INDUSTRIAL ELECTRONICS

INDUSTRIAL ELECTRONICS
Second Edition

James T. Humphries
Leslie P. Sheets

School of Technical Careers
Southern Illinois University at Carbondale

Breton Publishers
Boston, Massachusetts

PWS PUBLISHERS

Prindle, Weber & Schmidt ·🐿· Duxbury Press ·♠· PWS Engineering ·⚖· Breton Publishers ·⚙
20 Park Plaza · Boston, Massachusetts 02116

PWS Publishers is a division of Wadsworth, Inc.

Library of Congress Cataloging-in-Publication Data
Humphries, James T., 1946–
   Industrial electronics.

   Includes index.
   1. Industrial electronics.   I. Sheets, Leslie P.,
1941–   II. Title.
TK7881.H85   1986   621.381   85-29766
ISBN 0-534-06390-X

Printed in the United States of America
  2 3 4 5 6 7 8 9—90 89 88 87 86

ISBN-0-534-06390-X

Sponsoring editor: George J. Horesta
Production supervision: Technical Texts, Inc.
Production editor: Jean T. Peck
Copy editor: Marilyn Prudente
Text design: Mary S. Mowrey
Cover design: Sylvia Dovner
Cover photo: Donald E. Carrol/The Image Bank
Text illustration: Horvath & Cuthbertson
Composition: University Graphics, Inc.
Cover printing: New England Book Components, Inc.
Text printing and binding: Halliday Lithograph

Acknowledgments

   Chapter 1: Figure 1.50, National Semiconductor Corporation; Figures 1.71 and 1.72, RCA Solid State Division.
   Chapter 2: Figures 2.12 and 2.13, AT&T Communications.
   Chapter 3: Figures 3.1a and 3.4, Reliance Electric Company; Figure 3.5, reprinted from EDN, 1978 © Cahners Publishing Company; Table 3.6, reprinted from EDN, 1978 © Cahners Publishing Company.
   Chapter 4: Figures 4.12, 4.14a and b, 4.16, 4.17, 4.18, 4.19, 4.20, 4.21, 4.22, 4.23, 4.24, 4.25, Western Division IMC Magnetics Corporation; portions of the section on brushless DC motors are drawn from Chapter 6 of *DC Motors, Speed Controls, and Servo Systems,* 6th edition, 1982 Electro-Craft Corporation, Hopkins, MN; portions of the section on stepper motors are drawn from the *TORMAX Steppers and Control Manual,* Western Division IMC Magnetics Corporation, Maywood, CA.
   Chapter 5: Figure 5.1, Reliance Electric Company; Figure 5.1, G.M. Ferree, STC Photography; Figures 5.3 and 5.7, Bob White, STC Photography.
   Chapter 6: Figures 6.2, 6.3, 6.4, 6.5, Micro Switch, a Honeywell Division; Figure 6.6, Allen-Bradley Company; Figures 6.8 and 6.11, Micro Switch, a Honeywell Division; Figure 6.13, Chilton Company, a division of ABC, Inc; Figure 6.14, Ledex, Inc.; Figure 6.18, Square D Company; Figures 6.20 and 6.21, reprinted from *Machine Design,* May 19, 1983, copyright 1983, by Penton/IPC, Inc., Cleveland, OH; Figure 6.22, Sprecher & Schuh, Inc.; Figure 6.30, E G & G Wakefield Engineering; Figure 6.31, reprinted with permission, Signetics Corporation, copyright Signetics Corporation; Figure 6.52, copyrighted by and reprinted with permission of General Electric Company; Figure 6.53, Motorola Semiconductor Products, Inc.; Figure 6.54, Hamlin, Inc.; Figure 6.57, Simpson Electric Company; Figures 6.59 and 6.61, copyrighted by and reprinted with permission of General Electric Company.
   Chapter 7: Figures 7.3, 7.4, and 7.5, copyrighted by and reprinted with permission of General Electric Company; Figure 7.6, RCA Solid State Division; Figures 7.8 and 7.9, Furnas Electric Company; Figure 7.14, Reliance Electric Company; Figures 7.16 and 7.17, copyrighted by and reprinted with per-

(Continued on page 566)

To our wives, Marg and Joyce

PREFACE

Anyone involved in industrial electronics knows that industry is changing rapidly. Since an industrial electronics course tries to prepare students for entry into the workplace, an industrial electronics text should reflect those changes that are occurring in industry. We have made every effort to bring this text up-to-date to reflect what industry is demanding of our graduates and what instructors are teaching in their courses.

The goal of *Industrial Electronics* in the second edition is the same as in the first. It is to provide the student with an understanding of the basic components and systems used in industrial electronics. The book is intended for use in electronics programs offered in two- or four-year colleges.

The text avoids design questions. Instead the text focuses on the underlying concepts and the operation of electronic devices, circuits, and systems. We feel that if concepts are understood, designing circuits in most cases is not a problem. We definitely do not subscribe to the notion that the best way to understand electronic circuits is to design them.

The text is comprehensive. Experience has shown that a course in industrial electronics requires the coverage of a large number of topics. But how does a teacher cover these topics in one course? One solution is to use several textbooks for the course. Another is to supplement one text extensively with instructor-prepared materials. A third approach is to modify the course and teach only those topics covered in the textbook. However, we feel that all of these alternatives are unacceptable. Thus, we have purposely written a comprehensive text and have tried to include most of the topics you will wish to discuss.

Alternatively we have also written the book so that its presentation is as flexible as possible, should you not wish to present all the topics included. Although we cover most of the topics in the book in a one-semester course, some of the topics can be treated in other courses. Thus, the material in this book could easily be expanded to two semesters or two quarters, depending on your depth of treatment for each topic.

We require all our electronics students at the School of Technical Careers at Southern Illinois University, Carbondale, to take this course. We feel that every graduate of a two-year or four-year electronics program should have a basic understanding of both digital and analog circuitry. Many of our graduates report that their preparation in analog circuits and systems has been invaluable. They typically find themselves acting as liaisons between analog applications and people who have a predominantly digital background.

The text is not overly mathematical. We expect students to have a mathematics background no higher than algebra and trigonometry, and we have avoided any higher level of mathematics in our presentation. We realize, however, that mathematics is a concise way of representing concepts. Thus, we have included it where we feel that an adequate grasp of the concept demands a mathematical treatment.

We also expect students to have some background in basic digital gates and logic gained from an introductory course in digital electronics earlier in their electronics education. It may also be helpful, but not essential, for the student to have completed a technical physics course.

Users of the first edition of *Industrial Electronics* will notice significant differences in the text. We have dropped the first chapter that dealt with the introduction to the operational amplifier. Many users of the text said that this material was duplicated in introductory courses on amplifiers. For the same reason, the chapter on microprocessors was dropped. Users reported that most students taking a course in industrial electronics had already had a course in digital electronics and microprocessors. All remaining chapters have been retained in the second edition. Most of them have been expanded to include more in-depth coverage of the devices and circuits covered. For example, in the chapter on process control, simple circuits have been added to aid the student in understanding the industrial process. As most of the added circuits contain op amps and other readily available components, the circuits can be built and studied in a laboratory setting, both as applications of integrated circuits and as examples of process controls.

Several new chapters also have been added. One new chapter covers brushless and stepper DC motors. These devices are becoming more popular in industry and need special treatment. Also, a chapter on special linear integrated circuits has been added. Most courses teach the op amp as part of a course on amplifiers. Few texts on op amps, however, cover such devices as the phase-locked loop and the analog-to-digital and digital-to-analog converter. Users of the first edition will notice that the chapter on programmable controllers has been greatly expanded. This reflects the emphasis on programmable

controllers in industry and in the school industrial electronics laboratory. As a preface to the subject on programmable controllers, a chapter on sequential control has been added. We feel that the changes made in this second edition of the text have made it a useful and state-of-the-art addition to the industrial electronics classroom.

Learning aids have been built into the presentation. We have included questions at the end of every chapter that give immediate feedback on student understanding. Where appropriate, problems are also included. Users of the first edition will notice that the numbers of problems at the end of chapters have, in most cases, been doubled. In addition, at the end of the text, bibliographies are included for each chapter. You may wish to use these references as supplemental reading or study assignments if you cover certain topics in more depth. Data sheets have also been included at the back of the book for reference. They present detailed information about some of the IC chips used in the text. We hope that you will alert students to them so that they will use this information to gain greater understanding of the IC chips and circuits we have used as examples. They may want to use these same IC chips in projects of their own design, and these data sheets should help.

We hope you will find this textbook to be a useful teaching aid in your program. An ancillary instructor's guide is available without charge from the publisher for the convenience of instructors who adopt this book for classroom use. Also, a book of laboratory experiments is now available as an accompaniment to the second edition. We have tried to include experiments that will reinforce the concepts taught in the text material. Every effort has been made to use inexpensive, generic components and circuits for experiments. In many cases, equipment already on hand can be adapted for use in these experiments. The laboratory manual is available as a separate publication and is keyed to reading assignments in the text.

Publishing a textbook is the result of a coordinated effort that involves many people. We would like to express our thanks to the following individuals: to Bob White and his students for their excellent photographic work; to Arthur Barrett, David Longobardi, Thomas Dunsmore, and Eddy E. Pollock, who reviewed the manuscript for the first edition; to James M. Northern, James B. Wertz, Gerald K. Thomas, and Albert B. Barnett, who provided many helpful recommendations for this revision; to Frank Mann, Arthur Barrett (again!), and especially Frederick F. Driscoll, who reviewed the manuscript for the second edition and offered many suggestions for improvement; to George J. Horesta for words of encouragement at just the right moments; to Sylvia Dovner and her staff at Technical Texts, Inc. for the almost Herculean task of making our writing readable; to our colleagues here at Southern Illinois University in Carbondale for their support; to our wives, Marg and Joyce, for typing the manuscript; and to our children for understanding why we had to be away from home so many evenings and weekends.

James T. Humphries
Leslie P. Sheets

INTRODUCTION

Industrial electronics can be defined as the control of industrial machinery and processes through the use of electronic circuits and systems. Each of the topics in this text has been carefully chosen to help you, the future technician, survive in such an environment. We feel that knowledge gained by studying these topics will make you better prepared for entry-level employment as an electronics technician.

Although many topics have been included in this text, there are many additional topics you will need to know to function successfully in industry. As a starting point we assume that your previous electronics courses have given you a firm grounding in alternating and direct current theory, the functions of electrical and electronic components, and mathematics through algebra and trigonometry.

The topics in this text are built around a very general process control system, since we believe the electronics technician should be a generalist. Furthermore, it is not possible to cover in one text all the circuits and systems you are likely to encounter in industry. The range of industrial applications is simply too broad. Therefore, we have concentrated on some basic circuit and component concepts with appropriate examples. Once you have mastered the basic circuit and component functions, you will be ready to put these parts together into a functional system. That is, your knowledge of the functions of subparts and subsystems will help you understand how the overall system functions.

An example of a very basic process control system is illustrated in Figure I.1. Each block in the diagram represents a division of the elements within a process control system. Note that it is sometimes difficult to separate physically one block or topic from another in a real system. And, of course, not all blocks will be used in every control system; however, all are likely to be encountered in industry.

Each block in Figure I.1 contains the associated chapter in which that topic is discussed. The chapters by themselves may seem disconnected from one another, but they obviously are related when you consider the complete system.

The process in Figure I.2 is a hypothetical example. It shows a conveyor belt and a motor driver that must keep a constant speed on the belt regardless of the load. This example is related to the system of Figure I.1 in the following manner: The process in Figure I.1 is the transportation of the load from one point to another at a constant speed. As the load on the belt changes, some type of speed sensor is used to detect changes in speed and convert this change into an electric output. If the controlling system is some distance from the belt, telemetering can be used to transport the information from the sensor to the controller.

Before the controller can operate on this sensor output, it may require conditioning of some type to get the electric signal in the correct range or form. The controller can then make a decision, on the basis of the results of the in-

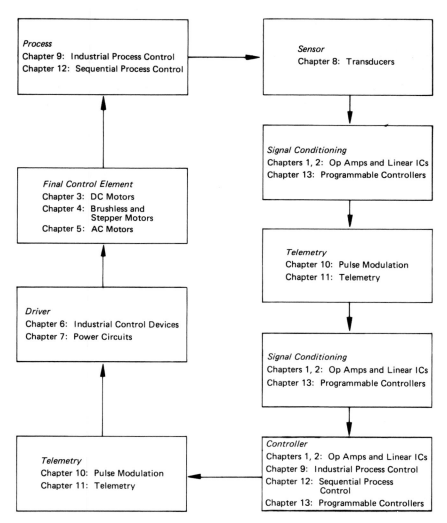

Figure I.1   General Process Control System

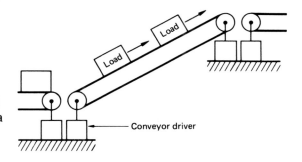

Figure I.2   Conveyor Belt System Used to Illustrate a Simple Process and Its Control

coming data, as to whether the belt should have more or less power than it had previously or the same power. This decision is then telemetered back to the belt drive circuitry, which will apply the appropriate amount of power, depending on the results of the controller's decision. The final control element, the motor, will then be provided with the appropriate power necessary to keep the belt operating at a constant speed. This simple process illustrates the concept of process control and the basis for the inclusion of the topics in this text.

CONTENTS

5
AC MOTORS • 144

6
INDUSTRIAL CONTROL DEVICES • 161

7
POWER CONTROL CIRCUITS • 208

8

TRANSDUCERS • 249

9

INDUSTRIAL PROCESS
CONTROL • 312

10

PULSE MODULATION • 340

11

INDUSTRIAL TELEMETRY AND DATA COMMUNICATION • 364

12

SEQUENTIAL PROCESS CONTROL • 395

13

PROGRAMMABLE CONTROLLERS • 415

1

OPERATIONAL AMPLIFIERS FOR INDUSTRIAL APPLICATIONS

OBJECTIVES

On completion of this chapter, you should be able to:

- Describe the characteristics and applications of the voltage follower;
- Compare the integrator and differentiator, and indicate how the ideal circuits are made practical;
- Calculate the output voltage of a logarithmic amplifier (log amp) with a given input;
- List the advantages of the active op amp filter over the passive devices;
- List and explain the Barkhausen criteria;
- Describe the operation of the comparator with and without hysteresis;
- Differentiate among the phase-shift, Wien-bridge, and twin T oscillators;
- Compare the op amp and the current-differencing amplifier (CDA);
- Describe the operation of the CDA, and draw the schematic diagram for the CDA;
- Describe the operation of the operational transconductance amplifier (OTA), and draw the OTA schematic diagram;
- Describe an application for the op amp, current-to-voltage converter, and draw the schematic diagram of that application;
- List the general procedures for troubleshooting op amp circuits.

INTRODUCTION

In this chapter we consider applications of the two basic operational amplifier (op amp) configurations. Several of these circuits perform higher-level mathematical functions. We will examine how the op amp can be used to subtract, integrate, differentiate, and generate logarithms. All these circuits are used in analog computers.

A common application of op amps in audio circuits is the filter. We will see in this chapter how op amp filters are superior to other types at certain frequencies. We will also consider the op amp oscillator. Other devices that perform similar functions but are not strictly considered op amps are also included in our presentation.

The versatility of the op amp is seen in the variety of its applications. It is used in signal interfacing, signal conditioning, impedance matching, active filtering, and many other applications.

VOLTAGE FOLLOWER

The *voltage follower,* a special case of the noninverting amplifier, is a circuit in which the output duplicates, or follows, the input in phase and amplitude. Let us take the noninverting amplifier of Figure 1.1, change the input resistor R_i to infinity, and decrease the feedback resistor R_F to 0 Ω, as shown in Figure 1.2.

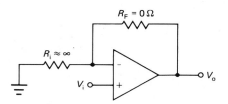

Figure 1.2 Developing the Voltage Follower from the Noninverting Amp

What is the gain of this circuit? We can use the following equation:

$$A_V = \frac{R_F}{R_i} + 1$$

where

A_V = voltage gain

R_i = resistor across which input voltage is developed

In this equation we substitute zero for R_F and let R_i approach infinity. The resulting gain is 1, as shown:

$$A_V = \frac{R_F}{R_i} + 1 = \frac{0}{\approx\infty} + 1 = 1$$

Now, the circuit can be redrawn as in Figure 1.3, which shows the voltage follower in its most common representation.

What are its advantages? The voltage follower is primarily used as a *buffer amplifier,* an amplifier that isolates the preceding stage from

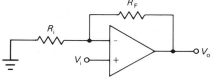

Figure 1.1 Basic Noninverting Op Amp Circuit

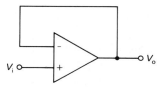

Figure 1.3 Voltage Follower

the following stage. It has a very high input impedance, even higher than a device with no feedback. Negative feedback causes input impedance to increase and output impedance to decrease. The voltage follower's high impedance ensures that the input circuit is not loaded down. And, of course, the voltage follower has unity gain with no phase shift between input and output. The voltage follower does not change the input in either phase or amplitude. The output follows the input—hence, the name, voltage follower.

DIFFERENCE AMPLIFIER

If we combine the inverting amplifier and the noninverting amplifier, we have the *difference* (or differencing) *amp*. It is shown in Figure 1.4. The output produced results from a combination of the input sources, V_{i1} and V_{i2}. From the principle of superposition, the contribution of V_{i1} to the output (V_{o1}) can be represented as follows:

$$V_{o1} = -\left(\frac{R_F}{R_i}\right)V_{i1}$$

In this equation we consider V_{i2} shorted to ground; thus, the circuit becomes an inverting

op amp. The contribution of V_{i2} to the output (V_{o2}) is determined by grounding V_{i1}, thus producing a noninverting op amp. The output (V_{o2}) produced by V_{i2} is as follows:

$$V_{o2} = \left(\frac{R_F}{R_i} + 1\right)V_2$$

For V_{i1} shorted to ground, V_2 is defined (from the voltage divider equation) as follows:

$$V_2 = \left(\frac{R_B}{R_A + R_B}\right)V_{i2}$$

where

   R_A = resistance between input signal and noninverting terminal

   R_B = resistance between noninverting terminal and ground

Suppose we combine all these factors into one equation, that is, by superposition:

$$V_o = V_{o2} + V_{o1}$$

Then:

$$V_o = \left(\frac{R_F}{R_i} + 1\right)\left(\frac{R_B}{R_A + R_B}\right)V_{i2} - \left(\frac{R_F}{R_i}\right)V_{i1}$$

The first expressions on the right side of the equation are for the noninverting amp; the expressions following the minus sign are for the inverting amp.

If all resistors are equal, this rather formidable equation can be simplified as follows:

$$V_o = V_{i2} - V_{i1}$$

Further, if $R_i = R_A$ and $R_F = R_B$, then the output voltage is as follows:

$$V_o = \left(\frac{R_F}{R_i}\right)(V_{i2} - V_{i1})$$

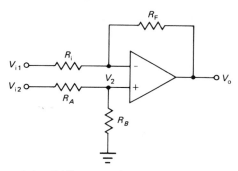

Figure 1.4   Difference Amp

INSTRUMENTATION AMPLIFIER

An interesting application of the difference amp circuit involves the idea that neither one of the op amp inputs is at ground potential. The difference amp can then be used to connect inputs that are floating to a grounded load without grounding either source. The instrumentation amplifier is such an application.

An *instrumentation amp* is an amplifier that is generally used to amplify DC and low-frequency signals produced by a transducer. (Transducers are discussed in Chapter 8.) These signals are typically only a few millivolts in amplitude, while common-mode noise may be several volts. The instrumentation amp has good common-mode rejection characteristics. Common-mode rejection is the ability of the amplifier to reject noise common to both inputs. Since noise will be common to both inputs, the amplifier of Figure 1.4 is frequently used when a signal may be masked by noise or when a signal is very small.

One problem inherent in the amplifier of Figure 1.4 is its input impedances. The input impedances to this amplifier may be unequal. And they are relatively low for use in instrumentation applications when the inverting and noninverting gains are not equal. This problem is reduced by making the gains large and equal; that is, $R_F = R_B$ and $R_i = R_A$, with R_F and R_B large. The input impedances then become much higher but may still be unequal. The amplifier can be improved by adding a voltage follower to each input. Such an amplifier is then called an instrumentation amp. Desirable characteristics of very high input impedance and excellent common-mode rejection combine in the instrumentation amp of Figure 1.5.

The gain of this circuit is expressed by the following equation:

$$A_V = \left(\frac{R_3}{R_2}\right)\left(\frac{2R_1 + R_A}{R_A}\right)$$

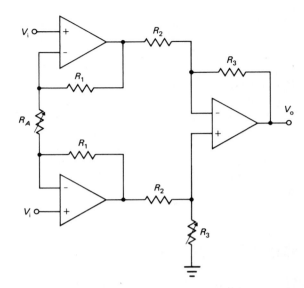

Figure 1.5   Instrumentation Amp

where

R_2 = resistance between voltage follower and output op amp

R_3 = feedback resistance to inverting terminal of output op amp and resistance to ground at noninverting terminal of output op amp

The instrumentation amplifier finds use in industry where signals are masked by noise and where it is desirable not to load the sensor unduly. Some manufacturers make an instrumentation amp within a single chip. The LM725 is an example. A single chip is desirable because, in that case, the amplifiers and resistors of the instrumentation amp can be more closely matched. Close matching improves instrumentation amp performance.

INTEGRATOR

Operational amplifiers can be arranged so as to perform the mathematical function of integra-

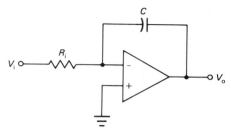

Figure 1.6   Ideal Op Amp Integrator

tion. Such a device is called an *integrator*. *Integration* is the process of summing the input signal over time or, in analytical geometry, of finding the area under a curve. The op amp integrator is illustrated in Figure 1.6. Notice that this circuit looks like the inverting amplifier except that the feedback resistor (R_F) has been replaced by a capacitor (C).

Integration of DC

To get a better idea of how an integrator works, we start with a DC input. Consider the circuit in Figure 1.7 with switch S_1 closed and switch S_2 open.

Assuming current is constant for the moment, the charge (Q) on any capacitor is related to the current (I) flowing through it and the time (t) spent charging it:

$$\Delta Q = I \, \Delta t$$

where

ΔQ = change in charge, measured in coulombs

Δt = change in time, measured in seconds

In a similar fashion, we can say that the change in voltage (ΔV) across the capacitor is a function of its change in charge (ΔQ) and the capacitor's capacitance (C):

$$\Delta V = \frac{\Delta Q}{C}$$

Substitution gives us the following equation:

$$\Delta V = \frac{I \, \Delta t}{C}$$

This equation tells us that if we know the capacitance of the capacitor, the amount of the current flowing through it, and the time of charging, we can predict the change of voltage across it.

What factors determined the current flowing through R_i and R_F? We found that input voltage (V_i) and the input resistance (R_i) alone determined this current. So, if we now substitute these factors into the preceding equation and simplify, we get the following:

$$\Delta V = - \frac{V_i \, \Delta t}{R_i C}$$

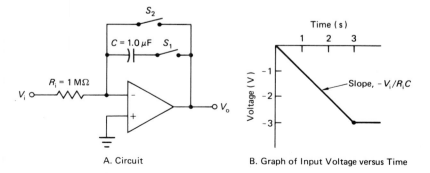

Figure 1.7   Integrator Example

A. Circuit

B. Graph of Input Voltage versus Time

Let us now apply these equations to a specific example. Figure 1.7A has values of 1 MΩ for R_i and 1 μF for C. A positive voltage is applied to the input for a certain time interval.

When switch S_1 is closed, current flows through the capacitor at a constant rate. If the capacitor is initially uncharged, how can we determine what the voltage on the capacitor will be after 3 s? We must know how fast the capacitor charges. The rate of change of voltage on the capacitor can be determined by rearranging the preceding equation and is measured in volts per second:

$$\frac{\Delta V_o}{\Delta t} = \frac{-V_i}{R_i C} = \frac{-1 \text{ V}}{(1 \text{ M}\Omega)(1 \text{ μF})} = -1 \text{ V/s}$$

In our example the rate of change of output voltage is -1 V/s. When, at the end of 3 s, we open the switch S_1, we should have a voltage of -3 V across the capacitor. This result is graphically illustrated in Figure 1.7B. Notice that the slope of the line in Figure 1.7B is $-V_i/R_i C$. Switch S_2 can be used to discharge the capacitor (C).

An interesting variation of this circuit results when switch S_2 in Figure 1.7A is replaced with a silicon-controlled rectifier (SCR). SCRs are discussed in Chapter 6. What we have then is a sawtooth wave generator, as illustrated in Figure 1.8. The reference voltage (V_{ref}) on the

SCR gate must be well within the range of 0 V to the $-V$ supply of the op amp.

Integration of a Square Wave

We have discussed the integrator's operation with a positive DC placed on its input. Now, instead of placing a constant positive DC voltage at the input of the integrator, let us alternate positive and negative DC voltages (otherwise known as a *square wave*). The output, then, instead of ramping negative, ramps positive and negative, a waveform known as a *triangle wave*. The situation is illustrated in Figure 1.9.

In this example a triangle wave output results from a square wave input. The triangle wave output is established by the following equation:

$$V_{o(p-p)} = -\frac{V_{i(pk)}t}{R_i C} = \frac{1 \text{ V}_{pk} \times 100 \text{ μs}}{(10 \text{ k}\Omega)(0.0022 \text{ μF})}$$
$$= 4.54 \text{ V}_{(p-p)}$$

where

$V_{o(p-p)}$ = peak-to-peak output voltage

$V_{i(pk)}$ = peak input voltage

Figure 1.9 shows a *practical integrator*. Resistor R_A has been added to prevent the ampli-

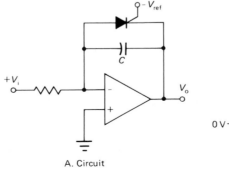

Figure 1.8 Op Amp Integrator Combined with an SCR

A. Circuit

B. Sawtooth Waveform

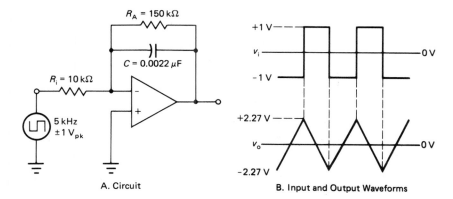

Figure 1.9 Practical Op Amp Integrator with Square Wave Input

A. Circuit

B. Input and Output Waveforms

fier from integrating the DC offset voltage. If this small voltage is integrated over a sufficiently long time, it drives the amplifier into saturation and the amplifier no longer operates as an integrator. Resistance R_A should be approximately 10 times R_i. Another way to see the effect of this resistor is to think of the amp's low-frequency gain as very high because of the capacitor's high reactance X_C. The resistor R_A limits this gain at low frequencies to R_A/R_i. But since the offset voltage is small, the DC output is also small.

Integration of a Sine Wave

The integrator gain equation for a sine wave input is similar to the inverting amplifier gain equation:

$$\frac{V_o}{V_i} = -\frac{R_F}{R_i} = A_V$$

Replacing R_F and X_C, we have the following equation:

$$A_V = \frac{X_C}{R_i} \angle 90°$$

where

$\angle 90°$ = notation that indicates a 90° phase shift occurs

A phase shift is expected across a capacitor. When X_C divided by the op amp, open-loop gain equals R_i, the phase shift between the input sine wave and the output waveform is approximately 45°. As frequency increases, the phase shift approaches 90°. In this region where X_C divides by op amp, open-loop gain is much smaller than R_i, the circuit in Figure 1.9 operates as an integrator. The output of the integrator with a sine wave input is a cosine wave, as shown in Figure 1.10.

In calculus the integral of a sine wave is a negative cosine wave. Thus, the phase relationship between input and output is 90°. But notice in Figure 1.10 that the output leads the input by 90°. We expect the output to lag the input since the integrator is, in fact, a lag net-

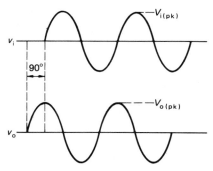

Figure 1.10 Integrator Input and Output Waveforms

work. A *lag network* is a circuit in which the output voltage lags the input voltage. But the integrator of Figure 1.9 uses the inverting amp (180° lag) and the integrator lag network (90° lag), which work in conjunction to provide a phase-leading relationship, output to input.

In Figure 1.10 to determine the maximum amplitude of the output cosine wave, $V_{o(pk)}$, we expand the preceding equation, drop the angle notation, and multiply it by the peak of the input sine wave, $V_{i(pk)}$:

$$V_{o(pk)} = \frac{V_{i(pk)}}{2\pi f R_i C}$$

where

  f = frequency of input sine wave

This new equation indicates that as the input frequency increases, the output amplitude of the cosine wave decreases. In Figure 1.9 if the input signal is a 1 V peak, 5 kHz sine wave, the output cosine wave amplitude is 1.45 V peak.

DIFFERENTIATOR

Interchanging the resistor and capacitor in the integrator produces a *differentiator,* which is shown in Figure 1.11. Mathematically, the *differentiation function* is the inverse of integration.

The differentiator is not as frequently used as the integrator. We noted that the integrator

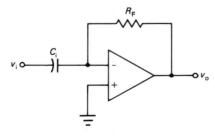

Figure 1.11   Ideal Op Amp Differentiator

(for sine waves) was a basic lag network, that is, a low-pass filter. Recall that a *low-pass filter* passes all frequencies below a given frequency. The inverse is true of the differentiator: It is a basic high-pass filter. A *high-pass filter* is a circuit that passes all frequencies above a given frequency. Thus, the differentiator is much more likely to oscillate or become unstable than the integrator is, and, at the same time, it will amplify any noise that is present on the signal.

The mathematical concept of differentiation is simple. Instead of producing an output that is proportional to the input voltage (as the inverting amp does), the differentiator output is proportional to the slope of the input voltage. If the slope of the input voltage is zero, the differentiator output is 0 V. If the slope of the input is constant, the output is constant. If the slope of the input is infinite (as in the edges of a square wave), the output is an infinite voltage (ideally).

Note the circuit in Figure 1.12A. Resistor R_F and capacitor C form a lead network (or high-pass filter). A *lead network* is a circuit in which the output voltage leads the input voltage. Series resistor R_s holds the high-frequency gain to $-R_F/R_s$. The input and output waveforms are shown in Figure 1.12B.

Differentiation of a Sine Wave

As in the integrator, a sine wave input to the differentiator gives a cosine wave output. Mathematically, a differentiated sine wave gives a cosine wave. The difference between integrator and differentiator output due to a sine wave input is in phase angle and amplitude. The differentiator, with its 180° phase inversion, causes the output voltage to lag the input by 90°. This relationship is diagramed in Figure 1.12B.

The output amplitude is expressed by the following equation:

$$V_{o(pk)} = 2\pi f R_F C V_{i(pk)}$$

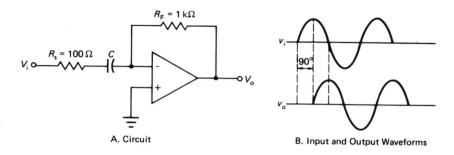

Figure 1.12 Differentiator with a Sine Wave Input and Negative Cosine Wave Output

A. Circuit

B. Input and Output Waveforms

In this equation, as the input frequency increases, so does the output amplitude.

Differentiation of a Square Wave

Notice what happens when we change the input from a sine wave to a square wave, as shown in Figure 1.13. The output of a differentiator with such an input is a series of positive and negative spikes. Why? Note that the slope of the AB portion of the square wave of Figure 1.13 is zero. We expect to get no output. The slope of the BC portion is infinity (or close to it). The output quickly rises to its maximum value and then falls back to zero. Thus, the output waveform is a series of positive and negative spikes.

Differentiation of a Triangle Wave

Let us examine what would happen if a triangle wave were the input for the differentiator. In calculus the operation of differentiation is the inverse of integration, as we mentioned earlier. Integration produces a triangle output with a square wave input. Thus, the differentiator produces a square wave output with a triangle input.

The input-output relationship with a practical differentiator is shown in Figure 1.14. The slope of the input from A to B and C to D is a constant positive value. The output at those times is a constant negative voltage. This output value changes only if the input slope changes.

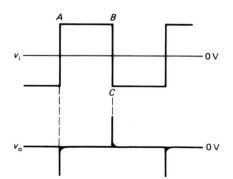

Figure 1.13 Square Wave Input to a Differentiator and Spiked Waveform Output

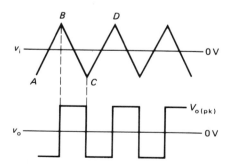

Figure 1.14 Triangle Wave Input to a Differentiator and Square Wave Output

To determine the amplitude of the output square wave of Figure 1.14, we use the following equation:

$$V_{o(pk)} = -R_F C \left(\frac{\Delta V_i}{\Delta t} \right)$$

The change in input voltage with respect to time $(\Delta V_i / \Delta t)$ is the slope of the input waveform.

The verification of the equations in the differentiator section is left for you to do.

LOGARITHMIC AMPLIFIER

A circuit frequently seen in analog computers is the logarithmic amplifier (log amp). In a log amp, the output is proportional to the logarithm of the input. The *log amp* takes advantage of the logarithmic relationship between the current through a semiconductor device and the voltage across it. Two such nonlinear devices are the semiconductor diode and the bipolar junction transistor (BJT). The circuit configurations for these log amps are illustrated in Figures 1.15A and 1.15B.

In the circuit shown in Figure 1.15A, the voltage across the diode (V_D) is logarithmically related to the current through the diode (I_D) by the following relationship:

$$V_D = A \log_B \frac{I_D}{I_r}$$

where

$\log_B$ = logarithm to a specified base (B)

A = constant of proportionality, which depends on the base of the logarithm

I_r = theoretical reverse leakage current of diode

Likewise, in Figure 1.15B the collector current (I_C) divided by its reverse leakage current (I_{Cr}) is logarithmically related to the voltage from base to emitter (V_{BE}) by the following relationship:

$$V_{BE} = A \log_B \frac{I_C}{I_{Cr}}$$

Practical Log Amp

In actual use of the log amp, a more practical circuit such as the one in Figure 1.16A would be utilized. Here, the more practical equation is as follows:

$$V_{o1} = -V_{BE} = A \log_B(V_i) + K \qquad \textbf{(1.1)}$$

where

V_{o1} = output of log amp

V_{BE} = voltage from base to emitter of transistor

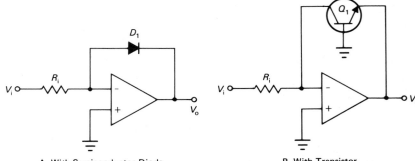

Figure 1.15 Ideal Log Amp

A. With Semiconductor Diode   B. With Transistor

A = constant of proportionality

K = offset value

The value of resistance for both R_1's in Figure 1.16A is given by the following expression:

$$\frac{V_{i(max)}}{I_{C(max)}} \le R_1 \le \frac{V_{i(min)}}{I_b} \qquad (1.2)$$

where

$V_{i(max)}$ = maximum input voltage

$V_{i(min)}$ = minimum input voltage

$I_{C(max)}$ = maximum collector current of transistor used

I_b = input bias current of op amp

For the 741, I_b is approximately 80 nA.

For evaluation of A and K in Equation 1.1, the following equations may be used:

$$A = \frac{V_{o2} - V_{o1}}{\log(V_{i2}) - \log(V_{i1})} \qquad (1.3)$$

$$K = V_{o1} - A \log(V_{i1}) \qquad (1.4)$$

where

V_{i1} = first applied input voltage

V_{i2} = second applied input voltage

V_{o1} = output voltage of log amp due to V_{i1}

V_{o2} = output voltage of log amp due to V_{i2}

An example might be helpful at this point to demonstrate the evaluation of A and K. Assume the input voltage applied will range over three different values: 1 V, 2 V, and 3 V. For these three inputs, three output voltages (V_{o1} in Figure 1.16A) result: 0.60 V, 0.62 V, and 0.64 V (hypothetically).

In Equation 1.3, V_{o1} is 0.60 V, V_{o2} is 0.62 V, V_{i1} is 1 V, and V_{i2} is 2 V. (Notice that the negative signs were dropped from the output voltage measurements.) Substituting these values into Equation 1.3 results in a value of 0.0664 for A. In Equation 1.4, V_{o1} is 0.60 V, A is 0.0664, and V_{i1} is 1 V. Substituting these values into Equation 1.4 results in a value of 0.60 for K.

Now, the procedure will be repeated for the second group of values for V_i and V_o. For Equation 1.3 and the second group of values, V_{o1} is now 0.62 V, V_{o2} is 0.64 V, V_{i1} is 2 V, and V_{i2} is 3 V. Substituting these values into Equation 1.3

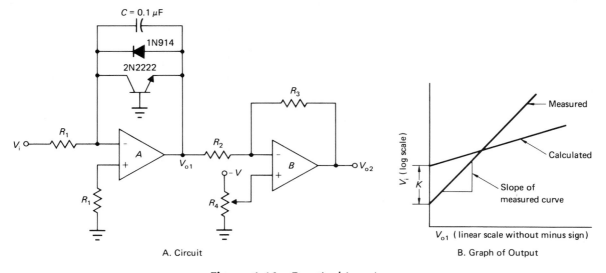

A. Circuit   B. Graph of Output

Figure 1.16 Practical Log Amp

results in a value of 0.114 for A. In Equation 1.4, V_{o1} is now 0.62 V, A is 0.114, and V_{i1} is 2 V. Substituting these values into Equation 1.4 results in a value of 0.586 for K.

After all values for A and K have been calculated, they can be averaged in order to obtain values that can be used in the amplifier circuit following the log amp in Figure 1.16A.

After A and K are determined, they are used with the second amplifier (B amplifier) in Figure 1.16A. The noninverting terminal of amplifier B is set to the value of K, and the gain of amplifier B is set to the reciprocal of A. When a voltage is input at V_i, the logarithm of that voltage is now output at V_{o2}.

As a precaution for the circuit of Figure 1.16A, be sure to use only positive input voltages in the range for which you determined the R_1 value in Equation 1.2. In addition, the transistor in the feedback circuit of the log amp (2N2222 in Figure 1.16A) can be any small-signal, high-speed transistor. If the transistor's response is a straight line on the graph in Figure 1.16B, the circuit of Figure 1.16A should produce an accurate logarithmic output. Whichever transistor you choose, though, will be heat-sensitive. For best results, either keep the transistor at a constant temperature or use a temperature-stabilized circuit. The bibliography for this chapter given at the back of the text lists good sources for such designs.

Antilog Amp

The *antilog amplifier* performs the inverse function of the log amp. This operation is done by simply reversing the position of the semiconductor and the input resistor. The antilog amp is shown in Figures 1.17A and 1.17B.

Designing and working with log amps can be a tedious and exacting job. For this reason several manufacturers have designed and built log amps within a single IC chip.

Applications of Log Amps

Applications of log amps are numerous. For example, they are used for compressing large voltage ranges into smaller ones. Also, since many transducers have a logarithmic transfer function, the log amp is used to compensate for this nonlinear response.

Probably the most frequent use of the log amp is in analog computers. Recall that multiplying and dividing may be accomplished by adding and subtracting the logarithms of numbers and then obtaining the antilogarithm. Since log amps perform these operations, they can be used in multipliers and dividers in computers. The block diagram of a multiplier is shown in Figure 1.18.

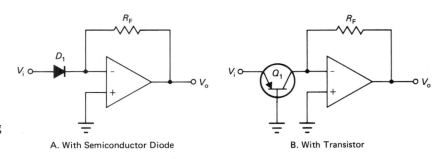

Figure 1.17   Ideal Antilog Amp

A. With Semiconductor Diode   B. With Transistor

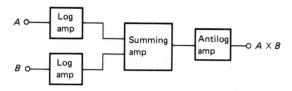

Figure 1.18 Multiplying with Log and Antilog Amps

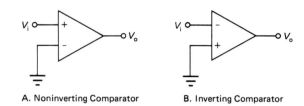

A. Noninverting Comparator   B. Inverting Comparator

Figure 1.19 Comparators (Zero-Crossing Detector)

COMPARATOR

The simplest op amp circuit is the *comparator,* which is an op amp without feedback. With no feedback, the op amp's gain is very high, equal to its open-loop gain. Figure 1.19 shows an example of an inverting and a noninverting comparator.

Since the gain of a comparator is very high, a voltage of only a few microvolts causes the amplifier to saturate. Therefore, the output of this amplifier is normally either $+V_{sat}$ (saturation voltage) or $-V_{sat}$. The op amps in Figure 1.19 compare the input against a reference, which in this case is 0 V. Such comparators are also called *zero-crossing detectors* because the output changes when the input crosses the 0 V point. Whenever the input voltage is above or below 0 V, the amplifier saturates.

A more general form of the inverting and noninverting comparators is shown in Figure 1.20. In these comparators, the op amp compares the input to a reference voltage (V_{ref}). Figure 1.21 shows a sine wave input to a comparator like the one shown in Figure 1.20B. Figure 1.20B (the noninverting comparator) has a 5 V peak sine wave applied with a reference of $+3$ V. We see that when the input rises above $+3$ V, the output goes to $+V_{sat}$. When the input falls below $+3$ V, the output goes to $-V_{sat}$.

The comparators we have just discussed are susceptible to triggering by noise voltages, alone or in combination with a signal source. The basic comparator can be improved by add-

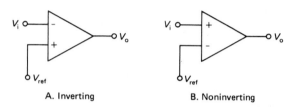

A. Inverting   B. Noninverting

Figure 1.20 General Forms of Comparators

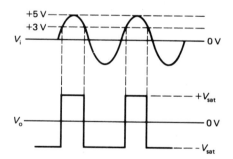

Figure 1.21 Input and Output Waveforms for Noninverting Comparator

ing positive feedback, as shown in Figure 1.22A. This circuit is called a comparator with hysteresis. *Hysteresis* is the property by which the output of a device is dependent on the history of the input movement and the present direction of the input movement. Originally applied to magnetic circuits, the term *hysteresis* now has been applied to other systems that have that type of action.

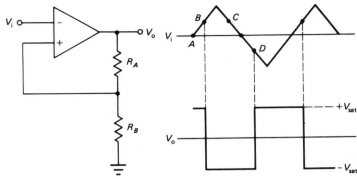

Figure 1.22 Comparator
with Hysteresis

A. Circuit, with Feedback   B. Input and Output Waveforms

In Figure 1.22A, the noninverting terminal receives a reference from the voltage divider formed by resistors R_A and R_B. Note that from point A to B on the input waveform in Figure 1.22B, the output is at $+V_{sat}$ ($+10$ V). When the input voltage rises above $+1$ V, the amplifier's output changes to $-V_{sat}$ (-10 V). At this time, the *trip point,* or reference voltage, at the noninverting terminal is no longer $+1$ V. It is -1 V. Now, the output of Figure 1.22B does not change at point C but, instead, at point D.

The reference voltage (trip point) is calculated from the following equation:

$$V_{ref} = \left(\frac{R_B}{R_A + R_B} \right) (\pm V_{sat})$$

For example, suppose $R_A = 9$ kΩ and $R_B = 1$ kΩ. Then, the circuit of Figure 1.22, when at $-V_{sat}$ (-10 V), has the following trip point:

$$V_{ref} = \left(\frac{1 \text{ k}\Omega}{9 \text{ k}\Omega + 1 \text{ k}\Omega} \right) (-10 \text{ V}) = -1 \text{ V}$$

ACTIVE FILTER

The operation of many electronic systems requires that certain bands of frequencies be passed and others attenuated. A device that does this operation is called a *filter.*

Classification of filters depends on point of view. If the components from which the filter is made are the basis of classification, then we say filters are either passive or active. A *passive filter* uses capacitors, resistors, and inductors. *Active filters* employ amplifiers along with the components mentioned in the passive category. Inductors usually are not used at audio frequencies since they can be simulated with amplifiers, capacitors, and resistors. Also, inductors are relatively expensive, large, and heavy. Figure 1.23 shows this type of classification of filters and the frequencies at which they are used.

Since an in-depth treatment of active filters would be quite lengthy, only an introduction is given in this text. For more information on active filters, consult the bibliography for this chapter (at the end of the book) or other reference material.

As we will see, the frequency behavior of any given amplifier can be controlled by adding reactive components to it in different configurations. Such active filters are reliable, low in cost, and easy to tune. In addition, when op amps are used, they give not only voltage gain but also high input impedance and low output impedance characteristics.

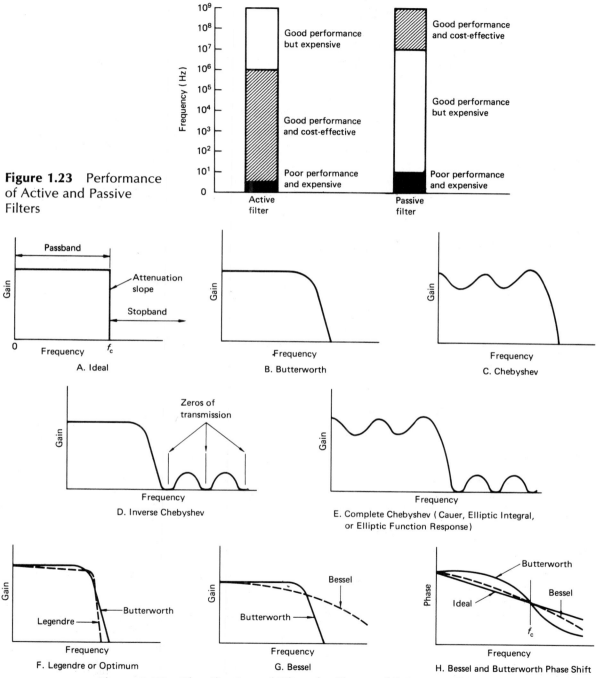

Figure 1.23 Performance of Active and Passive Filters

Figure 1.24 Classification of Filters by Shape of Response Curves

Table 1.1   Comparison of Filter Types

Name of Filter Type	Distinguishing Characteristic	Remarks
Butterworth	Maximally flat amplitude response	Most popular general-purpose filter
Chebyshev	Equal-amplitude ripples in passband	Attenuation slope steeper than in Butterworth near cutoff
Inverse Chebyshev	Equal-amplitude ripples in stopband	No passband ripple; zeros of transmission in stopband
Complete Chebyshev (also called Cauer, elliptic function, elliptic integral, or Zolatarev)	Equal-amplitude ripples in both passband and stopband	Zeros of transmission in stopband
Legendre	No passband ripple, but steeper attenuation slope than in Butterworth	Not maximally flat
Bessel (also called Thomson)	Phase characteristic nearly linear in pass region, giving maximally flat group delay	Good for pulse circuits because ringing and overshoot minimized; poor attenuation slope

Filters are also classified according to the shapes of their response curves. Figure 1.24 shows some of these curves. Table 1.1 compares these filter types.

Probably the most common classification of filters is related to how they treat frequencies. These classifications are low-pass, high-pass, bandpass, and band-reject (or notch).

Low-Pass Filter

A *low-pass filter* is defined as a circuit that passes all frequencies below a certain frequency while attentuating all higher frequencies. A simple low-pass Butterworth filter with its bandpass characteristic is shown in Figure 1.25.

As shown in Figure 1.25B, the output of the filter is constant until the *cutoff frequency* (f_c) is approached. After f_c is passed, the output voltage decreases, or rolls off, at a rate of -6 dB per octave, or -20 dB per decade. (Recall that a decade is a tenfold increase or decrease in frequency, and an octave is a twofold increase or

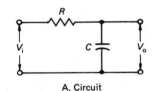

A. Circuit

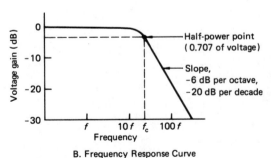

B. Frequency Response Curve

Figure 1.25   Low-Pass Filter

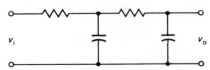

Figure 1.26   Second-Order, Low-Pass Butterworth Filter

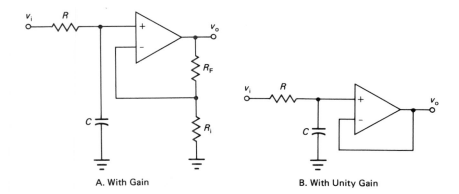

Figure 1.27 First-Order, VCVS Low-Pass Filter

A. With Gain

B. With Unity Gain

Table 1.2 Phase Shifts at Half-Power Point

Order of Filter	Phase Shift at f_o or f_c*	Roll-Off (dB per Decade)
1	−45°	−20
2	−90°	−40
3	−135°	−60
4	−210°	−80

*f_o = oscillation frequency for oscillators; f_o = cutoff frequency for filters.

decrease in frequency.) This filter is called a *first-order filter* because its transfer function is a linear equation and because it has only one *RC* filter combination.

Figure 1.26 shows a second-order Butterworth filter. Notice that there are two *RC* networks here. This filter exhibits a roll-off of −12 dB per octave, or −40 dB per decade. A third-order Butterworth filter has three *RC* networks and roll-offs of −18 dB per octave or −60 dB per decade; and so on. Table 1.2 shows the phase shifts at the half-power (0.707) points for these filters.

Let us now examine some low-pass active filters, as shown in Figure 1.27. These filters are first-order, *voltage-controlled, voltage source* (*VCVS*) *filters,* recognized by their use of the noninverting amplifier configuration. The only difference between the two amplifiers in Figures 1.27A and 1.27B is the gain.

Figure 1.28 shows a second-order, VCVS, low-pass filter. Components R_1 and C_1 make up one low-pass *RC* network, while R_2 and C_2 make up the other. For the resistors and capacitors as labeled in Figure 1.28, the equations for cutoff frequency f_c and voltage gain A_V are as follows:

$$f_c = \frac{1}{2\pi \sqrt{R_1 C_1 R_2 C_2}}$$

and

$$A_V = \frac{R_F}{R_i} + 1$$

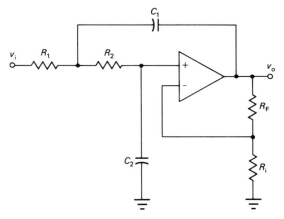

Figure 1.28 Second-Order, VCVS Low-Pass Filter with Gain

Thus far, we have shown examples of VCVS filters only. Figure 1.29 gives an example of a second-order, multiple-feedback, low-pass Butterworth filter. In this filter, identified as a *feedback filter,* the gain is fixed by the ratio of R_3 and R_1. The cutoff frequency equation is similar to the preceding equation. This circuit does have a disadvantage: Its gain and frequency characteristics cannot be separately controlled, as in the VCVS filters.

Higher-order filters (above second-order) may be constructed by cascading lower-order filters. For example, cascading two second-order filters creates a fourth-order filter.

High-Pass Filter

A *high-pass filter* is defined as a circuit that passes all frequencies above a certain frequency while attenuating all lower frequencies. The high-pass, first-order Butterworth filter and its associated graph are shown in Figure 1.30. The graph is similar to that of the low-pass filter. As frequency decreases, the output of the filter is constant until f_c is approached. Once f_c is passed, the output rolls off at -6 dB per octave (-20 dB per decade).

Active high-pass filters resemble low-pass filters, with the positions of capacitors and resistors interchanged. For instance, compare the low-pass filter in Figure 1.29 with the high-pass filter in Figure 1.31. For the resistors and ca-

pacitors shown in Figure 1.31, the equations for cutoff frequency f_c and voltage gain A_V are as follows:

$$f_c = \frac{1}{2\pi \sqrt{C_2 C_3 R_1 R_2}}$$

and

$$A_V = \frac{C_3}{C_1}$$

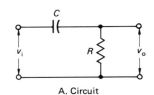

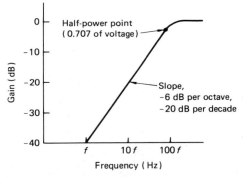

A. Circuit

B. Frequency Response Curve

Figure 1.30 High-Pass, First-Order Butterworth Filter

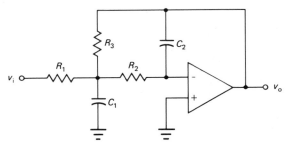

Figure 1.29 Second-Order, Multiple-Feedback, Low-Pass Butterworth Filter

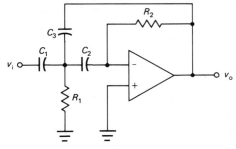

Figure 1.31 Second-Order, Multiple-Feedback, High-Pass Filter

Bandpass Filter

A *bandpass filter* is defined as one that passes a band of frequencies while attenuating all others. A simple way to construct a bandpass filter is to cascade a high-pass and a low-pass filter if their frequency response curves are made to overlap. The result of such a union is illustrated in Figure 1.32, with a graph showing gain versus frequency.

An example of a second-order, multiple-feedback bandpass filter is shown in Figure 1.33. For the resistors and capacitors shown in Figure 1.33, the equations for cutoff frequency f_c and voltage gain A_V are as follows:

$$f_c = \frac{1}{2\pi C_1} \sqrt{\frac{R_1 + R_2}{R_1 R_2 R_3}}$$

and

$$A_V = \frac{R_3}{2R_1} \qquad (\text{at } f_c)$$

Note the difference between the filter of Figure 1.33 and the high-pass filter of Figure 1.31. The capacitor C_1 of Figure 1.31 has been replaced by resistor R_1, thus making Figure 1.33 a bandpass filter. Generally, the multiple-feedback filter shown in Figure 1.33 is used for narrow-bandwidth applications ($Q > 10$) (recall that Q represents quality factor). Where wider bandwidths are required ($Q < 10$), high-pass and low-pass filters are cascaded.

Band-Reject (Notch) Filter

The *band-reject* (*notch*) filter is defined as a circuit that attenuates a specified band of frequencies while passing all others. Like the bandpass filter, the band-reject filter is constructed by connecting high-pass and low-pass filters in series (cascade) or in parallel. The parallel connection is most commonly used. It is sometimes called the *twin T filter*.

An example of the twin T and its associated bandpass characteristic is shown in Figure 1.34. Note that C_1, C_2, and R_3 form the high-pass section, and R_1, R_2, and C_3 form the low-pass section.

An active version using the twin T filter is shown in Figure 1.35. In this version, $C_1 = C_2$, $C_3 = 2C_1$, $R_1 = R_2$, and $R_3 = R_1/2$. The gain for this filter is 1. For the resistors and capacitors as labeled in Figure 1.35, the equation for the cutoff frequency is as follows:

$$f_c = \frac{1}{2\pi R_2 C_2} \qquad \qquad \textbf{(1.5)}$$

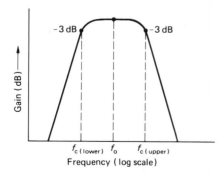

Figure 1.32 Frequency Response Curve for Bandpass Filter

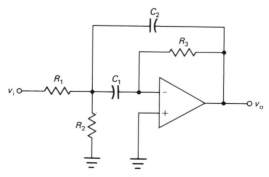

Figure 1.33 Second-Order, Multiple-Feedback Bandpass Filter

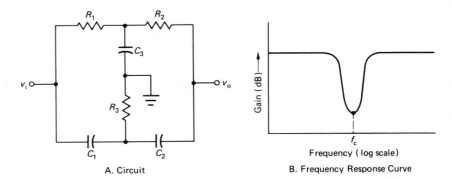

Figure 1.34   Band-Reject .Filter

A. Circuit

B. Frequency Response Curve

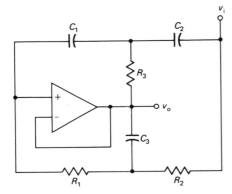

Figure 1.35   Twin T Band-Reject Filter

GYRATOR

While not strictly a filter circuit, the gyrator may be classified along with filters because of its unique reactive properties. The *gyrator* acts as a simulated inductor. We have already discussed the disadvantages of inductors at audio frequencies. At frequencies below radio frequencies (RF), the gyrator can simulate the behavior of an inductor. The gyrator is basically a two-terminal (or single-port) device. When a capacitor is connected to one port, the device rotates electrically, or gyrates, the capacitor so that it looks like an inductor at the other port. Two examples of gyrators are shown in Figure 1.36.

Gyrators may be used to replace inductors in low-frequency circuits. However, one terminal of the replaced inductor must be grounded for these circuits to work properly.

CAPACITANCE MULTIPLIER

Capacitance may be multiplied in a similar fashion to the creation of inductance. The circuit shown in Figure 1.37 will multiply the capacitance of C_1 according to the following relationship:

$$C_m = C_1 \left(\frac{R_1}{R_3} \right)$$

The resistance of R_1 should be as large as possible, since the impedance appears in series with the effective capacitance, C_m.

OSCILLATOR

An *oscillator* is a circuit that generates an alternating current (AC) waveform with only a direct current (DC) input. Commonly, the frequency of alternation is determined by the oscillator's components.

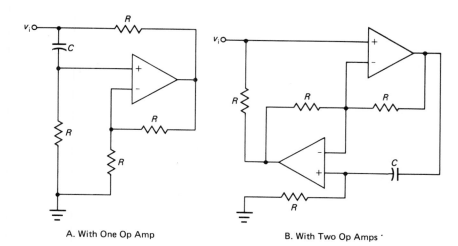

Figure 1.36 Gyrator
(Simulated Inductor)

A. With One Op Amp   B. With Two Op Amps

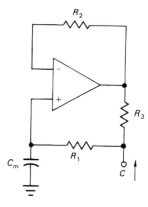

Figure 1.37 Capacitance Multiplier

Oscillators are quite widely used in industrial electronics for both analog and digital applications. Because most oscillators require an amplification device of some kind, op amps can be used to form basic oscillator circuits.

Sine Wave Oscillator

Sine wave oscillators, whether they use op amps or other amplification devices, must meet certain criteria to sustain oscillations. These criteria are sometimes called the Barkhausen criteria. The *Barkhausen criteria* are as follows:

1. The overall gain of the circuit (called *loop gain*) must be 1 or greater. Thus, losses around the circuit must be compensated for by an amplifying device.
2. The phase shift around the circuit (input to output and back to the input) must be 0° (or 360°).

The three common types of sine wave oscillators are phase-shift, Wien-bridge, and twin T.

Phase-Shift Oscillator. To illustrate the Barkhausen criteria, we consider the phase-shift oscillator shown in Figure 1.38. The feedback network of the phase-shift oscillator ensures that its own 180° phase shift, when added to the same shift for the inverting amp, equals 360°. Thus, one part of the Barkhausen criteria is met. The phase-shift network attenuates by a factor of about 30. The amplifier must then have a gain of at least 30 to satisfy the remaining Barkhausen criterion.

With the resistors and capacitors as labeled in Figure 1.38, the equation for the frequency

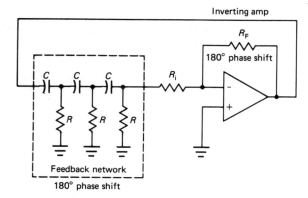

Figure 1.38   Phase-Shift, Sine Wave Oscillator

of oscillation (f_o) of this phase-shift oscillator is as follows:

$$f_o = \frac{1}{2\pi RC\sqrt{6}}$$

Wien-Bridge Oscillator. Another common sine wave oscillator is the Wien-bridge oscillator, shown in Figure 1.39. The *Wien-bridge oscillator* operates on the balanced bridge principle. When the impedance of the R_1C_1 branch equals the impedance of the R_2C_2 branch, the feedback voltage is in phase with the output voltage. Again, when the gain of the amplifier is sufficient to replace losses around the circuit, both Barkhausen criteria are met.

For the resistors and capacitors as labeled in Figure 1.39, the frequency of oscillation of the Wien-bridge oscillator is as follows:

$$f_o = \frac{1}{2\pi\sqrt{R_1C_1R_2C_2}}$$

Twin T Oscillator. The third sine wave oscillator is the *twin T oscillator* of Figure 1.40, so named because it uses the twin T configuration discussed in the section on notch filters.

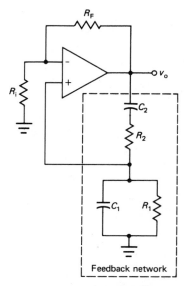

Figure 1.39   Wien-Bridge Oscillator

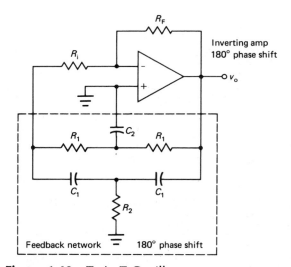

Figure 1.40   Twin T Oscillator

The feedback network gives a 180° phase shift only to the frequencies around f_o. The twin T network has losses around 25. Hence, if the amplifier's gain is at least that much, the circuit oscillates. For the resistors and capacitors as labeled in Figure 1.40, the frequency of oscillation of the twin T oscillator is as follows:

$$f_o = \frac{1}{2\pi R_1 C_1}$$

when $C_1 = 2C_2$ and $R_2 = R_1/2$.

Square Wave Oscillator

Figure 1.41 shows a square wave oscillator. When power is applied to this circuit, a small positive or negative voltage appears across R_A because of the presence of DC offset voltage. Since, at the first instant, capacitor C is uncharged, it applies 0 V to the inverting terminal. The small offset voltage appearing across R_A drives the amplifier into saturation. Capacitor C starts to charge and continues charging until it reaches the voltage across R_A. When this

voltage is exceeded, the oscillator output changes from one output saturation voltage state to the other, changing the reference voltage across R_A. The capacitor now discharges and then charges in the opposite direction. Again, when the voltage across the capacitor reaches the new reference voltage, it changes state. Thus, the process repeats itself.

As we see, this circuit is no more than a relaxation oscillator with an op amp comparator. A *relaxation oscillator* is an RC circuit with a capacitor that is charged and discharged (relaxed) to provide the basis for the oscillation. The output voltage oscillates between $+V_{\text{sat}}$ and $-V_{\text{sat}}$, producing a square wave.

For the resistors and capacitors as labeled in Figure 1.41, and when $R_A = 10R_B$, the frequency of oscillation of the square wave oscillator is as follows:

$$f_o = \frac{5}{R_1 C}$$

The square wave oscillator can be used to produce a triangle wave output by cascading it to an integrator, as shown in Figure 1.42.

MISCELLANEOUS OP AMP APPLICATIONS

Throughout the text we show applications of op amps in conjunction with other components

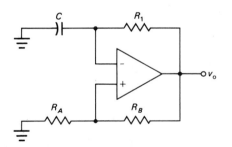

Figure 1.41 Square Wave (Relaxation) Oscillator

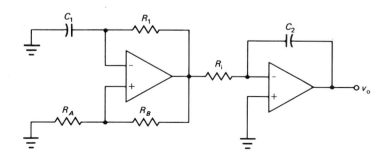

Figure 1.42 Triangle Wave Generator Using a Square Wave Oscillator and an Integrator

and devices. We have chosen seven applications of op amps to discuss here to further show the flexibility and versatility of op amps.

Op Amp Supply

Figure 1.43 illustrates a method by which a dual power supply can be made from a single supply input, one op amp, two transistors, a few resistors, and a few capacitors. The circuit is basically a high-gain amplifier with a very high input impedance. Resistors R_1 and R_2 form a voltage divider that references the non-inverting terminal at about half the input voltage. Feedback current from the junction of the Q_1 and Q_2 emitters to the input of the op amp keeps the output voltage evenly balanced.

LED Overvoltage Indicator

Figure 1.44 shows a circuit that is used to indicate the presence of an overvoltage by turning on an LED (light-emitting diode). By an adjustment of the gain of the op amp (R_F/R_i) and by proper choice of zener diodes, this circuit can be used to show an overvoltage visually. In this case, if the input voltage goes over $+3$ V, diode D_4 conducts, turning on D_2. If the input voltage goes lower than $+3$ V, diode D_3 conducts, turning on D_1.

Precision Half-Wave Rectifier

Diodes in a precision half-wave rectifier usually do not turn on fully until about 0.7 V is dropped across them. Refer to Figure 1.45. Be-

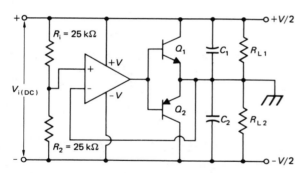

Figure 1.43   Op Amp Power Supply Circuit

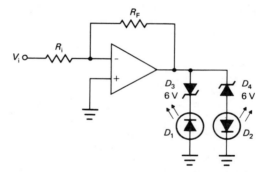

Figure 1.44   Op Amp LED Overvoltage Indicator

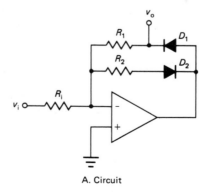

Figure 1.45   Precision Half-Wave Rectifier

A. Circuit

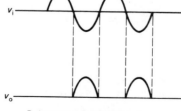

B. Input and Output Waveforms

cause the diodes are in the feedback circuit of the op amp, they turn on as soon as the input goes slightly above or below 0 V. They do so because of their high resistance before they turn on. Very high feedback resistance makes gain high. The 0.7 V needed to turn on the diodes occurs with a very small input voltage.

Op Amp Voltage Regulator

The op amp is also used as a low-current voltage regulator. Figure 1.46 illustrates such an application. Changes in the unregulated input result in a very small change across D_1 (provided it is biased correctly). Therefore, the output voltage remains relatively constant with changing load demands. The output voltage is a function of the zener voltage times the gain $(R_F/R_i + 1)$. Since zener voltage, R_F, and R_i remain constant, the output remains constant.

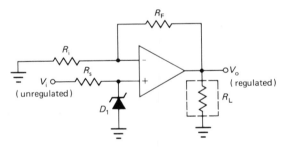

Figure 1.46   Op Amp Voltage Regulator

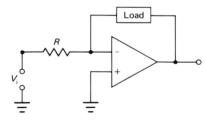

Figure 1.47 Op Amp Voltage-to-Current Converter

Op Amp Voltage-to-Current Converter

Some applications require a current source rather than a voltage source. Examples include analog meter movements, resistance measurement circuits, and driving inductive loads such as relay coils. Figure 1.47 shows a device that provides a current source, the voltage-to-current converter. In such a circuit the amount of feedback or load current does not depend on the resistance of the load. Load current depends on the input voltage and current. This effect is important, for example, in metering circuits. Only the input voltage being measured should affect the current through the meter. Any changes in the current through the meter because of meter resistance or voltage changes are undesirable.

Op Amp Current-to-Voltage Converter

Some transducers, particularly light- and heat-sensitive ones, give an output current proportional to the input variable (light or heat). To work properly with this current output device, we need an amplifier that can process input current rather than input voltage. The op amp shown in Figure 1.48 converts input current to an output voltage.

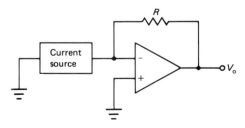

Figure 1.48 Op Amp Current-to-Voltage Converter

In this circuit almost no current flows in the inverting terminal. So, any input current from the source will flow through the feedback resistance. Because of the virtual ground, the voltage drop produced by the feedback resistor is also the output voltage. The output voltage is then determined by the input current, providing that the feedback resistor stays constant.

Op Amp Sample-and-Hold Circuit

A *sample-and-hold* (S/H) *circuit,* as its name implies, is used to sample a voltage at a particular time and hold it constant for another specified time interval. Figure 1.49 shows just such a circuit designed with an op amp.

In the S/H circuit, when a sample control voltage is applied to the switch, the switch turns on. The capacitor charges owing to the input voltage applied across it by the switch. The capacitor continues to charge until the switch is turned off. Since there is very little resistance to current flow in the charging circuit, the capacitor charges instantly to the input voltage.

After the switch is turned off, the input voltage stays on the capacitor. The voltage on the capacitor is then a measurement of the voltage at the input at the instant the sample was taken. The time when the switch is on is called the *sample period.* When the switch is off, the time is called the *holding period.*

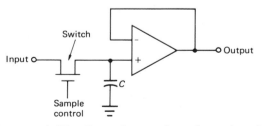

Figure 1.49 Op Amp Sample-and-Hold Circuit

LOW-VOLTAGE OP AMP

In 1979, National Semiconductor Corporation introduced a new device into the op amp world, the LM10. The LM10 is a low-voltage op amp. Its two main advantages are a voltage supply range from as low as 1.1 V to as high as 40 V and the ability to function in a floating or conventional mode. A block diagram of the LM10 is shown in Figure 1.50. Performance characteristics of the LM10 are similar to those of the LM108 op amp. The LM10, though, has a high output drive capability in addition to built-in thermal overload circuitry.

BRIDGE AMPLIFIER

The op amp makes an excellent *bridge amplifier.* Figure 1.51 illustrates a differencing amplifier driven by a bridge. Let us assume that R_2 is adjusted to give no difference in potential between points A and B. When this adjustment is made, we say that the bridge is balanced. At this point the output of the differencing amplifier should be zero. The resistor R_7 may have to be adjusted for 0 V out of the op amp. Let us further suppose that resistor R_1 decreases resistance so that a +0.5 V potential is felt from A

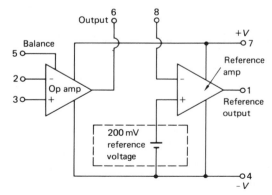

Figure 1.50 Block Diagram of LM10 Low-Voltage Op Amp

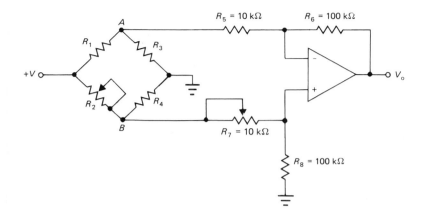

Figure 1.51 Differential Amplifier Driven by a Bridge

to B. Since the gain of the differential amplifier is 10, the output voltage would be -5 V.

The major problem with this circuit is that the differencing amplifier could load down the bridge, causing inaccuracies in the measurement. We could solve this problem by connecting the bridge to the instrumentation amplifier in Figure 1.52. The entire circuit is shown in Figure 1.52.

The gain of the instrumentation amplifier is expressed as follows:

$$A_V = -\left(\frac{R_3}{R_2}\right)\left(\frac{2R_1 + R_A}{R_A}\right)$$

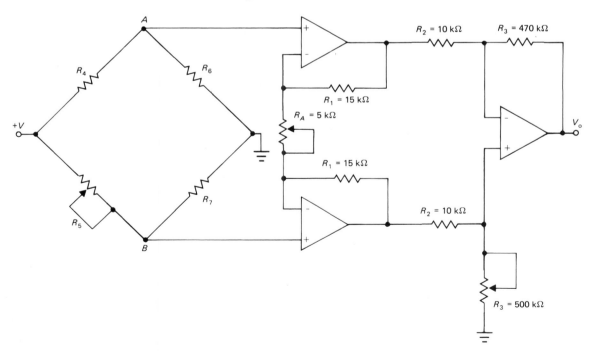

Figure 1.52 Instrumentation Amplifier Driven by a Bridge

The gain of the circuit in Figure 1.52 is, then, the following:

$$A_V = -\left(\frac{470 \text{ k}\Omega}{10 \text{ k}\Omega}\right)\left[\frac{2(15 \text{ k}\Omega) + 5 \text{ k}\Omega}{5 \text{ k}\Omega}\right]$$

$$= -329$$

Let us suppose that the bridge showed a difference in potential between A and B of -25 mV. The gain is expressed as follows:

$$A_V = \frac{V_o}{V_i}$$

Therefore, the output voltage will be:

$$V_o = A_V V_i = (-329)(-0.025)$$

$$= +8.23 \text{ V}$$

In addition to gain, this circuit has good common-mode rejection characteristics. Amplifiers with good common-mode rejection characteristics will cancel any noise that happens to be at the input while they amplify the input signal.

CLIPPER

It may be necessary to limit the output of an op amp stage to a certain maximum voltage. Some devices may be damaged with an input voltage higher than a specified maximum voltage. The *clipper,* shown in Figure 1.53, is a circuit that limits its own output voltage. Let us assume that each zener has a 5.4 V zener voltage (reversed-biased) and a 0.6 V forward-biased potential. With a positive input, the output would go negative, up to a maximum of -6.0 V. The voltage can go no higher than that because D_2 will clamp the output to -5.4 V, and D_1 will clamp at -0.6 V. Together, they clamp the output voltage to -6.0 V. Much the same action is repeated with a negative input. With a negative going input voltage, the output voltage will go positive, only as far as $+6.0$ V. At this point the D_2 would have a forward-biased drop of $+0.6$ V, and D_1 would have a reverse-biased drop of $+5.4$ V. Together, the total output voltage can rise no higher than $+6.0$ V.

WIDEBAND AC VOLTMETER

Most multimeters have AC voltmeters that have limited value in measuring high frequencies and low AC voltages. The circuit illustrated in Figure 1.54 is a sensitive AC voltmeter, capable of measuring signals as low as 15 mV at frequencies up to 500 kHz. The gain of the noninverting amplifier is adjusted by selecting R_1 through R_6, positioned as R_i in the noninverting amplifier. As R_i is decreased, the gain increases. The AC voltage gain will be equal to the sum of R_{10} plus the impedance of the bridge rectifier, divided by the range selector resistor chosen by the range selector switch. The input voltage is applied to the high-impedance noninverting terminal of the op amp. The AC is amplified by the amplifier applied to the bridge rectifier. The bridge rectifier converts the AC output voltage to DC and applies the DC to the meter. Resistor R_8 adjusts the output voltage of the op amp for zero, compensating for the input offset voltage. The capacitor C_3 is a decoupling

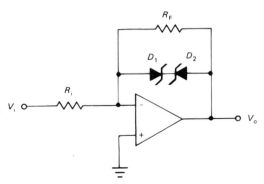

Figure 1.53 Op Amp Clipper

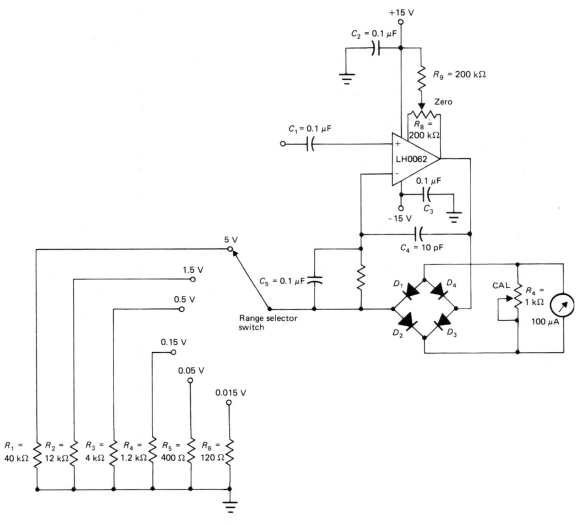

Figure 1.54   Sensitive AC Voltmeter

capacitor that prevents spurious oscillations, thereby improving amplifier stability.

This circuit uses the LH0062 high-speed, field effect transistor (FET) op amp. This device has a gain-bandwidth product of 15 MHz with a slew rate of 75 V/μs. With external compensation the slew rate may be extended to 120 V/μs.

OP AMP CURRENT DRIVERS

Low-power op amps, such as the 741, have a limited current output capability. In the 741 the output current is generally restricted to about 25 milliamperes (mA). If more current is desired for a particular application, two choices are available. First, a power op amp may be chosen. Such op amps can handle output powers up to several watts; however, power op amps are sometimes expensive and difficult to procure. A less expensive choice for larger current is to drive a transistor or a power MOSFET with an op amp. A circuit using the transistor is shown in Figure 1.55. The operational amplifier forces any input voltage at the noninverting terminal to be felt at the inverting terminal. If a positive 2 V potential is connected to the input at point A, the same potential is felt at point B. The transistor would draw a current equal to the following:

$$I = \frac{V_i}{R}$$

If the resistor were a 10 Ω resistor, the approximate current drawn would be as follows:

$$I = \frac{+2}{10} = 0.2 \text{ A} = 200 \text{ mA}$$

The circuit would then be supplying 200 mA to the relay coil, energizing it. The op amp, alone

and unassisted, may never have been able to generate this much current.

Another power op amp circuit, in this case a lamp driver, is shown in Figure 1.56. When the input voltage is below the reference voltage, the output is held low by diode D_1. The transistor will not be conducting. When the input voltage goes above the reference, the output of the comparator goes high, turning on the transistor. Current will then flow through the lamp, L_1, turning it on. The resistor R_2 limits the surge current in the circuit to a safe level until the filament in the bulb heats up. Resistor R_1 determines the base drive to the transistor. If the base drive needs to be increased, the resistance of R_1 will need to be decreased.

CURRENT-DIFFERENCING AMPLIFIER (CDA)

The *current-differencing amplifier* (CDA), or Norton amplifier, developed in the early 1970s,

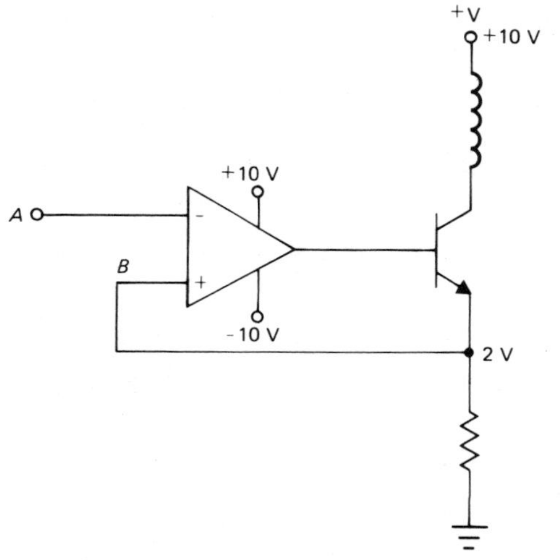

Figure 1.55 Op Amp Driving Transistor for Higher Currents

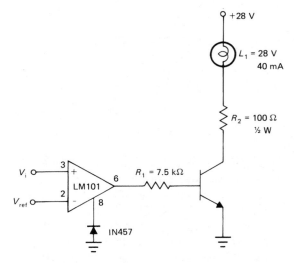

Figure 1.56 Op Amp Lamp Driver

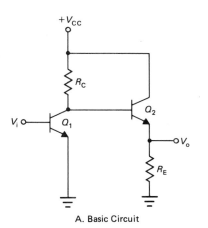

Figure 1.57 Inverting
Section of CDA

A. Basic Circuit

B. With Constant-Current Sources (Transistors)
Replacing Resistors

answered a need in the linear device field. Industry required a device that would be linear yet compatible with digital circuitry, and one that would operate from a single power supply. The CDA meets these needs and is relatively inexpensive as well. Four CDAs are housed within a 14-pin, IC, dual in-line package (DIP).

The basic CDA circuit is shown in Figure 1.57. The inverting section of the basic CDA is shown in Figure 1.57A. In Figure 1.57B, the resistors R_C and R_E have been replaced by transistors acting as constant-current sources. Since constant-current sources have very high dynamic resistances, the gain of this circuit reaches 60 dB.

Figure 1.58 illustrates the addition of a circuit that provides a noninverting input. This addition makes a differential input device. Transistor Q_5 and diode D_1 form a *current mirror stage*. Since the base-emitter junction of Q_5 and the cathode-anode junction of D_1 are matched, the application of a bias voltage causes both devices to conduct the same amount of current. Or we say that the current in D_1 is *mirrored* in Q_5. The CDA then tries to keep the difference between the two currents equal to zero. The same concept appears in the op amp, which tries to keep the difference in

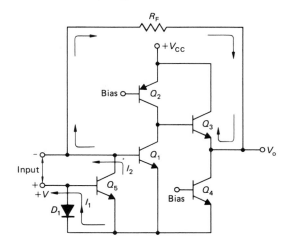

Figure 1.58 Simplified CDA Schematic Showing Inverting and Noninverting Terminal Inputs

potential between the two inputs equal to zero. The feedback resistor, the only component external to the IC in Figure 1.58, provides a path for current from Q_5 to the output transistor Q_3.

Biasing the CDA

Figure 1.59 shows the biasing arrangement used most for CDAs. When $+V$ is connected to

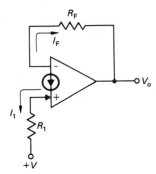

Figure 1.59 Current Mirror Biasing the CDA

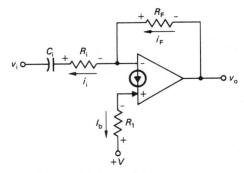

Figure 1.60 CDA Inverting Amp

R_1, current flows (I_b). The CDA mirrors the same amount of current through R_F as flows through R_1. The current through R_1 can be found from the following equation:

$$I_1 = \frac{+V - 0.7}{R_1} \approx \frac{+V}{R_1}$$

The output voltage likewise is as follows:

$$V_o = I_F R_F + 0.7 \approx I_F R_F$$

Since $I_1 = I_F$ (from the mirror principle), we have the following equation:

$$V_o = \left(\frac{+V}{R_1}\right) R_F$$

The output DC voltage in quiescent conditions is determined by the bias supply ($+V$), R_F, and R_1. For centered operation (where $V_o = +V/2$), $R_1 = 2R_F$. If $2R_F$ is substituted into Equation 1.5, then $V_o = +V/2$ (centered operation).

CDA Inverting Amplifier

The CDA inverting amplifier is similar in many ways to the comparable op amp circuit. The gain equation is identical. Figure 1.60 shows the basic CDA inverting amplifier. It is generally used in AC applications because the DC bias may be affected by the previous stage if there is no coupling capacitor.

Let us suppose that a positive-going input signal is applied to this circuit. Current i_i (the current due to the AC voltage applied to the input) flows as illustrated. The current mirror, which tries to keep the same current flowing out of the inverting terminal, causes the total current in R_F to decrease by an amount equal to i_i. So the change in R_F current, denoted as i_i, is then equal but opposite to i_F, which is the AC component of the total current in R_F. The gain equation is as follows:

$$A_V = -\frac{v_o}{v_i}$$

where

v_i = AC input voltage

v_o = AC output voltage

Thus, we can state v_o and v_i in terms of currents and resistances:

$$\frac{v_o}{v_i} = \frac{-i_F R_F}{i_i R_i}$$

Since $i_i = i_F$, this equation reduces the gain equation to the following:

$$A_V = \frac{v_o}{v_i} = \frac{-R_F}{R_i}$$

The output is now reduced from the $+V/2$ level by the amount A_V times v_i.

Figure 1.61A illustrates a basic inverting amplifier using a CDA. Current through R_1 (which is designated I_1) is calculated from the current equation mentioned earlier as follows:

$$I_1 = \frac{+V - 0.7\text{ V}}{R_1} = \frac{+10\text{ V} - 0.7\text{ V}}{2\text{ M}\Omega}$$

$$= 4.6\ \mu\text{A}$$

From our previous discussion, we know that I_3 equals I_1, where I_3 is the current through R_3. The output voltage, then, is as follows:

$$V_o = I_F R_F + 0.7\text{ V} = 4.6\text{ V} + 0.7\text{ V}$$

$$= 5.3\text{ V} \approx \frac{+V}{2} = 5\text{ V}$$

If we inject a 0.1 V peak signal into this amplifier, the output voltage change, where $A_V = -R_F/R_1 = -10$, is as follows:

$$v_o = A_V v_i = 10 \times 0.1\text{ V} = 1\text{ V}_{pk}$$

Thus, a 0.1 V peak input produces a 1 V peak output riding on a +5 V DC level. This relationship is diagramed in Figure 1.61B.

CDA Noninverting Amplifier

The circuit in Figure 1.62 connects a CDA so that it operates as a noninverting amplifier. The gain equation for this amp is the same as that of the inverting amp (except for the sign):

$$A_V = \frac{R_3}{R_2}$$

As in the inverting amplifier, the output voltage is +5 V, with about 4.6 μA flowing through R_1 and R_3 in the no-input-signal condition. The gain is also identical ($A_V = 10$). A 0.1 V peak input produces a 1 V peak output with no phase difference.

CDA Comparator

The CDA comparator is illustrated in Figure 1.63. The output of the device in Figure 1.63A is 0 V as long as the input voltage V_i is below the reference voltage V_{ref}. When the input voltage rises above V_{ref}, the output goes to saturation voltage and stays there as long as the input remains above the reference. Comparators such as these find extensive application in digital systems.

Figure 1.61 CDA Inverting Amp with AC Input

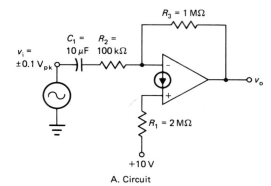

A. Circuit

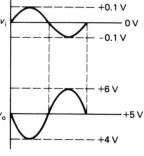

B. Input and Output Waveforms

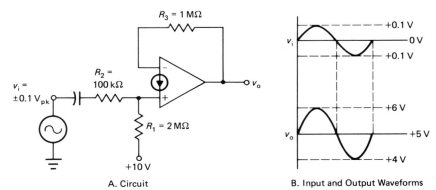

Figure 1.62 Noninverting CDA Amp with AC Input

A. Circuit

B. Input and Output Waveforms

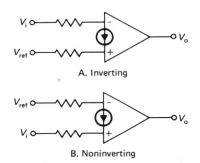

A. Inverting

B. Noninverting

Figure 1.63 CDA Comparator

Another type of comparator is shown in Figures 1.64A and 1.64B. In digital electronics this device is called a *Schmitt trigger*. In analog electronics it is usually called a *comparator with hysteresis*. A comparator with hysteresis often is needed to deal with the spurious switching caused by noise. As in an op amp comparator with hysteresis, the CDA circuit changes the trip point every time it is triggered.

The schematic shown in Figure 1.64A is an inverting comparator with hysteresis. Let us assume that the supply voltage is $+10$ V, and that when the output goes high, it goes to $+10$ V. Also, we will assume that when the output goes low, it will go to 0 V. The $+10$ V input at the reference will draw 5 μA of current. If the input is at 0 V, no current will be drawn from the inverting terminal. In this case the current in the

noninverting terminal is higher than the current in the inverting terminal, and the output will be high. A high output will draw 1 μA of current through R_3. The total current drawn from the noninverting terminal is 6 μA. To make the output go low, the current in the inverting terminal needs to go higher than that in

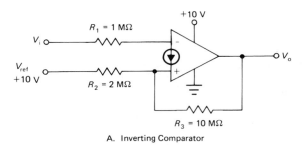

A. Inverting Comparator

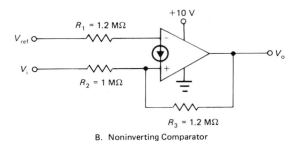

B. Noninverting Comparator

Figure 1.64 CDA Comparators with Hysteresis

the noninverting terminal. The output will trip and go low when the input voltage draws more than 6 μA. This will happen when the input voltage goes above 6 V. When the input voltage goes above 6 V, the output voltage goes low. This 6 V input potential is called the upper trip voltage. Six microamperes is the upper trip current. When the output voltage goes low, the only current drawn from the noninverting terminal is 5 μA. The trip voltage and current have changed. To trip the comparator and make the output go high again, the current in the inverting terminal will have to go below the 5 μA in the noninverting terminal. This will happen when the input voltage goes below 5 V. This potential is called the lower trip voltage. Five microamperes is the lower trip current.

Note the difference between the inverting comparator with hysteresis, shown in Figure 1.64A, and the noninverting comparator with hysteresis, shown in Figure 1.64B. We can see that the input and the reference have been switched. There are no other differences. In the noninverting comparator, a high output results from the current in the noninverting terminal exceeding the current in the inverting terminal. As in the inverting comparator, an input voltage causes the current to be drawn from the input terminal, tripping the comparator. The output voltage will go high when the input voltage exceeds a predetermined reference voltage. This reference voltage is called the upper trip point. When the input goes lower than a reference voltage called the lower trip point, the output voltage goes low.

Upper and lower trip points in the noninverting comparator are calculated in the same manner as for the inverting comparator. When the output is low (0 V), the input current must go above 8.33 μA. This 8.33 μA current will be drawn when the input voltage goes above 8.33 V, driving the output high ($+10$ V). When the output goes high, 5 μA will be drawn through the 2 MΩ R_3. The input voltage V_i must go below 3.3 V before the output will switch again

and go low. In this application, then, the upper trip voltage is 8.33 V and the lower trip voltage is 3.33 V. This variation in trip points is called hysteresis.

CDA Voltage Follower

One of the simplest applications of the CDA is the buffer, or isolation amplifier, pictured in schematic form in Figure 1.65. A positive DC voltage placed at the input will draw current from the noninverting terminal. That same current will be mirrored in the inverting terminal and converted to a voltage by the 1 MΩ feedback resistor. The output voltage should then follow the input in voltage and phase. This circuit does have some limitations, however. The input must be greater than $+0.6$ V to turn on the mirror current transistor in the CDA. This disadvantage may be overcome by other biasing techniques.

CDA Oscillator

The CDA is a device that has shown considerable flexibility in application, as does the normal op amp. The diagram in Figure 1.66 shows the CDA used as a square wave oscillator. Capacitor C_1 charges and discharges through R_1 between the limits set by R_2, R_3, and R_4. You will recognize this circuit as a modification of the comparator with hysteresis just discussed.

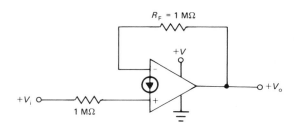

Figure 1.65   CDA Buffer Amp

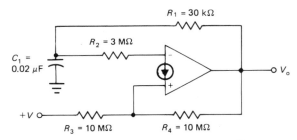

Figure 1.66   CDA Square Wave Oscillator

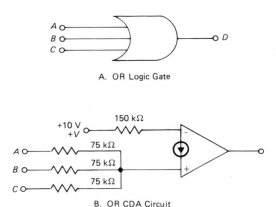

Figure 1.67   CDA OR Gate Function

When the output is low, the capacitor will discharge through R_1. As it discharges, the voltage across it will be decreasing, as will the current drawn from the inverting terminal. When the current in the inverting terminal goes below the current in the noninverting terminal, the output will go high. This will occur (when $+V = +10$ V) when the capacitor voltage goes below 3.33 V. The output then goes high and the capacitor starts to charge. When the output goes high, the current drawn from the noninverting terminal is doubled, from 1 μA to 2 μA. The capacitor voltage must go to twice the voltage ($+6.67$ V) to trip the CDA and make the output go low. With the component values given, the CDA square wave generator should oscillate at 1 kHz. With the component values given, the output waveform should be relatively symmetrical. If an unsymmetrical output is needed, the ratios of R_2, R_3, and R_4 can be varied.

CDA Logic Circuits

Some industrial applications require decision-making circuits, such as those normally provided by digital logic chips like the 7400 TTL series. In many applications the $+5$ V power supply is not available to power standard logic chips. The CDA can be used in place of logic chips in low-speed digital and switching applications. The larger voltage swing and the slower

speed can be an advantage in some industrial applications.

You will recall the digital logic OR function performed by the device schematically represented in Figure 1.67A. High voltages at inputs A, B, or C in this circuit will give a high output at D. The equivalent CDA circuit is shown in Figure 1.67B. If $+V$ is equal to $+10$ V, the current drawn from the inverting terminal will be approximately 66.7 μA. With no input, this voltage holds the output low, since the current in the inverting terminal is higher than that in the noninverting terminal (0 A). A $+10$ V input voltage at either A, B, or C will draw $+10$ V/75 kΩ, or 133 μA. The output voltage will go high since the current in the inverting terminal is higher than that in the noninverting terminal. The Boolean expression for this circuit is $D = A + B + C$.

This device can handle up to 50 inputs, each with its own 75 kΩ resistor, because the output current capability is 10 mA. This capability is called *fan-out*. A NOR function can be created by switching the inputs. In this case the output will be held high until an input pulls it low.

Another logic function provided by the CDA is symbolized by the schematic diagram in Figure 1.68A. We get a high output at D

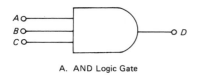

A. AND Logic Gate

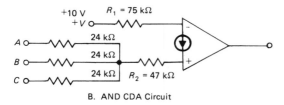

B. AND CDA Circuit

Figure 1.68 CDA AND Gate Function

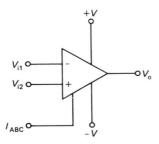

Figure 1.69 Schematic Diagram of the OTA

when highs are present at *A*, *B*, and *C*. The CDA version of this circuit is shown in Figure 1.68B. With a + *V* of +10 V, the input voltage supply will pull 133 μA from the inverting terminal. This will hold the output low with no input signal, as in the OR gate just discussed. With an input at *A* or *A* and *B*, there is not enough current to bring the total current in the noninverting terminal higher than in the inverting terminal. The only combination of input voltages that will cause high enough current in the noninverting terminal to trip the comparator is *A*, *B*, and *C*. The output will go high if and only if a high is present at *A*, *B*, and *C*. A NAND gate may be constructed by reversing the inputs, just as a NOR gate was created by switching inputs of an OR gate.

We have seen from this discussion that the CDA can substitute for the op amp in many applications. In fact, most of the op amp configurations can be duplicated by the CDA.

OPERATIONAL TRANSCONDUCTANCE AMPLIFIER (OTA)

A *transconductance amplifier* is a circuit or component that changes an input voltage into an output current. An *operational transconductance amplifier* (OTA) is an IC transconductance amplifier. In many ways the OTA resembles the op amp device discussed earlier. It possesses a differential voltage input, has a high input impedance, and high open-loop gain. The most striking difference in the two devices can be seen in the schematic diagram in Figure 1.69. Note the addition of a terminal labeled I_{ABC} in the OTA. This terminal is a control input that is called the *amplifier bias current control*. Many of the OTA's parameters can be controlled by varying the current in this terminal.

Table 1.3 shows the changes in OTA parameters with an increase in I_{ABC}. Notice that the parameter *transconductance* is given in Table 1.3 instead of voltage gain. Recall that transconductance is the change in output current divided by the change in input voltage, as expressed by the following equation:

$$g_m = \frac{I_o}{V_i}$$

where

g_m = transconductance in microsiemens, μS (formerly called micromhos)

I_o = output current

V_i = input voltage

The OTA, as its name implies, differs in regard to its output characteristics. Recall that the

Table 1.3 Changes in OTA Parameters with Increased I_{ABC}

Parameters	Changes
Input resistance	Decrease
Output resistance	Decrease
Input capacitance	Decrease
Output capacitance	Decrease
Transconductance	Increase
Input bias current	Increase
Input offset current	Increase
Input offset voltage	No change
Slew-rate	Increase

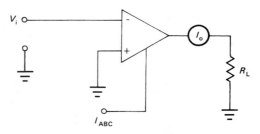

Figure 1.70 OTA Gain Control

op amp's output was defined in terms of voltage. In contrast, the OTA's output is defined in terms of current. In this regard, the OTA resembles an FET (field effect transistor) or vacuum tube. Therefore, transconductance is the suitable parameter to express this relationship, being the change in output current divided by the change in input voltage.

Like FETs and vacuum tubes, the OTA can be used as a voltage amplifier with the addition of load resistance or impedance. The voltage gain in this case equals the (mutual) transconductance (g_m) times the load resistance $(g_m R_L)$. Or, we could say that the output voltage will equal the transconductance times the output resistance times the input voltage as follows:

$$V_o = g_m V_i R_L$$

With an output resistance, the OTA will behave much like the normal op amp.

One of the simplest applications involving the OTA is in a gain control, such as shown in Figure 1.70. The V_{ABC} input voltage controls the transconductance and output current. An increase in V_{ABC} and I_{ABC} will cause transconductance to increase, thereby increasing output current and voltage. In OTAs like the CA3080 manufactured by RCA, the transconductance can be calculated if the supply voltage and I_{ABC}

are known. The equation that relates these factors is as follows:

$$g_m = \frac{300}{V} I_{ABC}$$

where 300 is a constant. For example, let us say that we had a supply voltage of ± 15 V with an I_{ABC} of 20 μA. The circuit transconductance is found by substituting these values into the preceding equation:

$$g_m = \frac{300}{V} I_{ABC} = \frac{300}{15 \text{ V}} (20 \ \mu A) = 400 \ \mu S$$

By substitution we can construct an equation that will allow us to predict the output voltage if we know the I_{ABC}, V_i, and R_L. The equation that allows us to make this prediction is as follows:

$$V_o = \frac{300}{V} I_{ABC} V_i R_L$$

where

V_o = output voltage
V = supply voltage
V_i = input voltage
R_L = load resistance

When using this equation we will assume that all the output current will flow into R_L, as shown in Figure 1.70. Using this formula, let us

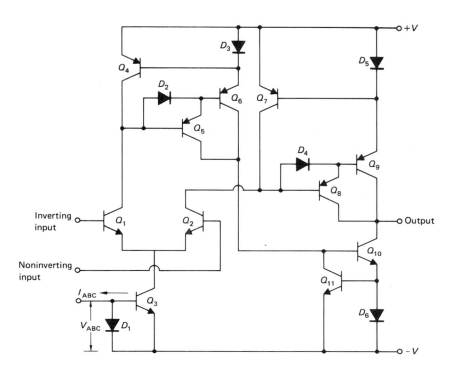

Figure 1.71 Internal Construction of the OTA

calculate the output voltage with an input of 0.5 V, an I_{ABC} of 100 μA, and an R_L of 10,000 Ω.

$$V_o = \frac{300}{V} I_{ABC} V_i R_L$$

$$= \frac{300}{15\ V} (100\ \mu A)\ (0.5\ V)(10,000\ \Omega)$$

$$= 10\ V$$

We should see an output voltage of approximately 10 V from this OTA. As expected, the OTA, because of its transconductance properties, has an output impedance much higher than that of the standard op amp.

Figure 1.71 shows the input stage of the OTA. Notice that it is a differential amplifier, just like the op amp. Transistor Q_3 and diode D_1 form the current source for the differential amplifier. This circuit differs from the op amp circuit in that the collector current of Q_3 can be changed by varying the current in the I_{ABC} ter-

minal. Changing this current varies the transconductance of the differential amplifier.

Numerous applications exist for OTAs. It is probably best suited to telemetry and communications use by virtue of the unique control provided by the I_{ABC} terminal. The OTA is also used in measurement circuits in a sample-and-hold configuration. Other applications include nonlinear mixing, gain control (including automatic gain control), waveform generation, and multiplication.

An example of an OTA application is the circuit shown in Figure 1.72, which has an amplitude modulator. Here, the output current is equal to the transconductance times the input voltage. In this amplifier, the level of the unmodulated carrier signal is controlled by the average current into the I_{ABC} terminal. The carrier frequency is amplitude-modulated when the modulating signal V_m causes changes in I_{ABC}. As the input-modulating signal (V_m) goes positive,

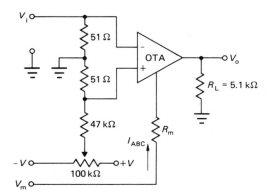

Figure 1.72 Amplitude Modulation Using an OTA

the current in I_{ABC} increases, thus increasing the transconductance of the OTA. As V_m goes negative, the current in I_{ABC} decreases, decreasing the OTA's g_m. The output, developed across the 5.1 kΩ resistor, is the familiar amplitude-modulated waveform.

PROGRAMMABLE OP AMPS

Most op amps have characteristics that the user has little or no control over. For instance, the input bias current for an op amp is generally fixed for each device according to its internal construction. *Programmable op amps,* however, allow the user to change some of these characteristics. Often, circuit designers are forced to make trade-offs when choosing a particular device for a job. The programmable op amp gives the engineer more flexibility in fitting an op amp to a particular application.

The programmable op amp has a special terminal similar to the I_{ABC} terminal on the transconductance amp. The control terminal on the programmable op amp is called the *set terminal.* Current flow from this terminal can change an op amp's slew rate, gain-bandwidth product, input offset current and voltage, and noise.

PROTECTING OP AMPS

Operational amplifiers, like other ICs, are susceptible to damage from too much voltage in the wrong places. Often, technicians make mistakes when breadboarding, or constructing prototypes of, circuits with op amps. In these cases damage may result to a device that is expensive or difficult to obtain. It is a good idea to try to protect the op amp from damage. Several commonly used methods of op amp protection are shown in Figure 1.73. Remember that, although we will be demonstrating these techniques on op amps, they can be used with any linear IC.

Most op amps will be destroyed if too high a voltage is placed across the input terminals. Figure 1.73A shows one input protection method using zener diodes connected back to back. The voltage at the input can go no higher than the zener voltage V_Z plus 0.6 V, regardless of the polarity. An alternate input protection method is shown in Figure 1.73B. This method uses normal semiconductor diodes D_2 and D_3 and operates in a similar fashion to the zener diodes. The difference is that the normal silicon

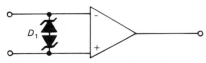

A. Input Protection with Zener Diodes

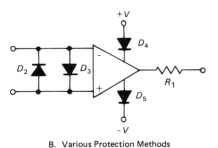

B. Various Protection Methods

Figure 1.73 Op Amp Protection Methods

diodes clamp the input voltage to ± 0.6 V. The input can go no higher than this.

Most op amps have built-in short circuit protection and, as such, need no additional short circuit protection. For example, the 741 op amp is short circuit protected, limiting output current to about 25 mA. For devices that do not have such protection, however, a 200 Ω series output resistor (R_1) will limit any short circuit current that may be drawn. The device will then be protected at the output.

One of the easiest mistakes to make in breadboarding is to cross the positive and negative supplies. Most ICs, including the op amp, will not tolerate this condition. The device will be destroyed. Protection for this kind of fault is found in diodes D_4 and D_5. Diode D_4 protects against a negative voltage being connected to the positive supply terminal. If a negative voltage is connected to the positive supply terminal, diode D_4 will be reverse-biased and no current will flow. This will protect the op amp from damaging reverse voltages. The negative supply is likewise protected from positive voltages by reverse biasing D_5.

TROUBLESHOOTING OP AMPS

Once a problem has been isolated to an IC op amp, the op amp itself must be checked. Fault analysis of any kind can be broken down into three categories: no output, low output, or distorted output. A simple check of the op amp's output determines in which of these general areas a malfunction is occurring.

No Output

A no-output condition may be caused by any of the following faults:

- Lack of power supply—check the op amp terminals for proper supply voltages.
- Lack of input voltage—check for the proper input signal or voltage.

- Saturated op amp—check for undesired DC inputs that may be causing saturation, check for a shorted input resistor or an open feedback resistor, and check for proper grounding of components and terminals.
- Malfunctioning op amp—take out the op amp and check it in an inverting amp configuration or replace it. (The inverting mode of operation ensures that the op amp is working in order to give an inverted output. A short or open in the noninverting mode may cause it to appear to be working poorly.)

Low Output

A low-output condition may be caused by any of the following faults:

- Low power supply—check the op amp terminals for proper supply voltages.
- Low input voltage—check the op amp input for proper signal or voltage levels.
- Change in component values—check the input and feedback resistances for changes in resistance.
- Malfunctioning op amp—replace it with a good unit.

Distorted Output

A distorted-output condition may be caused by any of the following faults:

- Distortion in the power supply or low power supply—check the op amp terminals for proper supply voltages.
- Distorted input—check the input to the op amp for distortion.
- Malfunctioning op amp—replace the op amp with a good unit.
- Operation outside of published specifications—check the slew rate, gain bandwidth, and so on.

Troubleshooting Breadboarded Circuits

As a technician, you may either breadboard circuits of your own or be required to check breadboarded circuits. In addition to observing the troubleshooting hints just mentioned, you should check closely your circuit's wiring. Both beginner and expert often do not wire a circuit correctly. Make sure that the IC pinout from the data sheet is observed.

CONCLUSION

We only have scratched the surface of op amp applications in this chapter. But we have seen that the IC op amp has applications perhaps undreamed of by those who developed it in the 1960s. Since this device is a major analog circuit building block, it is important that you have a firm grasp on its operation.

QUESTIONS

1. The voltage follower is a special application of the _____ amplifier. Its input impedance is considered to be _____.
2. The difference amplifier is a combination of the inverting and noninverting amplifiers. As such, it has _____ inputs.
3. The instrumentation amplifier is a _____ amplifier with each input connected to a voltage follower. It amplifies difference-mode signals while _____ common-mode noise.
4. The output of the integrator is a voltage that is produced by _____ the input signal over time. The output of a differentiator is proportional to the _____ of the input.
5. In the log amp, the output is proportional to the _____ of the input. The mathematical operation of _____ can be accomplished with log amps.
6. A filter that passes everything above a specified cutoff frequency is called a _____ filter. The roll-off of a first-order filter is _____ per decade.
7. The CDA is compatible with _____ logic, since it can operate on one 5 V supply.
8. In the OTA, the output _____ is proportional to the input voltage.

9. Describe the gain and input impedance of the voltage follower and some applications of this circuit.
10. Describe the advantages of the difference amplifier when compared with single-input circuits.
11. Contrast the integrator and differentiator in terms of function and input and output waveforms.
12. What is the major advantage of a comparator with hysteresis over one without it?
13. List the four different types of pass filters, and give an example of each in an active filter.
14. What are the advantages of active filters over passive ones?
15. Describe the function of a gyrator.
16. Describe the need for and construction of a CDA.
17. What is the phase relationship between input and output in the circuit shown in Figure 1.62?
18. Compare the characteristics and applications of op amps, CDAs, and OTAs.

PROBLEMS

1. Suppose the difference amp in Figure 1.4 has the following resistances: $R_F = 20$ kΩ, $R_i = 10$ kΩ, $R_A = 5$ kΩ, and $R_B = 5$ kΩ.
   a. Calculate V_o with $V_{i1} = +1$ V and $V_{i2} = -2$ V.
   b. Draw the resultant output waveform caused by applying a 2 V peak input sine wave at each input (assume both signals are in phase).

2. Suppose the instrumentation amp in Figure 1.5 has $R_1 = 45$ kΩ, $R_2 = 10$ kΩ, and $R_3 = 100$ kΩ, with R_A adjusted to 10 kΩ. With a differential input of 1 V, what is the output voltage?

3. The integrator in Figure 1.9 has a 1 kHz, 1 V peak input sine wave. What is the output waveform amplitude? How does the output waveform compare with the input in phase and waveshape?

4. The integrator in Figure 1.9 has a 2 kHz, 0.5 V peak square wave input. Draw the output waveform in the correct phase with the input, and indicate the output amplitude.

5. Draw the output waveform, indicating proper phase and amplitude, with the inputs to Figure 1.12 as follows:
   a. 2.5 kHz, 1 V peak sine wave.
   b. 1.0 kHz, 0.5 V peak triangle wave.

6. Draw the output waveform of the comparator circuit shown in Figure 1.20A with a 5 V peak sine wave input and a V_{ref} of -1 V (assume $\pm V_{sat} = \pm 10$ V).

7. Calculate upper and lower trip points (V_{ref}) for the circuit shown in Figure 1.22 (assume $\pm V_{sat} = \pm 15$ V, $R_A = 4$ kΩ, and $R_B = 1$ kΩ).

8. Assume the following resistance and voltage values in Figure 1.60: $R_i = 100$ kΩ, $R_F = 470$ kΩ, $R_1 = 1$ MΩ, and $+V = +20$ V. Calculate (a) A_V, (b) $V_{o(DC)}$, (c) $V_{o(AC)}$ with a 2 V peak input sine wave, and (d) I_b.

9. With the supply voltage shown in Figure 1.61, a DC voltage of 7.5 V is measured at the output of the CDA.
   a. What is the ratio of R_3 to R_1 that produces this voltage?
   b. With this bias point, what is the maximum peak voltage that can be applied at the input without distorting the output? *Hint:* How far can the output voltage go in a positive direction?

10. Assume the following resistance and voltage values in Figure 1.62: $R_1 = 470$ kΩ, $R_2 = 50$ kΩ, $R_3 = 220$ kΩ, $+V = 5$ V, and the input sine wave is $+0.1$ V peak. Calculate (a) A_V, (b) $V_{o(DC)}$, (c) $V_{o(AC)}$, and (d) I_b.

11. In the instrumentation amplifier circuit in Figure 1.52, R_2 and R_3 are increased to 1 MΩ. What is the gain of this stage?
   a. With this new resistance, calculate the output voltage with a 55 mV bridge potential at A and B.
   b. If $R_4 = R_6 = 1000$ Ω and $R_5 = R_7 = 500$ Ω, what change in R_5 resistance will give a 55 mV potential difference between A and B?

12. In Figure 1.51, if the diodes are 10.4 V zeners, what is the output voltage clamped to? Assume forward bias of $+0.6$ V.

13. The op amp in Figure 1.55 is driving a transistor so that a 3 V potential is found at the emitter. If the emitter resistance is 100 Ω, how much current is the transistor drawing? Assume $I_C = I_E$. If the transistor beta is 50, how much output current will the op amp need to drive the transistor to get a 3 V potential at the emitter?

14. The lamp driver in Figure 1.56 has an op amp output of $+10$ V. How much current is the op amp providing to drive the transistor? What is the voltage drop across the resistor R_1? If the DC beta of the transistor is 50, what is the approximate collector current?

15. The comparator in Figure 1.64A has the

following resistances: R_1 = 1.2 MΩ, R_2 = 2.5 MΩ, and R_3 = 10 MΩ. Assume V_o = +10 V. Find (a) upper and lower trip voltages, and (b) upper and lower trip currents.

16. The comparator in Figure 1.64A has the following resistances: R_1 = 1.0 MΩ, R_2 = 1.5 MΩ, and R_3 = 1.5 MΩ. Assume V_o = +10 V. Find (a) upper and lower trip voltages, and (b) upper and lower trip currents.

17. The OTA in Figure 1.55 has a supply voltage of ±10 V and an I_{ABC} of 100 μA. Find (a) the OTA transconductance, (b) the output voltage with an input of 10 mV and a load resistor of 1000 Ω, and (c) the I_{ABC} needed to give a 1 V potential at the output with a 120 mV input and a load resistance of 1000 Ω.

2

LINEAR INTEGRATED CIRCUITS FOR INDUSTRIAL APPLICATIONS

OBJECTIVES

On completion of this chapter, you should be able to:

- Describe the function of the voltage-to-frequency converter (VFC) and calculate its output frequency;
- Describe the function of the phase-locked loop (PLL);
- Define and calculate PLL capture and lock ranges;
- Describe the function of the frequency-to-voltage converter (FVC) and calculate its output voltage;
- Calculate the resolution of a digital system;
- Describe the analog-to-digital (A/D) conversion process and describe the operation of two analog-to-digital converters (ADCs);
- Describe the digital-to-analog (D/A) conversion process and describe the operation of two digital-to-analog converters (DACs);
- Describe the operation of the analog switch;
- Define current sourcing and current sinking.

INTRODUCTION

In Chapter 1 we discussed the op amp and some of its applications. As we have seen, few linear ICs can match the op amp for versatility in design. In addition to being versatile, it is inexpensive and reliable. There are applications, however, that are difficult to achieve with op amps. Because of this, IC designers have developed more specialized ICs, some of which we will be discussing in this chapter. In this chapter we will examine voltage-controlled oscillators, phase-locked loops, voltage-to-frequency converters, frequency-to-voltage converters, analog-to-digital converters, and digital-to-analog converters. These devices are used widely in industrial applications. You also will see applications for them in later chapters.

VOLTAGE-TO-FREQUENCY (V/F) CONVERSION

The *voltage-to-frequency converter* (VFC) is a device that changes analog voltages or currents to an AC frequency. The output signal frequency usually is proportional to the input voltage or current. The input is normally a constant or varying DC potential. The output frequency continuously tracks the input signal, responding directly to the input voltage. A block diagram shows the concept of the VFC, with a variable DC input voltage and a variable output frequency.

A common VFC is the NE/SE566 IC made by Signetics. The 566 is a VCO with two buffered outputs. A block diagram of the 566 is shown in Figure 2.1. Note that the two buffered outputs, pins 3 and 4, provide both a square wave and a triangle wave. An external resistor (R_1) and capacitor (C_1) determine the oscillator's frequency. The input voltage at pin 5 also will determine, with the resistor and capacitor, the frequency of oscillation. By changing the input voltage, the output frequency can be varied over a ten-to-one range with a frequency that is linearly proportional to the voltage.

The block diagram in Figure 2.1 can show us the basic operation of this VCO. The capacitor C_1 charges through the current source and

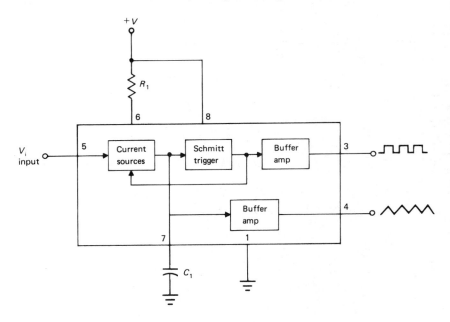

Figure 2.1   Voltage-Controlled Oscillator (VCO) Functional Block Diagram

through resistor R_1. The current source provides a constant charging current. How fast the capacitor charges depends on the resistance of R_1, the capacitance of C_1, and the input voltage controlling the current source. With a constant capacitance and resistance, an increase in the voltage to the current source will increase the rate at which the capacitor charges. When the capacitor reaches a set point, it starts to discharge at a linear rate through R_1 and controlled by the current source. A capacitor charging and discharging at a constant rate displays a triangle wave voltage signal across itself. A voltage follower–buffer amp will provide a triangle wave output to pin 4 of the 566 VCO. The buffer amp keeps the output from being loaded down and changing the oscillator frequency and waveshape. The triangle wave is also connected to the input of a Schmitt trigger (a comparator with hysteresis). The output of the comparator is a square wave. The square wave is sent to a buffer amp for isolation from loading effects in the output.

A circuit schematic diagram of a 566 VCO is shown in Figure 2.2. Resistors R_2 and R_3 form a voltage divider to provide a bias for pin 5, the control voltage input. The voltage on this pin must be between the values $+V$ and three-quarters of $+V$ for the device to operate properly. For example, if the $+V$ is $+10$ V, the voltage at pin 5 must be between $+10$ V and $+7.5$ V. This is a design requirement for the 566 IC. As mentioned before, the resistor R_1 and capacitor C_1 help set the frequency out of the device. Another design requirement for this VCO relates to the size of R_1. Resistor R_1 must be between 2 kΩ and 20 kΩ, or the device will not perform as specified.

The output frequency at pins 3 and 4 is approximated by the following formula:

$$f_o = \frac{2[(+V) - (V_i)]}{(R_1)(C_1)(+V)} \tag{2.1}$$

With the resistors as shown in Figure 2.2 and with a $+12$ V supply, the approximate out-

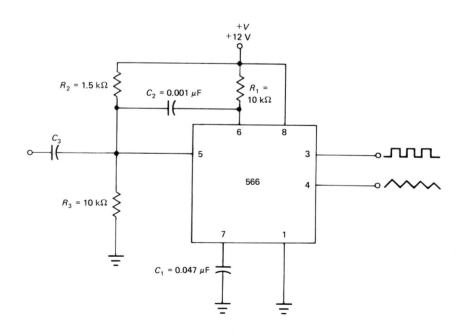

Figure 2.2 Basic VCO Circuit

put frequency is as follows:

$$f_o = \frac{2[(+12 \text{ V}) - (+10.43 \text{ V})]}{(10 \text{ k}\Omega)(0.047 \text{ }\mu\text{F})(+12 \text{ V})} = 557 \text{ Hz}$$

The voltage V_i is calculated by the voltage divider formula to solve for the drop across R_3:

$$V_i = \left(\frac{R_3}{R_2 + R_3}\right)(+V)$$

$$= \left(\frac{10{,}000}{1500 + 10{,}000}\right)(+12) = +10.43 \text{ V}$$

The VCO output frequencies at pins 3 and 4 should be approximately 557 Hz. Let us see what happens to the output frequency if the resistance of R_2 increases to 2 kΩ:

$$f_o = \frac{2[(+12 \text{ V}) - (+10 \text{ V})]}{(10 \text{ k}\Omega)(0.047 \text{ }\mu\text{F})(+12 \text{ V})} = 709 \text{ Hz}$$

Note that when the resistor R_2 increases to 2 kΩ, the voltage at V_i decreases from 10.43 V to 10.0 V. This decrease in voltage causes an increase in the output frequency from 557 Hz to 709 Hz.

From this discussion we can see one way to carry information about a sensor that changes resistance. A *sensor* (discussed in a later chapter) is a device that undergoes a change in a characteristic when an environmental parameter changes. For example, some resistors change their resistance as a function of temperature. We could substitute a temperature-sensitive resistance for either R_2 or R_3. The output frequency would then be directly related to the temperature changes around the resistor.

In other applications a voltage is applied to the input through capacitor C_3. The amplitude changes applied to the capacitor C_3 will be converted to frequency changes at the output. Using a value of 1.5 kΩ for the resistor R_2, the output frequency will be 557 Hz with no input

voltage. This frequency is called the *center* or *rest frequency*.

If we connect a sine wave generator to the input with a frequency of 1 kHz, the output frequency will change at a rate of 1 kHz, as shown in Figure 2.3. Note that as the input signal's amplitude increases, the output square wave frequency decreases. When the input voltage goes less positive, the output frequency increases. We can apply Equation 2.1 to solve for the output frequency at the input voltage peaks. With a ± 1 V peak input signal, we know that the voltage at pin 5 will be $+10.43$ V plus 1 V and $+10.43$ V minus 1 V. The input voltage at V_i will be $+9.43$ V and $+11.43$ V at the input negative and postive peaks, respectively.

At a V_i of $+9.43$ V the frequency would be:

$$f_o = \frac{2[(+V) - (V_i)]}{(R_1)(C_1)(+V)}$$

$$= \frac{2[(+12 \text{ V}) - (+9.43 \text{ V})]}{(10 \text{ k}\Omega)(0.047 \text{ }\mu\text{F})(+12 \text{ V})} = 911 \text{ Hz}$$

At a V_i of $+11.43$ V the frequency would be:

$$f_o = \frac{2[(+V) - (V_i)]}{(R_1)(C)_1)(+V)}$$

$$= \frac{2[(+12 \text{ V}) - (+11.43 \text{ V})]}{(10 \text{ k}\Omega)(0.047 \text{ }\mu\text{F})(+12 \text{ V})} = 202 \text{ Hz}$$

We can see from this illustration that when the input voltage goes from $+10.43$ V to $+11.43$ V, the output frequency goes from 557 Hz, the rest frequency, to 202 Hz. When the input voltage goes down to $+9.43$ V, the output frequency goes from 557 Hz to 911 Hz. This process is called frequency modulation (FM).

We may need to look at our circuit in a different way for some applications. We may need to know how much input voltage would produce a given range in output frequencies. If we label the high output frequency f_h and the low

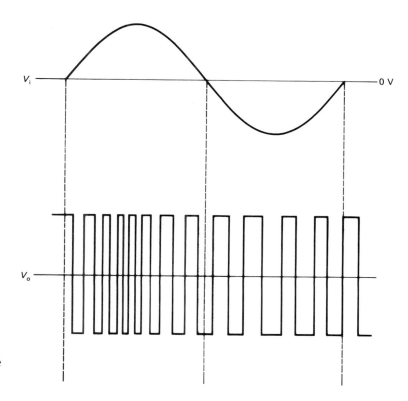

Figure 2.3 VCO Basic
Input and Output Voltage
Waveforms

frequency f_l, we can say that the range (Δf) is the difference between them, as expressed by the following equation:

$$\Delta f = f_h - f_l$$

Using Equation 2.1, we know that the minimum voltage (V_{min}) in will produce f_h, the highest frequency out:

$$f_h = \frac{2[(+V) - (V_{min})]}{(R_1)(C_1)(+V)}$$

The lowest frequency, f_l, is produced by the maximum voltage in (V_{max}):

$$f_l = \frac{2[(+V) - (V_{max})]}{(R_1)(C_1)(+V)}$$

By substitution, we can arrive at an equation for Δf as follows:

$$\Delta f = \left(\frac{2[(+V) - (V_{min})]}{(R)_1(C_1)(+V)} \right)$$
$$- \left(\frac{2[(+V) - (V_{max})]}{(R_1)(C)_1(+V)} \right)$$

Simplifying, we get:

$$\Delta f = \frac{2(V_{max} - V_{min})}{(R_1)(C_1)(+V)}$$

If we think of $V_{max} - V_{min}$ as ΔV, we can solve for ΔV:

$$\Delta V = \frac{(\Delta f)(R_1)(C_1)(+V)}{2}$$

As an example, let us suppose we want to know what change in input voltage (ΔV) would give a 300 Hz change in output frequency (Δf). Using the preceding equation we get:

$$\Delta V = \frac{(300 \text{ Hz})(10 \text{ k}\Omega)(0.047 \text{ }\mu\text{F})(+12 \text{ V})}{2}$$

$$= 0.846 \text{ V}$$

According to our calculation, we can get a 300 Hz change in the output frequency with a 0.846 V change in the input voltage.

In some cases the output of the 566 must be compatible with digital transistor-transistor logic (TTL). It is preferable to use the circuit shown in Figure 2.4 to interface with TTL. TTL requires a current sink of more than 1 mA. This circuit provides current sinking capability to 2 mA when the switch is down. When the switch is up, a saturated transistor is inserted between

the 566 and the logic circuitry. This circuit is used for TTL circuits that need a fast fall time (<50 ns) and a large current sinking capability.

VCOs, such as the NE566, are used in the FM modulation process, because the output frequency is a very linear function of the input voltage amplitude. Other applications include tone generators, signal generators, pulse generators, and function generators. An example of a negative ramp pulse generator is shown in Figure 2.5. In the ramp generator portion of this circuit, the transistor is turned on by the output of pin 3. The transistor quickly discharges the capacitor at the end of its charging cycle. The transistor also charges the timing capacitor (C_1) at the end of the discharge period. Because the circuits are reset quickly, the temperature stability of the ramp generator is excellent. The period t is $\frac{1}{2}f_o$, where f_o is the 566 free-running frequency in normal operation. The period of

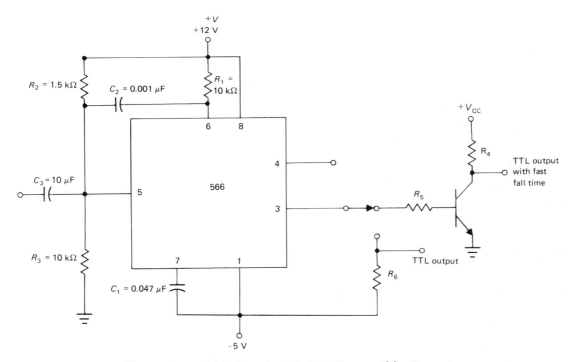

Figure 2.4   VCO Circuit with TTL Compatible Output

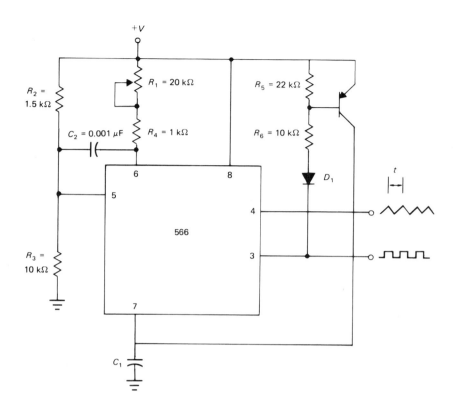

Figure 2.5 VCO Negative Ramp Pulse Generator

the output waveform, therefore, is given by the following equation:

$$t = \frac{(R_t)(C_1)(+V)}{5(+V - V_c)}$$

where

t = period of the output waveforms

R_t = resistance connected between pin 6 and $+V$

PHASE-LOCKED LOOPS

A *phase-locked loop* (PLL) is a group of electronic circuits arranged in a feedback system, as shown in block diagram form in Figure 2.6. The PLL is composed of three basic circuits: the phase comparator or detector, the low-pass filter, and the VCO. The VCO is the same circuit that we discussed previously, a free-running oscillator that has the capability to lock or synchronize with an incoming frequency. The VCO center or rest frequency is determined by an external timing resistor and capacitor and by the input voltage. The capacitor and resistor are normally fixed in the PLL, while the input voltage is variable.

The VCO output is connected to the input of the phase comparator. The *phase comparator,* as its name suggests, compares the phase of the VCO signal (f_c) to the phase of the input signal (f_o). The output of the phase comparator is an error correction voltage. The size of this voltage is a function of the phase and frequency differences between the two input signals. The

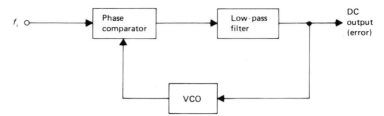

Figure 2.6 Phase-Locked Loop (PLL) Functional Block Diagram

error voltage is then filtered by the low-pass filter. The filtered output of the low-pass filter is the input to the VCO. When the input phase changes, this indicates that the input frequency is changing. The phase comparator output voltage increases or decreases just enough to keep the VCO frequency the same as the input frequency. The average voltage applied to the VCO is a function of the input signal.

PLL Capture and Lock Ranges

The PLL has three basic states: the free-running state, the capture state, and the locked or tracking state. In the *free-running state,* the VCO is some distance away from the input frequency f_c. In this state the input frequency is too far away from the VCO frequency to have any effect. Changes in the incoming frequency do not affect the VCO frequency; the VCO frequency does not change. If the input frequency is brought closer to the VCO frequency, a point is reached where the low-pass filter lets through enough voltage to have an effect on the VCO. Remember that the phase comparator puts out a voltage that is proportional to the difference between the two input frequencies, f_o and f_c. The error voltage is now large enough to cause the VCO to move toward the input frequency, f_c. If we look at the input to the VCO at this time, we see a waveform such as that shown in Figure 2.7. As the capture process starts, a small sine wave appears. This is the *difference frequency* or *beat frequency* between the VCO and f_c. Note that the negative half of the waveform

is slightly larger than the positive half. This condition is the DC component of the beat frequency. This DC component is what actually drives the VCO toward the lock state. Every successive cycle causes the VCO to move closer to the locked state, where the VCO and input frequency will be equal. Note that as the VCO moves closer to the input signal, the beat frequency decreases. This decrease allows the low-pass filter to pass a larger DC voltage to the VCO. At the same time, the closer the VCO moves toward the locked state, the longer it lingers near that state. This behavior extends the negative part of the sine wave and reduces the positive alternation. This feedback process continues until the VCO locks on the incoming frequency and the beat or difference frequency is zero.

This positive feedback behavior makes the VCO *snap* into lock when the input frequency is close enough. We can define *close enough* by the term capture range. The *capture range* of a PLL is the frequency range centered around the free-running VCO, over which the device can get a lock on the input signal. The low-pass filter plays an important role in the capturing process. Note that the low-pass filter is made up of an internal resistor and an external capacitor. If the input signal frequency is too far away from the VCO, the beat or difference frequency will be too high to pass through the filter. The low-pass filter will attenuate it. The input frequency is said to be out of the capture range.

Once the VCO has locked on the input frequency, the low-pass filter does not restrict the PLL. The VCO is free to track a signal past the

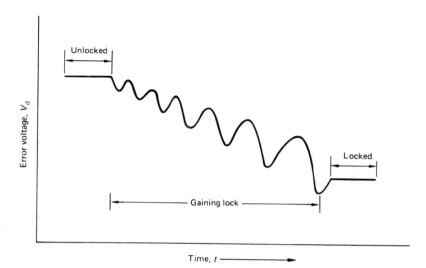

Figure 2.7 Error Voltage versus Time during PLL Capture

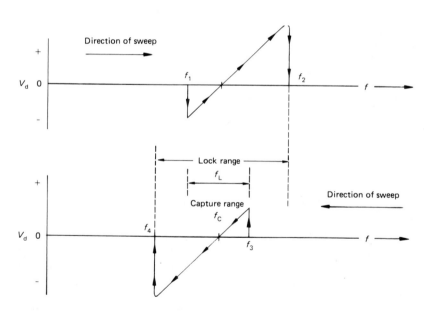

Figure 2.8 PLL Capture and Lock Ranges

capture range. The low-pass filter does, however, affect the speed at which the VCO can follow the input frequency. If the input frequency changes too fast, the VCO will stop following the input; it will become unlocked. As you might expect, there is a limit to the distance that the VCO can track the input frequency.

This limit is called the lock range. The *lock range* is the range of input frequencies over which the PLL will remain in lock. It is always larger than the capture range.

We can get a better picture of the capture and lock ranges of the PLL if we examine the diagram in Figure 2.8. This diagram is a plot of the DC voltage input to the VCO versus the input frequency. The input frequency is assumed to be a sine wave, the frequency of which is being swept over a wide frequency range. In Figure 2.8 the input frequency starts low and is being gradually increased. The PLL is not in lock at this time. Note that the PLL does not respond to the input signal until it reaches frequency f_1. This frequency is the lower edge of the capture range. At f_1 the PLL locks on the input signal, causing the error voltage to jump in a negative direction. This negative voltage is the input voltage the VCO needs to bring itself into lock with the input signal. From f_1 the error voltage increases in a linear fashion as the input frequency is increased. If we were to observe the VCO, we would see it track along with the input frequency. The VCO will track until the upper edge of the lock range is reached at f_2. At this point the PLL loses lock, and the error voltage goes back to zero. When the PLL loses lock, we also should see the VCO go back to its free-running frequency.

We will now sweep the input frequency down, observing the effect on the error voltage. The result is pictured in the bottom portion of Figure 2.8. Note that the same cycle repeats itself, only at different frequencies. The PLL recaptures the input frequency at f_3. This is the upper edge of the capture range. Note the change in the error voltage as we decrease the input frequency. The error voltage decreases, decreasing the VCO and keeping it in lock. When f_4 is reached, the VCO loses lock on the input frequency, and VCO goes back to its free-running frequency. Note that the capture range is between f_1 and f_3. The lock range is between f_2 and f_4.

Further examination of Figure 2.8 reveals several things. First, the PLL is selective in the frequencies it can lock on. It will capture only frequencies inside the capture range of the device. Second, the VCO has a very linear V/F transfer characteristic. This is useful when an application demands a linear correspondence

between input frequency (AC) and output voltage (DC). The linear V/F transfer characteristic is an important factor when the PLL is used to demodulate FM signals.

The VCO in the PLL performs the same functions as the VCO discussed earlier in this chapter. When the PLL is locked on to a frequency, the VCO input voltage is a function of the input frequency. When the VCO is used in demodulation applications, the VCO V/F transfer characteristic must be linear. If the output frequency of the VCO is not linear, distortion will result in the demodulation process. How much the VCO frequency changes for a given input DC voltage change is called *conversion gain*. The conversion gain equation is:

$$K = \frac{\Delta f}{\Delta V_e}$$

where

K = conversion gain in Hz/V

Δf = change in VCO frequency

ΔV_e = change in VCO DC input voltage

The conversion gain of a PLL can be found by taking the change in VCO control voltage (ΔV) for a given percentage frequency change and multiplying by the rest frequency. The VCO responds almost instantly to an input control voltage change.

The NE565 IC PLL is one of the more popular IC PLLs. It is a general-purpose PLL that can be used up to frequencies of 1 MHz. Compared to other PLLs, it has an exceptionally wide lock range (up to $\pm 60\%$ of the VCO free-running frequency). Its sensitivity, however, is 1 mV, which is lower than that of the other 560 series PLLs. The sensitivity of the device indicates how large an input signal must be before the VCO can lock on and track. Unlike the other members of the 560 series of PLLs, the 565 does not have a zener at the power supply input to keep the input voltages regulated. When using the 565, you should use a well-regulated power supply.

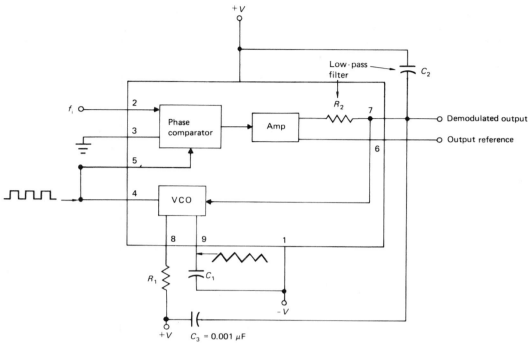

Figure 2.9 Basic PLL Circuit

The diagram in Figure 2.9 shows a circuit that uses the 565 to demodulate an FM input. This input may come from a VCO driven by an input sine wave, such as we discussed earlier in this chapter. In this circuit the average DC output of the phase comparator is directly proportional to the input frequency (f_c). As mentioned before, the VCO can lock on and track over a range equal to ±60% of its free-running frequency with a linearity of 0.5%.

The capacitor C_2 and resistor R_2 make up the low-pass filter. As you can see from the diagram, C_2 is an external capacitor and R_2 is an internal resistor. This internal resistance (in the 565) is about 3600 Ω ±20%. This configuration of an internal resistor and an external capacitor is typical of the Signetic 560 series PLLs. The capture range depends on the size of this resistor, as we shall see later. Note that R_1 and C_1

are part of the timing network that defines the VCO free-running frequency (f_o). The free-running VCO frequency is defined by the following formula:

$$f_o = \frac{1.2}{4R_1C_1}$$

where

f_o = VCO free-running frequency in Hz

The VCO free-running frequency should be adjusted to be in the center of the input frequency range. As in the VCO there is no limitation on the size of C_1, but R_1 must be between 2 and 20 kΩ. A good value is about 4 kΩ.

Note that pins 4 and 5 are shorted. Shorting these pins places the VCO output to one of the phase comparator inputs. Pin 6 is the reference for the demodulated output voltage. This pin provides a potential that is close to the potential of the demodulated output at pin 7. A resistor connected between pins 6 and 7 will reduce the gain of the output stage without changing the DC voltage at the output significantly. The lock range can then be decreased with little change in the free-running frequency. By using a potentiometer connected across pins 6 and 7, the lock range can be decreased from $\pm 60\%$ of f_o to $\pm 20\%$ of f_o (with a supply voltage of ± 6 V). The capacitor C_3, connected between pins 7 and 8, eliminates possible oscillations in the control current source.

The lock range for this circuit is defined as follows:

$$f_L = \pm \frac{8f_o}{V_{CC}}$$

where

f_L = lock range in Hz

f_o = free-running VCO frequency in Hz

V_{CC} = total supply voltage between $+V$ and $-V$ supply voltages

The capture range depends on the lock range and is given by the following equation:

$$f_C = \pm \left[\frac{f_L}{2\pi(3600)C_2} \right]^{1/2}$$

The schematic diagram in Figure 2.10 shows a PLL with resistor and capacitor values. The free-running frequency is calculated as follows:

$$f_o = \frac{1.2}{4R_1C_1} = \frac{1.2}{4(10,000\ \Omega)(0.01\ \mu\text{F})}$$
$$= 3000\ \text{Hz}$$

When not locked on the input frequency, the VCO should be oscillating at about 3000 Hz.

Knowing the free-running frequency, we can now calculate the PLL lock range:

$$f_L = \pm \frac{8f_o}{V_{CC}} = \pm \frac{8(3000\ \text{Hz})}{12}$$
$$= \pm 2000\ \text{Hz}$$

Finally, we can calculate the capture range of the PLL:

$$f_C = \pm \left[\frac{f_L}{2\pi(3600)C_2} \right]^{1/2}$$
$$= \pm \left[\frac{2000\ \text{Hz}}{2(3.14)(3600)(10\ \mu\text{F})} \right]^{1/2} = 94\ \text{Hz}$$

Many areas in modern industrial electronics employ the versatile IC PLL. The PLL has been used for a number of years in its discrete form in sophisticated military and aerospace applications. The IC PLL was introduced in the 1970s and has become increasingly popular in the area of telemetry and data communications. The low-cost IC PLL has spurred many applications in consumer areas as well as industrial. In industry PLLs are used for data communications, frequency dividers and synthesizers, Touch-Tone® telephones, and motor controls.

Frequency Shift Keying

The PLL in Figure 2.11 demodulates a data communications format called frequency shift keying (FSK). In this format binary information is carried by an oscillator that shifts between two frequencies. A VCO usually is driven by a binary data signal. One frequency represents the *high state* or *binary one* (called a *mark*), and another frequency represents the *binary zero* or *low state* (called a *space*). Before

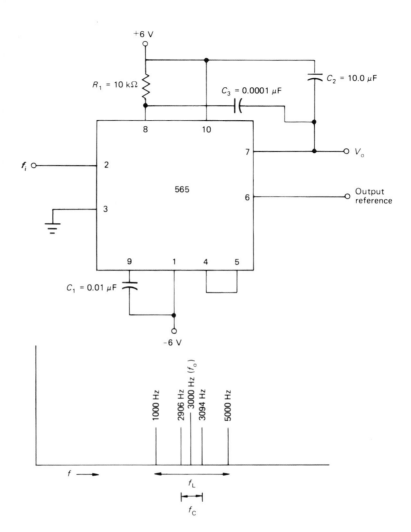

Figure 2.10   PLL Circuit
Example

the PLL came on the scene, FSK demodulation was done by bulky audio filters and large, unreliable electromagnetic relays.

FSK demodulation is achieved when the PLL locks on and tracks the incoming frequency shift. In the case of the circuit in Figure 2.11, the PLL is set to receive FSK signals of 1070 Hz and 1270 Hz. When either frequency appears at the input, the PLL locks on and tracks the carrier as it shifts between the two frequencies. A corresponding two-level DC

voltage will appear at the output of the PLL (pin 6). The low-pass filter capacitor C_2 in conjunction with the additional three RC networks filters out all unwanted AC components. All that is left is two clean DC voltage levels at the input to the op amp. The resistor R_1 adjusts the free-running frequency so that the DC voltage level at the output is the same as the voltage at pin 6. The output signal can be made logic-compatible by connecting a voltage comparator between the output and pin 6.

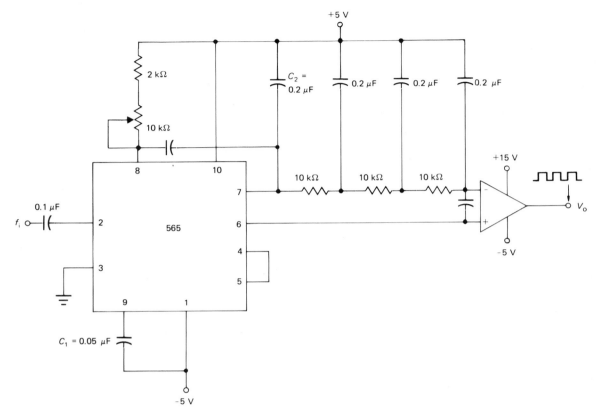

Figure 2.11 PLL FSK Decoder

Tone Decoder

The *tone decoder* (TDC) is a specially designed PLL. As its name implies, it is used in applications where it demodulates or decodes different frequency tones. Any frequency entering its detection band will cause the TDC output to go low, supplying current to a load. As in the PLL the capture range and the center frequency can be adjusted independently by external components. A block diagram of an NE567 is illustrated in Figure 2.12.

Note that several blocks in this device are similar to the PLL we just discussed (compare with Figure 2.6). The VCO, the phase detector

(called the I phase detector here), and the low-pass filter all behave in the same way they would in a PLL. In fact, this device is a PLL with several components added. Any input signal within the capture range will cause the VCO to lock on and track. The VCO will then be oscillating at the same frequency as the input signal. The I phase detector gives a zero output voltage only when the input signal frequency is the same as the free-running frequency of the VCO. The Q phase detector (Q stands for quadrature) gives a maximum voltage output when the I phase detector is zero. The high voltage out of the Q phase detector exceeds the reference voltage on the noninverting input to the

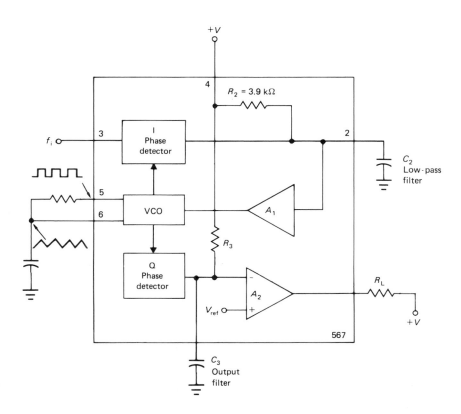

Figure 2.12 Tone-Decoder (TDC) Functional Block Diagram

comparator A_2. When the input frequency is at or near the VCO free-running frequency, the comparator output will go low, causing current to flow in R_L. If the input signal frequency increases or decreases, going out of the detection band, the output of the Q phase detector will decrease. When the input frequency moves far enough away in either direction, the input voltage to the comparator will be lower than the reference voltage. The comparator output will go high, interrupting current flow to the load. The TDC can then act as a frequency detector. Whenever a frequency goes within the detection band of the TDC, current will flow through a load, such as a light-emitting diode (LED).

The circuit shown in Figure 2.13 is a basic TDC application. When the input frequency goes within the detection band, the output will go low, causing current to flow in R_2. The center frequency of the detection band is given by the following equation:

$$f_o = \frac{1.1}{R_1 C_1}$$

For this circuit the center frequency is:

$$f_o = \frac{1.1}{R_1 C_1} = \frac{1.1}{(11,000 \ \Omega)(0.01 \ \mu\text{F})}$$
$$= 10,000 \text{ Hz} = 10 \text{ kHz}$$

where

f_o = center frequency

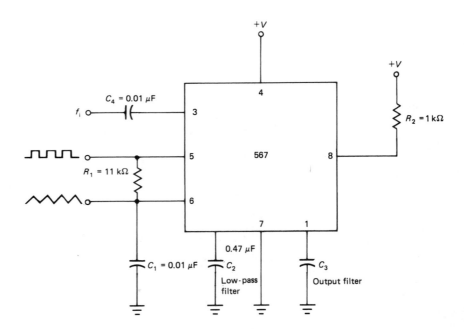

Figure 2.13   Basic TDC
Circuit

The bandwidth (BW) in hertz is given by the following equation:

$$BW = 10.7 \left[\frac{V_i f_o}{C_2} \right]^{1/2} \tag{2.2}$$

where

BW = detection bandwidth in Hz

V_i = input voltage in V

f_o = VCO free-running frequency in Hz

C_2 = capacitance of the low-pass filter capacitor in μF

Using this equation on the circuit in Figure 2.13, we can solve for the BW in hertz, with an input signal of 0.1 V:

$$BW = 10.7 \left[\frac{(0.1 \text{ V})(10{,}000 \text{ Hz})}{0.47} \right]^{1/2}$$

$$= 493 \text{ Hz}$$

Any frequency that enters into this 493 Hz detection band centered around 10,000 Hz will cause the TDC output to go low. This BW equation may be seen in another form as follows:

$$BW = 1070 \left[\frac{V_i}{C_2 f_o} \right]^{1/2} \tag{2.3}$$

where

BW = detection bandwidth in %

This equation will give the BW as a percentage of f_o. In our example we can calculate this percentage as:

$$BW = 1070 \left[\frac{0.1 \text{ V}}{(0.47)(10{,}000 \text{ Hz})} \right]^{1/2}$$

$$= 4.93\%$$

As a check, compare our two answers. We found with Equation 2.2 the detection bandwidth was 493 Hz with an f_o of 10,000 Hz. A bandwidth of 4.93% of f_o from Equation 2.3 is 0.0493 times 10,000 Hz, or 493 Hz.

As in the PLL, resistor R_1 should be between 2 and 20 kΩ. The NE567 TDC f_o limits

Touch-Tone Decoder

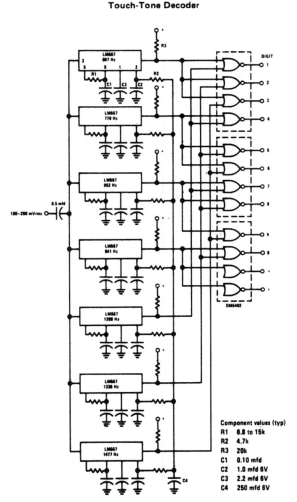

Component values (typ)

R1	6.8 to 15k
R2	4.7k
R3	20k
C1	0.10 mfd
C2	1.0 mfd 6V
C3	2.2 mfd 6V
C4	250 mfd 6V

Figure 2.14   Touch-Tone® Decoder

range from 0.01 Hz to 500 kHz. Detection bandwidths can be adjusted to 14% of f_o. The output can sink up to 100 mA.

An application of the NE567 is shown in Figure 2.14. This circuit is used in a Touch-Tone® decoding application. The Touch-Tone® telephone dialing system uses a combination of two frequencies every time a dialing button is depressed. The decoding circuit, as shown in

the figure, uses seven TDCs tuned to the following frequencies: 697 Hz, 770 Hz, 852 Hz, 941 Hz, 1209 Hz, 1336 Hz, and 1477 Hz. When the number 1 is pressed, the transmitter will generate a 697 Hz tone and a 1209 Hz tone. When these two tones are received by the TDCs that are tuned to their frequencies, the outputs of the two TDCs will go low. The output to the top NOR gate will go high when each input goes low. The output of the top NOR logic gate will give an output that will enable the switching circuit to dial a 1 digit. Note that if the TDCs are tuned properly, no other NOR gate's output will go high with a 697 Hz and 1209 Hz input to the circuit. Note also the pull-up resistors between the outputs of the TDCs and the inputs to the NOR gates. These resistors will hold the gate inputs high until the output of the TDC pulls it low.

FREQUENCY-TO-VOLTAGE (F/V) CONVERSION

Previously in this chapter we have covered VFCs, like the VCO. VFCs change analog input voltages and currents to variable-frequency output signals. The output signal is normally proportional to the DC input voltage change. *Frequency-to-voltage converters* (FVCs) perform the inverse function. They change an input frequency to a proportional change in output voltage. (Strictly speaking, the PLL falls into this classification. It gives a proportional DC output voltage change for a given input frequency change.) FVCs offer economical solutions to a wide variety of industrial problems where an analog frequency needs to be converted to an analog voltage. Some examples of FVC applications are motor speed controls and tachometers, power line frequency controls, and speedometers.

One of the simplest FVCs is the LM2907,

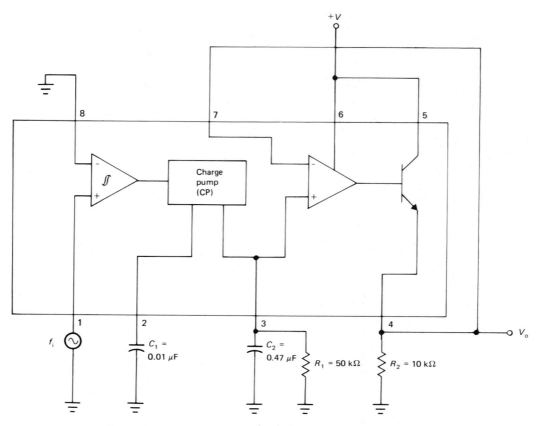

Figure 2.15   Functional Block Diagram of LM2907 FVC

made by National Semiconductor Corporation. This device has four main parts, as shown in Figure 2.15: the input Schmitt trigger, a charge pump, an op amp, and an output driver. The input Schmitt trigger converts the input signal into a square wave and applies it to the charge pump. Giving the Schmitt trigger 15 mV of hysteresis prevents false triggering. The charge pump alternately charges and discharges C_1 with a constant 200 μA current. The capacitor C_1 is charged to a preset voltage at a constant rate and then discharged. During the time that C_1 is being discharged, the charge pump charges capacitor C_2 with a 200 μA current. Capacitor

C_2 will then be charged by the same current as C_1 for the same amount of time. Capacitor C_2, however, charges to twice the average voltage across C_1. Resistor R_1 (connected to pin 3) provides a discharge path for C_2. The actual average DC voltage found at pin 3 to ground is equal to:

$$V_3 = V_{CC}R_1C_1f_i$$

where

V_3 = voltage to ground at pin 3

V_{CC} = supply voltage

f_i = input frequency in Hz

If the op amp–driver section has a gain of one, the output at pin 4 should be as follows:

$$V_o = V_{CC}R_1C_1f_iK \qquad (2.4)$$

where

V_o = F/V output voltage at pin 4

K = gain of the op amp–driver (usually 1)

When working with an LM2907 FVC, it is sometimes necessary to know how high the input frequency can go and still give a proportional output voltage. The maximum input frequency is given by the following equation:

$$f_{i(max)} = \frac{I_2}{C_1V_{CC}}$$

where

$f_{i(max)}$ = maximum input frequency in Hz

I_2 = constant current used to charge capacitor C_1 (usually 200 μA)

Using these equations, we can calculate the output voltage from our FVC with an input frequency of 1000 Hz and a supply voltage of +12 V as follows:

$$\begin{aligned}
V_o &= V_{CC}R_1C_1f_iK \\
&= (+12\text{ V})(50{,}000\ \Omega)(0.01\ \mu\text{F}) \\
&\quad \times(1000\text{ Hz})(1.0) \\
&= 6\text{ V}
\end{aligned}$$

This circuit should give an output voltage of +6 V with an input frequency of 1000 Hz. Its maximum input frequency, using a minimum (worst case) estimate of the C_1 charge current of 150 μA from the data sheet, will give:

$$f_{i(max)} = \frac{I_2}{C_1V_{CC}} = \frac{150\ \mu\text{A}}{(0.01\ \mu\text{F})(12\text{ V})}$$

$$= 1250\text{ Hz}$$

The ripple voltage on capacitor C_2 can be calculated from the following equation:

$$V_{r(p-p)} = \left(\frac{V_{CC}}{2}\right)\left(\frac{C_1}{C_2}\right)\left(1 - \frac{V_{CC}f_iC_1}{I_2}\right) \qquad (2.5)$$

where

$V_{r(p-p)}$ = peak-to-peak ripple voltage in V

Using this equation we can calculate the ripple output voltage as follows:

$$\begin{aligned}
V_{r(p-p)} &= \left(\frac{V_{CC}}{2}\right)\left(\frac{C_1}{C_2}\right)\left(1 - \frac{V_{CC}f_iC_1}{I_2}\right) \\
&= \left(\frac{12\text{ V}}{2}\right)\left(\frac{0.01\ \mu\text{F}}{0.47\ \mu\text{F}}\right) \\
&\quad \times \left[1 - \frac{(+12\text{ V})(1000\text{ Hz})(0.01\ \mu\text{F})}{150\ \mu\text{A}}\right] \\
&= 0.0255\text{ V} = 25.5\text{ mV}
\end{aligned}$$

When designing circuits with the LM2907, certain limitations are placed on the sizes of R_1, C_1, and C_2. First, R_1 should not be too much less than 100 Ω. Since the output impedance of the charge pump is about 10 Ω, too large a resistance for R_1 will interfere with the linearity of the charge pump. At a minimum the resistance of R_1 should be greater than or equal to the maximum voltage at pin 3, divided by the minimum current flowing in pin 3 (about 150 μA):

$$R_1 > \frac{V_{3(max)}}{I_{3(min)}} \qquad (2.6)$$

where

$V_{3(max)}$ = maximum full-scale output voltage required in the application

$I_{3(min)}$ = minimum C_2 charge current

Having chosen R_1, we may choose the capacitor C_1 according to the following equation:

$$C_1 = \frac{V_{3(max)}}{R_1V_{CC}f_{i(max)}}$$

where

$V_{3(max)}$ = maximum full-scale output voltage required in the application

$f_{i(max)}$ = maximum or full-scale input frequency

This equation was derived from our original output voltage Equation 2.4 (assuming $K = 1$):

$$V_o = V_{CC}R_1C_1f_iK$$

Since the capacitor C_1 also acts as internal frequency compensation for the charge pump, its value should be kept to at least 100 pF. The size of capacitor C_1 may also be found by solving the following equation for C_1:

$$f_{(max)} = \frac{I_2}{C_1V_{CC}}$$

so

$$C_1 = \frac{I_2}{f_{i(max)}V_{CC}}$$

We can also solve Equation 2.4 for R_1, if we have calculated or chosen C_1 first:

$$R_1 = \frac{V_{3(max)}}{C_1V_{CC}f_{i(max)}}$$

Solving for C_2 in the ripple voltage Equation 2.5, we can calculate a value for the capacitor C_2, provided we know the maximum acceptable ripple voltage. Thus:

$$C_2 = \left(\frac{V_{CC}}{2}\right)\left(\frac{C_1}{V_{r(p-p)}}\right)\left(\frac{V_{CC}f_iC_1}{I_2}\right)$$

Let us use these equations to solve an applications problem. Our application involves a tachometer (Figure 2.16) that will measure a maximum of 3000 revolutions per minute (r/min) on an analog meter movement. The ripple voltage should be no higher than 3% of the full-scale meter reading. With a 10 V full-scale voltage, the ripple should not be greater than 3% of 10 V, or 0.3 V_{p-p}. We are using a 1 mA meter movement. This means that 1 mA will cause a full-scale meter deflection. If the maximum shaft rotation is 3000 r/min, then the number of revolutions per second (r/s) of the frequency is 3000/60, or 50 Hz. We can now solve for C_1 as follows:

$$C_1 = \frac{I_2}{f_{i(max)}V_{CC}} = \frac{150 \ \mu A}{(50 \ Hz)(12 \ V)}$$
$$= 0.25 \ \mu F$$

Assuming a 2 V drop across the output transistor, our maximum output voltage swing will be 10 V. Knowing this, we can solve for the resistance of R_1:

$$R_1 = \frac{V_{3(max)}}{C_1V_{CC}f_{i(max)}}$$
$$= \frac{10 \ V}{(0.25 \ \mu F)\ (12 \ V)\ (50 \ Hz)} = 66.7 \ k\Omega$$

This should satisfy the maximum requirement set earlier (Equation 2.6) of less than 100 kΩ. Does it satisfy our minimum requirement? Using a minimum of 150 μA for the charge current of C_2 in pin 3, we make the following calculation:

$$R_1 \geq \frac{V_{3(max)}}{I_{3(min)}} \geq \frac{10 \ V}{150 \ \mu A} = 66.7 \ k\Omega$$

We have indeed satisfied the criteria, as the resistance of R_1 is equal to this value.

Finally, C_2 may be calculated. Since the ripple is worst at low frequencies, let us assume

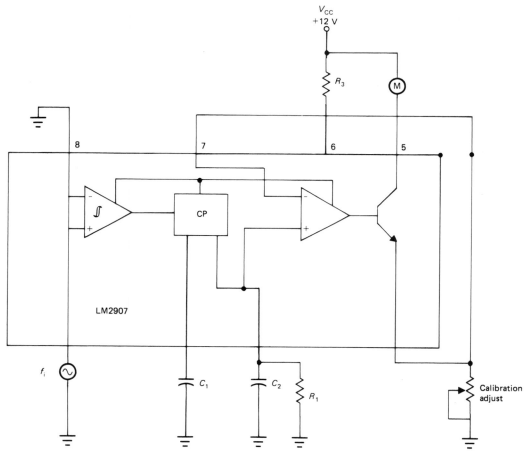

Figure 2.16   LM2907 Used as a Tachometer

that the lowest frequency to be measured is 60 r/min, or 1 Hz:

$$C_2 = \left(\frac{V_{CC}}{2}\right)\left(\frac{C_1}{V_{r(p-p)}}\right)\left(1 - \frac{V_{CC}f_iC_1}{I_2}\right)$$

$$= \left(\frac{12\text{ V}}{2}\right)\left(\frac{0.25\ \mu\text{F}}{0.3\text{ V}}\right)$$

$$\times \left[1 - \frac{(12\text{ V})(1\text{ Hz})(0.25\ \mu\text{F})}{150\ \mu\text{A}}\right]$$

$$= 4.9\ \mu\text{F} = 5\ \mu\text{F}$$

The LM2907 FVC has many uses in addition to the tachometer applications just discussed. For example, in Figure 2.17 the FVC is being used to extend the capture and lock ranges on a PLL. Normally, the IC PLL capture and lock ranges are relatively narrow, a considerable device limitation. The PLL capture and lock ranges can be extended by the FVC. The output of the PLL's low-pass filter and the FVC are summed at the summing junction, which may be an op amp summing amplifier. The output of the summing junction is applied to the PLL VCO input. The LM2907 here converts the input frequency to an output voltage that puts the VCO initially at the correct frequency to match the input frequency. The phase detector closes the gap between the VCO and the input frequency by exerting control through the summing point. If properly adjusted, the PLL can lock over a wide range with the help of the FVC.

A position-sensing application can be seen in Figure 2.18. Note that in this application the timing resistor R_1 has been removed, and $C_2 = 200C_1$. The output current produces a staircase effect instead of a DC voltage. If the input signal comes from a device that senses a passing notch on a rotating shaft, then the output will step up each time the notch is passed. When a specified number of steps is reached, the output can be turned on. In this circuit the staircase voltage is compared to a reference voltage on the internal comparator, and the output current is turned on when this voltage level is reached. In the circuit in Figure 2.18, this should happen after 100 input pulses are received. The output may turn on a relay that could remove power from the rotating device. Applications of this kind of circuit can be found in automated packaging operations and in line printers.

ANALOG AND DIGITAL CIRCUITS

Industrial electronics systems are composed of many subsystems. These subsystems have digital as well as linear electronic circuitry in them. As microprocessors get less expensive and more powerful, they will become increasingly popular in industrial circuits and systems. Since analog and digital circuitry are fundamentally incompatible, circuits called interfaces must bridge the gap between these two basically different types of circuits. The *interface* provides the necessary function of tying the parts of a system together. Such circuits usually are neither purely digital nor purely analog in their makeup. They usually contain both digital and analog circuitry.

In this section, we will deal with the circuits that will convert from digital signals to analog signals and from analog to digital. Recall that the digital signal is made up of bits (the abbreviation for BInary digiT). A *bit* is a voltage level that is either high or low, a 1 or a 0, never a value in between. These high and low levels are grouped together in clumps called words. A *word* is a group of bits. A word may, for example, be made up of four bits. Any number

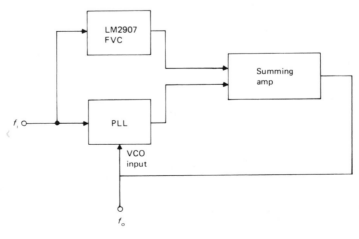

Figure 2.17 LM2907 Used to Extend Capture Range of PLL

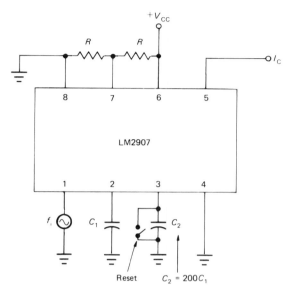

Figure 2.18   LM2907 Used as Position Sensor

can be represented as a series of bits. For example, the number 3 in the decimal numbering system is represented in the binary system as 0011, read zero-zero-one-one.

To convert an analog signal to a digital 1 means converting a voltage or current to a series of 0s and 1s to a series of high and low voltages. The series of bits must faithfully represent the analog value of voltage or current. If, in the conversion process, the binary output is not an accurate representation of the analog input, the system becomes unreliable.

It is important in the binary numbering system to be aware of the terms most significant digit and least significant digit. In the decimal numbering system, we have 10 digits (0–9) that can be used in combination to represent any number. For example, the number 15 is made up of two digits, a 1 and a 5. The 5 is in the units place and the 1 is in the tens place. Each digit in the ones place represents only one unit. As such, the digit in the ones place is called the *least significant digit* (LSD). The number 1 in

the tens place is called the *most significant digit* (MSD), since each digit is worth 10 units. To extend the concept further, we will analyze the number 745. The MSD is 7 and the LSD is 5.

In digital terminology, the significant digits are called significant bits. Thus, the most significant bit is abbreviated MSB. Likewise, the least significant bit is LSB.

In the binary numbering system the same basic rules apply to the significance of bits. The difference is that in the binary numbering system we have only two digits, 0 and 1. In this system, for example, the binary number 1000 reperesnts the number 8. Bit 1 in the leftmost position is the MSB, and the 0 to the far right is the LSB. Note that the number 8 is represented by four bits. In computer terminology this group of bits is called a four-bit word. The highest number we can represent with a four-bit word is 1111, corresponding to the decimal number 15. To represent numbers larger than decimal 15, we would need to have words made up of more than four bits. For example, a word with eight bits can represent decimal numbers up to 255.

In the measurement or representation of analog voltages, a voltage may take on any value between two extremes. For example, a voltage with a range of 0–5 V may take any value between 0 V and 5 V. This is not true of digital representations of voltage values. For example, if we were to convert the voltage range of 0–5 V, we would find that the digital representation of that range would be spread over that range in discrete steps or increments. Table 2.1 shows a 5 V range converted to a digital word of four bits. Note from this table that our four-bit word represents 16 individual items of data. We can state this another way by saying that the resolution of our four-bit system is one part in 15. *Resolution* is defined as the degree to which values of a quantity can be distinguished or separated. Given as a percentage, a resolution of 1 in 15 would be equal to 6.66%. Note also in our 5 V–four-bit system that each

Table 2.1 Conversion of a Range of 0–5 V to Four-Bit Digital Words

Decimal Value	Binary Representation
0.000	0000
0.333	0001
0.667	0010
1.00	0011
1.33	0100
1.67	0101
2.00	0110
2.33	0111
2.67	1000
3.00	1001
3.33	1010
3.67	1011
4.00	1100
4.33	1101
4.67	1110
5.00	1111

bit has a value of 0.33 V, or one-third of a volt. In this system we cannot represent a voltage between 2.00 and 2.33.

The lack of resolution is a limitation on the application of analog-to-digital conversions. If greater accuracy is needed, a word with greater numbers of bits is required. For example, an eight-bit system has a resolution of 1 in 255, or 0.392%. If an analog range of 5 V were broken down in 255 parts, each bit would be equal to about 20 mV instead of 0.33 V for a system using a four-bit word. Even greater accuracy is achieved by using a 12-bit word. A 12-bit word system has a resolution of one part in 4095, with a 0.024% resolution. In a system with a 12-bit word a voltage range is broken down into 4095 steps or increments. Even with the increased resolution a 12-bit system offers, however, a digital representation of an analog voltage is never exact.

DIGITAL-TO-ANALOG (D/A) CONVERSION

Several different types of circuits can convert digital signals or representations back into analog form. A basic *digital-to-analog converter* (DAC) is shown in Figure 2.19. Note that an analog output results with a four-bit digital signal input. Although we are using a four-bit word system in this example (and in other examples to follow), you should know that we are using this merely to simplify our illustrations. Actually, four-bit systems rarely are used in industry due to the lack of resolution. Words with 8, 10, 12, and 16 bits are more popular, since they have acceptable resolution for most industrial applications.

Binary Weighted Ladder DAC

An actual DAC is shown in Figure 2.20. This particular circuit is known as the *binary weighted ladder*. Note that we have a simple inverting summing amplifier in this circuit. Voltages V_1 to V_4 are either a high or low voltage. These voltages are applied to each of the four respective inputs to the op amp, and each of the four inputs is weighted. For example, the input V_1 is applied across an 80 kΩ resistor. This input gets an attenuation of about one-tenth at the output. Now notice that the input V_2 gets less attenuation by a factor of 2. This is appropriate since the digit associated with V_2 has

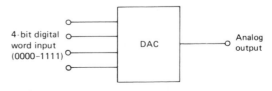

Figure 2.19 DAC Functional Block Diagram

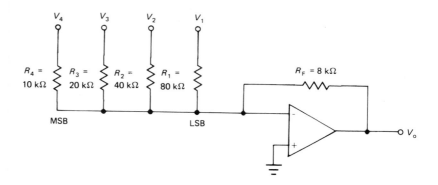

Figure 2.20 Binary
Weighted Ladder DAC

more significant position than the input V_1. Going to the MSB (at V_4), we see that this input has a gain factor of 1, eight times that of the LSB input.

To see how this works in practice, we will put in a 0001 digital signal (with $+V = 5$ V). To accomplish this input, we will make V_1 high (equal to $+5$ V) and the other inputs at ground potential, or 0 V. Since all the other inputs are equal to 0 V, the summing amp has only one input voltage. This input voltage is given an attenuation equal to 8 kΩ/80 kΩ = 0.1. A 5 V input will give a 0.5 V output.

Let us see what happens if we put a 0011 in our DAC. Voltages V_1 and V_2 each will be 5 V, and V_3 and V_4 each will be 0 V. The voltage at V_2 will be attenuated by a factor of 0.2. This will give an output of 1.0 V from V_2. Added to the 0.5 V output from V_1, the total is 1.5 V. If we were to connect a voltmeter at the output of the amplifier, we would see an output of 1.5 V with a 0011 input.

Table 2.2 shows the DAC output voltages from 0000 to 1111. We can see from this table that the output voltage ranges from 0 V to 7.5 V in 0.5 V steps or increments. The range may be changed by adjusting the size of the feedback resistor. Decreasing the size of the feedback resistor will decrease the range and decrease the size of each individual step or increment.

Table 2.2   DAC Input and Output Voltages

Binary Input	Analog Output
0000	0.0
0001	0.5
0010	1.0
0011	1.5
0100	2.0
0101	2.5
0110	3.0
0111	3.5
1000	4.0
1001	4.5
1010	5.0
1011	5.5
1100	6.0
1101	6.5
1110	7.0
1111	7.5

The weighted ladder approach just described works well for four- to six-bit systems. At higher word lengths, however, the spread in resistive values needed in the inputs becomes much greater. In eight-bit words, the final resistor in the LSB position may be 10 MΩ or more. Resistors spread over so wide a range will change their resistances by different amounts

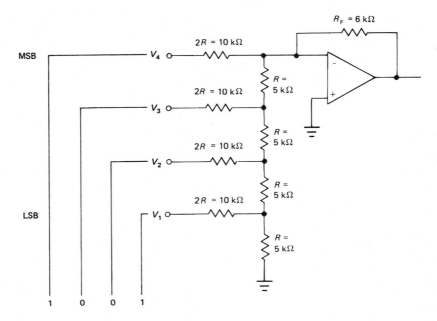

Figure 2.21   R-2R Ladder
DAC

with temperature changes, introducing errors. The R-2R ladder solves this problem by allowing lower-value resistors to be used.

R-2R Ladder DAC

The R-2R ladder is so named because resistors are either R (in this case 5 kΩ) or 2 times R (10 kΩ) as illustrated in Figure 2.21. Relative weighting of the bit inputs is done using a Thevenin voltage source. The analysis is a somewhat complicated exercise in Thevenizing and little will be gained by such an analysis here. The output voltage of this circuit is given by the following equation:

$$V_o = -R_F \left(\frac{V_4}{2R} + \frac{V_3}{4R} + \frac{V_2}{8R} + \frac{V_1}{16R} \right)$$
$$\text{MSB} \longrightarrow \text{LSB}$$

The output voltage with an input of 0010 is as follows:

$$V_o = -6 \text{ k}\Omega \left[\frac{0 \text{ V}}{2(5 \text{ k}\Omega)} + \frac{0 \text{ V}}{4(5 \text{ k}\Omega)} \right.$$
$$\left. + \frac{+5 \text{ V}}{8(5 \text{ k}\Omega)} + \frac{0 \text{ V}}{16(5 \text{ k}\Omega)} \right]$$
$$= 0.75 \text{ V}$$

Table 2.3 shows the DAC output voltages from 0000 to 1111. We can see from this table that the output voltage ranges from 0 V to 5.625 V in 0.375 V steps or increments. The R-2R system requires only two values of resistors for its D/A conversion. These resistors are usually precision resistors.

Although DACs can be constructed out of discrete devices, it is usually more economical to use an IC specially dedicated to D/A conversion. Such an IC DAC is the DAC0800 manufactured by National Semiconductor Corporation. The output may be in either current or voltage. In addition, the voltage and current can increase when the input goes from 00000000 to 11111111, or it may decrease. The

Table 2.3 *R-2R* Ladder DAC Input and Output Voltages

Binary Input	Analog Output
0000	0.0
0001	0.375
0010	0.75
0011	1.125
0100	1.5
0101	1.85
0110	2.25
0111	2.625
1000	3.0
1001	3.375
1010	3.75
1011	4.125
1100	4.5
1101	4.875
1110	5.25
1111	5.625

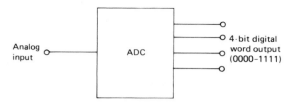

Figure 2.22 ADC Functional Block Diagram

current ranges from 0 mA to 1.992 mA, or from 1.992 mA to 0 mA, depending on the output selected. The output voltages range from 0 V to −9.960 V, or −9.960 V to 0 V, depending on the output selected.

DACs are found in many industrial applications, such as cathode ray tube (CRT) character generator and display drivers for computers, programmable power supplies and function generators, and audio tracking, digitizing, and decoding.

ANALOG-TO-DIGITAL (A/D) CONVERSION

An *analog-to-digital converter* (ADC) converts analog voltages or currents to a series of digital bits—just the opposite of what occurs in D/A conversion. A simplified block diagram of the A/D conversion process is shown in Figure 2.22. As in the DACs, digital word lengths are commonly 8, 10, 12, and 16 bits in length. Analog-to-digital converters (ADCs) are somewhat more complicated and difficult to understand than DACs. Also, since most ADCs use DACs as functional blocks, the DACs were presented first in our discussion.

Three A/D conversion principles are used in modern ADCs: digital ramp or counting, successive approximation, and parallel or flash. The counting ADC is the easiest to understand and will be presented first.

Counting ADC

A block diagram of the *counting ADC* is found in Figure 2.23. If we start the up-down counter at zero, the counter will have a digital output that increases every time a clock pulse is received. These digital bits are converted back into an analog form and compared to the input in the comparator. If the analog input voltage is greater than the DAC's output, the counter counts up. When the counter counts upward, the output of the DAC increases in a stepwise fashion. If the analog input voltage is less than the DAC's output, the counter counts down. When the counter counts downward, the output of the DAC decreases in a stepwise fashion, bringing the digital output in line with the analog input. You will recognize this as a feedback system that keeps the digital output approximately equal to the analog input.

The major disadvantage of this type of ADC is its slow speed. The conversion time for

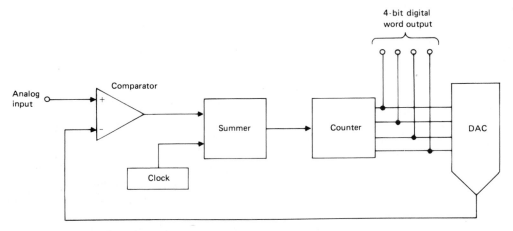

Figure 2.23 Counting ADC

this device is virtually instantaneous for 0 V, but is equal to 2^n clock periods, where n is the number of bits in the digital word. Thus, a 12-bit DAC would take 2^{12} or 4096 clock pulses to accomplish the conversion. This would be too long a time interval for many applications. Because of low conversion speed, the counting type of ADC is not used very widely, especially in 10- and 12-bit ADCs.

Successive Approximation ADC

More popular is the *successive approximation ADC,* shown in simplified block diagram form in Figure 2.24. As in the counting ADC, let us assume that the successive approximation register output (SAR) is 0000. The operation of this device may be compared to a guessing game. For example, suppose I have a number

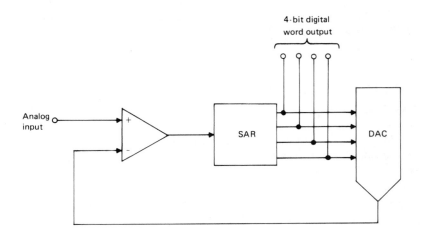

Figure 2.24 Successive Approximation ADC

in mind between 1 and 16. What would be the most efficient strategy for you to use to guess the number if you can only ask questions that can be answered with a yes or no? You could start out at 1 and ask, "Is the number greater than 1?" If I answer no, you can conclude that the number is 1. If my answer is yes, you could continue successively from 1 to 2 to 3, and so on, asking the same question. You would have to ask 15 questions, however, if my chosen number is 16. A better way to approach the question is to split the range of possible numbers in half every time you ask a question. For example, you could start out by asking, "Is the number greater than 8?" (Eight is halfway between 1 and 16.) I would answer yes because my chosen number is 9. Then you would reply, "Is it greater than 12?" (Note that in asking this question, you have divided the range between 8 and 16 in half.) I would answer no to this question. You would again respond by dividing the remaining range in half (you know now that the number is greater than 8 but less than 12) by asking, "Is it greater than 10?" I would reply no and you would know that the number is greater than 8 but less than 10. Of course, the number 9 fits that description.

This simple illustration demonstrates almost exactly what happens in the successive approximation ADC. The control logic in the SAR sets the output to 0000 and changes the MSB from 0 to 1, making the output of the SAR 1000. This SAR output gets converted to an analog voltage and applied to the comparator. The comparator, then, asks the question, "Is the input greater than the DAC output?" If it is, the comparator will answer yes by putting out a high voltage. The SAR will move the next MSB from 0 to 1 if the input is high. The same question is asked, and the SAR responds by making trial outputs. If the answer to the first question is no, the comparator output will be low. The SAR will then change the MSB back to a 0 and move on to the next MSB, making it high, 0100. The process continues until the cor-

rect binary word is determined. Generally, the successive approximation ADC is faster than the counting ADC and so is more popular.

Flash ADC

The fastest of the three ADCs mentioned is the parallel or flash converter, shown in Figure 2.25. The *flash converter* is a series of comparators with a decreasing reference voltage applied to each inverting input. The voltage at A_1 inverting input is higher than the voltage at A_2. The reference voltages continue to decrease as we move down the diagram, due to voltage divider action. If the analog input voltage V_i is less than the reference voltage V_{ref}, the comparators with inverting terminal voltages lower than V_i will have high outputs. The comparators with inverting terminal voltages higher than V_i will have low outputs. For example, if V_i is greater than the voltage at point A but less than the voltage at point B, outputs of comparators A_3 through A_7 will be high, and outputs of comparators A_1 and A_2 will be low. Since the outputs of these comparators are not in binary form, the comparator outputs must be converted into binary form by a decoder, usually made up of a network of logic gates. The converter in Figure 2.25 is a three-bit word ADC. Note that it has seven comparators to provide this three-bit output. In general, this type of converter needs $(2^n - 1)$ comparators, where n is the number of bits in a word, For example, an eight-bit converter will require $(2^8 - 1)$ or 255 comparators. The number of comparators expands exponentially as the number of bits in the word increases. The extra complexity and numbers of comparators in this circuit makes this an expensive choice for A/D conversion. Note that no clock is required in this application.

A six-bit flash ADC made by TRW has a conversion time of 33 nanoseconds (ns) and can provide up to 30 million samples per sec-

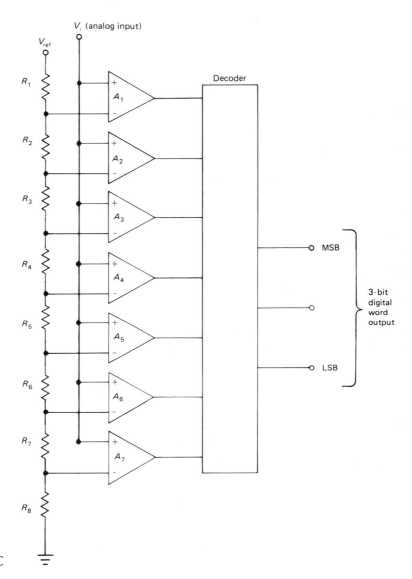

Figure 2.25 Parallel ADC

ond. This chip, the TDC1014J, accepts an input up to 1 V peak to peak and is supplied in a 24-pin DIP with an optional evaluation board available. TRW has recently introduced an ADC that has nine-bit output with 511 comparators, decoding logic, and an ECL compatible output on a single chip.

ANALOG SWITCHES

A *switch* is an electronic component that has two stable states: an on state and an off state. The on state typically is a low-resistance condition, where power is controlled by the switch but not consumed by the switch. The ideal switch is one that has 0 Ω resistance when on and infinite ohms resistance when off. The mechanical switch comes closest to this ideal resistance ratio. Mechanical switches are used in applications where current demands are high. In applications where the switch must be operated at a frequency more than 1 kHz, however, a mechanical switch is not suitable. The bipolar transistor can be used as a switch in these applications. Recall that when no base current flows in a transistor, the device is considered to be an open circuit from emitter to collector. Applying a base voltage will cause the device to conduct current flow from collector to emitter, corresponding to the on state. Many manufacturers make a special switching transistor that has low on state resistance and high switching speed. In general, we may say that most transistors have difficulty with switching speeds above 100 kHz. Higher speeds can be realized with FETs.

Discrete junction field effect transistor (JFET) and metal-oxide semiconductor field effect transister (MOSFET) devices can be used as switches in much the same way as the transistor. Applying a high enough voltage to the gate of a depletion-mode MOSFET, for example, will turn off the device. A low voltage will turn it back on. In enhancement MOSFETs, however, the opposite is true. A high voltage on the gate will turn on the MOSFET and a low gate voltage will turn it off. The IC MOSFET is more popular today than the discrete MOSFET for many switching applications. An example

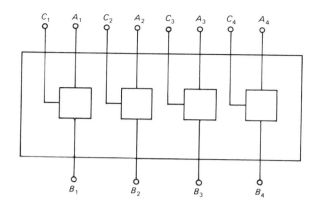

Figure 2.26   CMOS Analog Switch

of an IC MOS (metal-oxide semiconductor) switch is the 4066 CMOS (complementary MOS) switch, shown in block diagram form in Figure 2.26.

In the 4066 (and similar switches) when the control line is connected to a high potential (V_{DD}), the resistance between A and B is low (about 100 Ω). When the control line is connected to a low voltage (V_{SS}), the switch is disabled and a high impedance is established between A and B. This type of control is called *active HIGH enable* in digital terminology. A high voltage on the control line turns on the switch and a low voltage turns it off.

The on state and off state resistances normally vary with applied signal voltages and temperature. Table 2.4 summarizes typical parameter values for different types of switches.

Table 2.4   Typical Analog Switch Parameters

	Mechanical Switch	Transistor	JFET	MOSFET
On State Resistance	10^{-2}	10	30	80
Off State Resistance	10 pA	100 pA	100 pA	100 pA
Commutation Rate	1 kHz	100 kHz	10 MHz	50 MHz

SOURCING AND SINKING ICs

An important topic generally left out of textbooks is sourcing and sinking as related to the IC output. In ICs two types of output circuits are commonly used: current sourcing and current sinking. The names come from the position of the load in the output circuit. In *current sinking*, shown in Figure 2.27A, we see the load connected between the output and the $+V$ supply. In this type of output circuit the load is isolated from ground when the output is high. Current will flow through the load only when the output is at a lower potential than $+V$. If the output is high (at or near $+V$), no current flows through the load. If the output goes low, a difference in potential will exist between the supply and the output of the IC, and current will flow. The IC switches or *sinks* the output current from ground. This output is most often associated with the *open collector* configuration. Recall that the open collector uses a pull-up resistor. If the IC output is driving a TTL (or similar) input, as shown in Figure 2.27B, the pull-up resistor helps make a solid voltage level at the output. This is necessary since the open

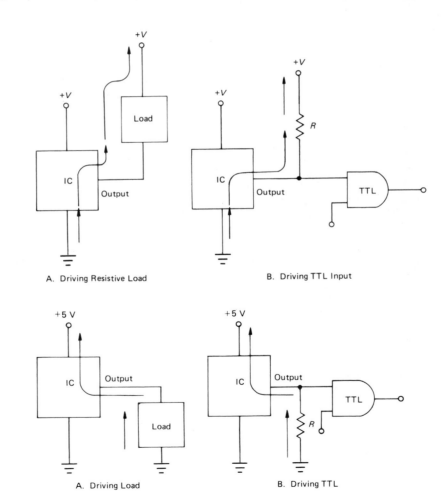

Figure 2.27 Current Sinking an IC Output

A. Driving Resistive Load

B. Driving TTL Input

Figure 2.28 Current Sourcing an IC Output

A. Driving Load

B. Driving TTL

collector output is floating. The pull-up resistor also provides better immunity from noise and faster output rise and fall times.

In *current sourcing,* the load is connected between the output and ground. When the output goes high, a difference in potential exists between the output and ground. Current will then flow through the load, as illustrated in Figure 2.28A. This circuit is called current sourcing because the IC provides a *source* of power to the load. An IC with a current source output

will normally be low, and go high when power is to be applied to the load. When the IC output is applied to a TTL device (or one with a similar input), a resistor called a pull-down resistor is connected as shown in Figure 2.28B. The pull-down resistor is used for reasons similar to those for which the pull-up resistor is used in the current sink output. The current sink, however, is an open collector output, and the current source is an open emitter.

PROBLEMS

1. In the VCO in Figure 2.2, $R_1 = 15$ kΩ and $C_1 = 0.01$ μF.
   a. Calculate f_o.
   b. Calculate the voltage at pin 5 if R_2 is 1.8 kΩ and R_3 is 10 kΩ.
   c. Calculate the output frequency with the input in part b.
   d. Calculate the output frequencies at the peak voltages with a ± 0.1 V peak input.
   e. With the resistances stated in part b, what change in voltage will give a 200 Hz change in output frequency?
2. The PLL in Figure 2.10 has the following component values: $R_1 = 15$ kΩ, $C_1 = 0.015$ μF, and $C_2 = 4.7$ μF. Find (a) f_o, (b) the capture range, and (c) the lock range.
3. A PLL is needed to have a capture range of 2000 Hz, a lock range of 100 Hz, and a free-running frequency of 5000 Hz. Choose component values that will give these parameters.
4. The tone decoder in Figure 2.13 has the following component values: $R_1 = 10$ kΩ, $C_1 = 0.015$ μF, and $C_2 = 4.7$ μF. Find (a) f_o, and (b) the BW in Hz.
5. Design a tone decoder to have a BW of 10% with an f_o of 1500 Hz.
6. The LM2907 FVC in Figure 2.15 has the following component values: $C_1 = 0.01$ μF, $R_1 = 100$ kΩ, $R_2 = 10$ kΩ, and $V_{CC} = +12$

V. With $K = 1$, calculate (a) the voltage output with a 200 Hz input frequency, (b) the maximum input frequency if I_2 is 150 μA, and (c) the ripple voltage on C_2.
7. Design an LM2907 tachometer to give the following parameters: We wish to measure a maximum 2000 Hz frequency with a ripple voltage no higher than 2%.
8. A six-bit system covers a voltage range of 0–5 V. Calculate (a) how much voltage each bit is worth, and (b) the resolution of the system in parts and percentage.
9. In the DAC in Figure 2.20, the R_F is 10 kΩ.
   a. Calculate the output voltage with 1101 input when $1 = +5$ V.
   b. Calculate the output voltage with 1111 input when $1 = +5$ V.
   c. Find the value of R_F that will give an output of approximately 0–5 V from 0000 to 1111.
10. In the DAC in Figure 2.20, the R_F is 10 kΩ and $R = 8$ kΩ.
    a. Calculate the output voltage with 1101 input when $1 = +5$ V.
    b. Calculate the output voltage with 1111 input when $1 = +5$ V.
    c. Find the value of R_F that will give an output of approximately 0–5 V from 0000 to 1111.

3

WOUND-FIELD DC MOTORS AND GENERATORS

INTRODUCTION

In these days of rapid electronic advancement, technicians, especially digital electronic technicians, frequently overlook the importance of the electric motor as a key system element. In fact, in some educational systems, digital electronics students are not required to take industrially related courses, such as courses related to motors and generators. In this chapter on wound-field DC motors and generators, we have attempted to correct that situation by presenting a condensed version of the topic. Of course, some topics are omitted here, but the important principles are presented in some detail.

Many applications in industry and in process control (see Chapter 9) require electric motors as critical functional elements. Because such motors are electromechanical rather than purely electrical, their response sets the entire system's performance limits. So, technicians who wish to be completely functional in the industrial electronics environment must be familiar with the varieties of motors and generators. In addition, technicians must understand basic operating principles if they are to perform intelligent troubleshooting (and system design).

In discussions involving both DC motors and DC generators, it is convenient to refer to a generalized electric machine, because much of what is said about electric motors is equally applicable to generators. The generalized electric machine is referred to as a *dynamo,* a machine that converts either mechanical energy to electric energy or electric energy to mechanical energy. When a dynamo is driven mechanically by a power source, such as a gas or diesel engine or a steam or water turbine, and provides electric energy to a load, it is called a *generator.* If electric energy is supplied to the dynamo, and its output is used to provide mechanical motion or *torque* (a force cting through a distance), it is called a *motor.* Generators are rated at the kilowatts they can deliver without overheating at a rated voltage and speed. Motors are rated at the horsepower they can deliver without overheating at their rated voltage and speed.

DYNAMO CONSTRUCTION

Figure 3.1A shows a simplified, exploded view of a DC dynamo. Figure 3.1B shows the dynamo in cross section.

The dynamo construction consists of two major subdivisions, the *rotor (armature),* or rotating part, and the *stator,* or stationary part. We will discuss these two parts in detail in the following sections.

Rotor

The rotor consists of an armature shaft, armature core, armature winding, and a commutator.

The *armature shaft* is the cylinder on which the rotor components are attached. The armature core, armature winding, and commutator (mechanical switch) are attached to the armature shaft. This entire assembly rotates. Sometimes, fins are attached to the shaft to provide cooling of the dynamo.

The *armature core* is constructed of laminated (thin-sheet) layers of sheet steel. These layers provide a low-reluctance (magnetic resistance) path between the magnetic field poles. The laminations are insulated from each other and attached together securely. The core is laminated to reduce *eddy currents,* which are circular-moving currents. The grade of sheet steel used is selected to produce low loss due to hysteresis (lagging magnetization). The outer surface of the core is slotted to provide a means of securing the armature coils.

Two basic armature, or field pole, forms are available in the construction of the dynamo: salient (standing out from the surrounding material) and nonsalient (minimal projection) poles. Figure 3.2 illustrates these forms for the armature of the dynamo.

The *armature winding* consists of insulated coils, which are insulated from each other and from the armature core. The winding is embedded in the slots in the armature core face, as shown in Figure 3.1B.

The *commutator* consists of a number of wedge-shaped copper segments that are assembled into a cylinder and secured to the armature shaft. The segments are insulated from each other and from the armature shaft, generally with mica. The commutator segments are soldered to the ends of the armature coils. Because of the armature rotation, the commutator provides the necessary switching of armature current to or from the circuit external to the armature.

The armature of the dynamo performs the following four major functions:

1. It permits rotation for mechanical generator action or motor action.

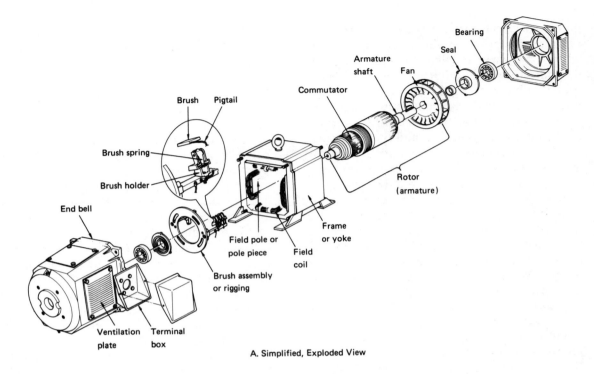

Brush   Pigtail

Brush spring

Brush holder

End bell

Commutator

Armature
shaft

Fan

Seal

Bearing

Rotor
(armature)

Field pole or
pole piece

Field
coil

Frame
or yoke

Brush assembly
or rigging

Ventilation
plate

Terminal
box

A. Simplified, Exploded View

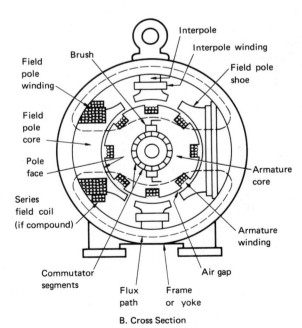

Field
pole
winding

Brush

Interpole

Interpole winding

Field pole
shoe

Field
pole
core

Pole
face

Series
field coil
(if compound)

Armature
core

Armature
winding

Commutator
segments

Flux
path

Frame
or yoke

Air gap

B. Cross Section

Figure 3.1   DC Dynamo

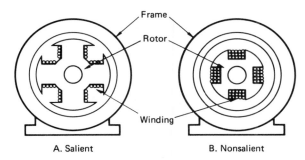

A. Salient   B. Nonsalient

Figure 3.2   Pole Construction

2. Because of rotation, it provides the switching action necessary for commutation.
3. It provides housing for the armature conductors, into which a voltage is induced or which provide a torque.
4. It provides a low-reluctance flux (magnetic lines) path between the field poles. Recall that *flux* or *flux lines* are invisible magnetic lines of force.

Stator

The stator consists of eight major components: a field yoke, field windings, field poles, air gap, interpoles, compensating windings, brushes, and end bells. (See Figure 3.1.)

The *field yoke,* or *frame,* is made of cast or rolled steel. The yoke supports all the parts of the stator. It also provides a return path for the magnetic flux produced in the field poles.

The *field windings* are wound around the pole cores and may be connected either in series or in parallel (shunt) with the armature cir-

cuit. The windings consist of a few turns of heavy-gauge wire for a series field, or many turns of fine wire for a shunt field, or both for a compound (both series and shunt field) dynamo.

The *field poles* are constructed of laminated sheet steel and are bolted or welded to the yoke after the field windings have been placed on them. The *pole shoe* is the end of the pole core next to the rotor. It is curved and is wider than the pole core. This widening of the pole core spreads the flux more uniformly, reduces the reluctance of the air gap, and provides a means of support for the field coils.

The *air gap* is the space between the armature surface and the pole face. The air gap varies with the size of the machine, but it generally is between 0.6 and 0.16 cm.

The *interpoles,* or *commutating poles,* are small poles placed midway between the main poles. The interpole winding consists of a few turns of heavy wire, since interpoles are connected in series with the armature circuit. The interpole magnetic flux, therefore, is proportional to the armature current.

The *compensating windings,* shown in Figure 3.3, are used only in large DC machines. They are connected in the same manner as the interpole windings but are physically located in slots on the pole shoe.

Brushes rest on the commutator segments and are the sliding electrical connection between the armature coils and the external circuit. Brushes are made of carbon of varying degrees of hardness; in some cases, they are made of a mixture of carbon and metallic copper. The

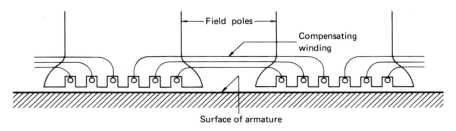

Figure 3.3   Location of Compensating Winding in the Face of Field Poles

brushes are held in place by springs in brush holders, the entire assembly being called the *brush rigging*. Electrical connection between the brush and brush holder is made with a flexible copper braid called a *pigtail*.

End bells support the brush rigging as well as the armature shaft. The end bells generally determine the type of protection the dynamo receives; that is, the construction can be open, semiguarded, dripproof (as in Figure 3.1), or totally enclosed. Semiguarded types have screens or grills to protect rotating or electrically live parts and to provide ventilation. A dripproof housing protects the equipment from dripping liquids. The totally enclosed construction is completely sealed and must dissipate the heat buildup by radiation from the enclosing case.

Figure 3.4 shows a cutaway view of a completely assembled DC machine, in this case, a DC motor. Notice the dual brushes in each brush rigging, provided for high currents. Also, the side access covers are of dripproof construction.

Dynamo Classification

The construction shown in Figures 3.1 and 3.4 is known as a wound-field dynamo. There are many other classifications of dynamos. Figure 3.5 charts selected DC motor classifications and shows a rich variety from which to choose. Each class of DC motors has its advantages, depending on the desired applications. The types of DC motors shown in Figure 3.5 will be discussed briefly later; a table of their characteristics is given at the end of the chapter. However, the conventional permanent-magnet (PM) and

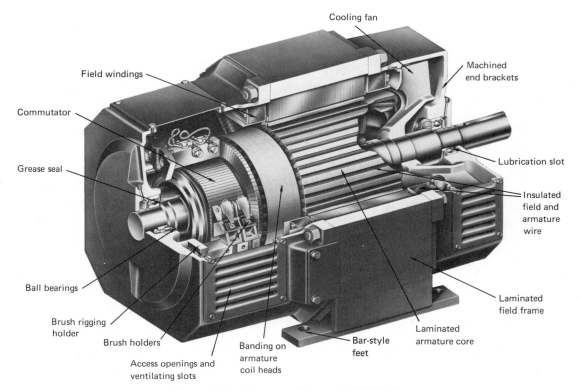

Figure 3.4 Cutaway View of DC Motor

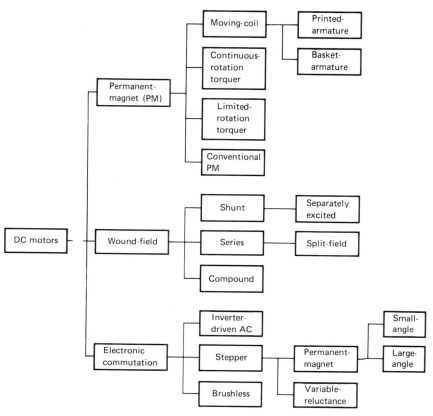

Figure 3.5   Selected DC Motor Types

wound-field DC motors are the subject of most of the discussion in this chapter.

Unlike DC motors, DC generators are primarily of the conventional permanent-magnet and wound-field types. Consequently, they will be the only types discussed.

BASIC PRINCIPLES OF THE DC GENERATOR

The simplest generator that can be built is an AC generator, shown in Figure 3.6. The basic generator principles are most easily explained through the use of the elementary AC generator. For that reason the AC generator will be discussed first. The DC generator will be developed next.

Elementary AC Generator

The elementary generator of Figure 3.6 consists of a single wire loop (the armature) placed so that it can be rotated in a stationary magnetic field. The magnetic field is shown in Figure 3.6 with arrows going from the N (north) pole to the S (south) pole. The pole pieces are shaped and positioned as shown to concentrate the magnetic field as close as possible to the wire loop. The rotation of the loop will produce an induced voltage, or emf (electromotive force), in the loop. Brushes sliding on the slip rings connect the loop to an external circuit resistance, allowing current to flow because of the induced emf. *Slip rings* are metal rings around the rotor shaft. Each ring attaches to the sepa-

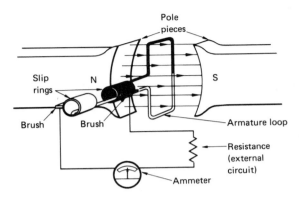

Figure 3.6   Elementary AC Generator

rate ends of the rotor coil. If the brushes were not contacting the slip rings but the loop were rotating, an emf would still appear at the slip rings, but no current would flow.

The elementary generator produces a voltage in the following manner: Assuming the armature loop is rotating in a clockwise direction, its initial or starting position is as shown in Figure 3.7A. This position is called the 0° position, or the *neutral plane*. At the 0° position, the armature loop is perpendicular to the magnetic field, and the black and white portions of the loop are moving parallel to the field. At the instant the conductors are moving parallel to the magnetic field, they do not cut any lines of force, and no emf is induced in the loop. The meter in Figure 3.7A indicates zero current at this time.

As the armature loop rotates from position A to position B (see Figure 3.7F), the sides of the loop cut through more and more lines of force, at a continually increasing angle. At 90° (position B) the sides of the loop are cutting through a maximum number of lines of force and at a maximum angle. The result is that between 0° and 90°, the induced emf in the conductors builds up from zero to a maximum value. Notice that from 0° to 90° rotation, the black side of the loop cuts down through the field while the white side cuts up through the field. The induced emf's in the two sides of the rotating loop are series adding. Thus, the resultant voltage across the brushes (the armature voltage, V_A) is the sum of the two induced voltages, and the meter in Figure 3.7B indicates a maximum value.

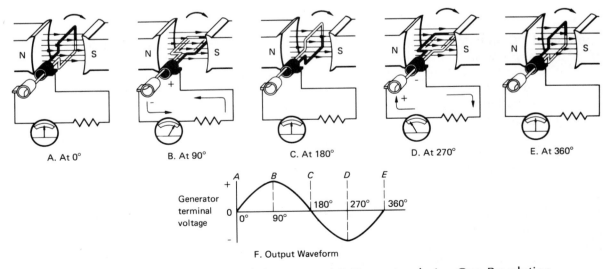

A. At 0°   B. At 90°   C. At 180°   D. At 270°   E. At 360°

F. Output Waveform

Figure 3.7   Output Voltage of Elementary AC Generator during One Revolution

As the armature loop continues rotating from position *B* to position *C,* the sides of the loop that were cutting through a maximum number of lines of force at position *B* now cut through fewer lines. At position *C* (180°) they are again moving parallel to the magnetic field. As the armature rotates from 90° to 180°, the induced voltage will decrease to zero, just as it had increased to a maximum value from 0° to 90°. The meter again indicates zero, as shown in Figure 3.7C.

From 0° to 180° the sides of the loop were moving in the same direction through the magnetic field. Therefore, the polarity of the induced voltage remained the same. However, as the loop starts rotating beyond 180°, from position *C* through *D* to position *E,* the direction of the cutting action of the sides of the coil through the magnetic field reverses. The black side cuts up through the field as the white side cuts down. As a result, the polarity of the induced voltage reverses, as shown in the graph in Figure 3.7F.

Elementary DC Generator

A single-loop generator with each end of the loop connected to a segment of a two-segment metal ring is shown in Figure 3.8. The two segments of the split metal ring are insulated from each other. The ring is called the commutator. The commutator in a DC generator replaces the slip rings of the AC generator and is the main difference in their construction. The commutator mechanically reverses the armature loop connections to the external circuit at the same instant that the polarity of the voltage in the loop reverses. Through this process, the commutator changes the generated AC voltage to a pulsating DC voltage, as shown in the graph of Figure 3.8F. This action is known as *commutation.*

We can follow the rotation of the loop in Figure 3.8 as we did in Figure 3.7. We see that as the segments of the commutator rotate with the loop, they contact opposite brushes. Also, the direction of current flow through the

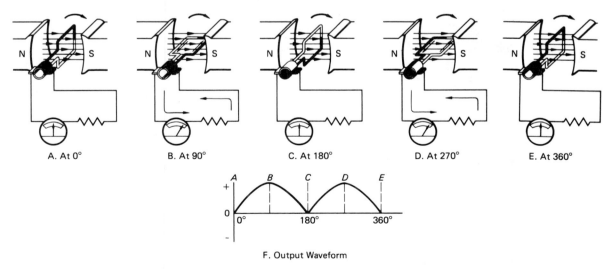

A. At 0°   B. At 90°   C. At 180°   D. At 270°   E. At 360°

F. Output Waveform

Figure 3.8   Elementary DC Generator and Effects of Commutation

brushes and the meter is the same. The voltage developed across the brushes is pulsating and unidirectional, peaking twice during each revolution between zero and maximum. This variation is called *ripple*.

The pulsating voltage of a single-loop DC generator is unsuitable for most applications. Therefore, in practical generators more armature loops and more commutator segments are used to produce an output voltage with less ripple.

Additional Coils and Poles

The effects of additional coils may be illustrated by the addition of a second coil to the armature. The commutator must now be divided into four parts since there are four coil ends, as illustrated in Figure 3.9A. Since there are four segments in the commutator, a new segment passes each brush every 90° instead of every 180°. This action allows the brush to

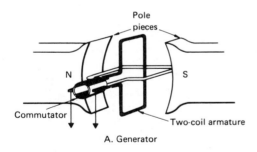

A. Generator

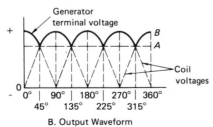

B. Output Waveform

Figure 3.9 Effects of Additional Coils in DC Generator

switch from the white coil to the black coil at the instant the voltages in the two coils are equal.

The graph in Figure 3.9B shows the ripple effect of the voltage when two armature coils are used. As the figure shows, the ripple is limited to the rise and fall between points *A* and *B* on the graph. With the addition of more armature coils, the ripple effect is reduced further. Decreasing ripple in this way increases the effective voltage of the generator output.

Practical generators use many armature coils. They also use more than one pair of magnetic poles. The additional magnetic poles have the same effect on ripple as the additional armature coils have. In addition, the increased number of poles provides a stronger magnetic field. This effect, in turn, allows an increase in output voltage, because the coils cut more lines of flux per revolution.

Electromagnetic Poles

Nearly all practical generators use electromagnetic poles instead of the permanent magnets found in the elementary generator. The electromagnetic field poles consist of coils of insulated copper wire wound on soft iron cores, as shown in Figure 3.1. The main advantages of using electromagnetic poles are (1) increased field strength and (2) control over the strength of the field. By variation of the input voltage to the field windings, the field strength is varied. By variation of the field strength, the output voltage of the generator is controlled.

Generator Voltage Equation

The average generated emf, in volts, of a generator may be calculated from the following equation:

$$\text{emf} = \frac{pN_{\text{cond}}\Phi n_{\text{A}}}{10^8 \times 60b_{\text{A}}} \tag{3.1}$$

where

p = number of poles

N_{cond} = total number of conductors on armature

Φ = flux per pole

n_{A} = speed of armature, in revolutions per minute

b_{A} = number of parallel paths through armature, depending on type of armature winding

For any given generator, all the factors in Equation 3.1 are fixed values except the flux per pole Φ and the speed n_{A}. Therefore, Equation 3.1 can be simplified as follows:

$$\text{emf} = C_{\text{emf}}\,\Phi n_{\text{A}} \qquad (3.2)$$

where

C_{emf} = all fixed values or constants for a given generator

DC MOTOR

Stated very simply, a *DC motor* rotates as a result of two magnetic fields interacting with each other. These two interacting fields are the field due to the current flowing in the field windings, and the field produced by the armature as a result of the current flowing through it. As shown in Figure 3.10A, the loop (armature) field is both attracted to and repelled by the field from the field poles. This action causes the armature to turn in a clockwise direction, as shown in Figure 3.10B.

After the loop has turned far enough so that its north pole is exactly opposite the south field pole, the brushes advance to the next commutator segment. This action changes the direction of current flow through the armature loop and, therefore, changes the polarity of the armature field, as shown in Figure 3.10C.

Torque

Torque is a twisting action on a body, tending to make it rotate. Torque (T) is the product of the force (F) times the perpendicular distance (d) between the axis of rotation and the point of application of the force:

$$T = F \times d$$

The torque is in newton-meters, the force is in newtons, and the distance is in meters. The unit of measure of torque in the British system of units is the pound-foot.

Recall from physics that work is also defined as a force acting through a distance. However, work requires that the motion be in the direction of the force; torque does not. A distinction between work and torque in the British system of measurement is made by the units: Torque is in pound-feet, and work is in foot-

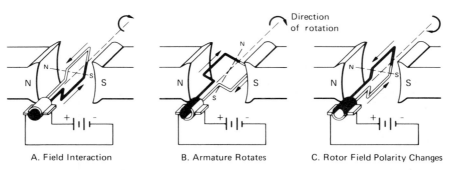

A. Field Interaction   B. Armature Rotates   C. Rotor Field Polarity Changes

Figure 3.10   Armature Rotation of Elementary DC Motor

pounds. In the metric system torque is measured in dyne-centimeters or newton-meters, and work is measured in ergs (which are the same units as dyne-centimeters) or joules (the same as newton-meters).

The metric measurement for torque with units of newton-meters (Nm) will be used in this text. However, conversion to the British system with units of pound-feet (lb-ft) can easily be made with the following conversion factor:

$$0.73756 \text{ lb-ft} = 1 \text{ Nm}$$

Motor Torque Equation

The torque for the DC motor can be calculated from the following equation:

$$T = \frac{pN_{cond}\Phi I_A}{2\pi b_W} \tag{3.3}$$

where

I_A = current in external armature circuit

b_W = number of parallel paths through winding

For any given motor, all the factors in Equation 3.3 are fixed values except the flux per pole Φ and the armature current I_A. Therefore, Equation 3.3 can be simplified as follows:

$$T = C_T \Phi I_A \tag{3.4}$$

where

C_T = all fixed values or constants for a given motor

Ideal DC Machine

In many problems involving the behavior of a DC dynamo as a system component, the machine can be described with satisfactory accuracy in terms of an idealized model having the following properties:

1. The stator has salient poles, and the air gap flux distribution is symmetrical.
2. The armature can be considered as a finely distributed (evenly spread) winding.
3. The brushes are narrow, and commutation occurs when the coil sides are in the neutral zone.

If these conditions exist, then the air gap flux Φ is linearly proportional to field current I_{field}. Then, Equations 3.2 and 3.4 can be simplified further, as follows:

$$\text{emf} = K_{emf}I_{field}n_A \tag{3.5}$$

$$T = K_T I_{field} I_A \tag{3.6}$$

where

K_{emf} = constants C_{emf} and Φ/I_{field}

K_T = constants C_T and Φ/I_{field}

Equations 3.5 and 3.6 are very useful and easily measured. The procedure for evaluating K_T will be shown now. The evaluation of K_{emf} will be given later in this chapter in the section titled "Evaluating Variables."

To determine K_T we use the following procedure: Attach a lever to the armature shaft, and secure a scale to the end of the lever. Apply

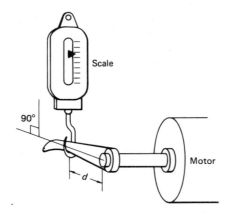

Figure   3.11   Arrangement   for   Measuring Torque

and measure the armature current and the field current, and note the reading on the force scale. The arrangement is shown in Figure 3.11. In Figure 3.11 the scale should be perpendicular to a line through the motor shaft and the lever on the shaft. The distance d is from the center of the shaft to the point of scale attachment.

The torque is the scale reading times the distance in Figure 3.11. The value of K_T is then determined from the following equation:

$$K_T = \frac{T}{I_{field}I_A}$$

This value of K_T should remain relatively constant for this motor. Factors that could affect the value of K_T are those factors that alter the motor in any way, such as heating of the wire in the motor. When the wires in the armature and field heat because of the current flow, the resistance increases and, therefore, reduces the current (and field strength) for a constant voltage. Also, as the armature turns faster, the air friction increases, causing K_T to change. The value of K_T should not change greatly, however.

After K_T has been determined, any value of torque can be approximated by measuring the values of armature and field currents and solving Equation 3.6 for torque T. Equation 3.6 holds whether the motor armature is rotating or stationary. For purposes of calculation, we will use a value of 0.5 Nm/A^2 in this chapter for K_T of a series motor, unless indicated otherwise. This value of K_T is for a representative ⅛-horsepower (hp) motor.

Counter emf in the Motor

In the development of the torque and voltage equations, recall that the motor had armature current and field current as inputs and torque as an output. Look at Equation 3.5 for emf. Observe that when the armature speed (n_A) and field current (I_{field}) interact, a voltage is gener-

ated. This action occurs in the motor as well as the generator. As armature and field currents are applied to the motor to produce torque and armature speed, the speed and the field current produce a voltage in the armature of the motor. This self-generated voltage opposes the external armature voltage that was applied to turn the motor. This self-generated and opposing voltage is termed *counter electromotive force* and is abbreviated cemf. The equation for determining cemf in the motor is exactly the same as the equation for determining emf in the generator:

$$\text{cemf} = K_{emf}I_{field}n_A \qquad (3.7)$$

Conveniently, the cemf generally does not equal or exceed the value of armature voltage (V_A) applied to the motor. If the cemf did exceed the applied armature voltage, the motor would be acting like a generator and would drive current back into the line. If a voltmeter were placed across the armature, it would indicate the larger of the two voltages, applied armature voltage or cemf.

Counter Torque in the Generator

Likewise, in the generator the rotation of the generator causes a current (I_A) to flow in the armature. Then, torque is produced in the generator because of armature current (I_A) and field current (I_{field}) interaction. This torque is in opposition to the external torque turning the armature and is termed *counter torque* (cT). The counter torque equation for the generator is the same as the torque equation for the motor:

$$cT = K_T I_{field}I_A$$

Counter torque produced by the generator is generally less than the torque being supplied by the device that is turning the generator. From the previous development, we see that both the torque and voltage equations are in operation in both the motor and the generator.

Let us now apply these equations to the basic motor and generator configurations and observe the results.

DYNAMO CONFIGURATIONS

A cross-sectional view of a dynamo is shown in Figure 3.12A. Schematic diagrams of the various ways to connect the field and armature in the dynamo are shown in Figures 3.12B through 3.12F. The armature is drawn as a circle representing the end view of the armature, with two squares representing the brushes, and the field is represented by the standard inductor symbol.

The general configurations of DC dynamos take their names from the type of field excitation used. When the dynamo supplies its own excitation, it is called a *self-excited dynamo* (Figures 3.12C through 3.12F). If the field of a self-excited dynamo is connected in parallel with the armature circuit, it is called a *shunt-connected dynamo* (Figure 3.12C). If the field is in series with the armature, it is called a *series-connected dynamo* (Figure 3.12D). If both series and shunt fields are used, it is called a *compound dynamo.* Compound dynamos may be connected as a *short shunt,* with the shunt field in parallel with the armature only (Figure 3.12E), or as a *long shunt,* with the shunt field in parallel with both the armature and series fields (Figure 3.12F). The series and shunt machines show the electrical placement of the in-

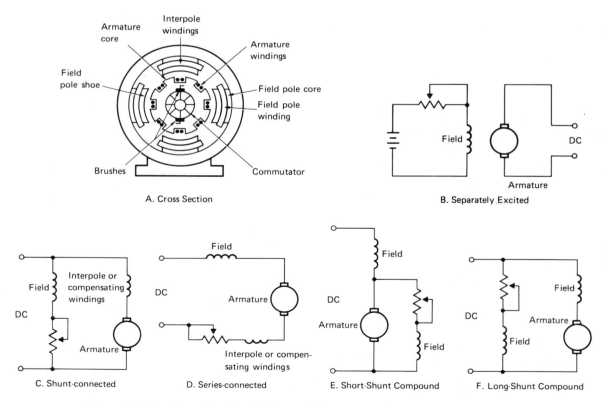

Figure 3.12   Cross-Sectional View of Dynamo and Dynamo Connections

terpole or compensating winding; their physical placement was discussed earlier.

Field potentiometers, as shown in this figure, are adjustable resistors placed in the field circuits to provide a means of varying the field flux, which controls the amount of emf generated or motor torque produced. These resistors are not placed in the armature circuit if that can be avoided, because the current is higher in the armature than in the field. In keeping with basic control system principles, it is more efficient and economical to control the low-power circuits, which, in turn, control the higher-power circuits. The current control of the lower-power field circuit has a profound effect on the output of the higher-power armature circuit, whether it is generating emf or producing torque.

CONSTANT LINE VOLTAGE

The following analyses are somewhat long and involved, but we feel this part of the chapter on DC motors and generators is the most important. In the following pages, we will produce the DC motor *characteristic curves,* which are graphs of the motor's electrical and mechanical responses. These curves are derived from applications of Equations 3.5 and 3.6, Ohm's law, and Kirchhoff's law. To truly understand DC dynamo principles, you must be very proficient with the characteristic curves and the equations just mentioned.

The first topic we deal with concerns the series motor with constant line voltage. This condition can occur when the torque load of a motor varies while the line voltage remains fixed. The characteristic curves developed help illustrate what happens in the motor.

Series Motor

The series-connected motor is the first machine considered for analysis. A simplified series motor is shown in Figure 3.13.

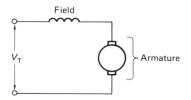

Figure 3.13   Basic Series-Connected Motor

Schematic. The field may be split into two windings, in actuality, one on each side of the armature. But in the schematic, it is generally drawn as one inductor. The symbol V_T in this figure and others in this text represents the line voltage. It symbolizes the connection made to the motor or generator external to its case. The field and the armature together are the schematic representation of the series motor.

Measuring Field and Armature Resistance. The DC resistances of the field and armature are important factors and must be measured. If the field and armature leads do not come out of the motor, some disassembly of the machine may be necessary to obtain these measurements. In this text a ⅛ hp machine will be used for illustration purposes.

As shown in Figure 3.13, the series motor has all the armature current flowing through the field. Therefore, the field windings should be made of heavy conductors in order to carry the armature current without heating or dropping an excessive amount of line voltage. A typical value of field resistance in this motor might be 50 Ω, as measured with an ohmmeter.

The resistance of the armature may be measured either with an ohmmeter or by a method called the locked-rotor–voltage-current method. If the ohmmeter method is used, the armature should be rotated *slowly* while measuring resistance. The ohmmeter reading will fluctuate owing to the slight amount of voltage and current being generated in the armature and to the fluctuating resistance caused by the brushes sliding from one commutator segment

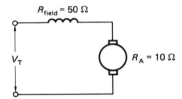

Figure 3.14 Series-Connected Motor with Measured Resistances

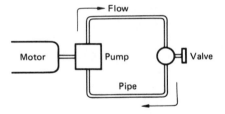

Figure 3.15 Hydraulic Pump Arrangement for Providing a Load to Motor

to another. The armature resistance may be on the order of 10 Ω.

In the *locked-rotor–voltage-current method,* a voltage is applied to the motor while the armature is held stationary (locked rotor). The armature voltage and current are measured, and the armature resistance is determined by applying Ohm's law to these measurements. The test voltage should not be applied for a long time, because of the heating of the conductors and possible damage to the insulation.

For the motor just described, we now have the circuit shown in Figure 3.14, where R_{field} is the field resistance and R_A is the armature resistance.

Evaluating Variables. The next procedure is to apply a line, or terminal, voltage (V_T) and evaluate the value of K_{emf} in the emf equation. Some provision should be made for applying an armature load or external counter torque to the motor since n_A is one of the variables we want to control. This load can be provided by having the motor drive a hydraulic pump or some other rotational device such as a generator.

The hydraulic pump arrangement is illustrated in Figure 3.15. Here, the pump circulates hydraulic fluid in the pipe. If more load is required on the motor, the valve is closed further to restrict the fluid flow. This restricted flow produces an increased back pressure on the pump, which, in turn, produces an increased load on the drive motor. This loading method would work well for motors of ¼ hp or less.

When considering the torque and emf equations and the schematic of the series motor, we see that there are at least four variables: V_T, I_A or I_{field} (since the currents are the same in a series circuit), torque T, and speed n_A. In most system analyses it is instructive to hold one of the variables constant while allowing a second variable to change, and observe the results on the remaining variables. For example, in Ohm's law if V_T is held constant and resistance R is increased, then current I must decrease in order to keep the equation balanced. This technique will be applied to the series motor system.

For the first case we will hold V_T constant at 100 V while the torque is varied, and observe the effects on I_A and n_A. The value of K_T was determined previously to be 0.5 Nm/A². We will arbitrarily pick an n_A of 1000 r/min and an I_A of 0.5 A for illustration purposes. These values would, of course, be measured in actual practice.

For the circuit of Figure 3.14, then, where I_A and I_{field} are equal (because it is a series circuit), the voltage drop across the field (V_{field}) is as follows:

$$V_{field} = I_{field}R_{field} = (0.5 \text{ A}) (50 \text{ Ω})$$
$$= 25 \text{ V} \qquad (3.8)$$

The voltage drop across the field winding is due to the resistance of the wire. There is no reactive component of impedance in the field.

Figure 3.16 is a convenient way of representing the armature of the motor as an equiv-

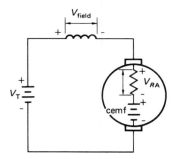

Figure 3.16 Voltage Distribution around Series Motor Circuit

alent circuit. The cemf in the armature in Figure 3.16 is a function of field current, I_{field}, and armature speed, n_A (see Equation 3.7). In the motor current flows from the line into the armature against the cemf. Applying Kirchhoff's law to the series motor yields the following equation:

$$V_T = \text{cemf} + I_A R_A + I_{field} R_{field} \qquad (3.9)$$

The line voltage must be balanced by the IR drops in the armature and the field and the cemf at all times. If the cemf were ever to become larger than line voltage, then current would flow back into the line against the line voltage.

The voltages V_{RA} (equal to $I_A R_A$, the drop across R_A) and V_{field} are voltage drops, while cemf is a voltage rise. It should be noted that a voltmeter placed across the armature would measure V_A (75 V), not the cemf generated.

At this point, the constant K_{emf} in the cemf equation (3.7) can be calculated. Rearranging the cemf equation yields the following:

$$K_{emf} = \frac{\text{cemf}}{I_{field} n_A} = \frac{70 \text{ V}}{(0.5 \text{ A}) (1000 \text{ r/min})}$$
$$= 0.14 \text{ V min/A r} \qquad (3.10)$$

This value of K_{emf} should remain relatively constant as long as the motor is not taken apart and physically altered.

Torque can also be calculated, as follows:

$$T = K_T I_{field} I_A$$
$$= (0.5 \text{ Nm/A}^2) (0.5 \text{ A}) (0.5 \text{ A})$$
$$= 0.125 \text{ Nm}$$

Now, we have completed the calculations for one set of data. Additional sets of data need to be calculated and graphed so that we can visualize what is occurring when the torque of the motor changes but the line voltage remains fixed. This procedure can be repeated for different values of torque. These calculations have been done and are tabulated in Table 3.1. The table also shows that as n_A decreases to zero, torque and armature current increase to some maximum value.

Locked Rotor. The condition when n_A is zero is called a *locked rotor*. When n_A is zero, voltage is applied to the motor, and the rotor or armature behaves as if locked. In this condition maximum armature current and torque are produced because the opposing cemf is zero. This condition also occurs momentarily each time the motor is started from a dead stop.

When power is first applied to the motor, the armature is stationary; maximum current and torque are produced. As the motor starts to turn, the armature speed produces the opposing cemf, which, in turn, reduces current and, therefore, torque. As armature speed continues to increase, the armature current and torque are further reduced, until a stable operating condition is reached in which the torque produced by the motor just balances the torque required by the load to turn it.

Another name for the locked-rotor torque is *stall torque*. The motor is stalled until it produces enough torque to overcome the torque demanded and begins to rotate. The maximum rate of change of speed occurs when power is first applied. The rate of change in speed decreases until a stable speed is reached, at which time the rate of change is zero. This curve of

Table 3.1 Series Motor, V_T Constant

V_T (V)	Torque (Nm)	$I_A (I_{field})$ (A)	V_{field} (V)	V_A (V)	V_{RA} (V)	cemf (V)	n_A (r/min)
100	—	—	—	—	—	—	decreasing to 0
100	0.926	1.36	68	32	13.6	18.4	96.6
100	0.500	1.00	50	50	10.0	40.0	286.0
100	0.463	0.962	48	52	9.6	42.0	314.0
100	0.245	0.70	35	65	7.0	58.0	592.0
100	0.125	0.50	25	75	5.0	70.0	1000.0
100	0.080	0.40	20	80	4.0	76.0	1357.0
100	0.045	0.30	15	85	3.0	82.0	1952.0
100	0.020	0.20	10	90	2.0	88.0	3143.0
100	0.005	0.10	5	95	1.0	94.0	6714.3
100	0	0	0	V_T	0	V_T	∞

armature speed versus time, shown in Figure 3.17, looks very much like the capacitance charge curve. In a typical motor this action occurs within seconds.

Zero Torque. The last line in Table 3.1 shows the results of the torque demand going to zero. The cemf becomes equal to the line voltage (V_T), and no current flows in the armature or field. Of course, these conditions can exist only in the ideal machine. The real machine has friction and other losses that limit the maximum armature speed.

Characteristic Curve. A graph of torque, current, and armature speed (from Table 3.1) is shown in Figure 3.18. These curves are the characteristic curves, or responses, for the series motor.

As shown in Figure 3.18, armature speed is by no means a linear function of torque. In fact, as T (demand) decreases, n_A increases dramatically. A dangerous condition can occur when the torque demand becomes very low and the

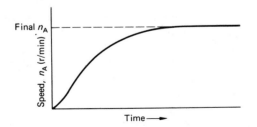

Figure 3.17 Time Response of Motor Starting from Standstill

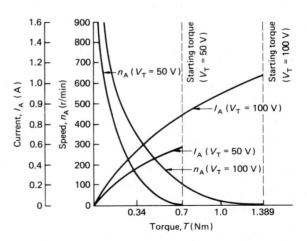

Figure 3.18 Characteristic Curves for Series Motor at Constant Voltage

speed very high. This condition is generally referred to as *runaway*.

Two sets of graphs are shown in Figure 3.18: a set for $V_T = 100$ V and a set for $V_T = 50$ V. The resulting curves show that the characteristic curves are similar in shape. These curves are also referred to as *constant–line voltage curves*. A different set of curves results if the line voltage is allowed to vary and the torque is held constant (they are shown later).

Shunt Motor

In the shunt-connected motor, shown in Figure 3.19, the field is connected in parallel, or shunt, with the armature. Field current and armature current are not the same.

Since the field winding no longer carries armature current, it can be made of fine wire and contain many turns. Thus, the shunt motor field has higher resistance than the series motor field. A typical value of shunt field resistance for a ⅛ hp motor may be 1000 Ω.

Evaluating Variables. Again, line voltage is held constant at 100 V, the torque required is varied, and armature current (I_A) and speed (n_A) are observed. The value of K_T is evaluated, as done previously; it is found to be 0.2 Nm/A². An initial set of values, $n_A = 1000$ r/min and $I_A = 6.25$ A, is measured.

The field voltage drop (V_{field}) is the same as the line voltage (V_T) and armature voltage (V_A) since they are all in parallel. To determine field current (I_{field}), we apply Ohm's law:

$$I_{field} = \frac{V_{field}}{R_{field}} = \frac{100\ V}{1000\ \Omega} = 0.1\ A$$

Then from Equation 3.6:

$$\begin{aligned}T &= K_T I_{field} I_A \\ &= (0.2\ \text{Nm/A}^2)\,(0.1\ A)\,(6.25\ A) \\ &= 0.125\ \text{Nm}\end{aligned}$$

To determine cemf, we first evaluate the internal voltage drop due to R_A:

$$\begin{aligned}V_{RA} &= I_A R_A = (6.25\ A)\,(10\ \Omega) \\ &= 62.5\ V\end{aligned} \tag{3.11}$$

The cemf, then, is as follows:

$$\begin{aligned}\text{cemf} &= V_T - V_{RA} = 100\ V - 62.5\ V \\ &= 37.5\ V\end{aligned} \tag{3.12}$$

At this point, K_{emf} for the shunt motor can be evaluated using Equation 3.10:

$$\begin{aligned}K_{emf} &= \frac{\text{cemf}}{I_{field} n_A} = \frac{37.5\ V}{(0.1\ A)\,(1000\ \text{r/min})} \\ &= 0.375\ \text{V min/A r}\end{aligned}$$

Knowing the value of K_{emf} now allows us to determine any value of n_A for the shunt motor. First, we select a value of torque required of the motor—for example, 0.1 Nm—and solve for armature current (I_A):

$$I_A = \frac{T}{K_T I_{field}} = \frac{0.1\ \text{Nm}}{(0.2\ \text{Nm/A}^2)\,(0.1\ A)} = 5\ A$$

Then:

$$V_{RA} = I_A R_A = (5\ A)\,(10\ \Omega) = 50\ V$$
$$\begin{aligned}\text{cemf} &= V_T - V_{RA} = 100\ V - 50\ V \\ &= 50\ V\end{aligned}$$
$$\begin{aligned}n_A &= \frac{\text{cemf}}{K_{emf} I_{field}} = \frac{50\ V}{(0.375\ \text{V min/A r})\,(0.1\ A)} \\ &= 1333\ \text{r/min}\end{aligned} \tag{3.13}$$

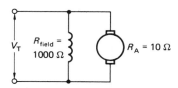

Figure 3.19 Basic Shunt-Connected Motor

Table 3.2   Shunt Motor, V_T Constant

V_T (V)	Torque (Nm)	I_{field} (A)	I_A (A)	V_{field} (V)	V_A (V)	V_{RA} (V)	cemf (V)	n_A (r/min)
100	0.2	.1	—	—	—	—	—	decreasing to 0
100	0.15	.1	7.50	100	100	75.0	25.0	667
100	0.133	.1	6.65	100	100	66.5	33.5	893
100	0.125	.1	6.65	100	100	62.5	37.5	1000
100	0.10	.1	5.0	100	100	50.0	50.0	1333
100	0.075	.1	3.75	100	100	37.5	62.5	1667
100	0.05	.1	3.33	100	100	33.3	66.7	1779
100	0.025	.1	1.25	100	100	12.5	87.5	2333
100	0	.1	0	100	100	0	V_T	approaches maximum

Table 3.2 shows the results of this procedure repeated for various values of torque. Similar to the results for the series motor, Table 3.2 shows that as armature speed decreases to zero, torque and armature current increase to some maximum value.

Maximum Armature Speed.   How is maximum armature speed determined in the shunt motor? From the cemf equation we have the following:

$$n_{A(max)} = \frac{cemf_{max}}{K_{emf}I_{field}}$$

$$= \frac{100 \text{ V}}{(0.375 \text{ V min/A r}) (0.1 \text{ A})}$$

$$= 2667 \text{ r/min}$$

Maximum speed for the shunt motor can be determined in another way without knowing line voltage. Look at the previous equation:

$$n_{A(max)} = \frac{cemf_{max}}{K_{emf}I_{field}}$$

where

$$cemf_{max} = V_T - I_A R_A$$

But $I_A = 0$ since cemf is maximum. Therefore, cemf $= V_T$.

Solving for I_{field} in terms of voltage and resistance gives the following:

$$I_{field} = \frac{V_T}{R_{field}}$$

Substituting yields the following:

$$n_{A(max)} = \frac{cemf_{max}}{K_{emf}I_{field}} = \frac{V_T}{K_{emf}(V_T/R_{field})}$$

$$= \frac{V_T R_{field}}{K_{emf}V_T}$$

V_T cancels. Thus:

$$n_{A(max)} = \frac{R_{field}}{K_{emf}} = \frac{1000 \text{ }\Omega}{0.375 \text{ V min/A r}}$$

$$= 2667 \text{ r/min} \qquad \textbf{(3.14)}$$

This equation shows that maximum armature speed of the shunt motor is a function not of line voltage but of circuit constants. In other words, theoretically, no matter what line voltage is applied, as torque demand goes to zero, speed will always go to the same value. This result is shown in Figure 3.20 for line voltage of 100 V and 50 V.

Doubling Line Voltage.   Observe in Figure 3.20 what occurs when line voltage is doubled.

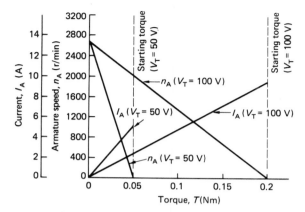

Figure 3.20 Characteristic Curves for Shunt Motor at Constant Voltage

Maximum torque produced is quadrupled because both I_A and I_{field} are doubled. Thus, some high-voltage motors produce disproportionately more available torque.

Comparing Series and Shunt Motors

As shown in Figure 3.20, armature current in the shunt motor is directly proportional to torque:

$$I_A = \frac{T}{K_T I_{field}}$$

(I_{field} stays constant as long as V_T is constant.)

But in the series motor (Figure 3.18), armature current is a second-order function of torque:

$$I_A = I_{field}$$

Therefore:

$$I_A^2 = \frac{T}{K_T}$$

This equation results in the parabolic-shaped curve of Figure 3.18.

CONSTANT TORQUE

When we examined the series and shunt motors in the previous sections, we observed that there were at least four variables: V_T, I_A or I_{field} (since they are the same in a series motor), torque T, and n_A. It is obvious that there are more variables than can be easily observed on a simple, two-dimensional graph. Our approach, then, is to hold one variable constant, vary another, and observe the effect on the remaining variables. In the previous sections we examined the effects on I_A and n_A as torque was varied and V_T was held constant. In the next investigation we will maintain torque at a constant value, vary the line voltage V_T, and again observe the effects on armature current I_A and speed n_A.

Series Motor

Since the same series motor will be used in this section as in the previous one, K_T (0.5 Nm/A²) and K_{emf} (0.14 V min/A r) will also be the same because we have not physically altered the motor. Therefore, from Table 3.1, when V_T = 100 V, torque T is 0.125 Nm and n_A is 1000 r/min.

The next step is to vary V_T while torque is held at 0.125 Nm. Armature and field currents will be constant at 0.5 A. From Equation 3.6:

$$T = K_T I_{field} I_A$$

Since the field and armature currents are equal, solving for current yields:

$$I_A = \sqrt{\frac{T}{K_T}} = \sqrt{\frac{0.125 \text{ Nm}}{0.5 \text{ Nm/A}^2}} = 0.5 \text{ A}$$

Therefore, from Equation 3.8 and Figure 3.14:

$$V_{field} = I_{field} R_{field} = (0.5 \text{ A})(50 \text{ }\Omega) = 25 \text{ V}$$

Also:

$$V_{RA} = I_A R_A = (0.5 \text{ A}) (10 \text{ } \Omega) = 5 \text{ V}$$

Setting V_T to 50 V in Figure 3.14 produces the following results:

$$V_A = V_T - V_{field} = 50 \text{ V} - 25 \text{ V} = 25 \text{ V}$$

And from Figure 3.16:

$$\text{cemf} = V_A - V_{RA} = 25 \text{ V} - 5 \text{ V} = 20 \text{ V}$$

Therefore, rearranging Equation 3.10 yields:

$$n_A = \frac{\text{cemf}}{K_{emf} I_{field}} = \frac{20 \text{ V}}{(0.14 \text{ V min/A r}) (0.5 \text{ A})}$$
$$= 286 \text{ r/min}$$

Table 3.3 shows the preceding procedure repeated for different values of V_T. From Table 3.3 we see that as the line voltage continues to increase, so does armature speed; no theoretical limit is reached. In practice, maximum armature speed is determined by how long the motor can hold together.

Minimum line voltage is reached when n_A goes to zero, or stalls. At zero armature speed, cemf is also zero:

$$n_A = \frac{\text{cemf}}{K_{emf} I_{field}} = \frac{0}{(0.14 \text{ V min/A r}) (0.5 \text{ A})}$$
$$= 0$$

Also:

$$V_A = V_T - V_{field}$$

Or after rearrangement:

$$V_T = V_A + V_{field} = (\text{cemf} + V_{RA}) + V_{field}$$
$$= (\text{cemf} + I_A R_A) + V_{field}$$
$$= 0 + (0.5 \text{ A}) (10 \text{ } \Omega) + 25 \text{ V}$$
$$= 30 \text{ V}$$

This equation shows that when the line voltage decreases to 30 V, the motor will stall if it still is required to produce 0.125 Nm of torque.

The graph in Figure 3.21 is the characteristic curve for the series motor with constant torque. Two armature speed curves have been drawn to show what happens at a higher, fixed torque. Notice that these lines are not parallel. Also, the higher-torque example does not start rotating until a line voltage of 30 V is exceeded.

Shunt Motor

In the shunt motor under varying line voltage but constant torque, there are five possible variables: V_T, I_A, I_{field}, T, and n_A. Field current in the shunt motor varies directly as the line voltage. If the torque is held constant and field current

Table 3.3 Series Motor, Torque Constant

V_T (V)	Torque (Nm)	$I_A(I_{field})$ (A)	V_{field} (V)	V_A (V)	V_{RA} (V)	cemf (V)	n_A (r/min)
stall voltage	.125	.5	25	decreasing to minimum	5	0	0
50	.125	.5	25	25	5	20	286
100	.125	.5	25	75	5	70	1000
150	.125	.5	25	125	5	120	1714
200	.125	.5	25	175	5	170	2429
∞	.125	.5	25	∞	5	∞	∞

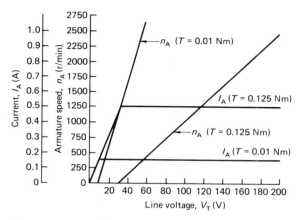

Figure 3.21 Characteristic Curves for Series Motor at Constant Torque

$$I_{\text{field}} = \frac{V_{\text{field}}}{R_{\text{field}}} = \frac{150 \text{ V}}{1000 \ \Omega} = 0.15 \text{ A}$$

$$I_A = \frac{T}{K_T I_{\text{field}}} = \frac{0.125 \text{ Nm}}{(0.2 \text{ Nm/A}^2)(0.15 \text{ A})}$$

$$= 4.17 \text{ A}$$

$$V_{RA} = I_A R_A = (4.17 \text{ A})(10 \ \Omega) = 41.7 \text{ V}$$

Recall that in a shunt motor, $V_A = V_T$ from Equation 3.12. Therefore:

$$\text{cemf} = V_A - V_{RA} = 150 \text{ V} - 41.7 \text{ V}$$

$$= 108.3 \text{ V}$$

$$n_A = \frac{\text{cemf}}{K_{\text{emf}} I_{\text{field}}} = \frac{108.3 \text{ V}}{(0.375 \text{ V min/A r})(0.15 \text{ A})}$$

$$= 1925 \text{ r/min}$$

changes, then armature current must vary inversely.

The calculations required to produce the characteristic curves for the shunt motor of Figure 3.19 are as follows: Using values for K_T of 0.2 Nm/A^2, for T of 0.125 Nm, and for K_{emf} of 0.375 V min/A r, increase the line voltage from 100 to 150 V. Then, we have the following calculations:

Table 3.4, which shows the results of repeating the above procedure for various values of line voltage, is another verification that line voltage does not determine maximum possible speed for the shunt motor. Maximum armature speed is determined as follows:

Table 3.4   Shunt Motor, Torque Constant

V_T (V)	Torque (Nm)	I_{field} (A)	I_A (A)	V_{field} (V)	V_A (V)	V_{RA} (V)	cemf (V)	n_A (r/min)
stall voltage	.125	decreasing to 0	increasing to maximum	stall voltage	stall voltage	increasing to maximum	decreasing to 0	decreasing to 0
90	.125	6.09	6.94	90	90	69.4	20.6	610
100	.125	0.1	6.25	100	100	62.5	37.5	1000
120	.125	0.12	5.21	120	120	52.1	67.9	1509
150	.125	0.15	4.17	150	150	41.7	108.3	2250
160	.125	0.16	3.91	160	160	39.1	120.9	2015
200	.125	0.2	3.125	200	200	31.3	168.8	2250
300	.125	0.3	2.08	300	300	20.8	279.2	2482
∞	.125	0	0	—	—	0	—	maximum

$$n_{A(max)} = \frac{R_{field}}{K_{emf}} = \frac{1000\ \Omega}{0.375\ \text{V min/A r}}$$

$$= 2667\ \text{r/min}$$

We also need to determine *stall voltage,* the voltage at which the motor stops turning. At stall, we have the following:

$$\text{cemf} = 0$$
$$V_T = V_{field} = V_{RA} = I_A R_A$$

or

$$I_{field} R_{field} = I_A R_A$$

Solving for I_A yields the following:

$$I_A = \frac{I_{field} R_{field}}{R_A}$$

Also:

$$T = K_T I_{field} I_A$$

Substitution gives us the following:

$$T = K_T I_{field} \left(\frac{I_{field} R_{field}}{R_A} \right)$$

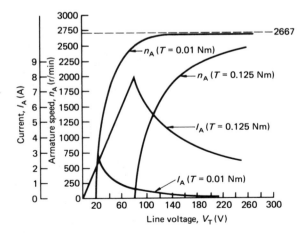

Figure 3.22 Characteristic Curves for Shunt Motor at Constant Torque

Solving for I_{field}, we have the following result:

$$T = \frac{K_T I_{field}^2 R_{field}}{R_A}$$

$$I_{field} = \sqrt{\frac{T R_A}{K_T R_{field}}}$$

$$= \sqrt{\frac{(0.125\ \text{Nm})\ (10\ \Omega)}{(0.2\ \text{Nm/A}^2)\ (1000\ \Omega)}}$$

$$= 0.07906\ \text{A}$$

Also:

$$I_A = \frac{T}{K_T I_{field}} = \frac{0.125\ \text{Nm}}{(0.2\ \text{Nm/A}^2)\ (0.07906\ \text{A})}$$

$$= 7.905\ \text{A}$$
$$V_T = I_A R_A = (7.905\ \text{A})\ (10\ \Omega) = 79.05\ \text{V}$$

The data in Table 3.4, along with lower-constant-torque data, are used to obtain the curves in Figure 3.22. Notice that as torque decreases, the curve becomes more angular. Thus, if the torque demand on the shunt motor is low, maximum speed is reached quickly with the application of a relatively small line voltage. In the ideal or theoretical motor with a torque demand of zero, maximum speed will be reached any time the line voltage exceeds 0 V. Figure 3.22 also shows that armature current is linear up to the point where the motor starts rotating. After that point, armature current decreases, owing to increased cemf.

MOTOR CONTROL

Now that we have completed the basic characteristic curves, we must discuss how to control the motor. Methods and circuitry for electronic motor control will be presented in Chapter 7, but a discussion of what parts of a motor lend themselves to control can be undertaken now.

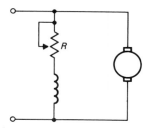

Figure 3.23 Shunt Motor with Field Current Control

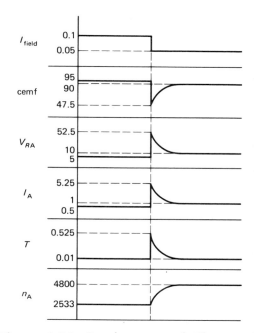

Figure 3.24 Synchrogram of Changes That Occur within Shunt Motor

A general rule for control systems is to place the control mechanism in the low-power part of the system so that the circuit can then control the higher-power part of the system. For example, a signal in a transistor amplifier is generally applied to the base because it is the low-power circuit that controls the higher-power output circuit. However, just as the transistor is not always base-driven, neither are all control systems controlled in the low-power circuitry. Much depends on the application and the component(s) used for the control.

In the shunt motor it is relatively easy to control the speed by controlling field current. Field current is generally much smaller than armature current and, as we will see, has a great effect on armature current. For example, consider the shunt motor in Figure 3.23. A variable resistor R has been placed in the field circuit to control field current. Let us see what effect a changing field current has on armature speed.

For convenience, we consider this motor to be the same as the one in the previous section pertaining to shunt motors with constant line voltage. With $V_T = 100$ V, Table 3.5 shows the

measured and calculated values. We will assume that the motor load needs the same amount of torque to turn it, no matter what speed it has.

Now, suppose resistor R is increased suddenly so that field current is cut in half. A synchrogram of this result is shown in Figure 3.24. A *synchrogram* is a vertical alignment of related graphs to show a time relationship.

We will assume that the speed of the motor does not change instantly whenever a change in torque is applied because of inertia of the ar-

Table 3.5 Measured and Calculated Values for Shunt Motor

V_T (V)	Torque (Nm)	I_{field} (A)	I_A (A)	V_{field} (V)	V_A (V)	V_{RA} (V)	cemf (V)	n_A (r/min)
100	.01	.1	.5	100	100	5	95	2533

mature. If at the instant field current changes, n_A is still 2533, then we have the following results. Rearranging Equation 3.13:

$$\text{cemf} = K_{emf} I_{field} n_A$$
$$= (0.375\,\text{V min/A r})(0.05\,\text{A})(2533\,\text{r/min})$$
$$= 47.5\,\text{V}$$

Rearranging Equation 3.12:

$$V_{RA} = V_T - \text{cemf} = 100\,\text{V} - 47.5\,\text{V}$$
$$= 52.5\,\text{V}$$

Rearranging Equation 3.11:

$$I_A = \frac{V_{RA}}{R_A} = \frac{52.5\,\text{V}}{10\,\Omega} = 5.25\,\text{A}$$

From Equation 3.6:

$$T = K_T I_{field} I_A$$
$$= (0.2\,\text{Nm/A}^2)\,(0.05\,\text{A})\,(5.25\,\text{A})$$
$$= 0.525\,\text{Nm}$$

With this increased torque produced, the motor will increase in speed. As the motor increases in speed, more cemf is produced, which, in turn, decreases armature current, causing reduced torque. A stable speed is reached when torque returns to its original value, as the following calculations show:

$$I_{A(final)} = \frac{T}{K_T I_{field}} = \frac{0.01\,\text{Nm}}{(0.2\,\text{Nm/A}^2)\,(0.5\,\text{A})}$$
$$= 1\,\text{A}$$
$$V_{RA(final)} = I_A R_A = (1\,\text{A})\,(10\,\Omega) = 10\,\text{V}$$
$$\text{cemf}_{final} = V_T - V_{RA} = 100\,\text{V} - 10\,\text{V}$$
$$= 90\,\text{V}$$
$$n_{A(stable)} = \frac{\text{cemf}}{K_{emf} I_{field}}$$
$$= \frac{90\,\text{V}}{(0.375\,\text{V min/A r})\,(0.05\,\text{A})}$$
$$= 4800\,\text{r/min}$$

These calculations show that a 0.05 A change in field current can cause a 0.5 A change in armature current. The surprising aspect is that when field current is reduced, armature speed is *increased*. Thus, if the shunt motor field is accidentally disconnected, the motor could go into runaway.

Observation of the speed curve in Figure 3.24 shows that the previous shunt motor maximum speed of 2667 r/min (Equation 3.14) has been greatly exceeded (4800 r/min). This result occurs because the field resistance must now include the control resistor's resistance:

$$R_{field} = \frac{V_T}{I_{field}} = \frac{100\,\text{V}}{0.05\,\text{A}} = 2000\,\Omega$$
$$n_{A(max)} = \frac{R_{field}}{K_{emf}} = \frac{2000\,\Omega}{0.375\,\text{V min/A r}}$$
$$= 5333\,\text{r/min}$$

Note: These last two equations were used to calculate *maximum* possible speed. Figure 3.24 is not representing maximum speeds.

The series motor, shown in Figure 3.25, has a control resistor in parallel with the field. In the series motor, as resistor R is decreased, the actual field current is reduced, thus reducing field strength. However, armature current is increased because the total field resistance is decreasing. The net effect is that torque is increased in the series motor, just as it was in the shunt motor previously discussed.

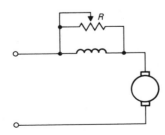

Figure 3.25 Series Motor with Control Resistor

CHARACTERISTIC CURVES

The characteristic curves mathematically derived in the previous sections show that the emf and torque equations do represent the motor's physical operation. It is not our intent, though, to have you, as a technician, become proficient in manipulating numbers. Rather, we expect you to perceive basic motor operation through the equations and characteristic curves. For instance, consider the following examples:

1. In the series motor with constant line voltage, what will happen to the armature speed if torque is decreased? From the curve in Figure 3.18, we see that speed will increase. You probably will not need to know how much the speed increases, but you should know if it increases nonlinearly.
2. If the field current in a shunt motor of constant line voltage is decreased, what happens to the motor armature speed? As discussed in the previous section, speed generally increases because cemf typically is larger than V_{RA}. This action is not easily observable in either Equation 3.6 or 3.7, since cemf is related to speed n_A as follows:

$$n_A = \frac{\text{cemf}}{K_{emf}I_{field}}$$

And since cemf and I_{field} both decrease, they both have counteracting effects on n_A or the characteristic curves.

SEPARATELY EXCITED MOTOR

The schematic for the separately excited motor is shown in Figure 3.26. The equations and characteristic curves for this motor are nearly identical to those for the shunt motor. The derivations are left as exercises.

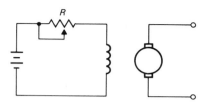

Figure 3.26   Separately Excited Motor

COMPOUND MOTORS

The compound motor has both a shunt and a series field. As a review, Figure 3.27 shows the two possible ways of connecting the fields: long shunt and short shunt. Of these two possibilities, the long shunt is most often used since the shunt field has a relatively constant current—and, therefore, field strength—as long as the line voltage remains constant. The long-shunt arrangement also produces a series field strength that is proportional to armature current.

In most cases the series winding is connected so that its field aids that of the shunt winding, as shown in Figure 3.28A. Motors of this type are called *cumulative compound motors*. If the series winding is connected so that its field opposes that of the shunt winding, the

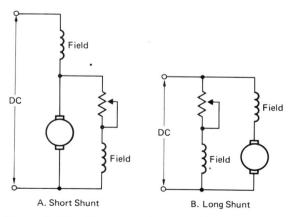

Figure 3.27   Compound Motor Configurations

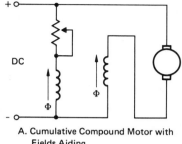

Figure 3.28 Compound Motors

A. Cumulative Compound Motor with Fields Aiding

B. Differential Compound Motor with Fields Opposing

motor is called a *differential compound motor*, as shown in Figure 3.28B. In these figures, Φ represents the flux per pole, and the arrows indicate relative direction of the flux.

The characteristics of a cumulative compound motor are between those of a series motor and those of a shunt motor. If the field strength of the series field is stronger than that of the shunt field, the characteristics are more like those of a series motor, and vice versa. Figure 3.29 illustrates the relationship between the motors.

As the figure indicates, the cumulative compound motor has higher starting torque than the shunt motor and better speed regulation than the series motor. Unlike the series motor, however, it does have a definite no-load speed.

In some operations it is desirable to use the cumulative series winding to obtain good starting torque. When the motor comes up to speed, the series winding is shorted out. The motor then has the improved speed regulation of a shunt motor.

The differential compound motor is seldom used because of two basic problems. One problem is speed instability under heavy load. The speed regulation is very good under light loads. But as the load increases, so does the speed of the motor. As the speed increases, the armature current also increases. The increased current causes the field strength of the series winding's opposing field to eventually exceed

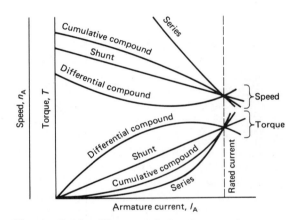

Figure 3.29 Characteristic Curves for Compound, Series, and Shunt Motors

the shunt's field strength, and the motor starts to run backward. As shown in Figure 3.29, the other problem is that starting torque is low for a relatively high armature current. That is, it uses energy in opposing magnetic fields when armature current is high.

GENERAL CONSIDERATIONS

Reversing Direction

Reversing the direction of any DC motor requires that the current through the armature with respect to the field be reversed. If the currents in the armature and field are both reversed, there will be no change in rotational direction. This situation is shown in Figure 3.30, where the direction of F (force) remains unchanged when both conductor and field currents are reversed.

To change the direction of a motor, we must reverse the armature rather than the field, for the following four reasons:

1. The field is more inductive than the armature and frequent reversals produce switch contact arcing and erosion.
2. In a compound motor both fields must be reversed. Otherwise, the motor will change from a cumulative compound motor to a differential compound motor.
3. The armature circuit connections are usually opened for various types of braking, as will be discussed later.
4. If the field-reversing circuit fails to close, runaway could result.

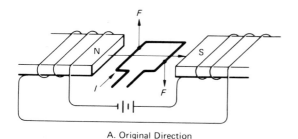

A. Original Direction

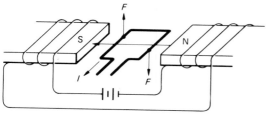

B. Direction Reversed

Figure 3.30 Magnetic Field and Armature Current Reversed

Motor Starters and Controllers

In large motors a motor starter is required to prevent excessive starting currents. A *motor starter* is a switching device that is intended to start and accelerate the motor. A *motor controller,* on the other hand, is a device that controls the power flow to a motor, usually resulting in some form of speed control. Controllers will be discussed later in Chapter 7.

A motor starter generally consists of a resistive bank inserted in the motor circuit to prevent full line voltage from being applied to the motor. Some method is used to reduce this resistance as the motor speeds up, thus keeping motor currents from reaching destructive levels. This reduction may be done manually or automatically.

Figure 3.31 shows a typical manual starter circuit. A spring on the starter switch holds it in the off position. But once in the run position, the switch will be held there by an electromagnet. It is up to the operator to provide the necessary time delay between switch positions to allow the speed to stabilize and prevent high currents. An automatic starter typically uses the same resistive configuration, but the switching is done automatically with time-delay relays.

Most starters and controllers include one or more devices known as *overload relays.* These devices protect the motor from overheating caused by loads above the rated value. There are four types of overload protection systems in

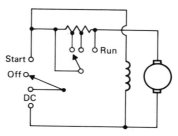

Figure 3.31 Manual Motor Starter Circuit

use: (1) thermal relays, (2) magnetic relays, (3) electronic relays, and (4) thermistors.

Thermal overload relays have a small heating element connected in series with the motor. Some consist of a bimetallic strip with a heating coil around it. As the current heats the strip, it opens and stops the motor. Other devices resemble fuses, which melt to break the circuit. Some method of resetting the circuit, either manually or automatically, is also provided.

Figure 3.32 shows the schematic symbol for the thermal relay. The question-mark shapes represent the heating elements. At the right in Figure 3.32 is the normally closed (NC) contact that the thermal relay would open.

Magnetic overload relays are actuated by an electromagnetic coil, much as the typical circuit breaker is. This relay would be ineffective against heat buildup due to a large number of successive starts and stops.

Electronic overload relays sense both the voltage applied and the current flow through the motor. They electronically simulate the iron loss and copper loss within the motor.

A *thermistor* is a resistor whose resistance changes greatly as the temperature increases. The temperature characteristic of most thermistors is negative; that is, as temperature increases, the thermistor's resistance decreases. However, in heat-sensing applications of motors, the thermistor with a positive temperature coefficient (discussed in Chapter 8) generally is used.

The thermistor is embedded in the windings of the motor and is monitored electronically to detect heat. When the windings get too hot, the power is shut off. Thermistors generally are not placed in the motor rotor because of the connection problem. Thus, it is unlikely that a thermistor would be used in a DC machine.

Stopping a Motor

Large motors with heavy loads may take a very long time to coast to a stop. The motor itself, however, can be used to slow the rotation. Two methods used for electromechanical braking are dynamic braking and plugging.

Dynamic braking is illustrated in Figure 3.33. Here, the armature of a shunt motor is disconnected from the line and connected to a resistor. The motor armature is still rotating within a magnetic field and therefore producing an armature voltage due to generator action. If the armature circuit is completed through a resistor, current will flow and a counter torque will be produced. This counter torque will slow the motor. As the motor slows down, the counter torque will decrease because the current is decreasing. Therefore, most of the slowing effect occurs at higher speeds, as indicated in Figure 3.34.

Plugging occurs when the armature is disconnected from the line and then is reconnected to the line in the opposite direction. This technique causes the motor to slow down very rapidly. When the motor comes to rest, it is automatically disconnected from the power line. Figure 3.35 is a simplified diagram illustrating plugging.

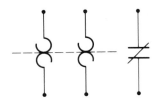

Figure 3.32 Schematic Symbol for Thermal Overload Relay

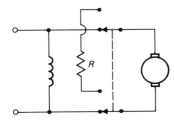

Figure 3.33 Dynamic-Braking Circuit

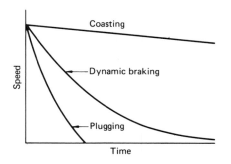

Figure 3.34 Time Relationship for Various Braking Methods

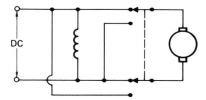

Figure 3.35 Plugging Circuit

DC GENERATOR

Classification

DC generators are classified by their method of supplying excitation current to the field coils. The two major classifications are separately excited and self-excited generators. Self-excited generators are further classified by the method in which the field coils are connected. Like the DC motor, the generator can be connected as series, shunt, or compound. The schematic diagrams for these generator configurations are identical to the corresponding motor schematics. These configurations will be discussed in more detail next.

Separately Excited Generator

A DC generator that has its field supplied by another generator, batteries, or some other out-side source is referred to as a *separately excited generator*. The circuit for this generator is shown in Figure 3.36.

The field excitation in this generator is the same as the excitation in the separately excited motor. However, the armature circuit here is different. Whereas the motor has an external voltage applied to the armature so that a torque is produced, the generator has an external torque supplied to rotate the armature, and a voltage (emf) is produced in the armature.

If the armature circuit is completed, current flows through the external load resistor R_L. The voltage distribution is illustrated in Figure 3.37, which shows that emf is the only voltage rise in the armature circuit. Thus, armature voltage must be less than—or, at best, equal to—the emf generated. Therefore, the generator armature voltage (V_{RL}) will also be lower than the emf generated. This result is an obvious departure from the motor characteristics, where the motor armature voltage is always higher than the cemf generated.

Now, let us look at an example that illustrates the development of the separately excited

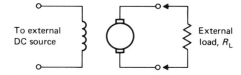

Figure 3.36 Separately Excited Generator Connection

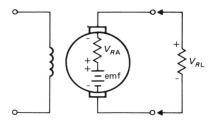

Figure 3.37 Voltage Distribution in Armature Circuit of Generator

characteristic curve. Assume a constant armature speed drive is turning the generator at 1000 r/min. Armature resistance (R_A) is measured at 10 Ω, load resistor (R_L) is 50 Ω, K_T is 0.5 Nm/A^2, armature current (I_A) is 0.5 A, and field current (I_{field}) is 0.5 A. With these values determined, the armature terminal voltage, or load resistor voltage, is easily found:

$$V_{RL} = I_A R_L = (0.5 \text{ A}) (50 \text{ Ω}) = 25 \text{ V}$$

Generated voltage (emf) can also be found, as follows:

$$\text{emf} = I_A R_A + I_A R_L$$

$$= (0.5 \text{ A}) (10 \text{ Ω}) + 25 \text{ V} = 30 \text{ V}$$

Also:

$$K_{emf} = \frac{\text{emf}}{I_{field} n_A} = \frac{30 \text{ V}}{(0.5 \text{ A}) (1000 \text{ r/min})}$$

$$= 0.06 \text{ V min/A r}$$

These calculations show that the emf of 30 V is greater than the armature terminal voltage of 25 V due to the internal drop across R_A. As the load resistor decreases, the armature current increases, and more voltage is dropped across R_A. If the load resistor goes to zero (or shorts), maximum armature current flows:

$$I_A = \frac{\text{emf}}{R_A} = \frac{30 \text{ V}}{10 \text{ Ω}} = 3 \text{ A}$$

This maximum current causes all the generated emf to be dropped internally, and the armature terminal voltage is zero. Figure 3.38 shows this result.

This same situation is encountered when a battery's terminals are shorted with a wire. A high current flows in the shorting wire, causing it to become hot. All the battery's open-circuit terminal voltage is dropped across its own internal resistance.

Maximum counter torque (cT) is also produced in the generator when the output is shorted since the following relation holds:

$$cT = K_T I_{field} I_A$$

Field current is constant and armature current is maximum. The characteristic curve is shown in Figure 3.39.

Some applications require the armature speed to vary but the output voltage to remain constant, such as in the automobile charging system. This result can be achieved by changing the field current as an inverse function of speed.

As an example, suppose the output voltage of 25 V of the previous generator is to be main-

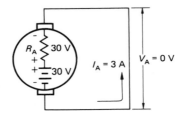

Figure 3.38 Voltage Distribution in Armature Circuit of Shorted Generator

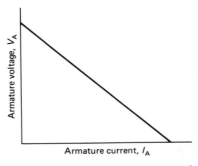

Figure 3.39 Characteristic Curve for Separately Excited Generator at Constant Armature Speed

tained but the speed increased to 2000 r/min. Let us find the required field current. As long as the load resistance remains constant, so must armature current and emf. Therefore, the emf equation can be used to solve for I_{field}, as follows:

$$I_{field} = \frac{emf}{K_{emf}n_A}$$

$$= \frac{30 \text{ V}}{(0.06 \text{ V min/A r}) (2000 \text{ r/min})}$$

$$= 0.25 \text{ A}$$

The automotive charging system is one that has a constantly changing speed drive for the generator but needs a constant output voltage for charging the battery. The automotive system can perform its task by using some method of sensing output voltage and adjusting field current to keep the voltage constant.

Self-Excited Generators

Shunt Generator. When the field windings are connected in parallel with the armature, as in Figure 3.40, the generator is shunt-connected.

Self-excited generators, of which the shunt generator is one, do not produce an output voltage when their armatures are rotated, for several reasons. Consider what happens when the generator in Figure 3.40 starts from 0 r/min. At 0 r/min, armature, field, and load currents are all zero. As the armature starts turning, is an output voltage produced? Ideally, no. Since no field current is flowing, there is no magnetic field across the armature. Remember that voltage is produced only when a conductor moves within a magnetic field.

An output voltage generally is produced, though, because of a small residual magnetism remaining in the soft iron of the field core. Figure 3.41 shows how this residual magnetism appears. Consider the circuit of Figure 3.41A, where field current starts from zero and increases to $I_{field(sat)}$ (field current saturation). As the armature speed is held constant, the generated voltage increases from 0 V to some voltage at point C on the curve in Figure 3.41B. Notice that the curve is approximately linear from point A to point B but then curves to point C. This entire curve is generally called the *magnetization curve;* it shows field strength as a function of field current.

So, field strength increases linearly with field current to point B. After that point, called the knee of the curve, it takes proportionately

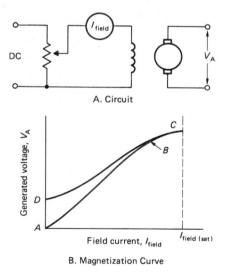

A. Circuit

B. Magnetization Curve

Figure 3.41   Determining Generator Magnetization Curve

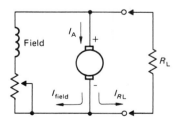

Figure 3.40   Shunt-Connected Generator

much more field current to produce the same increase in field strength. From point B to point C and beyond is called the saturation part of the curve. Realize, though, that there is no point reached where an increase in field current does not produce an increase in field strength.

As field current is decreased to zero, some magnetism is retained because some of the molecular domains (groups of molecules that act together) remain aligned in the direction they were when field current was present. This small magnetism is called *residual magnetism* and is a property of the soft iron field core called hysteresis.

Another cause for no generated output voltage arises in a self-excited generator when the field is connected in the wrong direction. Figure 3.42 shows a shunt-connected generator with residual magnetism. The solid lines represent the field due to the residual magnetism, and the dashed lines represent the field due to the generated field current. As the generator starts to rotate, current increases in the direction that opposes the residual magnetism. At some point, the field current will completely cancel out the field magnetism, and the output voltage will be zero.

A field resistance that is too large can also prevent voltage buildup in the generator. Figure 3.43 shows a graph of field current and voltage for various field resistances R_{field}. For the same field voltage, as field resistance decreases, field current increases.

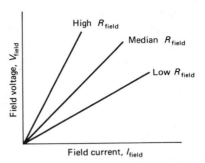

Figure 3.43 Voltage–Current Characteristics for Various Field Resistances

If this graph is superimposed on the magnetization curve (Figure 3.41), the graph in Figure 3.44 results. This graph shows that for a field resistance (R_{field}) and constant speed, a small initial voltage (V_{A1}) will be produced. As current starts to flow in the field, more voltage will be produced, generating more current, and so on. A stable armature voltage (V_{A2}) will be reached when the voltage and current produced are all dropped across the field, assuming no other losses. If the field resistance is decreased to a new value $R_{field(1)}$ (corresponding to the low R_{field} curve in Figure 3.43), armature voltage stabilizes at V_{A3}. If, however, the field resistance is increased above R_{field} to R_{crit} (the critical resistance), the output voltage drops to a value below V_{A2}. If the field resistance is increased

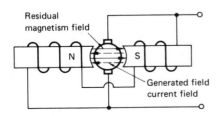

Figure 3.42 Field Connected in Incorrect Direction

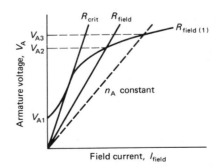

Figure 3.44 Magnetization and Field Resistance Curves Superimposed

above R_{crit}, the generator field will *unbuild* (or collapse). In other words, the output voltage will drop back down to V_{A1}.

In summary, the following four conditions may cause a self-excited generator not to build an output voltage:

1. There may be lack of residual magnetism. Residual magnetism may be lost by physical shock (such as being dropped on a floor), heat, vibration, or lack of use for a period of time. This condition can be corrected by *flashing the field,* that is, by applying a DC voltage to the field in the proper direction to reestablish the residual magnetism.
2. The field circuit may be reversed with respect to the armature. A simple test shows whether the field is reversed. Connect a voltmeter to the generator output. If the output increases slightly when the field is disconnected, then the field is reversed.
3. The field circuit resistance may be higher than the critical value. An open or a high-resistance field circuit may be the problem. Check the circuit with an ohmmeter.
4. An open or a high resistance may exist in the armature circuit. Again, check the circuit with an ohmmeter.

The characteristic curve for a shunt generator with varying speed is shown in Figure 3.45.

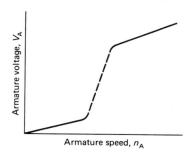

Figure 3.45 Characteristic Curve for Shunt Generator with Varying Armature Speed

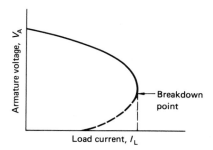

Figure 3.46 Characteristic Curve for Shunt Generator at Constant Armature Speed

This graph shows that the generator cannot build voltage until it reaches a speed that overcomes the resistance of the field and load. This point is similar to the critical resistance point in Figure 3.44.

Figure 3.46 shows the characteristic curve at a constant speed. As the load current increases, the armature voltage drops. At a point called the breakdown point, the current drain of the load and the field exceeds what the generator can supply, and so the generator unbuilds. Over most of the load current range, however, the armature voltage is relatively constant.

The nonlinearity of the magnetization curve prevents easy use of the emf and torque equations with self-excited generator circuits.

Series Generator. When the field is in series with the armature, as in Figure 3.47, the generator is series-connected.

The characteristic curve for the shunt generator at constant armature speed is shown in

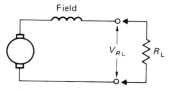

Figure 3.47 Series-Connected Generator

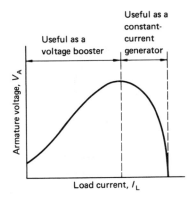

Figure 3.48 Characteristic Curve for Series Generator at Constant Armature Speed

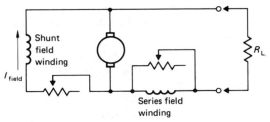

A. Overcompounded, Flat-compounded, and Undercompounded

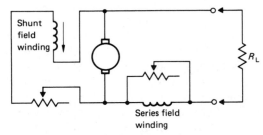

B. Differentially Compounded

Figure 3.49 Compound Generators

Figure 3.48. The curve shows that at zero load current (output open-circuited), no armature current can flow to build up field strength. Therefore, the output voltage (V_{A1}) will be a result of residual magnetism only. As the load resistance decreases, more field current can flow and more output voltage is generated. At the peak of the curve in Figure 3.48, an increase in field current does not increase output voltage because the field has become saturated.

Past the peak of the curve, no additional voltage is produced, but more voltage continues to be dropped across R_A and R_{field}. Thus, the output voltage eventually goes to zero. This dropping part of the curve can be useful in welding generators. In such generators the constantly changing arc length causes voltage to fluctuate, but the current must be constant in order to produce a consistent heat.

Compound Generator. As we have seen previously, terminal voltages associated with series-connected and shunt-connected generators vary in opposite directions with load current. If both a series and a shunt field were included in the same unit, it would be possible to obtain a generator with characteristics somewhere between the characteristics of these two types. The resulting device is a compound gen-

erator. The compound configurations and characteristic curves are shown in Figures 3.49 and 3.50.

If the number of turns in the series field is changed, three distinct types of compound generators are obtained: overcompounded, flat-compounded, and undercompounded. See Figure 3.49A.

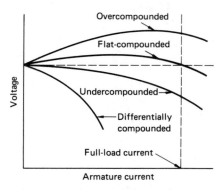

Figure 3.50 Characteristic Curves for Compound Generators

When the number of turns of the series field is more than necessary to give approximately the same voltage at all loads, the generator is *overcompounded.* Thus, the terminal voltage at full load will be higher than the no-load voltage. This feature is desirable when the power must be transmitted some distance. The rise in generated voltage compensates for the drop in the transmission line.

When the relationship between the turns of the series and the shunt fields is such that the terminal voltage is approximately the same over the entire load range, the unit is *flat-compounded.*

When the series field is wound with so few turns that it does not compensate entirely for the voltage drop associated with the shunt field, the generator is *undercompounded.* In this type of generator the voltage at full load is less than the no-load voltage.

An undercompounded generator in which the series and shunt fields are connected so as to oppose rather than aid one another is referred to as a *differentially compounded generator.* See Figure 3.49B. With this type of generator, the terminal voltage decreases rapidly as the load increases. Undercompounded generators are sometimes used in welding machines.

ARMATURE REACTION

As we noted earlier in the chapter, interpoles, sometimes called commutating poles, are small auxiliary poles placed midway between the main poles, as shown in Figure 3.51. They have a winding in series with the armature. Their function is to improve commutation and to reduce sparking at the brushes to a minimum.

To describe interpole function, we first must discuss *armature reaction,* which is the effect that the armature magnetic field has on the field distribution. Figure 3.52 illustrates how armature reaction is produced. Figure 3.52A shows the flux lines produced by the pair of

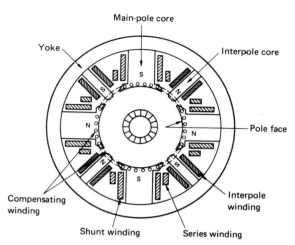

Figure 3.51 Arrangement of Various Field Windings on DC Machines

poles when no current is flowing through the armature coils. A vertical line through the field indicates the zero axis (neutral plane) of the field. Figure 3.52B shows the flux lines produced by current flowing through the armature coils alone. Figure 3.52C shows the resultant field of the two fluxes superimposed. Note that the zero axis of the resultant field is displaced, as indicated by the line designated "new neutral plane." This displacement results in a shift in the position of the old neutral plane. Shifting of the neutral plane results in sparking, burning, and pitting of the commutator. As the load current and the resulting armature reaction increase, this effect becomes more pronounced.

The short-circuiting effect can be counteracted by adding interpoles and compensating windings to the generator. Figure 3.53 shows the schematic diagram for a compound generator with interpoles (R_{pole}) and compensating windings (R_{comp}) added. The neutral plane, or load neutral, in the motor is shifted in the opposite direction with the same direction of armature rotation as shown in Figure 3.52. This example shows that motors and generators cannot be interchanged for best results.

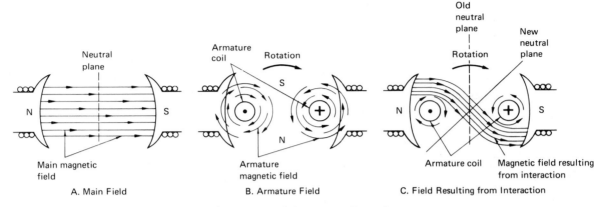

Figure 3.52 Armature Reaction

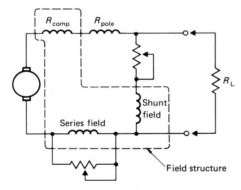

Figure 3.53 Compound Generator with Interpoles and Compensating Windings Added

POWER AND EFFICIENCY

Power, P, is defined as the rate of doing work. The original unit of power was developed from James Watt's determination of horsepower. Power is such an important and useful concept that the basic electrical unit of power, the watt, was named after him, even though he did not work with electricity. Watt defined 1 hp as 550 ft-lb/s. Conversion to metric or SI units uses the following equation:

$$550 \text{ ft-lb/s} = 1 \text{ hp} = 745.7 \text{ Nm/s}$$
$$= 745.7 \text{ W}$$

Torque and speed can also be related to power:

$$1 \text{ hp} = \frac{T \times n_A}{5252.1}$$

where torque is in pound-feet and speed is in revolutions per minute. In SI units the conversion is as follows:

$$1 \text{ W} = 1 \text{ Nm/s}$$

The *efficiency*, η, of any device, such as a motor or generator, is its output power (P_o) divided by its input power (P_i), when they are in the same units:

$$\eta\% = \frac{P_o}{P_i} \times 100$$

(η is the lowercase Greek letter eta.)

For example, a ⅛ hp motor operating at 110 V DC requires 1.02 A when operating at rated conditions. To determine the efficiency, we must find input and output power. Input power (P_i) is as follows:

$$P_i = IV = (1.02 \text{ A})(110 \text{ V}) = 112.2 \text{ W}$$

At rated conditions, the motor should be producing ⅛ horsepower:

$$\frac{1}{8} \text{ hp} = \frac{1}{8}(745.7) = 93.2 \text{ W}$$

Therefore:

$$\eta = \frac{P_o}{P_i} \times 100 = \frac{93.2 \text{ W}}{112.2 \text{ W}} \times 100 = 83.1\%$$

NAMEPLATE SPECIFICATIONS

The nameplate of a dynamo indicates the power, voltage, speed, and so forth, of the machine. These specifications, or nominal characteristics, are the values guaranteed by the manufacturer. The following information punched on the nameplate of a 100 kW generator is an example:

- Power—100 kW
- Speed—1200 r/min
- Voltage—250 V
- Type—compound
- Exciting current—20 A
- Class—B
- Temperature rise—50°C

These specifications show that the machine can deliver, continuously, a power of 100 kW at a voltage of 250 V, without exceeding a temperature rise of 50°C. It can then supply a load current of 100,000/250 = 400 A. It is a compound motor with both series and shunt windings. The shunt field current is 20 A. The class B designation refers to the class of insulation used in the machine. These ratings are designed by the manufacturer and should not be exceeded. Characteristics of selected DC motor types are given in Table 3.6.

CONCLUSION

The theory of operation of DC motors and generators is quite old and well established, compared with the theory of some electronic components such as the transistor and microprocessor. However, the electric motor is still the major dynamo used in industry. Because of its widespread use, all electronics technicians need to have a thorough understanding of the basic principles of DC motor and generator operation. With the development of new electronic motor control systems, new motor applications will continue to be created.

QUESTIONS

In Questions 1–5 fill in the missing words with increase(s) (I), decrease(s) (D), or stay(s) the same (S).

1. The torque on a series motor is held constant. The armature speed _____ and the armature current _____ as the applied voltage is decreased.
2. In a shunt-connected motor, if you hold line voltage constant and increase torque, then armature speed will _____, armature current will _____, and cemf will _____.

3. A separately excited motor has a fixed armature voltage. If the torque demanded of the motor remains constant and the field current is increased, then cemf will _____, armature current will _____, and armature speed will _____.
4. A shunt generator is operating at a constant speed and has unlimited power available from the prime mover. If the load current on the generator is increased, the output voltage of the generator will _____.

Table 3.6 Characteristics of Selected DC Motor Types

Motor Type and Basic Characteristics	Performance Ranges	Application Areas
CONVENTIONAL PERMANENT-MAGNET* A simple alternative to wound-field shunt. PM field plus wound armature. Linear torque-speed relationships in small units. Life limited by brushes in high-speed or severe applications. Readily controlled by transistors or SCRs.	Output from 1 W to a fraction of a horsepower. Time constants (to 63.6% of no-load speed) to less than 10 ms. Efficiencies from 60% to 70% in 10 W sizes. With new magnet materials, can deliver high peak powers (horsepower range).	For full range of inexpensive, good performance drive and control applications. With appropriate environmental precautions, suitable for military and aerospace use. Preferred as a high-performance, general-purpose servo motor.
LIMITED-ROTATION DC TORQUER No commutator wear or friction. Unlimited life. Infinite resolution. Smooth, cog-free rotation. No electromagnetic interference generation. Available as motor elements or fully housed.	Travel range typically to 120°. Torque from a few ounce-inches to greater than 40 lb-ft. Mechanical time constants from 10 to 50 ms.	Very high accuracy positioning or velocity control over a limited angle.
CONTINUOUS-ROTATION DC TORQUER Slow speed, high torque. Relatively low power output. Available as pancake-shaped components. Wide dynamic range. Large number of coils give smooth operation.	From tens of ounce-inches to hundreds of foot-pounds. Moderate mechanical time constants. Control to seconds of arc. Relatively expensive.	For direct coupling to load. For very precise control. Alternative to geared types.
MOVING-COIL, PRINTED-ARMATURE (IRONLESS ROTOR)† Similar to permanent-magnet DC units. Linear torque-speed characteristics. Smooth, noncogging rotation. Handles very high, short-duration peak loads. Fast response (less than 10 ms).	Outputs from less than 1 W to fractional horsepower. High efficiencies. Very low mechanical and electrical time constants.	Computer peripherals where smooth control and fast response are needed. Control applications needing high-response bandwidth, fast starting and stopping.
VARIABLE-RELUCTANCE STEPPER Brushless and rugged. High stepping rate dependent on driver circuitry. No locking torque at zero energization. Poor inherent damping. Low power efficiency. Can exhibit resonance. Operates open loop. Wide dynamic range. Easily controlled. Very reliable and low in cost in popular frame sizes.	Several hundred to thousands of pounds per second. Power output up to a few hundred watts.	Alternative to synchronous motor. Used in control applications where fast response rather than high power is the principal requirement. Interfaces well to digital computers.
SMALL-ANGLE, PERMANENT-MAGNET STEPPER Uses vernier principle to give very small stepping angles. High stepping rates. High cogging torques with zero input power. Efficiency usually very low.	Stepping rate from less than 100 lb/s to many thousands. Dependent on driver electronics. Power up to a few hundred watts. Single step takes a few milliseconds.	Useful in numerical control and actuator application where control is digital. Provides fast slewing and high-resolution tracking.
INVERTER-DRIVEN AC Operates from DC line using a switching inverter. Somewhat less efficient than AC induction motors; otherwise similar in performance. Single-phase (capacitor) or two-phase versions most common.	Outputs from less than 1 W to fractional horsepower. Efficiencies from 20% to 80% in larger models. Speeds up to 30,000 r/min and higher.	Use where DC is only power available. For universal applications where AC supplies vary widely, as in foreign applications. Use where brushes might not be sufficiently reliable, as in very high speeds or in severe environments. Variable-frequency versions used in accelerating high inertial loads.
BRUSHLESS DC PM units using electronic commutation of stator armature. Exhibits conventional DC motor characteristics, but torque modulation with rotation is higher. Lack of brushes gives reliability in difficult applications.	From less than 1 W to 1–2 hp. Relatively high time constants. Speeds to 30,000 r/min. Efficiencies to 80%. Voltages to hundreds of volts DC.	For brushless, long-life applications requiring superior efficiency and control. May be operated at very high altitudes or may be totally submerged.

116

Comparisons with Other Motors	Selection and Application Factors
Higher efficiency, damping, lower electrical time constants than comparable AC control motors, except in very low power applications. Far more efficient than stepper drives. More easily controlled than other motor types.	Select for safe operation with acceptable temperature rises. Check operating conditions for abrupt starts or reversals, which can d magnetize PM fields. Check for altitude or environmental effects on brushes, especially over 10,0(r/min. High stall currents are drawn in efficient or high-power motors. Low-output impedance electronic control required to utilize inherent mot damping.
Much simpler than continuous-rotation torquers with or without commutators.	Suitable for direct-drive, wide-band, high-accuracy mechanical control. Similar wide-angle brushless tachometers available. Requires high-power driving amps.
Supplies most precise control, smooth and accurate tracking for continuous-rotation applications. Requires higher-powered amps compared with geared units.	Stiff direct coupling to load preferred. Pulse-width modulation amps preferred for high control power.
Faster response than iron-rotor motors. Excellent brush life. Lower starting voltage, limited only by brush friction. Much more efficient than stepper motors.	Recommended for low-cogging, low-starting voltage, fast-response applic tions. Rotor heats up quickly. Thermal transients and heat removal can be important factors. Larger, high-performance units can be expensive. Low armature inductance permits commutation of very high current surges.
Power output and efficiency generally very low compared with DC control motors.	Care required in application. Performance dependent on electronic driver circuitry. Heat dissipation a possible problem. At certain pulse rates, resonances can occur, which reduce load-handlir capability. Load inertia reduces performance. Friction can improve damping. Damping, gearing, and mechanical couplings require special attention.
Efficiency of shaft power generation is low. More flexible than comparable means. Simple and inexpensive alternative to synchronous or wide-speed-range drives. Handles higher load inertia than variable-reluctance stepper and has better damping.	Choose where special control characteristics are preferred over efficiency. Check for resonances at all pulse rates. Gearing can require extra safety margins because of impacts inherent in stepp operation. Coupling compliances can help in accelerating load inertia, but addition resonances can be introduced. Driver circuit design is critical. Standard drivers available.
Less efficient than true brushless motors using electronic commutation. More complex, expensive, and noisier than brush-type DC motors. Less suitable for control than other DC types. Very long life with properly designed inverter.	Inverter can be separate or packaged with motor. High line circuit spikes. Electromagnetic interference generation, with bulky filter capacitors required f(suppression. SCR inverters preferred in higher-power uses, but transistor inverters are easi(to switch and more reliable. Power supply capacitors can be required and must withstand supply transient
More efficient, easier to control, generates less electromagnetic interference than inverter-type motors. Commutating transistors can be used for speed control, reversing current, and torque limiting without a separate controller, unlike other types. Delivers highest sustained output in a given package size.	Electronics can be packaged externally or within motor housing. High peak line currents. Bulky line filter required if electromagnetic interference is a problem. Power supply capacitors could be required. With properly designed electronics, life is limited only by bearings. Temperatures can set limits to some commutation sensors.

5. If the output of any operating DC generator were suddenly shorted, the torque required to keep the speed constant would _____.

6. Define these devices: (a) dynamo, (b) generator, and (c) motor.

7. Name the parts of a dynamo's rotor and stator.

8. Define torque. Name its units (British and SI).

9. When a generator supplies load current, the terminal voltage is not the same as the generated emf. Is it higher or lower? Why?

10. How may the direction of rotation of a DC motor be reversed?

11. Explain why the generator magnetization curve is not a straight line.

12. When a motor is in operation, why is the armature current not equal to the line voltage divided by the armature resistance?

13. What is an interpole? What is its purpose? How is its winding connected?

14. What is the effect of armature reaction in a motor?

15. What four factors may prevent the buildup of voltage for a self-excited shunt generator?

16. Why is it dangerous to open the field of a shunt motor running at no load?

17. When is a compound generator said to be (a) flat-compounded and (b) overcompounded?

18. Why is the speed regulation of a series motor poorer than that of a shunt motor?

19. Define efficiency.

20. Why should a series motor never be operated without load?

21. Describe what is meant by dynamic braking.

22. Draw schematic configurations for series, shunt, separately excited, and long-shunt and short-shunt compound dynamos.

23. Draw the constant-torque and constant-line voltage characteristic curves for the series and shunt motors and generators.

PROBLEMS

1. Find the armature speed of a series motor with 10 Ω field resistance, 5 Ω armature resistance, K_{emf} = 0.311 V min/A r, K_T = 1 Nm/A^2, an applied voltage of 200 V, and T = 0.09 Nm.

2. Given a shunt motor with a field resistance of 500 Ω, an armature resistance of 10 Ω, K_{emf} = 0.311 V min/A r, and K_T = 0.5 Nm/A^2. If the torque demanded is 0.2 Nm and the applied voltage is 100 V, what is the motor speed?

3. In the motor of Problem 2, suppose torque and armature current remain constant. What would the field resistance need to be in order to change the armature speed to 2500 r/min?

4. Given a separately excited motor with R_{field} = 500 Ω, I_{field} = 0.2 A, I_A = 0.45 A, R_A = 20 Ω, V_A = 100 V, and K_T = 0.5 Nm/A^2. The torque demand is constant for any n_A, and only the field voltage is reduced by half (V_A remains fixed). Find the new I_A, V_{RA}, cemf, and n_A.

5. In the motor of Problem 4, V_A and V_{field} are returned to what they were originally, and what the motor was turning suddenly locks up (stops turning). Find the armature current and torque produced by the motor. What is the power being dropped across R_A?

6. A separately excited generator has a field current of 0.3 A, with R_A = 5 Ω, load resistance (R_L) of 150 Ω, output voltage (V_{RL}) of 100 V, and an armature speed of 1500 r/min. Find the no-load voltage at 1000 r/min (R_L is infinite).

7. With the generator of Problem 6, what armature speed would be needed to give an output voltage of 150 V with a load resistance of 100 Ω?

8. With the generator of Problem 6, if you wished to keep the voltage across the 100 Ω load resistor at 150 V while changing the speed of the armature to 2000 r/min, the field current would have to be changed. What field current would be needed?

9. Use the generator of Problem 6. What is its output voltage when the load resistance goes to 0 Ω? What is the armature current?

10. Derive the characteristic curves for the separately excited motor of Figure 3.26. Use the appropriate motor values, similar to those used in the shunt motor characteristic curves of Figures 3.20 and 3.22.

4

BRUSHLESS AND STEPPER DC MOTORS

INTRODUCTION

In the previous chapter we discussed the popular wound-field DC motor. This device has stood the test of time as a versatile, reliable machine; it has been used in industry for many years. The wound-field DC motor has some disadvantages, however, which were mentioned in Chapter 3. One major disadvantage is the high maintenance costs incurred by the wearing of brushes and commutator segments. Another is the radio frequency emissions that are given off by the arcing at the brushes when the motor windings are commutated. The wound-field DC motor must also be connected to a DAC to be driven by a computer or by digital circuitry. These disadvantages in some measure spurred the development of motors that do not have these disadvantages. Two such devices are the brushless DC motor and the stepper motor that we will discuss in this chapter.

BRUSHLESS DC MOTORS

You will recall that in the conventional wound-field DC motor the commutator is divided into segments. Generally, the effect of winding inductance when the current starts and stops in the winding segments is reduced when the number of segments increases. Increasing the number of segments will improve performance. Another reason for increasing the number of winding segments is to control the torque ripple. The more windings there are, the lower the amount of torque ripple. Thus we find that a fractional horsepower DC motor may have anywhere from 7 to 32 commutator bars per armature, and an integral horsepower motor of 5–10 hp may have less than 100 bars. Larger motors may have more than 100 bars per armature. Although we can see the effect of commutation by connecting across the armature windings, we cannot see what is going on inside the armature. It is not easy to reach proper internal test points in a rotating structure like the armature. Much of the designer's effort in such motors goes into choosing the proper brush and commutator. A great variety of brush compositions have enabled designers to cope with the widely varying operational and environmental conditions that exist in modern technology. Even so, brush and commutator wear are the primary maintenance requirements for wound-field DC motors.

The problems associated with a commutator may be avoided if the switching is done by semiconductor devices instead of mechanically. Using semiconductor devices to commutate a DC permanent-magnet (PM) motor generates a new set of problems, however. A simple translation of the brush-type motor design to a brushless type is not practical. For ex-ample, if we were to take a 16-bar, two-pole DC motor and substitute semiconductor devices for the brush and commutator assembly, we would end up with 32 power transistors, two of which would be conducting at any one given time. This would result in an inefficient use of semiconductors—an efficiency factor of about 6%. Obviously the problem has to be solved in a new way to get cost-effective performance.

A DC motor in which the brushes and commutator have been replaced by functionally equivalent semiconductor switches is called a *brushless DC motor*. A brushless DC motor system should have the torque-speed characteristics of the conventional DC PM motor.

Brushless DC Motor Construction

The diagram in Figure 4.1 illustrates the speed-torque characteristics of a conventional PM DC motor. The brushless DC motor should have the same basic characteristics. When brushless DC motors are designed, one of the requirements to be met is the reduction of the number of semiconductor switches to as low a number as possible. The second design requirement is that, wherever possible, permanent-magnet materials should be used in the rotor or armature. This is necessary to eliminate the

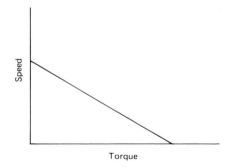

Figure 4.1 Speed-Torque Curve for a PM DC Motor

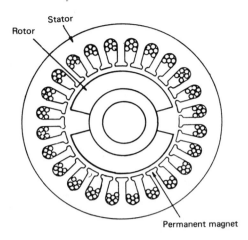

Figure 4.2 Cutaway View of Brushless DC Motor Assembly

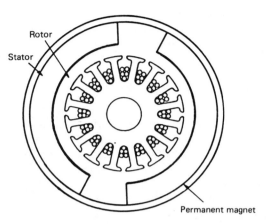

Figure 4.3 Cutaway View of Conventional PM DC Motor Assembly

need for slip rings in the rotor assembly. In a brushless DC motor it is usually most practical to provide a stator structure as shown in Figure 4.2 where the windings are placed in an external, slotted stator. The rotor consists of the shaft and a hub assembly with a PM structure. The picture shows a two-pole magnet in the rotor. Contrast Figure 4.3, which shows the equivalent cross-sectional view of a conventional PM DC motor. Figure 4.3 shows permanent magnets situated in the stator structure; the rotor carries the various winding coils. We can see that there are significant differences in winding and magnet locations. The conventional DC motor has the active conductors in the slots of the rotor structure, and, in contrast, the brushless DC motor has the active conductors in slots in the outside stator. The removal of heat produced in the active windings is easier in the brushless DC motor, since the thermal path to the environment is shorter. Since the PM rotor does not produce heat, the result is that the brushless DC motor is a more stable mechanical device from a thermal point of view.

In spite of the advantages just discussed, there are cases where a brushless DC motor uses the configuration shown in Figure 4.2. In these cases the roles of the two parts are reversed so that the PM outside structure rotates, and the wound lamination part is the stator. This design produces a high rotor moment of inertia, which is useful in applications where a high mechanical time constant is needed to smooth out any existing torque ripple. In such cases the thermal advantage of the design in Figure 4.3 is not used.

To illustrate the similarities and differences between the conventional and brushless DC motor systems, compare the sketches of the two in Figures 4.4 and 4.5. In Figure 4.4 we see the elements of a conventional DC motor and control. The bidirectional controller and driver stage is shown together with a power supply and control logic. The bidirectional controller enables the motor to turn in both directions. The control logic signals the driver circuit to turn on either Q_1 or Q_2, depending on which direction the rotor needs to turn. The equivalent brushless DC motor system is shown in Figure 4.5, where the main differences are seen to be

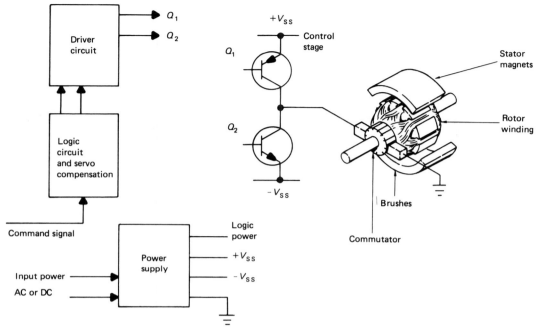

Figure 4.4 Main Parts of DC Motor Servo Control

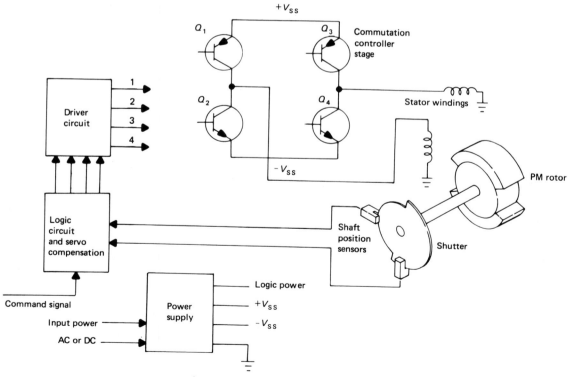

Figure 4.5 Main Parts of Brushless DC Motor Control

stator windings, PM rotor, four transistors instead of the two found in the conventional motor, and a shaft position encoder. The shaft encoder tells the control logic circuitry about the rotor's position. Remember that the logic signals control the commutation of the windings. It is vitally important, therefore, that the control logic knows where the rotor is so that proper commutation can take place. This is not a problem in the conventional mechanical commutator, since the commutation process is fixed by the position of the brushes and the commutator.

Brushless DC Motor Operation

One of the simplest practical brushless DC motor circuits is shown in Figure 4.6. This circuit is called a half-wave control circuit. The synchrogram of Figure 4.6 shows the *torque function*, $T(I_1)$, of each winding. This function shows each winding's contribution of torque at various shaft angles when a constant current I is flowing in the winding. The torque over the conduction angle then becomes the torque constant K of the winding. As we see in the synchrogram, each winding is used one-third of the time, and the logic control of the system is rather simple. The speed and torque output of the motor can be controlled by varying the power supply voltage. In the lower part of the diagram we see the same system with a reversed torque. The torque reversal is not achieved by reversing the power supply voltage as in a conventional DC motor. Instead it is done by shifting all logic functions by 180°. This example illustrates one of the basic differences between brush-type and brushless-type DC motors.

The example just given ignores one very important point in brushless motor commutation performance. The commutator is not able to deal with the inductive kick generated when the current stops in the motor coil windings.

The transistors in the circuit in Figure 4.6 will experience a forward voltage breakdown if current stops in the windings. This breakdown is caused by voltage produced by the stored energy in each winding. Such breakdown conditions can be tolerated in low-power systems in which the stored energy is low. If any significant amounts of current and voltage are handled in the half-wave system, however, a point is reached where breakdown conditions will cause damage to the semiconductor junctions. Other methods may need to be used to assure proper commutation of the inductive energy in each winding.

Instead of applying the remedy for the inductive energy problem to the circuit we have just reviewed, let us look at another type of circuit. The circuit in Figure 4.7 shows a two-phase brushless motor using two power supplies, $+V$ and $-V$. Note that we now have four power transistors and four diodes. Each half of the circuit controls its own winding, and the two halves are essentially independent of each other. The diagram in Figure 4.7 shows phase currents and logic signals for one direction of operation. The current waveforms in the diagram are intended to be typical of real-time response of current at a given shaft velocity. The diagram, therefore, shows not only current response with respect to rotor position, but also current versus time at a given shaft velocity. For instance, if we look at current I_{Q_1}, we see that it has an exponential initial increase to a steady-state value that is maintained until the 90° position has been reached. Then Q_1 is switched to the OFF condition. The stored energy is now dissipated through the power supply and returned through diode D_2. The exponential decrease is shown in I_{D_3}, while the current is rising in Q_2. Thus, torque is produced continually by the motor shaft as one stage is turned off and the next is turned on. The diodes provide an ideal path for the stored conducted energy. The energy is either stored in one of the filter capacitors or transferred to the other

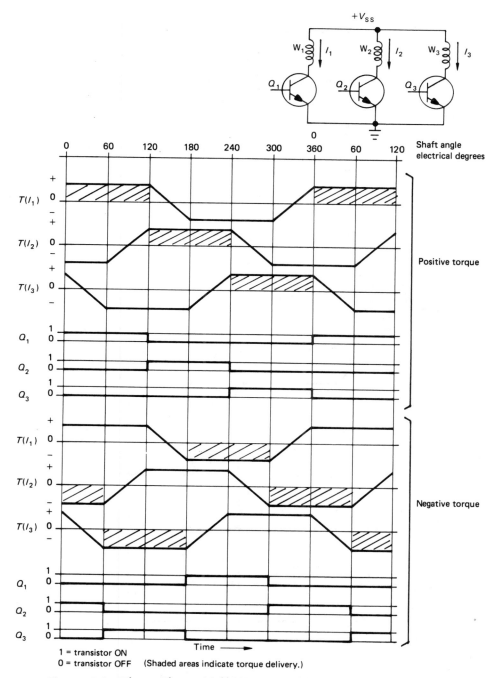

Figure 4.6 Three-Phase, Half-Wave Brushless DC Motor Controller

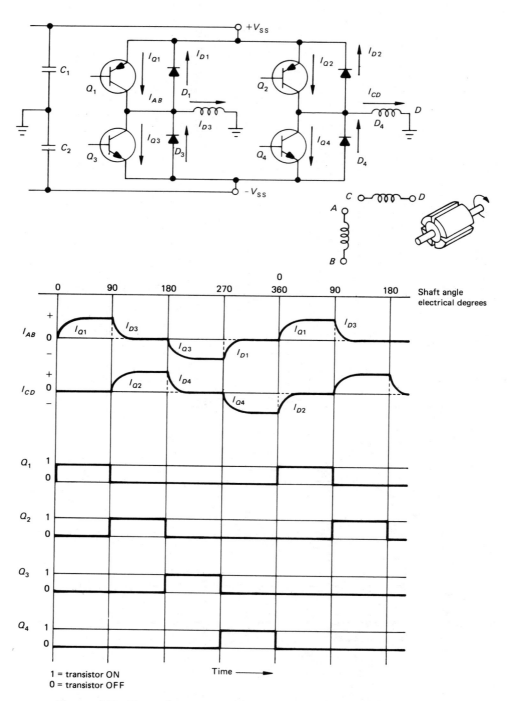

Figure 4.7   Two-Phase, Four-Pole Brushless DC Motor Controller

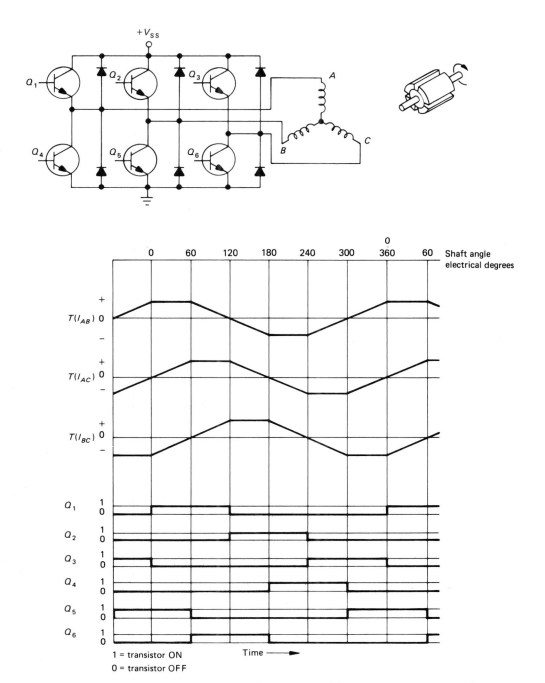

Figure 4.8   Three-Phase, Full-Wave Brushless DC Motor Controller

winding, depending on the logic sequence. If fast recovery diodes are used, commutation is carried out without any significant RFI (radio frequency interference) emission.

The diagram in Figure 4.8 shows an extended version of the two-phase, full-wave brushless motor system. The winding configuration is based on a "star" connection stator arrangement where each winding is oriented 120° from the other. The six transistors are connected to the end points of each stator leg, and thus form a three-phase, full-wave motor control. This system differs from the previous one in that conduction is always continuous in one leg when the other is being commutated. We can see that when Q_1 is energized between 0° and 60°, Q_5 is also conducting. Therefore, current is flowing from point A to point B. In the next sequence (60°–120°) Q_6 is energized, and the current will then flow from point A to point C. In the meantime the current through leg B declines to zero by conduction through D_2. The conduction angle per phase is 120°, as compared to 90° in the two-phase circuit. Since this circuit requires only a single power supply, it tends to be more compact than the two-phase circuit. Furthermore, since 67% of the available windings are used at any one time, compared to 50% in the two-phase circuit, the three-phase circuit is more efficient than the two-phase circuit of Figure 4.7.

We can now return to the half-wave, three-phase circuit illustrated in Figure 4.6, which had no way to handle the stored inductive energy. If we were to provide a diode path for each winding, such as shown in Figure 4.9A (typical of relay coil circuits), we would find that the inductive energy is well handled by the diode arrangement. The fly-back (or free-wheeling) diode is reverse-biased when the transistor is conducting. When the transistor is cut off, the diode conducts, being forward-biased by the reversed voltage in the coil when the field collapses. Because the diode is located across the coil and not across the transistors as in Figures

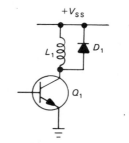

A. A Conventional Fly-Back Diode

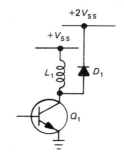

B. A Biased Fly-Back Diode

Figure 4.9   Fly-Back Diodes

4.7 and 4.8, we find that the cemf will also flow in the diode circuit when the polarity of the cemf is reversed (180°–360° in the coil). To prevent current flow due to the cemf during this "off" period of each cycle, diode conduction must occur to a voltage source that has a voltage at least two times the power supply voltage. One arrangement that accomplishes this is shown in Figure 4.9B. As the conduction of the transistor stops, the winding inductance will generate a voltage equal to $2V_{ss}$, which will allow the inductive transient to be dissipated in the power supply circuit. Because the diode is biased, when the cemf is reversed and could normally cause current flow, it cannot overcome the bias, and therefore no current flows. This type of circuit thus will allow for proper handling of the stored inductive energy, but at the expense of extra components and power dissipation in the bias supply.

Commutation-Sensing Systems

As mentioned previously the brushless DC motor system needs a method to let the control circuit know the position of the rotor. Commutation cannot be accomplished without this information. Several methods are available for sensing the rotor's position. The most commonly used methods are Hall effect sensors, electro-optical sensors, and RF sensors. The theory behind these sensors is explained in Chapter 8.

The *Hall effect sensing system* uses a sensor that detects the size and polarity of a magnetic field. The signals are amplified and processed to form logic-compatible signal levels. The sensors usually are mounted in the stator structure, where they sense the polarity and magnitude of the PM field in the air gap. The outputs of these sensors control the logic functions of the controller to provide current to the proper coil in the stator at the right time. The system can provide some compensation for the effects of armature reaction, which occurs in some motor designs. One drawback with such a location of the angular position sensor is that it is subject to stator temperature conditions. At times temperature variations can be rather severe in high-performance applications. It is not unusual, for instance, to allow a winding temperature to reach 160°–180°C for peak load conditions. Such a temperature change may affect the Hall effect switching performance and therefore affect the system performance.

The Hall effect device can, of course, be located away from the immediate stator structure. It may use a separate magnet for sensing angular position. In such a case the sensor is not necessarily subject to the severe operating conditions just mentioned. On the other hand, the sensor does not compensate for armature reaction problems in this position.

The second method for angle sensing is the *electro-optical switch.* This system most commonly uses a combination of an LED (light-emitting diode) and a phototransistor (discussed in Chapter 8). A shutter mechanism controls light transmission between the transmitter and the sensor. The sensor voltages can be processed to supply logic signals to the controller. The electro-optical system lends itself well to generation of precise angular encoding signals.

The third angle-sensing method, the *RF sensor,* is based on inductive coupling between RF coils. Several varieties of such devices have been used with varying degrees of circuit complexity. The angle-sensing accuracy of such devices depends on several design factors, which may limit their use in high-performance systems. In addition, their switching time depends on oscillator frequency and may cause switching time delays that can be undesirable at higher shaft speeds.

The three systems discussed have their own advantages and drawbacks, and the requirements of each application usually dictate which system is best.

Power Control Methods

So far we have discussed commutation control without reference to power control of the brushless DC motor. One method of power control is to *vary the supply voltage to the commutation system.* An example of this method is shown in Figure 4.10. The six switching transistors will control commutation at the proper time. The series-connected power transistor will handle velocity and current control of the brushless motor. This can be accomplished either by linear (Class A) control or by pulse-width or pulse-frequency modulation. If directional control is needed, the commutation sequence must be adjustable from 0° to 180°. In effect, then, we have a series regulator controlling the power supply voltage for a switching stage commutation controller.

Another way to control voltage and current

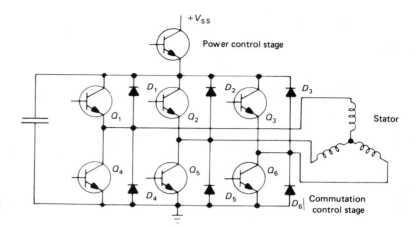

Figure 4.10 Series-Regulator Power Control for Three-Phase Brushless DC Motor

in the brushless motor is to *let the commutation transistors control the motor current* either by linear control means or by pulse-width or pulse-frequency modulation. (Again, modulation methods and circuits are covered in later chapters.) Such a control method results in better use of available semiconductor devices. Proper attention, however, must be paid to power dissipation in the controller stage.

In the case of *linear transistor control,* the control stage must be operated in a constant-current mode rather than constant-voltage mode. Otherwise the transistor stage held at a zero output voltage will tend to conduct cemf-induced currents during the inactive part of the cycle. This, in turn, causes a viscous damping effect that is detrimental to motor operation.

The *pulse-width or pulse-frequency control scheme* is well suited for control of voltage and current to a brushless motor. Since logic circuitry already in place is capable of switching the appropriate transistors on and off, the use of such control is easily accomplished. The switching rate must be within the switching capability of the power transistors so that undue dissipation losses do not occur during the transistor turn-off and turn-on times.

Any high-performance brushless DC motor will require some form of current limit control, either to protect the controller stage or to protect the magnetic circuit. With either of the

controller schemes discussed, it is easy to apply such current limit control. However, the switching-type controller can sustain current limit conditions without significant circuit power dissipation—as compared to a linear-type control system where the excess power is dissipated in the power transistors. For the reasons mentioned, pulse-width or pulse-frequency modulation is a superior control scheme for brushless DC motors in which any significant amount of DC power is being controlled.

Advantages and Disadvantages of Brushless DC Motors

Brushless DC motors can be used for a variety of applications. The advantages of state-of-the-art magnet materials have made possible brushless DC motor designs that have very high torque-to-inertia ratios. Since commutation is performed in circuit elements external to the rotating parts, there are no items that will suffer from mechanical wear, except the motor bearings. The brushless DC motor, therefore, will have a life expectancy limited only by mechanical bearing wear and the reliability of the electronic controller.

The brushless DC motor can be controlled with very efficient amplifier configurations. In cases of severe environmental conditions, the controller can be located remotely from the motor. The control system can interface easily with digital and analog inputs. Therefore, they are well suited for motion control requiring rotor movement in discrete steps. This kind of motor control is called *incremental motion control*. The motor and control also have a lower level of RF emission than the conventional DC motors and controls.

The brushless DC motor controller has, in general, a more complex configuration than the controller for an equivalent conventional DC motor, but may be similar in size and complexity to a closed-loop step motor controller. The commutation sensor system in some cases can provide some incremental motion applications, but usually an encoder system is added to the commutation system to suit the application.

Table 4.1 is a brief comparison of the conventional DC motor and the brushless DC motor. Because of the many available variations of the two kinds of motors, the comparison is a very general one. However, it may give

more insight into the basic characteristics of the two types of motors.

STEPPER MOTORS

The *stepper motor* is the only motor that is truly digital in its output. Most motor rotors, including those in AC motors, turn at a rate proportional to the voltage (or frequency) applied to them. The stepper motor, as its name implies, turns in discrete movements called *steps*. After the rotor makes a step, it stops turning until it receives the next command. Stepper motor operation can be likened to a series of electromagnets or solenoids arranged in a circle, as shown in Figure 4.11. When energized in sequence, the electromagnets react with the rotor, causing it to turn either clockwise or counterclockwise, depending on the input commands. The stepping angle, ψ, is determined by the design of the motor, but should not be greater than 180°. There are two large classes of steppers: mechanical, which rely on ratchet and pawl arrange-

Table 4.1   Brief Comparison of Conventional DC Motors and Brushless DC Motors

Type of Motor	Good Points	Problem Areas
Conventional DC PM Motors	• Provides control over a wide range of speeds • Capable of rapid acceleration and deceleration • Convenient control of shaft speed and position by servo amplifiers	Commutation (brushes) causes wear and electrical noise. This problem can be kept under control by selecting the best materials for each application.
Brushless DC Motors	• Provides control over a wide range of speeds • Capable of rapid acceleration and deceleration • Convenient control of shaft speed and position • No mechanical wear problem due to commutation • Better heat dissipation arrangement	More semiconductor devices are required than for the brush-type DC motor for equal power rating and control range.

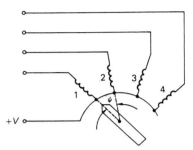

Figure 4.11   Series of Electromagnets Driving a Stepper Motor

ments or other linkages, and magnetic (true motors). Since magnetic stepper motors are more common, we will be dealing with them only. There are two basic types of magnetic stepper motors: permanent-magnet (PM) and variable reluctance (VR).

PM Stepper Motors

The *PM stepper motor* operates on the reaction between an electromagnetic field and PM rotor. In its simplest form, the PM unit consists of a two-pole PM rotor revolving within a four-pole slotted stator (Figure 4.12). Although the rotor is shown in the diagram as round and smooth, the actual PM rotor has teeth. The stator likewise is a toothed construction. Current is applied to each of the stator windings successively, creating a series of magnetic fields. The PM rotor, interacting with the magnetic reaction of the field, is pulled into alignment as each winding is energized. Each successive realign-

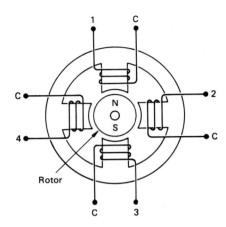

Figure 4.12   Construction of a Four-Phase (90°) PM Stepper Motor

ment produces a stepping motion of the rotor, in this case steps of 90°, since the four stator poles are 90° apart. By changing the excitation sequence to the windings, the motor can be made to operate either clockwise or counterclockwise.

Because of the rotor reaction and the relatively large step angle of the PM stepper, the rotor has little tendency to overshoot. *Overshoot* is the condition when the rotor goes past the pole that is attracting it. The pole, still attracting the rotor, pulls it back. The rotor then overshoots again (but not as much), this time in the opposite direction. This oscillating behavior continues until all the rotor energy is absorbed. The rotor then stops. The oscillating waveform shown in Figure 4.13 illustrates the

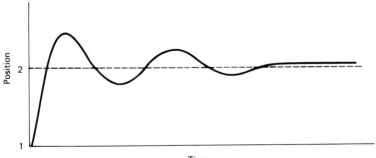

Figure 4.13   Stepper Motor Oscillations

oscillations *damping,* or dying out. Sometimes, when the number of oscillations cannot be tolerated, external means of damping must be used. In the case of the PM stepper, however, damping is rarely necessary.

PM Stepper Motor Parameters. It is very important that the proper amount of excitation voltages and currents be applied to the stator windings. The rotor may become demagnetized by excessive excitation beyond normal ratings. The best sources of information about proper excitation voltage and current are manufacturers' data sheets. For example, we would learn from a data sheet for an Airpax L82101-P2 stepper that it is designed to be operated with a stator voltage of 12 V. Some manufacturers give the rotor resistance and the maximum power that the rotor can dissipate instead of the current and voltage ratings. The same Airpax stepper has a stator resistance of 118 Ω.

Rotor inertia generally is higher in PM motors than in VR steppers. Recall from physics that *inertia* is the resistance a body possesses to starting motion or changing the direction of motion. Inertia is directly proportional to the mass of an object; the more mass a body has, the more inertia it possesses. *Rotor inertia* (also called *moment of inertia*) is generally given in g-m^2. The following units also are used for rotor inertia: kg-cm^2 and in the English system oz-in^2 and lb-in^2. PM motors permit higher rotor speeds than other types of motors and, conversely, lower stepping rates.

Another parameter usually seen on data sheets is the step angle. The *step angle* is defined as the specific amount of shaft rotation in degrees caused by a change in winding polarity. Step angles generally are larger in the PM stepper. Step angles range from 0.72° to 90°, with 7.5°–18° most common. PM steppers normally have 12 or 24 poles, permitting stepping angles of 3.75°, 7.5°, or 15°. Normally, only 45° and 90° steps are possible with a two-pole rotor. Increasing the number of poles results in smaller stepping angles and higher maximum stepping speeds.

Along with step angle, maximum number of steps per second (steps/s) is given in manufacturers' data sheets. This parameter replaces the maximum r/min rating of a DC or AC motor. The *maximum stepping speed* of a stepper motor is inversely proportional to the mass (and inertia) of the rotor; as mass and inertia increase, the maximum stepping speed decreases.

PM stepper motors exhibit an interesting and useful parameter called holding torque. Holding torque is produced when the stator windings are energized. When the stator is energized and the motor has stepped to its new position, the rotor will be held in place by the attraction between the two magnetic fields. *Holding torque* is defined as the torque needed to move the rotor a full step with the stator energized but at a standstill. The holding torque generally is larger than the running torque. The holding torque then acts as a powerful braking mechanism in holding a load.

Even when the stator field is not energized, it takes some torque to move the rotor, since the rotor produces a cemf when cutting the stator windings. This torque is called *detent torque* and is usually about one-tenth of the holding torque. This useful feature of PM steppers holds the load in the proper position, even when the motor is off. The position will not, however, be held as accurately as when the motor is energized.

The accuracy of a stepper motor is expressed either in a number of degrees or as a percentage per step taken. The *stepping accuracy* is the total error made by the stepper in a single step movement. For example, the accuracy of a motor may be ±6.5%. A 7.5° stepper motor will position a load to within ±0.5°. This error is said to be noncumulative, which means that the error does not build up with each step taken. The accuracy will be within ±6.5% whether one step or 1000 steps are taken.

VR Stepper Motors

VR (variable reluctance) steppers use a ferromagnetic multitoothed rotor with an electromagnetic stator similar to the PM stepper. A typical three-phase design (Figure 4.14) has 12 stator poles spaced 30° apart; the rotor has eight poles spaced at 45° intervals. The stator poles are energized sequentially by three-phase winding. When current is applied to phase one, the rotor teeth closest to the four energized (magnetized) stator poles are pulled into alignment. The four remaining rotor teeth align midway between the nonenergized stator poles. This is the position of least magnetic reluctance between rotor and stator field.

Energizing phase two produces an identical response. The second set of four stator poles magnetically attracts the four nearest rotor teeth, causing the rotor to advance along the path of minimum reluctance into a position of alignment. This action is repeated as the stator's electromagnetic field is sequentially shifted around the rotor. Energizing the poles in a definite sequence produces either clockwise or counterclockwise stepping motion.

The exact increment of motion (step angle) is the difference in angular pitch between stator and rotor teeth, in this case 30° and 45°, respectively, for a net difference of 15°. The VR stepper's step angles are small, making finer reso-

lution possible than with the PM type. Maximum stepping rates generally are higher than in the PM stepper. Also, because of the nonretentive rotor, VR steppers do not have detent torque when unenergized.

A typical VR stepper uses a stator with 12 fields. Poles are set about 30° apart and grouped for three-phase operation where each phase has four coils set 90° apart. The rotor has eight teeth spaced 45° apart. VR steppers have a maximum stepping speed of about 18,000 steps/s, much higher than the PM stepper can produce. At high speeds, however, the VR stepper tends to overshoot and must be damped.

Stepper Operation Modes

Based on our discussion of the theory of stepper motor operation, we can divide the motor operation into four modes: rest, stall, bidirectional, and slew. These modes of operation are illustrated by the speed-torque curves in Figures 4.15A and 4.15B for both VR and PM stepper motors.

The first mode of operation is the *rest mode*. Unenergized PM steppers, due to the interaction between the PM rotor and the stator, exhibit a resistance to movement called *detent*, or *residual*, *torque*. Sometimes called position memory, this characteristic is valuable when final position must be known in the event of electrical failure in a system. VR steppers do not exhibit this characteristic.

The second mode of operation is the *stall mode*. When a stator winding is energized, both VR and PM steppers resist movement. In data sheets this is sometimes called *static stall torque*.

The third mode of stepper operation is called the *bidirectional mode*. In the bidirectional mode the shaft continually advances and then stops momentarily (start-stop). The direction of rotation can be reversed instantaneously. Performance curves given in data

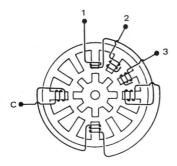

Figure 4.14 Three-Phase VR Stepper Motor

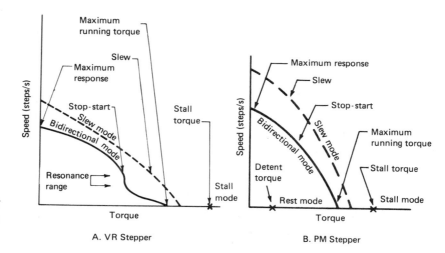

Figure 4.15 Stepper Motor Speed-Torque Curves

sheets indicate the maximum speed at which a given load may be run bidirectionally without losing a step. The maximum load that the motor can drive occurs at a rate of about 5 steps/s and is listed on some data sheets as *maximum running torque*. The maximum speed, in steps per second, that a stepper will run in bidirectional mode occurs with zero load, and is shown as *maximum stepping rate* or *maximum response* on the data sheets.

The fourth and last mode of stepper operation is called *slewing*. The stepper motor can be accelerated beyond its start-stop bidirectional range in a unidirectional slew mode. The rotor pulls into synchronism with the rotating stator field, much like a standard AC synchronous motor, to be discussed in Chapter 5. In the slew mode, the motor is beyond its bidirectional start-stop speed range; it cannot be instantaneously reversed and still maintain pulse and step integrity. Nor can the motor be started in this range. To attain slewing speed, whether from rest or from bidirectional mode, the motor must be carefully accelerated (or ramped). In a similar fashion, to stop or to reverse in the slew mode, the motor must first be decelerated carefully to some speed within its bidirectional capability. When ramping is sup-

plied for acceleration and deceleration, there is no loss of pulse-to-step integrity in slew mode.

Driving Stepper Motors

Early stepper systems used mechanical commutation switches to energize a stepper's stator windings in sequence. Applications typically required transmission of bidirectional shaft rotation for remote indicators such as repeater compasses.

The modern stepper motor usually is driven by a high-speed, solid-state circuit (the controller) that sequences commands to the motor for either clockwise or counterclockwise rotation, as shown in Figure 4.16. The subject of stepper motor driving circuitry will be discussed in more detail in later chapters.

Stepper Excitation Modes

Depending on the stator winding and performance desired, a stepper motor can be excited in several different modes: two-phase and two-phase modified; three-phase and three-phase modified; and four-phase and four-phase mod-

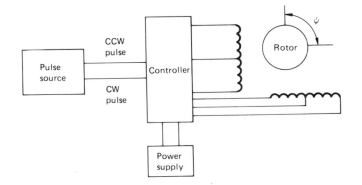

Figure 4.16 Block Diagram of a Stepper Motor Controller

ified. *Phase* refers to a stator winding, and *modified* means that two windings are driven simultaneously. All of these modes of excitation are compatible with most stepper controllers except the two-phase modes.

PM steppers normally are made with two stator windings that are center-tapped, as shown in Figure 4.17. When the center taps are connected to the controller, this stepper is considered to be a four-phase motor. Two-phase motors have the center taps either eliminated or open-circuited, and only the end taps are connected.

VR steppers, due to geometry and construction, lend themselves to a convenient Y-shaped (wye) winding arrangement, illustrated in Figure 4.18. They are excited in the three-phase or three-phase modified mode.

Two-Phase Mode. One entire phase of the motor, end tap to end tap, is energized at any given moment. Compared to standard four-phase excitation (the usual way of rating four-phase motors on data sheets), resistance is doubled; thus input current and power are halved. Heat dissipation is increased, since more copper is used in the motor winding. Because of the reduced input and greater heat dissipation, the output of the motor can be improved by as much as 10% over the standard four-phase mode of excitation. See Table 4.2 and Figure 4.19 for input sequence and rotor position.

Two-Phase Modified Mode. Both windings (that is, considering one winding to be end tap to end tap and ignoring the center taps shown in Figure 4.17) are energized simultaneously. Energy input in this mode is the same as in standard four-phase mode, but output performance is increased by about 40%. The controller is more complex and costly than a four-

Four-phase

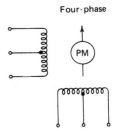

Figure 4.17 Four-Phase Stepper Motor with Two Center-Tapped Stator Windings

Three-phase

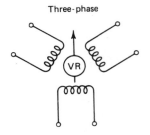

Figure 4.18 VR Three-Phase Stepper Motor with Y-Shaped Winding Arrangement

Table 4.2   Two-Phase Motor Input Sequence and Rotor Position

Excitation Mode	Energized Winding (Figure 4.19)	Rotor Position (Figure 4.19)	Motion Sequence
Two-phase commutation of $B+$ and $B-$	3–1	f	Index
	6–4	h	CCW
	1–3	b	CCW
	4–6	d	CCW
Two-phase modified commutation of $B+$ and $B-$	3–1 and 6–4	g	Index
	1–3 and 6–4	a	CCW
	1–3 and 4–6	c	CCW
	3–1 and 4–6	e	CCW

Table 4.3   Three-Phase Motor Energizing Sequence and Rotor Position

Excitation Mode	Energized Winding (Figure 4.20)	Rotor Position (Figure 4.20)	Motion Sequence
Three-phase commutation of $B+$ only	2–1	a	CCW
	3–4	c	CCW
	5–6	e	CCW
Three-phase modified commutation of $B+$ only	2–1 and 3–4	b	CCW
	3–4 and 5–6	d	CCW
	5–6 and 2–1	f	CCW

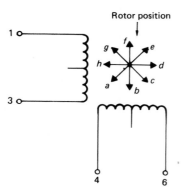

Figure 4.19   Rotor Position and Winding Orientation for Two-Phase Excitation Mode

Three-Phase Mode. Most VR steppers use three-phase windings. Windings are excited one phase at a time to obtain the ratings shown in data sheets. The energizing sequence and consequent motion are shown in Table 4.3. Figure 4.20 illustrates the rotor position and winding orientation.

Three-Phase Modified Mode. In this mode two adjoining phases of a three-phase motor are excited simultaneously. The rotor steps to a minimum reluctance position corresponding to the resultant of the two magnetic fields. Since two windings are excited, twice as

phase system for this mode. For critical applications maximum performance is obtained from minimal energy inputs.

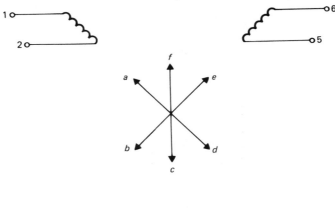

Figure 4.20 Rotor Position and Winding Orientation for Three-Phase Excitation Mode

much power as the standard mode (one phase at a time) is required, as shown in Table 4.3. The output is not significantly increased, but damping characteristics are noticeably improved.

Four-Phase Mode. In this mode of excitation each halfwinding (end tap to center tap as in Figure 4.17) is regarded as a separate phase, and phases are energized one at a time. The sequence of excitation is shown in Table 4.4; rotor position and winding orientation are

shown in Figure 4.21. Performance curves for four-phase motors shown on most data sheets use this mode of excitation. Although this mode is less efficient than others, the controller is uncomplicated, being a simple four-stage ring counter.

Four-Phase Modified Mode. Phases (halves) of different windings are simultaneously energized in this mode, as shown in Table 4.4. Since two phases are simultaneously excited, twice the input energy of single-phase ex-

Table 4.4 Four-Phase Motor Energizing Sequence and Rotor Position

Excitation Mode	Energized Winding (Figure 4.21)	Rotor Position (Figure 4.21)	Motion Sequence
Four-phase commutation of $B+$ only	2–1	f	Index
	5–4	h	CCW
	2–3	b	CCW
	5–6	d	CCW
Four-phase modified commutation of $B+$ only	2–1 and 5–4	g	Index
	2–3 and 5–4	a	CCW
	2–3 and 5–6	c	CCW
	2–1 and 5–6	e	CCW

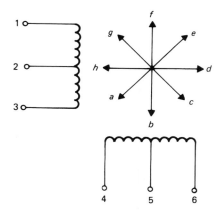

Figure 4.21 Rotor Position and Winding Orientation for Four-Phase Excitation Mode

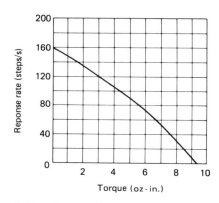

Figure 4.22 Speed-Torque Curve for Four-Phase Stepper Motor

citation is required. Torque output is increased by about 40%, and the maximum response rate is also increased, compared to the single-phase excitation.

Load Torque and Inertia

The maximum response rate of a stepper motor is measured in a no-load condition. As the torque load on the motor is increased, the maximum stepping or response rate will decline proportionally. The increased torque ratings are plotted on a continuous curve. The resulting curve describes the bidirectional operating areas of a stepper motor over its entire torque capacity range. The curve in Figure 4.22 is such a curve. Read from left to right, torque is shown to be greatest at 0 steps (pulses)/s, and response is highest at the no-load point. Operation at rates beyond those indicated on the curve in Figure 4.22 is in the unidirectional slow range discussed earlier.

Methods of Damping

A factor called resonance can be a problem with VR steppers and, occasionally, with PM motors, especially when load inertia is high. *Resonance* is the inability of the rotor to follow the step input command. Every object has a natural resonant frequency. When a motor's natural resonant frequency is reached, the motor will lose steps and oscillate about a point.

A brief description of the dynamic movement of a stepper motor will be helpful in further defining resonance. As shown in Figure 4.23, when the stepper goes from position 1 to position 2, the kinetic energy of the rotor must be dissipated, producing an oscillating motion. In the PM types, these oscillations are damped by the interaction of the rotor with the energized magnetic field, eddy currents, and hysteresis loss. However, in VR motors, which have no permanent magnet, eddy currents and hysteresis loss alone provide little damping effect. Under resonant conditions the rotor will oscil-

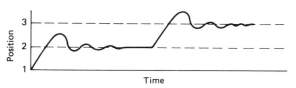

Figure 4.23 Stepper Motor Oscillations after Stepping

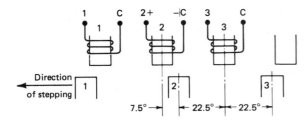

Figure 4.24 Resonance in a 15° Stepper Motor

late at random and lose pulse step integrity. Figures 4.24 and 4.25 graphically illustrate this situation in a 15° VR stepper. Rotor pole 2 is in a position 7.5° away from its final position. It is in line with energized stator winding 2. In this condition rotor poles 2 and 3 are the same distance away from stator winding 3, 22.5° apart on either side. If, at this precise moment, current is switched from stator winding 2 to winding 3, it is possible for the rotor to step in either direction regardless of programmed command.

In a 15° VR stepper, resonance typically may occur at several points, as shown in Figure 4.25. To eliminate resonance in VR steppers, the amplitude of oscillation must be kept within 7.5° deflection from each 15° step position. To do this, several methods of damping may be employed.

Slip-Clutch Damping. A *slip-clutch damper* is a mechanical device that uses a heavy inertia wheel sliding between two collars.

As the rotor moves, the inertia wheel resists movement and adds friction load to the system. This friction has the effect of reducing rotor speed and consequently decreasing overshoot and undershoot. The amount of friction is controlled by spring pressure, and wear is reduced by using Teflon discs to separate the steel members.

Most VR stepper motors are supplied with a rear shaft extension to accommodate this type of damper assembly. Figure 4.26 shows a typical slip-clutch damper installed on the rear shaft of a VR stepper motor. The disadvantage in using system friction to damp resonance is that resonance will vary as the system wears, slowing down response.

Resistive Damping. In *resistive damping*, external resistors are placed across the stator windings, as illustrated in Figure 4.27. The resistors allow rated current to pass through one phase while they limit current through the remaining two phases. As a result a slight reverse torque is applied to the rotor by the two windings not carrying rated current. This reverse torque prevents the rotor from accelerating quickly and limits overshoot. Resistive damping increases power consumption an average of 20% for most inertia loads.

Capacitive Damping. In place of resistors, capacitors can be used to apply reverse torque

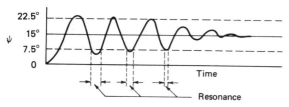

Figure 4.25 Graphic Representation of Stepper Motor Resonance

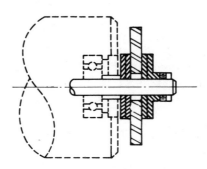

Figure 4.26 Slip-Clutch Damper

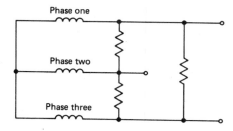

Figure 4.27 Resistive Damping

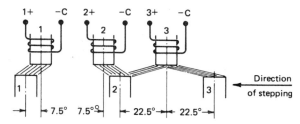

Figure 4.29 Two-Phase Damping

to the rotor in *capacitive damping,* as shown in Figure 4.28. At the moment phase one is de-energized and field two is energized, the capacitor on phase one slowly discharges. The discharging current applies retrotorque on the rotor. This action is repeated as the remaining phases are shifted. This damping method offers the advantage of lower power consumption than the resistive damping technique.

Two-Phase Damping. In the *two-phase damping* method, two stator windings are excited simultaneously. The excitation causes the rotor to step to a minimum reluctance position, midway between the stator poles, as shown in Figure 4.29. The step angle is not changed from single-phase excitation, but the final rotor position will be different. While advancing to this final step position, the two adjacent rotor teeth exert equal and opposing torque. In combination with the stator's magnetic field, this system

produces twice the damping effect of single-phase excitation. The disadvantage of two-phase damping is that power consumption is approximately double that of single-phase excitation. The speed and torque delivery, however, remain essentially unchanged.

Retrotorque Damping. The most satisfactory method of damping the stepper motor is through the use of reverse torque controller electronics. The *retrotorque controller* drives the stepper motor in single-phase mode and effects damping by supplying a pulse of power to the stator winding previously (last) energized. Both overshoot and undershoot tendencies are eliminated. Damping in this manner does not increase power consumption or noticeably affect performance. However, the greater complexity and increased expense of the retrotorque controller may be a factor for consideration.

Advantages and Disadvantages of Steppers

Compared to the usual closed-loop analog servo motors, stepper motors offer significant advantages:

1. Feedback is not ordinarily required when the stepper motor is applied properly. However, the stepper motor is perfectly

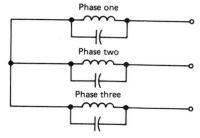

Figure 4.28 Capacitive Damping

compatible with feedback, whether analog or digital, whether for velocity or position, or both.

2. Error is noncumulative as long as pulse-to-step integrity is maintained. A stream of pulses can be counted into a stepper, and its final shaft position is known within a very small percent of one step. DC servo accuracy, in contrast, is subject to the sensitivity and phase shift of the loop.

3. Stepper motors do not have a null position. Bidirectional rotation is continuous.

4. Maximum torque occurs at low pulse rates. The stepper can readily accelerate its load. When the desired position is reached and command pulses cease, the shaft stops. There is no need for clutches and brakes. Once stopped, there is little tendency to drift. Indeed, PM steppers are magnetically detented in the last position. In sum, a load can be started in either direction, moved to a position, and it will remain there until commanded again.

5. A wide range of step angles is available off-the-shelf in the stepper motor line—1.8°, 7.5°, 15°, 45°, and 90° angles without special drivers and logic manipulation.

6. Inherent low velocity is available without gear reduction. For example, a Tormax model 200 stepper motor driven at 500 pulses/s turns at 150 r/min. Many AC and wound-rotor DC motors have difficulty turning at such a slow speed.

7. Steppers are true digital devices. They do not need a D/A conversion at the input as do conventional servos. Steppers may then be used with computers without any additional D/A conversions.

8. They offer close speed control and reversibility over a wide range.

9. Starting current is low.

10. The rotor inertia usually is low.

11. Multiple steppers driven from the same source maintain perfect synchronization, a problem in other types of DC motors.

These advantages are offset by certain disadvantages:

1. Efficiency is low. Much of the input energy must be dissipated as heat.

2. Loads must be analyzed carefully for optimum stepper performance. Inputs (pulse sources and controllers) must also be matched to the motor and load.

3. When load inertia is exceptionally high, damping may be required.

Table 4.5 summarizes typical parameter values for stepper motors. The category "Special PM Type" has been included. Special hybrid steppers have been produced that have im-

Table 4.5 Parameters for Stepper Motors

Type	PM	VR	Special PM Type
Sizes	5–32	8–25	23–34
Mode	2, 4	3, 4	4
Step angle (degrees)	45, 90	7.5, 15	1.8
Steps/revolution	8, 4	48, 24	200
Maximum steps/s (pps)	90–500	350–1025	250–600
Detent torque	Yes	No	Minimal
Maximum running torque (oz-in.)	0.17–24	0.15–12	25–100
Stall torque (oz-in.)	0.3–50	0.25–35	45–200
Resonance	Minimal	Some	Minimal

proved performance in some areas of operation. Table 4.5 reflects only typical values and not maximum values.

CONCLUSION

In this chapter we have seen two DC motors that present alternatives to the conventional wound-field DC motor: the stepper motor and the brushless DC motor. The stepper motor is a truly digital device. The rotor steps in discrete increments each time voltage is applied to the stator. The stepper is, therefore, the motor of choice in digital applications, such as may be required in microcomputer. The brushless DC motor is not a digital device. It is similar in function to the PM DC motor. The brushless DC motor is commutated by semiconductors, unlike the mechanical commutator of the wound-field and PM DC motors. Both the stepper motor and the brushless DC motor have specific applications for which their advantages give them a valuable place in the DC motor family.

QUESTIONS

1. List the parts of a brushless DC motor.
2. How does the rotor on a brushless DC motor get its electromagnetic field?
3. List several of the differences between the brushless DC motor and the PM DC motor.
4. Describe the commutation process in a brushless DC motor.
5. How are the semiconductors protected from damage in a brushless DC motor circuit?
6. List the sensors used in a brushless DC motor controller to detect the rotor position. What are their advantages and disadvantages?
7. Compare the advantages and disadvantages of brushless DC motors with those of PM DC motors.
8. Define the following terms as they relate to the stepper motor: (a) stepping angle, (b) maximum stepping rate, (c) rotor inertia,

(d) holding torque, and (e) residual or detent torque.
9. Describe the differences in construction between VR and PM stepper motors.
10. Describe the operation of the PM and VR stepper motors.
11. List and describe the four modes of stepper motor operation.
12. Draw the schematic diagram for a two-phase stepper motor, and describe how it is operated.
13. Describe the relationship between torque, speed, and rotor inertia in a stepper motor.
14. Describe the condition of resonance and how it is reduced in stepper motor controllers.
15. Explain the differences between resistive and capacitive damping in a stepper motor controller, and draw diagrams to illustrate both damping methods.

5

AC MOTORS

On completion of this chapter, you should be able to:

- Classify AC motors by horsepower and internal construction;
- Explain the concept of the rotating field and calculate its speed;
- Explain how torque is produced in an induction motor;
- Calculate the slip of an induction motor;
- List and describe the different methods of starting single-phase motors;
- Justify the need for special starting methods for synchronous motors.

INTRODUCTION

Alternating current (AC) has one property that allows it to be transported long distances by wire more efficiently than direct current. Alternating current circuits can be stepped up (transformed) in voltage and at the same time stepped down in current. AC transmission lines thus can carry high voltages at low currents and keep I^2R (power) losses much lower than the equivalent DC power lines can. Because of this property, most modern power-generating systems produce AC power. Consequently, the majority of motors used throughout industry are designed to use AC power.

There are other advantages to AC motors besides the wide availability of AC power. In general, AC motors of the same horsepower rating are smaller and, therefore, less expensive than DC motors. AC induction motors do not use brushes and commutators. Thus they are less prone to mechanical wear and sparking. This feature decreases maintenance requirements and the possibility of igniting explosive gases. AC motors are well suited to constant-speed applications; several AC motors can be made to run in synchronization at the same speed.

Figure 5.1   AC Motor

Industry uses AC motors in a wide variety of shapes, sizes, and power ratings. One type of AC motor in use is shown in Figure 5.1.

In this chapter we will examine the characteristics of AC motors, and what makes them different from DC motors. We will consider briefly the universal motor, which is basically a series DC motor. As its name implies, the universal motor can be operated with either DC or AC voltage. The remainder of the chapter will deal with the two other classifications of AC motors, the induction motor and the synchronous motor.

CLASSIFICATION OF AC MOTORS

AC motors may be classified in several ways. First, they may be classified in terms of their horsepower (hp) ratings. *Fractional hp motors* are motors rated under 1 hp. *Integral hp motors* are motors rated at 1 hp or more. Integral hp motors may range from 1 to 10,000 hp.

The number of phases (usually one or three) applied to the stator windings is another basis for classification. Single-phase motors are found in domestic, business, farm, and small-industry applications. Three-phase power generally is used in heavy industry.

In this chapter we will divide motors into three areas: (1) universal, (2) induction, and (3) synchronous. This classification is based on the internal construction of the motor. We cannot cover all aspects of AC motors in one chapter. Consequently, we will deal with the most important operating principles of the most common types of AC motors. For more information on AC motors, consult the bibliography for this chapter at the end of the book.

UNIVERSAL MOTOR

The *universal motor* operates equally well on AC or DC power. The construction of the universal motor is shown in Figure 5.2. Note that it is electrically the same as the series DC motor. Current flows from the supply through the field, through the armature windings, and back to the supply.

If you use the left-hand rule for coils, you can determine the directions of the field shown in the figure. The polarities of the armature and field flux oppose each other. If current is reversed, both fields reverse. The magnetic flux of the armature and the field still oppose one another. The motor, therefore, tends to turn in one direction no matter which way current flows. When 60 Hz AC is applied to the input terminals, the current (and the magnetic fields) changes direction 60 times per second.

Universal motors differ somewhat from DC motors in their physical construction. Recall that transformers have eddy currents and hysteresis losses. AC motors tend to have these same losses. To reduce these losses, AC motors are constructed with special metals, laminations, and windings, We can say, therefore, that a universal motor works equally well on AC or DC. A series DC motor, however, does not work well on an AC supply because DC motors are not constructed to reduce the losses mentioned.

The operating characteristics of the universal motor are similar to those of the DC motor. For example, starting torque is high in the universal motor. As in the DC motor, speed can become dangerously high if the load is taken off. Universal motors are normally fractional hp devices powered by single-phase AC.

PRINCIPLE OF THE ROTATING FIELD

Synchronous and induction motor construction can be divided into two basic parts: the rotor and the stator. Stators for both types of motors are similar in construction. So that eddy currents and hysteresis losses are reduced, the stator is made of many laminated steel discs. Coil slots are punched around the inside bore. The stator windings are wound around these slots. A typical single-phase stator winding is shown in Figure 5.3.

Since stators differ very little in induction and synchronous motors, the operating principle of rotating fields is basically similar for both. The principle of the rotating field is the key to the operation of synchronous and induction AC motors. The idea is relatively simple. Let us, for the moment, assume that the rotor is a permanent magnet. One way we can get the rotor to rotate is to place a magnet near it and move the magnet; see Figure 5.4.

Obviously, this technique is unsatisfactory. One of these two parts, the magnet or rotor,

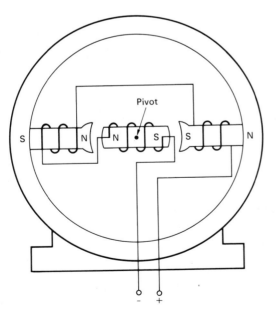

Figure 5.2 Universal Motor Construction

Figure 5.3   Single-Phase Stator Winding

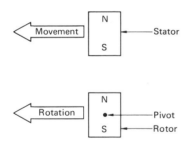

Figure 5.4   Rotating Rotor by Physically Moving Stator

must be stationary. Normally, the magnet that pulls the rotor around is a series of fixed windings called the stator windings. If these windings are fixed and cannot move physically, how can we get the rotor to rotate? The answer is to use for the stator a magnetic field that can be made to rotate electrically—the *rotating field.*

Let us see how this result can be obtained. Refer to Figure 5.5A. The principle of the rotating field is most easily seen in a two-phase system, as shown in the figure. But note that this principle is most frequently used in three-phase power systems.

Let us begin with a motor with two stator windings, each displaced from the other by 90°. Stator 1 pole pairs are vertical, and stator 2 pole pairs are horizontal. If the voltages applied to stator windings are 90° out of phase, then the currents that flow in those windings are 90° out of phase also. If the currents are out of phase by 90°, the magnetic fields produced by those currents are 90° out of phase.

To see how the fields rotate, refer to Figure 5.5B. Remember that phase 1 is applied to stator winding 1 and phase 2 is applied to stator winding 2. At time 1, note that the voltage at stator 1 is maximum while the voltage across stator 2 is zero. The current in stator 1 is then maximum, producing the vertical magnetic field shown. Note that it is north on top and south on the bottom. Also note that there is no horizontal field. The total resultant field is shown by the arrow.

At time 2, the current has decreased in stator 1 and increased in stator 2. Thus, there are two fields present, one in stator 1 and one in stator 2. The total resultant field is again shown by the arrow.

At time 3, the current in stator 1 has decreased to zero, while stator 2 current has reached a maximum. There is a strong horizontal field but no vertical field. The resultant field is again indicated by the arrow.

Looking at the total resultant field direction, we see a clockwise movement of the field. It has moved a total of 90°. The field starts out in the six o'clock position and ends in the nine o'clock position. In the figure, this rotation continues to one full cycle. When the two-phase voltages have finished one complete cycle, the resultant field has moved one complete 360° rotation.

We have, then, created a rotating field by doing two things. First, we placed two field windings (called the stator) at right angles to

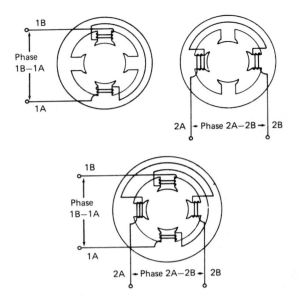

A. Two Phases Applied to Opposite Poles

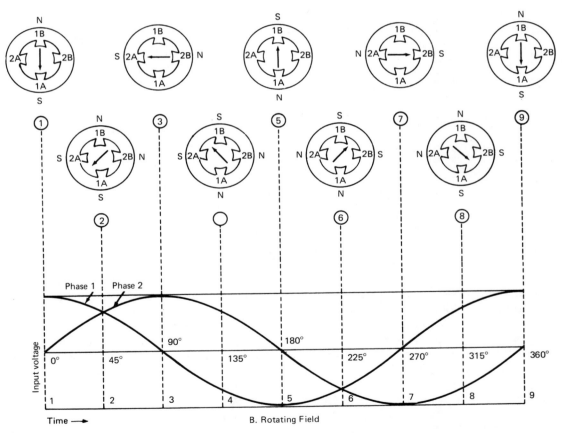

B. Rotating Field

Figure 5.5 Rotating Field in a Two-Phase System

each other. Second, we excited those windings with voltages 90° out of phase.

Two-phase motors are rarely used except in specialized equipment. We have discussed them here only as an aid to understanding the rotating field. This same principle of the rotating field is used, however, in three-phase and single-phase systems. In the three-phase system, the phases are 120° out of phase, not 90°. In the single-phase system, the required phase difference most commonly is produced by adding inductance or capacitance to one of the pairs of windings.

Note in Figure 5.5B that if the supplied current completes 60 cycles each second, the resultant field rotates at 60 r/s, or 3600 r/min. If we double the number of stator coils, the stator field rotates only half as fast. In a motor with a rotating stator field, the field travels one rotation per pole pair, per winding, for each cycle of applied voltage. If we double the number of poles for each winding, the magnetic field has more pole pairs to travel. Therefore, it takes twice as long to complete one rotation of all the pole pairs in one winding. This relationship can be stated in the following equation:

$$n_{st} = \frac{120f}{p}$$

where

n_{st} = stator speed in r/min

f = frequency of voltage applied to stator windings

p = number of magnetic poles per phase

For example, let us say we have a three-phase motor with four magnetic poles. If the frequency of the voltage applied to the stator is 60 Hz, what is the stator speed? We can calculate the stator speed as follows:

$$n_{st} = \frac{(120)(60 \text{ Hz})}{4} = 1800 \text{ r/min} \quad \textbf{(5.1)}$$

The speed at which the stator field moves is called the *synchronous speed.* This name is used because the field is synchronized to the frequency of the supply voltage at all times. Note that as we increase the number of poles, the stator speed decreases. Thus, we see that the speed of this type of motor can be changed by switching in different numbers of poles. However, this method of speed control is limited to a small number of discrete speed changes.

INDUCTION MOTOR

The driving torque of both DC and AC motors comes from the interaction of current-carrying conductors in a magnetic field. Recall that in the DC motor the armature moves and the field is stationary. Currents in both field and armature windings create the magnetic fields that produce the torque. The rotor current comes from the supply through the brushes and commutator. In the *induction motor,* electromagnetic induction produces the rotor currents.

Rotor Construction

Before we turn to the operation of the induction motor, we will examine the rotor construction in Figure 5.6. The rotor conductors are usually made of heavy-gauge copper. All these bars are connected at either end by copper or brass end rings. The end rings short-circuit the copper bar at both ends. No insulation is needed between the bars because the voltages generated are low and the bars represent the lowest resistance path for the current to flow. The core of the rotor (not shown in the diagram) is made of iron. This construction technique reduces air gap reluctance. It also concentrates the flux lines between the rotor conductors. Because of its unique appearance, this rotor is often called the *squirrel-cage rotor.* An actual squirrel-cage rotor is shown in Figure 5.7.

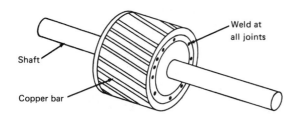

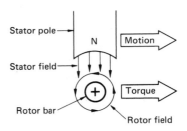

Figure 5.8 Interaction of Stator Field and Rotor Bars

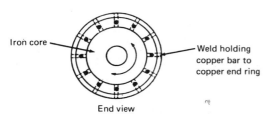

Figure 5.6 Squirrel-Cage Rotor Construction

Figure 5.7 Squirrel-Cage Rotor

Induction Motor Operation

Let us proceed now to the operation of the induction motor. Assume that we have a motor with a two-phase rotating field. If this field rotates over the stationary rotor windings, current is induced in the copper bars. One of these bars is shown in Figure 5.8.

The current generation follows the left-hand rule for generators. Remember that although the stator field (shown as a magnet) is moving right, the relative motion of the conductor is to the left. From the dot convention, current is generated, flowing into the page. This current flow generates a magnetic field around the conductor.

The interaction of the stator field and the rotor field produces a torque. The torque generated tends to move the conductor toward the right. We see, then, that the rotor tends to move in the direction of the stator rotation. Note that there is no physical contact between the rotor and stator. The only thing that links them is the electromagnetic field.

Since this motor has no physical connection between rotor and stator, it is relatively maintenance-free. There are no brushes to replace; since no brushes exist, there is no sparking to ignite explosive gases and vapors.

Slip. Note that it is impossible for the rotor of an induction motor to turn at the same speed as the stator. If it could turn at the same speed, no current would be produced in the rotor. No rotor current means no torque, so the motor will not run on its own. There must be relative motion between the rotor and the stator field for the motor to operate. The rotor, therefore, runs at a slower speed than the stator.

The difference between the rotor speed and the stator field speed is called *slip*. Slip (*s*) is normally expressed in equation form, as follows:

$$s = \frac{n_{st} - n_{rt}}{n_{st}}$$

where

n_{st} = stator speed in r/min

n_{rt} = rotor speed in r/min

In this equation slip will range between 0 and 1. Slip sometimes is expressed as a percentage. In that case, multiply the calculated value by 100. In this equation slip ranges between 0% and 100%. In actual loaded AC motors, slip ranges between 1% and 10%.

When the motor starts, the stator immediately goes to synchronous speed. At the first instant of starting, the rotor is stationary. The slip at starting is, therefore, 1, or 100%. Gradually, the rotor picks up speed, rotor currents decrease, and the rotor reaches a point where the torque produced equals the torque demanded by the load. The rotor speed ceases to increase but remains constant as long as this condition exists.

Rotor Reactance. As the motor gains speed, the rate at which the rotating field cuts the rotor conductors decreases. The amount of rotor voltage and current then decreases. We observe that the frequency of the rotor current varies directly with slip. As slip increases, so does rotor current frequency.

Recall the equation for inductive reactance (X_L) from basic electronics:

$$X_L = 2\pi f L$$

where

L = inductance

f = frequency of current flow through that inductance

In our example, if L is the inductance of the rotor and f is the stator frequency, we can add slip (s) to the equation because the reactance decreases as rotor frequency decreases. Our new equation for *rotor reactance* is, then, as follows:

$$X_{rt} = 2\pi f L s$$

where

X_{rt} = rotor reactance at slip s

Torque. Note that when the two speeds, rotor and stator, are close to each other, slip is nearly zero. As slip approaches zero, so does rotor reactance. When slip is near zero, the only impedance to rotor current flow is resistive. Thus, rotor current and voltage are in phase. At starting, however, slip is one, and the impedance is mostly reactive. In this case rotor current lags rotor voltage by about 90°.

With this value of slip (100%), the high rotor reactance produces a low power factor. Recall that the *power factor* of an AC circuit is the power delivered to or absorbed by the circuit, divided by the apparent power of the circuit. A resistor, because it does not have any reactance, has a power factor of one. A perfect coil or capacitor has a power factor of zero because it consumes no DC power.

In the near-zero-slip motor application, a low power factor means low torque due to the small rotor current. At the other end of the scale, when slip approaches one, torque is low because the power factor angle is large. The torque produced by an induction motor, then, varies approximately as the power factor of the rotor current. The torque equation is as follows:

$$T = K I_{st} I_{rt} \cos \theta$$

where

K = a constant (as in the DC motor)

I_{st} = stator current

I_{rt} = rotor current

$\cos \theta$ = power factor of the rotor current

With no load connected, the rotor of an induction motor rotates at a value close to synchronous speed. If the motor is loaded, slip increases, causing more rotor current to flow. Remember that rotor resistance is small. Under the loaded condition, more rotor current produces more torque output. This increased torque speeds up the motor to a value close to the value where it started. Induction motors seldom change speed more than 5% from no-load to full-load conditions. Therefore, they are classified as constant-speed devices.

There is a limit to the amount of torque the motor produces. If we keep increasing the torque demand, the motor produces more torque, up to a point. When the motor produces as much torque as it can (at a slip of about 25%), further torque demands simply decelerate the motor to a stop. The point at which this happens is called the *pull-out,* or *breakdown, torque.* This point is illustrated in the curve in Figure 5.9.

Figure 5.9 shows the torque and current curves for a three-phase induction motor with a squirrel-cage rotor. As load increases, slip and rotor reactance increase. The pull-out torque occurs at about 25% slip. Note that the torque at this point is about three times the normal full-load value.

Another important point to notice is the change in rotor current when the motor stalls.

It increases by a factor of five. As the rotor current increases, so does the stator current because of the induction or transformer action of the rotor and stator. The rotor can be considered the secondary and the stator the primary of a transformer. As the current in the secondary of a transformer increases, so does the primary current in order to keep the power in the two windings approximately the same. Thus, in the induction motor at stall, the stator current increases by approximately a factor of five over its full-load current. Current this large damages most motor windings. Designers normally protect motors from this kind of damage by installing circuit breakers of appropriate size.

Some tests require that the rotor of a motor be locked. In this case the stator voltage should be lowered to about 50% of its normal value. Lower stator voltage decreases locked-rotor current, thus protecting the motor.

Classifications of Squirrel-Cage Induction Motors

Squirrel-cage induction motors have several classifications, all based on differing types of rotor construction. Changes in the rotor construction produce differences in torque, speed, slip, and current characteristics. These motor types have been standardized by the National Electrical Manufacturers Association (NEMA). Examples of this classification scheme are NEMA motor types B, C, and D.

The NEMA type B motor is one with normal starting torque and current. A low rotor resistance produces low slip and high running efficiency. Generally, applications include constant-speed and constant-load requirements, where the motor does not need to be turned on and off frequently. Centrifugal pumps, blowers, and fans are examples of such applications. The type B is the most commonly used NEMA motor.

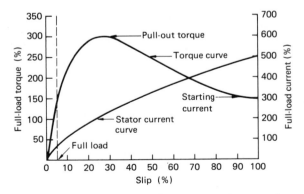

Figure 5.9 Torque and Current Curves for Three-Phase Induction Motor

The NEMA type C motor has the same slip and starting currents but a higher starting torque and poorer speed regulation than the type B. This motor is used with loads that are hard to start, such as compressors, conveyors, and crushers.

The NEMA type D motor has a high-resistance rotor. Slip increases and efficiency decreases in this motor. It is used in applications that require frequent starting, stopping, and reversing. Since type D has the highest starting torque, it is used to drive loads with very high torque demand, such as cranes, hoists, and punch presses. Because of its specialized rotor construction, the type D motor is the most expensive of the three.

Double Squirrel-Cage Construction

Another special type of rotor construction is illustrated in Figure 5.10. This rotor construction is known as the *double squirrel-cage rotor*. In this rotor the set of bars closest to the surface has a small cross-sectional area and a resistance of a few tenths of an ohm. The bottom bars have a much lower resistance, on the order of several thousandths of an ohm.

When the motor starts, the frequency induced into the bars is high. Since the lower bars have more inductance, the high frequency causes them to have a higher reactance than the top bars have. The current flow at starting, therefore, is higher in the top bars. Since these bars have a high resistance, they cause a reduction in the phase angle between the stator flux field and the rotor current. This reduction tends to increase starting torque. As speed picks up, the reactance of the bottom bars decreases, causing more current to flow in them. The effect of adding these extra bars is more torque to the motor under varying load conditions. The NEMA type C induction motor is an example of the double squirrel-cage construction.

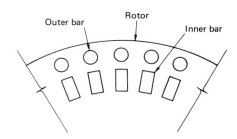

Figure 5.10 Double Squirrel-Cage Rotor Construction

Single-Phase Induction Motors

As their name implies, *single-phase motors* run when supplied with only one phase of input AC power. They are used in fractional hp sizes in small commercial, domestic, and farm applications. Single-phase AC motors are generally less expensive to manufacture in fractional hp sizes than three-phase motors. Of course, they eliminate the need to run three-phase power lines, which can be expensive and may not be necessary. Single-phase motors are used in blowers, fans, compressors in refrigeration equipment, portable drills, grinders, and so on.

The single-phase induction motor does not produce a rotating field and, therefore, has no starting torque. But if the rotor is accelerated, it behaves as if it were driven by a rotating field. With a single phase supplied to the stator, the stator field reverses at each alternation of the supply voltage. Thus, we can visualize the stator field as rotating in half-turn steps in either direction.

Because the single-phase induction motor does not produce any starting torque, it must depend on another system to start it. Consequently, single-phase induction motors are classified by the method used to start them. These classifications are discussed next.

Shaded-Pole Motor. One of the lowest-cost starting methods is the *shaded-pole motor.*

Shaded-pole motor construction is illustrated in Figure 5.11. Notice that a portion of each pole is surrounded by a copper strap, or shading coil. When current is increasing through the windings, the flux increases also. But as the flux travels through the portion of the pole piece shaded by the copper bar, the current induced into the copper strap builds up a field of its own. This field opposes the field buildup caused by the stator winding. The total flux is concentrated on the left side of the pole, as shown in Figure 5.12A.

As the applied voltage reaches a peak, the rate of change of flux is zero. At this time the flux is evenly spread over the pole piece, as shown in Figure 5.12B.

When the applied voltage decreases, the flux decreases also. At this time, however, the flux around the copper bar reverses direction, opposing the decrease in flux around the bar. This reversal produces a total flux concentration on the right side of the pole piece (Figure 5.12C).

Note that the total flux density shifts from left to right with respect to time. In other words, the flux moves toward the shaded pole. The rotor then follows this shift as if it were a rotating field.

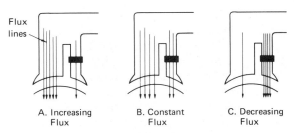

Figure 5.12   Movement of Flux in Shaded-Pole Motor

The shaded-pole motor produces starting torques in the range of 50% of full-load torque. It is manufactured in fractional hp sizes to about ⅟₂₀ hp, and it is not very efficient. A typical application of this motor is in a small fan or inexpensive phonograph.

Split-Phase Motor.   The *split-phase motor* has more starting torque than the shaded-pole motor. From Figure 5.13A we see that the split-phase motor has a stator with two sets of windings. One set, called the start winding, has a few turns of thin wire. It has a high resistance and a low reactance compared with the other windings, the run windings. The axes of these windings are displaced by 90°.

The current in the start winding (I_{start}) lags the line voltage by about 30°, as shown in Figure 5.13B. It is also smaller than the current in the run winding because the impedance of the start winding is larger. The current in the run winding (I_{run}) lags the applied voltage by about 45°. This phase shift produces a rotating field effect such as we discussed earlier in the chapter. The interaction of this field and the one produced by induction in the rotor causes a starting torque to be generated. This motor is sometimes called the *resistance-start motor*.

When the motor comes up to about 75% of its running speed, a centrifugal switch opens. A *centrifugal switch* is a device that disconnects the start winding in an AC motor at about

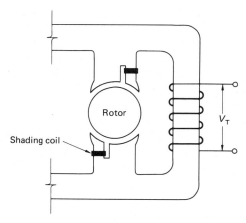

Figure 5.11   Shaded-Pole Motor Construction

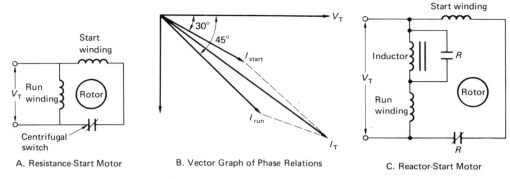

Figure 5.13   Split-Phase Motor

A. Resistance-Start Motor

B. Vector Graph of Phase Relations

C. Reactor-Start Motor

three-fourths of the full-load rotor speed. The centrifugal switch is mounted on the rotor and, as its name implies, is operated by centrifugal force. This switch disconnects the start winding. The motor continues to turn as long as a single phase is applied to the stator. Thus, the split-phase motor has good speed regulation. Starting torques range from 150% to 200% of full-load torque.

More starting torque is produced by adding an inductor in series with the run winding, as shown in Figure 5.13C. When the centrifugal switch opens, it not only disconnects the start winding but also shorts out the inductor. This motor is sometimes called the *reactor-start motor*. Direction of rotation is reversed by reversing either the start or run windings.

Applications of split-phase motors include washers, ventilating fans, and oil burners.

Another type of split-phase motor is the capacitor-start motor. This motor has more starting torque than the reactor-start motor because a capacitor is added in series with the start winding. The capacitor-start motor is illustrated in Figure 5.14A.

The capacitor produces a larger phase difference between currents in the start and run windings than the resistance-start motor does. The current in the start winding is about 90° out of phase with the current in the run winding, as shown in Figure 5.14B. This phase difference produces a larger starting torque than is available in the resistance-start motor. Starting torque can reach 350% of full-load torque.

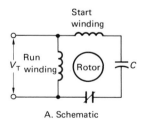

A. Schematic

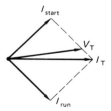

B. Phase Diagram

Figure 5.14   Capacitor-Start Motor

If the capacitor is disconnected by a centrifugal switch, the motor is called a *capacitor-start motor.* If the capacitor is left in the circuit, the motor is called a *capacitor-run motor.*

The capacitor in the capacitor-start motor must be a nonpolarized one. Sizes range from 80 μF for a ⅛ hp motor to 400 μF in a 1 hp motor. The capacitor-start motor normally is used in applications that demand high starting torque where DC motors cannot be used.

Wound-Rotor Induction Motor

Another type of induction motor, called the *wound-rotor motor,* is sometimes seen in industry, although not as often as the squirrel-cage. Its rotor construction is more like the DC motor armature than the squirrel-cage rotor construction. The windings are also similar to the stator windings. They are usually wye-connected, with the free ends of the windings connected to three slip rings mounted on the rotor shaft, as indicated in Figure 5.15. An externally variable wye-connected resistance is connected to the rotor circuit through the slip rings. In Figure 5.15, ϕ_1, ϕ_2, and ϕ_3 indicate the various phases in a three-phase system.

Rotor resistance is high at starting, producing high starting torque. The higher rotor resistance increases the rotor power factor, which accompanies the higher starting torque. As the motor accelerates, the variable resistance is reduced. When the motor reaches full-load speed,

the resistance is shorted out. The motor then acts like a squirrel-cage motor.

The advantages of the wound-rotor motor over the squirrel-cage motor are (1) smoother acceleration under heavy loads, (2) no excessive heating when starting, (3) good running characteristics, (4) adjustable speed, and (5) high starting torque. There are two major disadvantages of the wound-rotor motor: Its initial cost is high, and it has higher maintenance costs.

The wound-rotor motor is not used as much today as it was in the past. Its decline is due to improvements in speed control in other motors by varying stator frequency and to the development of the double squirrel-cage rotor.

SYNCHRONOUS MOTOR

The *synchronous motor* differs from the induction motor in several ways. A large polyphase synchronous motor needs a separate source of DC for its field. It also requires special starting methods, since it does not produce enough starting torque to start on its own. The rotor of the synchronous motor is essentially the same as an alternator rotor. Recall from the previous chapter that a DC motor may be run as a DC generator. Likewise, an AC generator (called an *alternator*) may be run as an AC motor.

Unlike the induction motor, the synchronous motor runs at synchronous speed. Also,

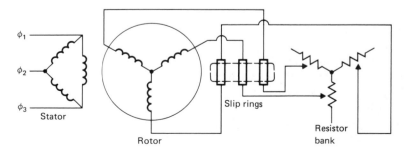

Figure 5.15 Wound-Rotor Induction Motor

the synchronous motor can be used to correct the low power factors caused by large inductive loads. As mentioned previously, synchronous motors range from fractional sizes to thousands of horsepower. Synchronous motors are found in a wider range of sizes than any other type of motor.

Synchronous Motor Operation

The basic operating principle of the synchronous motor may be described by visualizing the rotor as a magnet. Indeed, in fractional hp motors, the rotor may be a permanent magnet. If the stator field is sweeping around at synchronous speed, the magnet is literally yanked around in step with the revolving field.

Starting the Motor. The synchronous motor suffers from the disadvantage of not being able to start on its own. Refer to the simplified diagram in Figure 5.16. Current flowing in the direction shown causes interaction of the two fields. Before the rotor has a chance to turn, the current reverses, also reversing the field direction. Thus, total net torque produced over several cycles is zero. Special measures must be used to overcome this disadvantage.

One way to solve the problem is to add squirrel-cage winding slots to the rotor. In this way the motor can accelerate as an induction motor. When it approaches synchronous speed, the rotor is excited with DC. If the rotor is excited at the proper time, it locks into step with the stator field. The torque generated when the rotor locks into step with the stator field is called the *pull-in torque.*

Special sensing circuits are used to detect the precise instant at which excitation needs to be applied. If excitation is applied at the wrong time, the poles repel instead of attracting. The motor quickly comes to a stop. The motor also comes to a stop if too much torque is demanded by the motor. This torque, as mentioned previously, is called the pull-out torque.

Another method used to start the synchronous motor is to drive it to nearly synchronous speed with another motor, either mechanical or electrical. Such a motor is called a *pony motor.*

Loaded Operation. When the synchronous motor has no load connected, the rotor is positioned almost directly under the stator field, as shown in Figure 5.17A. If a load is connected, the rotor is pulled behind the stator, as shown in Figure 5.17B. At this time the motor is drawing lagging current. However, if we increase the amount of excitation, the strength of the rotor field increases. This increased strength produces a stronger attraction for the stator pole. If the excitation is strong enough, the rotor comes back under the stator pole. From this point, if we increase the excitation, the rotor moves ahead of the stator field. The motor then draws leading current. That is, the

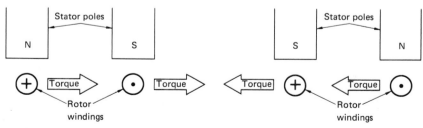

Figure 5.16 Synchronous Motor Starting

A. Torque Produced with Pole Polarity during One Half Cycle

B. Torque Produced during Opposite Half Cycle

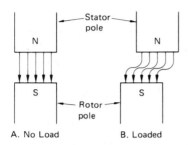

Figure 5.17 Pole Positions of Synchronous Motor

current and induced voltage lead the applied voltage. The motor then looks like a capacitor to the line.

Many industrial loads draw considerable inductive current, which degrades the power factor and increases utility rates. The synchronous motor can be adjusted to compensate for these inductive loads, thus improving power factors and efficiency.

Fractional hp Synchronous Motors

Up to this time we have been discussing integral hp synchronous motors. But there are many fractional hp synchronous motors in use today. Most do not use a separately excited rotor, however. It is too expensive and usually not necessary. Motors without a separately excited rotor are called *nonexcited synchronous motors*. There are two major types in use: reluctance motors and hysteresis motors.

Reluctance Motor. The *reluctance motor* uses differences in reluctance between areas of the rotor to produce torque. The reluctance principle utilizes a force that acts on magnetically permeable material. When such a material is placed in a magnetic field, it is forced in the direction of the greatest flux density. A diagram of the special rotor construction is shown in Figure 5.18.

The rotor has slots cut into it at regular intervals, as shown in the figure. These slots form salient poles (see Chapter 3). Proper motor operation demands that the number of salient poles be equal to the number of stator poles. The rotor also has bars, which cause it to accelerate as an induction motor. When the motor nears synchronous speed, the field magnetizes to the rotor and pulls it into step. The torque to drive the motor is generated because the reluctance is greater in the areas that do not have any bars. The areas of the rotor that have bars have a lower reluctance.

This motor has very low starting torque as well as low pull-out torque. Although the reluctance motor is normally found in fractional hp sizes, it is also encountered in integral sizes up to 100 hp.

Hysteresis Motor. Another interesting synchronous motor uses the hysteresis effect as a driving principle. The rotor of the *hysteresis motor* is made of a cobalt steel alloy. This alloy retains a magnetic field well and has high permeability.

Recall from basic electricity that when a magnetically permeable material is placed in a magnetic field, the domains in the metal align with the field. If the fields rotate, the domains still align themselves with the field, as shown in Figure 5.19. Power is consumed when the do-

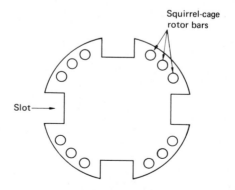

Figure 5.18 Rotor Construction of Reluctance Motor

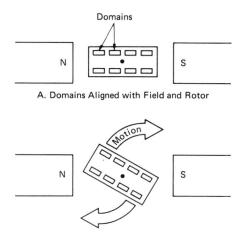

A. Domains Aligned with Field and Rotor

B. Domains Retain Alignment with Field

Figure 5.19 Rotor Behavior of Hysteresis Motor

mains maintain their alignment against the rotation of the rotor. This power loss is called hysteresis loss. During the time these domains are changing, starting torque is produced by another method, such as squirrel-cage bars. As the rotor approaches synchronous speed, current in the squirrel-cage bars decreases, producing less torque. At this time the rotor domains lock into alignment, resulting in a fixed magnetic polarity in the rotor. The retained magnetic field is now pulled into synchronization with the rotating field.

One interesting feature of the hysteresis motor is that the torque produced remains constant, regardless of load or speed changes. Consequently, it is used in clock and timer motors as well as audio turntables and tape drives.

CONCLUSION

The AC motor has special characteristics that differ from those of the DC motor. The induction motor, for example, is one of the most common motors in industry because it needs little maintenance. Since many AC motors are used in industry, technicians must have a general understanding of how they operate.

QUESTIONS

1. The universal motor is similar in construction to the _____-connected DC motor. The universal motor, then, can run on _____ as well as AC power.
2. Both induction and synchronous motors use the principle of the _____ stator field to generate torque. This principle requires that separate pairs of windings have applied voltages that differ in _____.
3. In the induction motor, energy is coupled from the stator to the rotor by _____ _____. Therefore, there is no _____ connection between rotor and stator.
4. Because only one phase is present, single-phase motors must have special circuitry to _____ them.
5. The rotors of synchronous motors run at _____ speed. Synchronous motors develop very little torque at _____. This problem is sometimes solved by placing squirrel-cage bars in the synchronous motor's _____.
6. Explain how the universal motor can be used in AC and DC circuits. Explain the difference in construction of a series DC motor and a universal motor.
7. Describe how a rotating magnetic field is created.
8. Is it possible to change the direction of an AC motor? If so, explain how.
9. What factors influence the rotor speed of an AC motor?

10. Describe how an induction motor produces torque.
11. Define slip in an induction motor.
12. Explain the NEMA classification system for induction motors.
13. Describe the construction of a double squirrel-cage motor. What effect does this construction have on motor operation?
14. Rank the types of single-phase induction motors in terms of the torque produced as a percentage of full-load torque.
15. Explain how starting torque is produced in the following engines: (a) split-phase, (b) capacitor-start, and (c) shaded-pole.
16. Describe the operation of the centrifugal switch.

17. Describe the method of rotor excitation in the wound-rotor motor.
18. List the advantages and disadvantages of the wound-rotor motor compared with the squirrel-cage motor.
19. At what speed does the synchronous motor run?
20. Why doesn't the synchronous motor develop enough torque to start on its own? How is the synchronous motor started?
21. What purpose does the synchronous motor serve other than the obvious one of turning mechanical loads?
22. Explain the method by which torque is produced in the hysteresis and reluctance motors.

PROBLEMS

1. Calculate the synchronous speed of a 12-pole motor with three-phase, 60 Hz power applied to its stator.
2. How many poles would a three-phase motor need to turn at a synchronous speed of 900 r/min?

3. If an induction motor's rotor is turning at 1750 r/min, what is the most likely stator synchronous speed? *Hint:* Examine Equation 5.1. On the basis of this speed, what is the slip?

CHAPTER
6
INDUSTRIAL
CONTROL DEVICES

OBJECTIVES

On completion of this chapter, you should be able to:

- Categorize individual control devices into one of five areas;
- Summarize the advantages and disadvantages of electromagnetic relays and reed relays;
- Summarize the advantages and disadvantages of solid-state relays;
- Identify the parts of the electromagnetic relay;
- Identify the schematic symbols of thyristors in common industrial use;
- Calculate the firing angle of an AC thyristor circuit;
- Calculate the junction temperature and power dissipation of a thyristor;
- Give examples of manually operated switches and their applications;
- Give examples of mechanically operated switches and their applications;
- List four methods by which thyristors are triggered;
- Describe the principles and applications of the solenoid;
- List two types of motor protection circuits and describe how they work;
- Contrast thermal and magnetic overload relays in terms of operation and applications;
- Contrast the electromagnetic relay and the solid-state relay in terms of operations and applications;
- List two methods of thyristor transient protection.

INTRODUCTION

Control devices lie at the heart of modern industrial systems. Very simply, a *control device* is a component that governs the power delivered to an electrical load. The load may be a motor or generator load, an electronic circuit, or even another control device. Indeed, every industrial system uses a control device. Since control devices are so common, a good background in the different types of control devices—how to recognize them and how they work—is essential.

Basically, control devices can be broken down into five categories: manually operated switches, mechanically operated switches, solenoids, electromagnetic switches (relays), and electronic switches (thyristors). A *switch* may be described as a device used in an electrical circuit for making, breaking, or changing electrical connections. In this chapter we discuss each of these basic categories in terms of the control devices in each category and the operating principles of these devices.

MANUALLY OPERATED SWITCHES

A *manually operated switch* is one that is controlled by hand. Many types and classifications of switches have been developed. Frequently, switches are classified by the number of poles and throws they have. The *pole* is the movable part of the switch. The *throw* of a switch signifies the number of different positions in which the switch can be set. The number of poles can be determined by counting the number of movable contacts. The number of throws can be found by counting how many circuits can be connected to each pole. Figure 6.1 shows several schematic diagrams of different combinations of poles and throws.

Toggle Switches

A *toggle switch* is an example of a manually operated switch. One type of toggle switch is shown in Figure 6.2. Toggle switches have many uses, most of them centered around energizing low-power lighting and other low-power electronic equipment. Electrical ratings are in terms of volts and amperes. These ratings should not be exceeded; otherwise, damage to the switch may result. If a switch needs to be replaced but an exact replacement cannot be found, substitute a switch with higher voltage and current ratings.

Push-Button Switches

Another common manually operated switch in industrial use is the *push-button switch*. This switch and its schematic diagram are shown in Figure 6.3.

Push-button switches can be broken down into two categories: momentary-contact and maintained-contact. In the *momentary-contact*

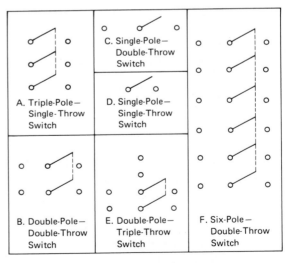

Figure 6.1 Switch Schematic Diagrams

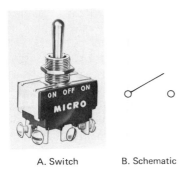

A. Switch   B. Schematic

Figure 6.2   Toggle Switch

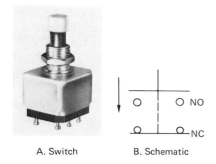

A. Switch   B. Schematic

Figure 6.3   Push-Button Switch

switch, actuation occurs only when the switch is pressed down. This push-button switch resets itself when released. (Resetting is the process of returning the switch back to its original state.) In the *maintained-contact switch,* as its name suggests, the switch is actuated even when the button is released. It must be pressed again for it to change states. This switch is most commonly used in turning electrical equipment on and off. A good example of its application is the dimmer switch in a car.

Note the abbreviations NO and NC in Figure 6.3B. These letters represent the electrical state of the switch contacts when the switch is not actuated. The abbreviation NC means that the switch contacts are normally closed; NO means that they are normally open.

Knife Switches

The *knife switch* is a manually operated switch used extensively for controlling main power circuits. The knife switch is shown in Figure 6.4. The knife switch typically is a bar of copper with an attached handle. The switch is actuated by pushing the copper bar down into a set of spring-loaded contacts. Normally the knife switch is enclosed in a box to prevent accidental contact with high voltage.

Rotary-Selector Switches

The *rotary-selector switch* is another common manually operated switch. This switch is shown in Figure 6.5. As the knob of the switch is ro-

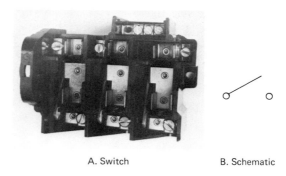

A. Switch   B. Schematic

Figure 6.4   Knife Switch

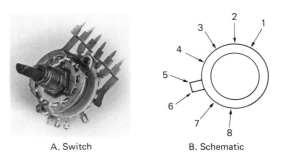

A. Switch   B. Schematic

Figure 6.5   Rotary Switch

tated, circuits can be opened and closed according to the construction of the switch. Some rotary switches have several layers, or wafers, as shown in Figure 6.5B. The addition of wafers makes it possible to control more circuits with a single change of position. The rotary switch is often used on test equipment and other multifunction low-power equipment. The range selector switch of a voltmeter is an example of the use of a rotary switch.

Manual Motor Starters

The *manual motor starter* is a special manually operated switch that is used to start or stop AC and DC motors. It is illustrated in Figure 6.6. A mechanical linkage usually transmits the movement of the toggle or handle to the actual switch contacts. The wiring diagrams, shown in Figure 6.7, do not show the operating mechanism, since it is not electrically controlled.

Figure 6.7A is a simplified diagram of a manual starter wired to control an AC motor. Note the three normally open (NO) switch contacts labeled S_1, S_2, and S_3. These switches are said to be normally open because this term re-

Figure 6.6   Manual Motor Starter

flects the state of the switch in its off position. Throwing the switch or pushing the start button mechanically closes these contacts, connecting the three-phase AC motor to the line. The contacts can be opened by pressing the stop button. The contacts can also be opened by tripping one of the two overload relays (OL). An *overload relay* is a device that will open a set of NC contacts when a certain preset temperature is reached. It will be discussed a little later in this chapter. When an overload relay trips (caused by a motor overheating), the starter mechanism unlatches and the S contacts open. This removes power from the motor. Usually the contacts cannot be reclosed until the starter mechanism has been reset. Resetting can be accomplished by pushing the stop or reset button, or by moving the handle to the reset position. The overload relay should be allowed to cool before resetting.

A simplified diagram of a manual starter wired to control a DC motor is shown in Figure 6.7B. Note the difference in wiring between this starter and the one used to control an AC motor. Only one overload relay is necessary, since only one path for current exists.

Manual starters generally are used in conveyors, compressors, pumps, blowers, fans, and small machine tools. They are low in cost, are simple to use and install, and do not give off the annoying AC hum often found in devices with a coil. The manual starter is used in situations where equipment can run continuously and automatic control is not required. As a general rule of thumb, if a motor has to be started and stopped more than 12 times an hour, week after week, manual control should not be used. They also are used as power disconnects and overload protectors where these two functions are not provided elsewhere. Manual starters normally are used on small AC and DC machines. If the motor is large, the designer will have no choice; another type of controller must be selected. In single-phase AC applications, manual starters are limited to 5 hp (3.7 kW) at voltages up to 230 V. In three-phase applications, 15 hp

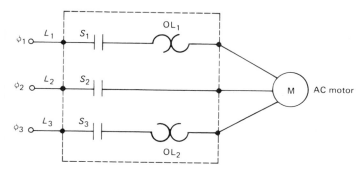

A. Starter Wired to Control an AC Motor

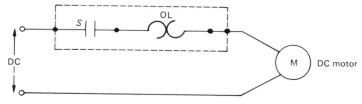

Figure 6.7 Simplified Diagrams of Manual Motor Starters

B. Starter Wired to Control a DC Motor

motors may run on manual switches to 600 V AC.

MECHANICALLY OPERATED SWITCHES

Mechanically operated switches are used extensively in industry. A *mechanically operated switch* (sometimes called an automatic switch) actuates automatically in the presence of some specific environmental factor, such as pressure, position, or temperature.

Limit Switches

The *limit switch,* shown in Figure 6.8, is a very common industrial device. It usually is actuated by horizontal or vertical contact with an object, such as a cam. Note in the figure that the actuation is accomplished by moving a lever. Often the lever is the roller type, as shown in Figure 6.8A. Or as in the case of the limit switch in Figure 6.8B, a wand type of arm is

Figure 6.8   Limit Switches

used. Note the difference between the wand and lever types. The lever may be actuated only by movement in one direction, but the wand may be actuated by movement in any direction.

Limit switches have two basic uses. First, they serve as safety devices to keep objects between certain physical boundaries. For instance, in a machine that uses a moving table, a limit switch can keep the table from moving past its safe physical limits by shutting off the table's drive motor. Second, limit switches are used to control industrial processes that use conveyors or elevators.

Mercury Switches

The *mercury switch* is nothing more than a glass enclosure containing mercury and several switch contacts placed so that the mercury will flow over and around them. Several examples of mercury switches are shown in Figure 6.9.

Because of the simplicity of the mercury switch, it is very versatile. It can be used as a position sensor; it can sense the amount of centrifugal or centripetal force and changes in speed, inertia, and momentum. In addition, it has little contact bounce (a major problem in relays and mechanical switches), is hermetically sealed against environmental impact, and has a relatively long life.

Most mercury switches are used with currents up to 4 A and voltages up to 115 V. When

actuated, the contact resistance can be less than a tenth of an ohm. As shown in Figure 6.9, the mercury switch comes in many different configurations. The glass envelope is usually filled with hydrogen gas to prevent arcing.

An application of a mercury switch is shown in Figure 6.10. When the mercury switch closes, trigger voltage is provided to the silicon-controlled rectifier (SCR), which turns on. The SCR is an electronic switch; its operation is discussed later in this chapter.

Snap-Acting Switches

A *snap-acting switch* is a switch in which the movement of the switch mechanism is relatively independent of the activating mechanism movement. In a snap-acting switch, as in a toggle switch, no matter how quickly or slowly the toggle is moved, the actual switching of the circuit takes place at a fixed speed. In other switches, like the knife switch, the speed with which the switch closes is determined by the mechanism driving it. In the snap-acting switch, the switching mechanism is a leaf spring that snaps between positions.

A *precision snap-acting switch* is one in which the operating point is preset and very accurately known. The operating point is the point at which the plunger or lever causes the

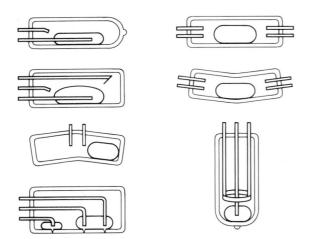

Figure 6.9 Mercury Switch Configurations

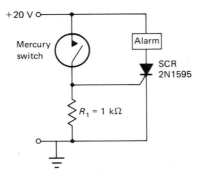

Figure 6.10 Mercury Switch Controlling SCR

switch to change position. The precision snap-acting switch is commonly called a micro-switch, which is actually a trade name of the Micro Switch Division of Honeywell Corporation. An example of a microswitch is shown in Figure 6.11. Frequently, limit switches, like those in Figure 6.8, contain snap-acting switches as their contact assembly. Much less arcing results from the rapid movement of the contacts.

The basic precision snap-acting switch is used in many applications as an automatic switch. It is most often used as a safety interlock to remove power from a machine when there is danger to the operator or technician.

Several different methods are used to actuate this type of switch. Some of the more common actuators and their uses are shown in Figure 6.12.

Motor Protection Switches

The motor is the most important source of rotating power in industry. We learned in earlier chapters that the motor tries to meet any torque demand placed on it. Unfortunately, the motor cannot detect unaided when a demand is too high; it is incapable of recognizing its own limitations. Motors must be protected from poten-

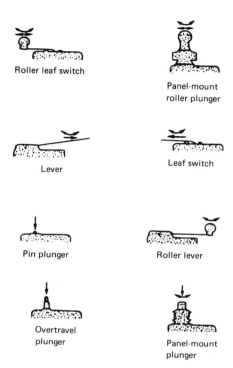

Figure 6.12 Common Actuators Used in Precision Snap-Acting Switches

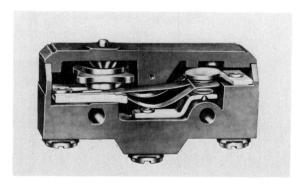

Figure 6.11 Cutaway Showing Construction of Precision Snap-Acting Switch

tially damaging conditions. Industrial motor protection is based on two conditions; excessive line current and high motor temperatures. Too much line current is called a *short circuit condition,* usually ranging from 10 to 100 times normal current flow. It is the more serious of the two conditions. If not interrupted, short circuits can damage motors, lines, and branch circuits, can cause fires, and can possibly injure personnel. Short circuits normally occur when physical contact is made between two or more line conductors or a line conductor and ground. The resulting damage caused by the short circuit may be thermal, physical (due to magnetic forces), or electrical arcing damage. Short circuits normally are prevented by fuses or circuit breakers.

Historically, *fuses* have been recognized as the best short circuit protection, due to their fast response time. The fuse, however, in some situations is not the best protection for a motor or a motor circuit. System designers must protect motor control circuits and feeder lines, as well as the motors themselves, according to National Electrical Code (NEC) requirements. Ordinary fuses may protect the motor from damage but may let through currents high enough to damage starters or other associated equipment. A good fuse should respond to an increase in current within a few milliseconds.

A device that is an improvement over the fuse is called a *motor short circuit protector* (MSCP). This device responds more quickly than a fuse, thus letting through a fraction of the energy a fuse lets through. When the MSCP opens, a trigger protrudes from the end of the device. The trigger indicates if the device has been opened by an overcurrent. The MSCP also is used in motor starters to remove power from the starter in the event of a short circuit. Power is removed when the trigger presses on a bar, which opens the power switch. MSCPs are available in ratings up to 480 V and 100,000 A.

A *circuit breaker* is a device that protects against short circuits by unlatching a breaker switch. The response times of circuit breakers range from 8 to 10 ms. For motor protection circuits this response may be too slow. If enough current gets through the circuit breaker before the breaker responds, some equipment may be damaged. To combat this slow response time, several manufacturers have developed a new design. Called the *blow-apart* design, this new type of breaker will respond to overcurrents in about 2 ms. The breaker arms, shown in Figure 6.13, are held together by a special calibrating spring. Current flows through each breaker arm creating like magnetic fields. When the repulsive force of the two electromagnetic breaker arms overcomes the force of the spring, the breaker arms blow apart. The faster the current increases, the faster the arms blow apart.

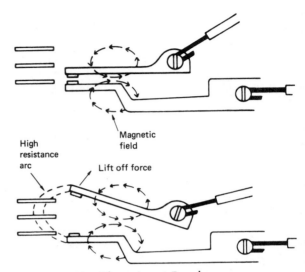

Figure 6.13   Blow-Apart Breaker

SOLENOIDS

The principle of the solenoid is an important one in industrial electronics. The *solenoid* is a device used to convert an electrical signal or an electrical current to linear mechanical motion. As shown in Figure 6.14, the solenoid is made up of a coil with a movable iron core. When the coil is energized, the core, or armature as it is sometimes called, is pulled inside the coil. The amount of pulling or pushing force produced by the solenoid is determined by the number of turns of copper wire and the amount of current flowing through the coil.

One factor often overlooked in solenoid design is the heating factor. Heat makes the coil less efficient at producing flux, lowering the flux density and the solenoid pulling force. Heat may be controlled by properly venting the enclosure, by using a heat sink, by forcing air past the solenoid, or by using a larger solenoid. The amount of heat produced by a solenoid is related to the duty cycle. The higher the duty cycle, the higher the temperature of the device. Solenoids in continuous duty applications may

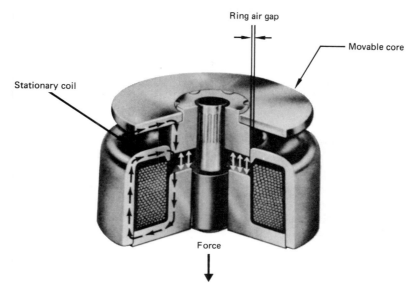

Figure 6.14 Linear Solenoid Construction

need a special circuit to reduce current and, therefore, power dissipation. Such a circuit is shown in Figure 6.15. This method of current reduction depends on using a solenoid that has a set of normally closed (NC) switch contacts. These contacts are placed across resistor R_1, as shown in the diagram. When the solenoid coil K_1 is energized, current flows through the normally closed switch contacts. With power applied, the armature will start to close. When the armature reaches the end of its travel, it opens the NC switch. Current through the solenoid will then be reduced, since the circuit has a higher total resistance.

Solenoids are used in a wide variety of applications, such as braking motors, closing and opening valves, shifting the position of objects, and providing electrical locking. The mechanical movement caused by the solenoid is usually between ½ and 3 in. Solenoids are available for use in either AC or DC circuits. When selecting a solenoid, the important considerations are the stroke length and force required, the length of time the solenoid is to be energized, how frequently the device will be turned off and on, and the ambient temperature conditions. You should bear in mind when working with sole-

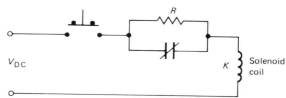

Figure 6.15 Circuit for Continuous Duty Solenoid Action

noids that too large a force in solenoid operation is almost as bad as not enough force. Too great a pull in the solenoid will cause it to pound itself into premature failure.

ELECTROMAGNETIC RELAYS (EMRs)

Relays are undoubtedly one of the most widely used control devices in industry. A *relay* is an electrically operated switch. Relays may be divided into two basic categories: control relays and power relays. There also are additional, more specialized relays.

Control relays, as their name implies, are most often used to control low-power circuits or other relays. Control relays find frequent use in automatic relay circuits, where a small electrical signal sets off a chain reaction of successively acting relays performing various functions.

Power relays are sometimes called contactors. The power relay is the workhorse of large electrical systems. The *power relay* controls large amounts of power but is actuated by a small, safe power level. In addition to increasing safety, power relays reduce cost, since only lightweight control wires are connected from the control switch to the coil of the power contactor.

The components of both power and control relays are the same; see Figure 6.16. Each has a coil wound around an iron core, a set of movable and stationary relay contacts (the switch), and the mounting. A switch usually is used to start or stop current through the coil. When current flows through the coil, a strong magnetic field is created. This electromagnet pulls the armature, which, in turn, moves the relay contact down to make electrical connection with the stationary contact. The physical movement of the armature occurs only when current

flows through the coil. Any number of sets of contacts may be built onto a relay. Thus, it is possible to control many different circuits at the same time.

Like switches, the relay contacts have maximum voltage and current ratings. If these ratings are exceeded, the life of the contact may be decreased greatly. Relays also have pull-in voltage and current ratings. *Pull-in current* is the amount of current through the coil needed to operate the relay. *Pull-in voltage* is the voltage needed to produce that current. Usually, the actual steady-state (continuous) voltage applied to the relay coil is somewhat higher than the pull-in voltage. This condition guarantees enough current so that the relay actuates when needed and holds under vibration.

Another rather specialized type of relay is the magnetic motor starter. Many industrial applications require the starter to be located some distance from the operator. This is not possible with the manual starter. Also, many processes require automatic control of motors that the manual starter cannot provide. *Magnetic starters* usually are relays that work on a variation of the solenoid principle, as shown in Figure 6.17. Current through the coil of an electromagnet generates a magnetic field. The field attracts the movable armature, pulling it up. The relay contacts are attached to the armature, closing when the armature is pulled in. Like the relays just mentioned, the magnetic motor starter has a pull-in voltage and current rating, which is larger than needed to keep the starter energized.

Magnetic motor starters usually have built-in motor overload protection circuits. (These protective switches are discussed later in this chapter.) The overload circuits are the main difference between a motor starter and a contactor. A contactor, being a general, high-power relay, does not have motor overload protection built in. Magnetic starters often are controlled by many of the switches we have discussed in this chapter, for example, limit switches, push buttons, and timing relays.

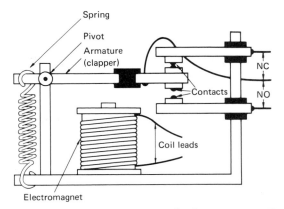

Figure 6.16 Basic Parts of Electromagnetic Relay

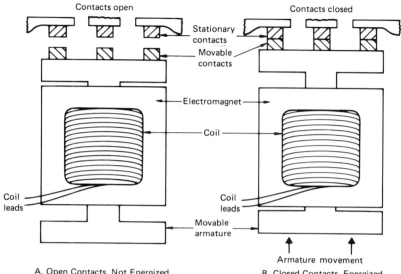

Figure 6.17 Magnetic
Motor Starter Operation

A. Open Contacts, Not Energized

B. Closed Contacts, Energized

You may recall from Chapter 5 on AC motors that to reverse the direction of a three-phase, squirrel-cage motor, we need only reverse any two of the power leads. Such a reversing task usually is done by a special starter called a *reversing starter*. A simplified diagram of the reversing starter is seen in Figure 6.18. You will notice that the starter actually contains two contactors, one for forward operation and the other for reverse operation. A mechanical interlock (not shown) keeps only one contactor on at a time.

Not all relays have the same construction as the armature relay of Figure 6.16. Another type of relay, called the *reed relay*, is also widely used in industry. The reed relay, like the armature relay, uses an electromagnetic coil, but its contacts consist of thin reeds made of magnetically permeable material. These contacts, called the reed switch, usually are placed inside the coil. A cutaway view of a reed relay is shown in Figure 6.19A.

When current flows through the relay coils, a magnetic field is produced. The field is conducted through the reeds, magnetizing them. As

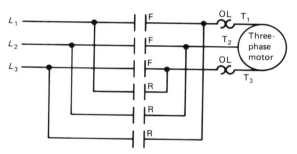

Figure 6.18 Reversing Starter

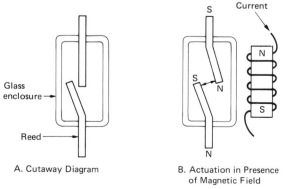

A. Cutaway Diagram

B. Actuation in Presence of Magnetic Field

Figure 6.19 Reed Relay

shown in Figure 6.19B, the magnetized reeds are attracted to one another, and the relay contacts are closed.

Reed relays are superior to armature relays in many ways. Reed relays are faster, more reliable, and produce less arcing than the armature relay. However, the current-handling capabilities of the reed relay are limited. In addition, because of the glass case that houses the reed relay contacts, the reed relay is more susceptible to damage by shock than the armature relay is.

Arcing and pitting can be further reduced in the reed relay by wetting the relay contacts with mercury. Since mercury is a liquid metal, it is drawn up the bottom relay contact by capillary action. The mercury completely covers the relay contact, thus protecting it from arcing, and extending contact life.

Timers

The proper timing of events in a process represents an important part of industrial electronics. In industrial processes certain actions must occur at a certain time interval after another operation occurs. For example, in a DC motor system there is often a time delay before full power is applied to the armature. This time delay gives the motor a chance to accelerate to a high speed. The actual time delay varies widely according to the application and the device doing the timing. Delays may range from as little as a microsecond to as long as a day.

Timing functions are performed by devices called *timers*. Two important classifications are delay timers and interval timers. As its name implies, a *delay timer* removes power from or applies power to a component or circuit after a certain preset time interval has elapsed. The *interval timer* applies or removes power for a certain amount of time.

The delay timer is the more common device. Delay timers can be classified into two basic groups: on-delay and off-delay. The on-delay timer is shown in Figure 6.20A. When the coil is energized, the timing mechanism causes time delay relay TDR_1 contact to close after a preset period of time. The red (r) lamp, P_1, will turn on when the TRD_1 closes. The timing function of TDR_1 contacts is called on-delay, timed-closed. Also, when the coil is energized, the TDR_2 contacts open after a set time delay. The green (g) lamp, P_2, will stay on after the coil is energized for an amount of time equal to the time delay. This set of contacts is called on-delay, timed-open. Deenergizing the TDR coil in Figure 6.20A will immediately open TDR_1 and close TDR_2 contacts.

The diagram in Figure 6.20B illustrates an off-delay timer. In this timer when the coil is energized, the TDR_1 contact is closed immediately. It remains on as long as the coil remains energized. The red lamp, P_1, will also turn on when the coil is energized, since power is ap-

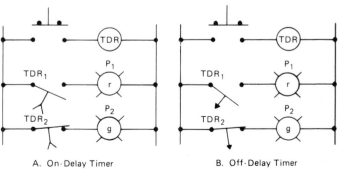

Figure 6.20 Delay Timers

A. On-Delay Timer   B. Off-Delay Timer

plied through TDR$_1$ contacts. TDR$_2$ contacts, on the other hand, will open and remain open as long as the coil is energized. This turns off lamp P$_2$. When the coil is deenergized, the delay mechanism will cause the TDR$_1$ contacts to open after a time delay. TDR$_1$ contacts in this relay are called off-delay, timed-open. After the same time interval after the coil loses power, the TDR$_2$ contacts, which were open, will close. The TDR$_2$ contacts are called off-delay, timed-open.

Overload Relays

A mechanical motor load imposes a certain torque demand on a motor. The motor tries to meet that demand by converting electrical power to mechanical energy. If the load demands more torque than the motor is designed to supply, the motor will draw above its rated current. A motor is considered to be *overloaded* when it draws too much current. You will recall from an earlier chapter that a large amount of current may be drawn by a motor at starting, due to the low cemf. Although this large starting current is permissible for a short time at starting, if it continues for any length of time, the motor will overheat. Overheating damages the insulation on windings and interferes with the normal lubrication process. Generally, if a motor draws more than its full load rated current, it is being overloaded. Actually, the term overload is almost a misnomer. The motor is not protected from an overload; rather, it is protected from the high motor temperatures that result from an overload.

A means must be designed to monitor the motor and protect it from excessive loads. The NEC specifically directs designers "to protect the motors, the motor control apparatus, and the branch-circuit conductors against excessive heating due to motor overloads or failure to start." Section 430 of the NEC details the requirements for motor circuit protection. One

device used to give a motor protection from intermittent or sustained overloads is called an *overload relay*. Overload relays are divided into two classifications: thermal and magnetic.

Magnetic Overload Relays.   The magnetic overload relay was one of the original methods of providing overload protection. It is still used, although it is not as popular as some of the thermal overload relays. It is used presently as a protection method for large DC motors where commutation, rather than heating, is the problem. The *magnetic overload relay* has a movable magnetic core located inside a coil. Through the coil passes all or a part of the motor current. As current flows in the coil, the core is pulled upward. When the core moves up far enough, it trips a set of contacts at the top of the relay. The contacts open, removing power to the motor. The movement of the core is controlled by a piston inside an oil-filled dashpot. The piston-dashpot arrangement is similar to a shock absorber in an automobile. The piston is slowed down by the resistance of the oil in the dashpot, producing a time delay. The time delay is adjusted by increasing the size of the holes in the piston. The tripping current is adjusted by moving the core on a threaded rod. Since both time-delay and tripping current can be adjusted, this relay is sometimes used to protect motors with long acceleration times or unusual duty cycles.

The magnetic overload relay, as we have seen, responds to the amount and duration of motor current. It does not measure the temperature of the motor directly. The magnetic relay does not respond to overloads lasting for very short time intervals. If too many overloads occurred, the motor windings could get dangerously hot; the magnetic relay would never trip. We would then need a method to detect motor temperatures directly. Sometimes this is done with thermistors embedded into the motor case. Thermistors with a positive temperature coefficient are used because of their extreme nonlinearity. However, these types of circuits are generally expensive and restricted to certain appropriate applications.

Thermal Overload Relays. A more economical device for general use in detecting motor temperatures is the externally mounted *thermal overload relay.* This device does not need to be in contact with the motor. It simulates the heating conditions that are occurring in the motor. It is adjusted so that the relay trips before the motor reaches a dangerous temperature. This thermal relay type has two basic parts. First, it has a resistive heating element that is heated when motor current passes through it. Second, it has a sensor-switch element that detects the temperature and opens the switch when the temperature is too high.

Many types of sensor-switch elements have been used in motor overload relays. The two that are used today are the bimetal strip and the eutectic alloy. The *bimetal strip,* explained in Chapter 8, deflects when the heater increases its temperature. A certain deflection corresponding to a preset temperature opens a set of contacts, removing power from the motor. The *eutectic alloy* type melts when it reaches a precise temperature. The alloy (similar to solder) releases a wheel when it melts. The turning wheel actuates a switch, which removes power from the motor.

Bimetal strips sometimes detect temperature changes directly on the motor case. An example of this kind of device is shown in Figure 6.21. Note that the protector senses the temperature at the end bell. The bimetal strip, when heated to a certain temperature, will snap open, removing power from terminals one and two.

For most motors, especially small motors, the simple types of protectors are economically justified. Thermal and magnetic overload devices protect the motor from damage due to excessive temperatures. In larger motors a current transformer senses the excessive current and turns on an electromechanical relay to remove power from the motor. Large motors and some small specialty motors require more protection than overload and short circuit protection can provide. For example, ground faults normally are not detected by fuses. A *ground fault* is defined as a current leakage path to ground. In motors the ground fault is usually caused by the breakdown of insulation on the motor windings. Among other things, ground faults can damage a motor's stator laminations and can also be a safety hazard. Ground faults are detected and dealt with by devices called *ground fault interrupters.* The ground fault interrupter detects current flowing in the ground lead and removes power from the circuit.

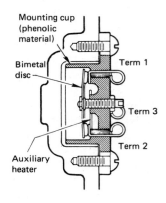

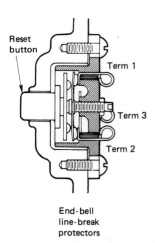

Figure 6.21 Bimetal Strip Overload Relay

Another form of protection needed in three-phase circuits is phase loss protection. If a three-phase motor loses a phase, the motor may still run (depending on the load), but the motor will definitely overheat, causing damage. *Phase loss detection circuits* are available to sense the loss of a phase and to de-energize the circuit if that loss occurs. The same problem of overheating occurs when one of the three-phase voltages increases or decreases. An increased or decreased phase voltage causes a phase imbal- ance, which also can be detected by appropriate circuitry.

Most of these functions have been accomplished in the past by EMRs with appropriate sensing circuits attached to detect the problems. Providing most of these forms of protection using electromechanical hardware is not as feasible today as it once was. Providing all of these functions we have mentioned (and some we have not) would be expensive and would require a large enclosure. It is more cost- and

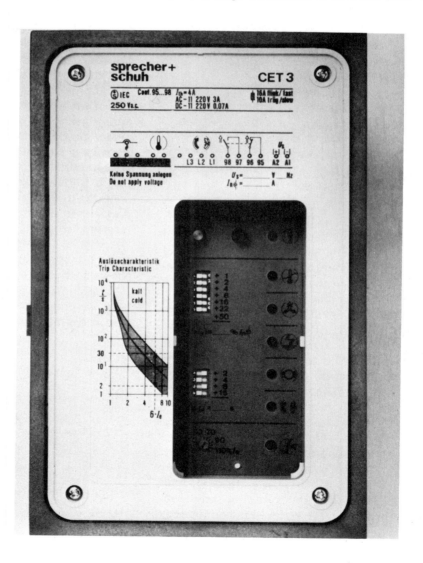

Figure 6.22 Solid-State Motor Protector

space-effective today to use a single *solid-state motor protector,* such as the one shown in Figure 6.22 manufactured by Sprecher & Schuh, Inc. This single solid-state unit provides protection against phase loss, thermal overload, ground fault, motor stall, phase reversal, and underload. The cost of this unit, when compared to its electromechanical counterparts, is considerably less. In addition to being less expensive, the solid-state motor protector is easier and less time-consuming to install. Most of these units are extremely flexible in allowing the operator to select the points at which power will be removed from the system. They are also very rugged and allow installation in environments hostile to EMRs.

SEMICONDUCTOR ELECTRONIC SWITCHES: THYRISTORS

While the EMR has advantages in numbers of circuits controlled, the *electronic switch* is faster, cheaper, and more energy efficient, and it lasts longer. The electronic switch does not exhibit contact bounce, a major problem in digital circuits. Also, the electronic switch is not acceleration-sensitive or vibration-sensitive, works better in hostile environments, and does not arc or spark. It is, therefore, explosion-proof. Because of these advantages, electronic switches, or thyristors, have replaced many EMRs and thyratron tubes in industry. In addition, thyristors are used in rectifiers, alarms, motor controls, heating controls, and lighting controls. Thyristors are used wherever large amounts of power are being controlled.

A *thyristor* is a four-layer, PNPN device. This PN sandwich, then, has a total of three junctions. As we will see later, this device has two stable switching states: the on state (conducting) and the off state (not conducting).

There is no linear area in between these two states, as there is in the transistor. That is, a switch is either on or off, never (we hope) in between.

Silicon-Controlled Rectifier (SCR)

The most common thyristor you will encounter as a technician is the *silicon-controlled rectifier* (SCR). It is widely used because it can handle higher values of current and voltage than any other type of thyristor. Presently, SCRs can control currents greater than 1500 A and voltages greater than 2000 V.

The schematic symbol for the SCR is shown in Figure 6.23A. Notice that the schematic symbol is very much like that of the diode. In fact, the SCR resembles the diode electrically since it conducts only in one direction. In other words, the SCR must be forward-biased from anode to cathode for current conduction. It is unlike the diode because of the presence of a gate (G) lead, which is used to turn the device on.

SCR switching operation is best understood by visualizing its PNPN construction, as shown in Figure 6.23B. If we slice the middle PN junction diagonally, as shown in Figure 6.24, we end up with an NPN transistor connected back to back with a PNP transistor. A

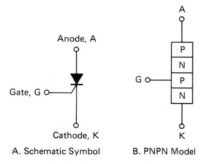

A. Schematic Symbol   B. PNPN Model

Figure 6.23   Silicon-Controlled Rectifier (SCR)

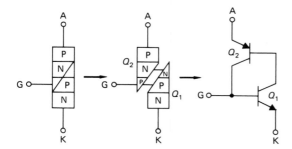

Figure 6.24 SCR Two-Transistor Analogy

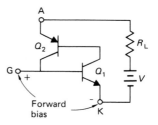

Figure 6.25 SCR Two-Transistor Analogy with Regenerative Feedback

forward-bias voltage between the gate-to-cathode leads turns on transistor Q_1. Transistor Q_1 conducts, turning on Q_2. Transistor Q_2 drives Q_1 further into conduction. This regenerative process, shown in Figure 6.25, continues until both transistors are driven into saturation. Only microseconds are needed to complete the process. This regenerative action is called *latching* because the transistors continue to conduct even if the gate-to-cathode, forward-bias voltage is removed.

Why does the SCR handle power efficiently? When it is in the off state, it does not draw much current. When it is on, the voltage across it is around 1 V, regardless of the current through the device. Recall the power equation:

$$P = IV$$

where

P = power dissipated by device

I = current through device

V = voltage across device

If the voltage remains at about 1 V, the power is approximately equal to the size of the current flow. For instance, if the voltage drop across the device stays at 1 V with 1000 A flowing through it, the power dissipated is only 1000 W. All the rest of the power is transferred to the load.

Turning on the SCR. The gate-to-cathode voltage required to turn on the SCR is referred to as the *gate turnon voltage* (V_{GT}). The amount

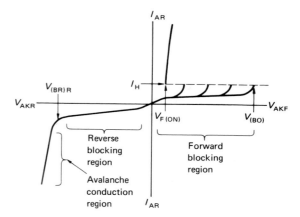

Figure 6.26 SCR Current-Voltage Characteristic Curve

of gate-to-cathode current needed to turn the device on is called *gate turnon current* (I_{GT}). Voltage V_{GT} usually ranges from 1 to 3 V, while I_{GT} ranges from 1 to 150 mA. The SCR can be turned on by methods other than the gate voltage. However, with the exception of the radiation method of triggering, little use is made of other triggering methods.

Refer to Figure 6.26, which shows the current-voltage behavior of an SCR with no gate current flowing and the gate grounded. When the SCR is forward-biased (positive voltage on the anode with respect to the cathode), very little current flows. This region is called the *forward blocking region*. However, if the *forward voltage* (V_{AKF}) is high enough, minority carriers will be accelerated to high velocities and create carriers in the gate-cathode junction. When enough carriers are generated, the device breaks into conduction, just as if a gate current were drawn. The voltage at which the device breaks over is represented by $V_{(BO)}$ and is called the *breakover voltage.*

Regardless of how the current flow starts, once it begins, it keeps on flowing. However, the diagram in Figure 6.26 also gives a clue as to how to shut off the SCR. Note the part of the curve labeled I_H. If anode current falls below this current, called the *holding current,* the device goes back into the off state.

Note that in the reverse direction, the SCR curve looks like a normal PN diode. Very little reverse current flows when the SCR is reverse-biased. This region is called the *reverse blocking region.* When the reverse breakdown voltage $V_{(BR)R}$ is reached, the device will quickly go into avalanche conduction, which usually destroys the device.

Although the most common method of turning on an SCR is gate voltage control, there are other ways to turn it on. SCRs are often turned on by radiation, particularly light. When radiation falls on the middle junction, it creates electron-hole pairs. If enough pairs are created, the device breaks into conduction. Such a device is called a *light-activated SCR* (LASCR).

The LASCR is similar in construction to the SCR, except that it has a window to let in light. The construction of the LASCR is shown in Figure 6.27A. The schematic symbol for the LASCR is shown in Figure 6.27B. Light is directed through the lens to fall on the silicon pellet. When light of great enough intensity falls on the middle PN junction, the device conducts. The LASCR is often used in industry to detect the presence of opaque objects. It is also used in cases where electrical isolation between circuits is needed.

There are two remaining methods by which SCRs are turned on. The first is called the *rate effect turnon method,* which operates in the fol-

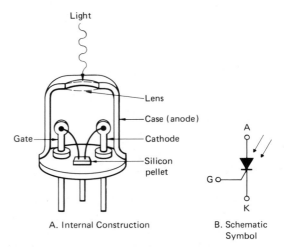

Figure 6.27   Light-Activated SCR (LASCR)

lowing manner. Recall that every transistor has PN junctions, which act as capacitors. If a quickly rising voltage is applied between the anode and gate of an SCR, charge current flows through the device. If this charge current is high enough, it can trigger the SCR into conduction. Most specification sheets specify this parameter in terms of volts per microsecond. It is called the *critical rate of rise,* or dV/dt. For example, the 2N1595 SCR has a critical rate of rise of 20 $V/\mu s$. In other words, if the anode voltage rises any faster than 20 V in 1 μs, the device turns on. The critical rate of rise also limits the maximum line frequency that can be connected to the device.

Engineers usually try to prevent rate effect turnon because the device turns on when they do not want it to. Firing by exceeding the critical rate of voltage rise is prevented by the addition of an *RC snubber circuit,* shown in Figure 6.28. The *RC* network delays the rate of the anode voltage rise.

Finally, the SCR can be turned on by temperature. As junction temperature increases, minority carrier current flow (leakage) increases. In fact, it doubles every 14°F (8°C). If

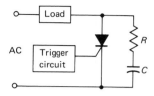

Figure 6.28 SCR Circuit with Decrease dV/dT

the temperature rises high enough, this increased current is enough to turn on the SCR. This thermal effect is usually prevented by using a heat sink.

Another problem with SCRs and temperature is the temperature dependence of some of the SCR parameters. Specifically, V_{GT}, I_{GT}, and $V_{(BO)}$ all decrease with increasing temperature. This decrease will cause a change in firing angle with temperature fluctuations in circuits where the firing angle depends on these parameters.

Recently, Motorola, the semiconductor manufacturer, made available a new type of SCR, the MOS SCR. The proposed schematic is shown in Figure 6.29. This device has the advantage of the high input impedance and fast turnon of the MOS family of devices with the switchlike action of an SCR. Because of the high input impedance and low gate current drain, the MOS SCR can be driven by standard CMOS and TTL logic. Figure 6.29 shows the MOS SCR directly controlled by a logic gate.

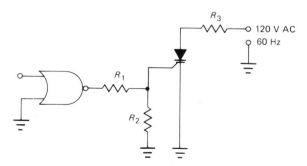

Figure 6.29 TTL Driving MOS SCR

Note the complete isolation between the 120 V AC line and the logic gate.

Another MOS device, called the *MOS-controlled thyristor* (MCT), has been developed by General Electric. Not only does this device have the high input impedance of the MOS devices, but it can be turned off easily with a gate pulse in a DC circuit. Prototype devices indicate that a 15 V gate signal will be required to turn off an MCT with a current density of 1000 A/cm^2.

Heat Sink for the SCR. The reliability and behavior of electronic equipment often depend on how well the semiconductor devices within are cooled. One manufacturer estimates that the life of a piece of electronic equipment is cut in half for every 10°C rise in temperature. Although possibly overstated, it is true, nevertheless, that a semiconductor's life and reliability are affected by its operating temperature. Also, the efficiency of a device is inversely proportional to the efficiency of the circuit. With the same input power, an increase in temperature means that less power is being delivered to the load and more wasted on generating heat. This will lower efficiency. For these two reasons, excessive heat should be removed from semiconductor devices.

Most SCRs are devices that handle large amounts of power. Consequently, the case of the SCR is often not large enough to dissipate the heat generated by the thyristor. When the case of the semiconductor is not effective in carrying away the heat from the junction, a heat sink is used. A *heat sink* is a device that, when attached to the semiconductor, will aid the device in getting rid of the heat it generates.

The maximum power that the thyristor can dissipate safely is a function of (1) the temperature at its junction, (2) the ambient (air) temperature, and (3) the size of its thermal resistance, R_θ. *Thermal resistance* is defined as the quotient of the temperature drop between two points and the power passing between them.

The temperature drop usually is measured between the point where heat is being generated and some other point of reference. In the semiconductor the heat usually is generated at the junction. The reference points are the case, the heat sink, and the ambient temperature. Thermal resistance is, then, a measurement of how well the device will conduct heat away from the junction. Its units of measure are degrees Celsius per watt (°C/W).

Power semiconductors and heat sinks are rated thermally by a statement of their thermal resistance. We will need to find out the correct size of heat sink to cool a semiconductor adequately for a particular application. Although we will be discussing the application of the heat sink with reference to the SCR, these principles apply to the cooling of any semiconductor device, for example, transistors, power MOS-FETs, and voltage regulators.

A diagram of a semiconductor with a heat sink and three thermal resistances is shown in Figure 6.30A. These resistances are considered to be in series, like electrical resistors. In fact, the behavior of the thermal resistances may be understood best by drawing the thermal circuit as an electrical one, as shown in Figure 6.30B.

1. *Junction-to-Case Thermal Resistance:* The first thermal resistance is between the junction and the case, symbolized by $R_{\theta JC}$. The heat created by the device starts its journey to the outside world at the junction. This thermal resistance is a function of the manufacture and design of the semiconductor itself and usually is reported in the manufacturer's data sheet. It cannot be changed or influenced by any factor, including heat sinks. Since some semiconductors come in different packages, it may be possible to obtain a package with a higher or lower thermal resistance. Table 6.1 shows some typical thermal resistances for selected semiconductor packages.

2. *Case-to-Sink Thermal Resistance:* The next thermal resistance the heat generated at

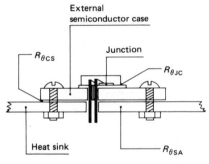

A. Semiconductor with Heat Sink

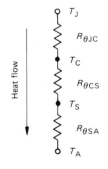

B. Thermal Resistance Equivalent Circuit

Figure 6.30   Semiconductor with Heat Sink

Table 6.1   Thermal Resistances for Selected Semiconductor Packages

Package	$R_{\theta JC}$ (°C/W)
TO-3	0.5–6.0
TO-3 (typical 2N3055)	1.5
TO-66	1.5–15
TO-66 (typical 2N3054)	7.0
TO-220	1.7–5
TO-220 (typical D44C8)	4.2
TO-92	175–200
TO-92 (typical 2N2222)	175
TO-5 (metal header)	20–40
TO-5 (typical 2N1595)	25
TO-5 (glass header)	30–50
TO-5 (typical 2N2322)	32

the junction encounters is between the case and the heat sink. This thermal resistance is symbolized by $R_{\theta CS}$. Unlike the junction-to-case thermal resistance, the case-to-sink thermal resistance can be controlled to some degree. The two major factors that affect this resistance are thermal grease and the presence of an electrical insulator. The thermal resistance across any two metals is a function of (1) the cross-sectional area between the two, (2) the condition of the surfaces, (3) how flat the surfaces are, (4) how much force is pressing the surfaces together, and (5) the thermal conductivity of the medium between the two metals.

To reduce the thermal resistance between the case and sink (thus being able to pass heat better), four factors must be kept in mind. First, keep the surfaces, in this instance the sink and semiconductor, as smooth and flat as possible. Second, keep the surface areas between the two as high as possible. Third, use thermal joint compound to keep the thermal conductivity of the medium between the surfaces as great as possible. *Thermal grease,* usually made of a silicon compound, fills the small voids that occur between surfaces that are not perfectly smooth. The thermal grease has a higher thermal conductivity (lower thermal resistance) than the air that would normally be filling the spaces between the metals. Fourth, where bolts, screws, or studs are used for mounting the sink, apply tightening torques according to the manufacturer's recommendations. Increased tightness beyond the recommended torque will increase thermal conductivity. Do not overtighten screws or bolts; stress limits may be exceeded and cause damage.

Sometimes electrical insulators are needed between the semiconductors and heat sinks. These insulators are necessary since collector or anode voltages often are present on the case. Bringing the case into electrical contact with the heat sink may cause a short circuit, since the sink is usually at ground potential. An appropriate insulator should be chosen, that is, one with the correct insulating properties and

the lowest thermal resistance. Table 6.2 shows the thermal resistances of certain selected insulators used with three semiconductor packages. Note that the thermal resistances given are based on the use of thermal grease and the proper tightening of screws and bolts.

The method most overlooked by technicians in reducing case-to-sink thermal resistance is the proper torquing of mechanical hardware. Exerting proper torque is the least expensive and most efficient method to reduce this thermal resistance. The thermal resistance of a TO-3 package may range from a low of 0.1°C/W with no insulator or thermal grease to 2°C/W with an insulator mounted dry. A semiconductor dissipating 15 W with a TO-3 case is 29°C hotter in the latter case (insulator mounted dry), a considerable temperature difference.

3. *Sink-to-Ambient Thermal Resistance:* The last thermal resistance in our heat transfer example is the resistance from the heat sink surface to the ambient (usually air). This resistance is symbolized by the designation $R_{\theta SA}$. It is probably the most important thermal resistance of the three in keeping the semiconductor junction cool. The smaller this thermal resistance, the cooler the device will be under

Table 6.2 Thermal Resistances of Insulators

Insulator	Package	R_θ (°C/W)
Beryllium oxide	TO-3	0.22
	TO-220	1.4
Mica	TO-3	0.8
	TO-220	5.2
Plastic	TO-3	0.8
	TO-220	5.2
Silicone rubber	TO-3	1.2
	TO-220	7.9
Bare package	TO-3	0.15–0.25
	TO-220	0.55–0.65

Note: All measurements taken with thermal grease in use except for silicone rubber.

steady-state conditions. Or, another way to put it, the lower $R_{\theta SA}$ is, the more power the device can dissipate without exceeding maximum junction temperatures. This thermal resistance may be expressed as a function of several parameters. A discussion of each follows.

First, the sink-to-ambient thermal resistance is a function of the convection heat transfer coefficient. This parameter is affected by the amount of air moving past the surface of the sink. If the air is still at the surface, this condition is called natural convection. More heat can be carried away if air is forced to move past the surface. The amount and type of convection change the heat transfer coefficient. Generally, the larger the amount of airflow, the larger the heat transfer coefficient, and the lower the thermal resistance.

Another factor that affects the sink-to-air thermal resistance is emissivity. You know from your studies in physics that a blackbody absorbs and gives off heat better than any other substance. For this reason heat sinks are usually made with substances that have good emissivity. Some substances, such as lacquer and oil paint, approximate the blackbody standard and improve the heat transfer coefficient. The factor called emissivity, usually symbolized by the Greek letter epsilon (ϵ), is the ratio of the emissive power of a body compared to the emissive power of a standard blackbody. The emissivity of a standard blackbody is equal to one. Materials that do not emit as much heat have an emissivity less than one. So, emissivity is a ratio, usually given in decimal form, between zero and one. For example, lacquer of any color has an emissivity of 0.8 to 0.95, and oil paint of any color has an emissivity of 0.92 to 0.96. You should be aware that the term blackbody has little to do with the optical color of a material. Bodies of any color can have high emissivities and be referred to as blackbodies. For example, anodized aluminum has an emissivity of 0.8, regardless of its optical color. Emissivity is affected more by the type of sur-

face. A matte or dull surface will have a higher emissivity than a shiny, specular surface.

Another factor that determines the sink-to-air thermal resistance is the area of the sink. The heat given off by the sink is directly proportional to the amount of heat sink area. The larger the amount of sink area, the more effective it is in dissipating heat. Greater surface area will lower the sink-to-ambient thermal resistance. In most applications involving natural convection, the only way to decrease the sink-to-ambient thermal resistance is to increase the surface area of the heat sink.

Looking back at the drawing of the thermal system shown in Figure 6.30B, we can see that as the heat flows from the junction, it encounters the three resistances just discussed: the junction-to-case thermal resistance, $R_{\theta JC}$; the case-to-sink thermal resistance, $R_{\theta CS}$; and the sink-to-ambient thermal resistance, $R_{\theta SA}$. As shown in the diagram these resistances are considered to be in series. We can find the total thermal resistance between the junction area and the ambient condition by adding these resistances:

$$R_{\theta JA} = R_{\theta JC} + R_{\theta CS} + R_{\theta SA}$$

The following basic equation describes the heat transfer in this system:

$$P_{D} = \frac{T_{J} - T_{A}}{R_{\theta JC} + R_{\theta CS} + R_{\theta SA}}$$

where

P_{D} = power dissipated by the semiconductor in W

T_{J} = junction temperature in °C

T_{A} = ambient temperature in °C

$R_{\theta JC}$ = junction-to-case thermal resistance

$R_{\theta CS}$ = case-to-sink thermal resistance

$R_{\theta SA}$ = sink-to-ambient thermal resistance

We can use this equation to solve for the maximum power a junction can be called upon to dissipate safely without exceeding a specified junction temperature. Let us consider an example at this point. Let us assume that we are using an SCR that happens to be in a TO-3 package. In the data sheet for this device we find it has an $R_{\theta JC}$ of 1.5°C/W. From the heat sink manufacturer's data we find an $R_{\theta SA}$ of 2.5°C/W and an $R_{\theta CS}$ of 0.3°C/W. How much power will the SCR be able to dissipate and not exceed a junction temperature of 200°C if the ambient temperature is 50°C? Using the power dissipation equation, we can calculate the resultant maximum power dissipation as follows:

$$P_D = \frac{T_J - T_A}{R_{\theta JC} + R_{\theta CS} + R_{\theta SA}}$$
$$= \frac{200°C - 50°C}{1.5°C/W + 0.3°C/W + 2.5°C/W}$$
$$= 34.9 \text{ W}$$

In this system the semiconductor will safely dissipate 34.9 W with a junction temperature of 200°C.

Rearranging this equation, we can solve for the junction temperature, given the thermal resistances, ambient temperature, and the power dissipated. Using the thermal resistances from our example and a semiconductor dissipating 20 W at an ambient temperature of 50°C, we can calculate the junction temperature:

$$T_J = (P_D R_{\theta JA}) + T_A$$
$$= (20 \text{ W} \times 4.3°C/W) + 50°C$$
$$= 136°C$$

Furthermore, we can solve for the particular size heat sink necessary for a specific application. Assume we have an application that requires a TO-220 package, which we are calling upon to dissipate 5 W. We wish to keep the junction temperature to a maximum of 150°C. In the data sheet for the TO-220 we find it has an $R_{\theta JC}$ of 3.0°C/W. With the semiconductor mounted on an insulator and using thermal

grease, we have an $R_{\theta CS}$ of 1.0°C/W. Solving the original equation for $R_{\theta SA}$, we get:

$$R_{\theta SA} = \frac{T_J - T_A}{P_D} - (R_{\theta JC} + R_{\theta CS})$$
$$= \frac{150°C - 50°C}{5 \text{ W}}$$
$$\quad - (3.0°C/W + 1.0°C/W)$$
$$= 13.2°C/W$$

We would need a heat sink with a thermal resistance of 13.2°C/W to meet this application.

Referring again to Figure 6.30A, we can see that we cannot make an actual measurement of the junction temperature to see if we are exceeding the maximum junction temperature specifications. However, we can measure the case temperature with a temperature probe. Since case temperature is a more practical measurement, some manufacturers give a maximum allowable case temperature as a function of the current drawn by the device. In the case of the SCR, the information usually is in the form of a graph of maximum allowable case temperature plotted against the on-state anode current. The case temperature may be solved for using the following equation:

$$T_C = (P_D R_{\theta CA}) + T_A$$

Let us calculate the case temperature, using the data from our example:

$$T_C = (P_D R_{\theta CA}) + T_A$$
$$= [P_D(R_{\theta CS} + R_{\theta SA})] + T_A$$
$$= [5 \text{ W}(1.0°C/W + 13.2°C/W)]$$
$$\quad + 50°C$$
$$= 121°C$$

The case temperature in this application would be 121°C. This temperature could then be compared to the maximum allowable case temperature graph to see if it is excessive.

In some instances bolting a semiconductor to the chassis of a piece of equipment may provide an acceptable heat sink. Or, if a commercial heat sink is not available, a sink may be manufactured from a flat piece of copper or aluminum. In either case the thermal resistance would need to be known. The nomogram in Figure 6.31 gives an estimate of the thermal resistance of a flat piece of vertically or horizontally mounted copper or aluminum. Let us say we have a 25 in.^{2} vertically mounted piece of copper $3/16$ in. thick. Using a straightedge, draw a vertical straight line down from the 25 on the surface area scale to find a thermal resistance of about 2°C/W on the scale for vertically mounted copper $3/16$ in. thick.

In the foregoing calculations, we have assumed a steady-state condition, with the SCR (or semiconductor) conducting all the time. As we will see in the next chapter, this assumption will not hold in certain kinds of power control circuits in AC applications. In most cases, however, the manufacturer will provide information in data sheets that will allow the estimation of the maximum case temperature under different dynamic conditions. Thermal resistance calculations in those situations where the SCR is on for only a portion of the AC cycle are difficult and complex, and are outside the scope of this text.

Installation, Mounting, and Cooling Considerations. Before installing power semiconductors, it is important to understand why so much emphasis is placed on using proper mounting techniques. To obtain maximum efficiency and greater reliability in the use of power semiconductors, the heat developed by power dissipation at the junction must be removed as fast as possible. As stated earlier there are usually three distinct obstacles in the path of this heat transfer: (1) interface and material between the semiconductor element and the semiconductor device mounting surface; (2) interface between the device mounting surface and the heat sink; and (3) transfer of heat through the heat sink to the ambient, which might be natural convection air, forced air, oil, or water. These obstacles are the thermal resistances previously discussed: (1) $R_{\theta JC}$, (2) $R_{\theta CS}$, and (3) $R_{\theta SA}$. Thermal transfer is depicted using an electrical analog in Figure 6.30B.

The $R_{\theta JC}$ generally is a function of device construction and design and is determined by

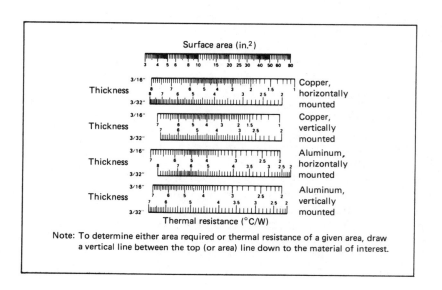

Figure 6.31 Heat Sink Nomogram

the semiconductor manufacturer. Except for disc-mount devices that require the user to apply the correct mounting force to assure the proper $R_{\theta JC}$, the user has no control over $R_{\theta JC}$ once a device has been selected. The user's only option then is to select semiconductors that offer low junction-to-case thermal resistances. The other two thermal resistances, however, vary quite a lot. The user must carefully consider the various methods and techniques available to ensure long life and efficiency.

Primary considerations in obtaining a low $R_{\theta CS}$ are the degree of flatness and surface finish of the device and heat sink mating surfaces, the use of a thermal joint compound, and the proper application of force (torque). Two suggestions worth highlighting are: (1) use a torque wrench to insure proper mounting force, and (2) use thermal joint compounds sparingly. Only a thin film of thermal joint compound is needed between the device and heat sink mounting interface. Most technicians apply too much compound. Remember when using insulating hardware that the $R_{\theta CS}$ will increase about tenfold.

The final obstacle in the heat removal path is the $R_{\theta SA}$. The semiconductor user can minimize this value by choosing the proper heat sink and most effective cooling method. A semiconductor can be no better than its heat sink. A wide range of heat sink materials, configurations, sizes, and finishes are available that will meet almost any space, cost, and heat dissipation requirement.

The most cost-effective and popular air-cooled heat sinks are made from aluminum extrusion. For higher output current at less cost, forced convection is the answer. Forced air can frequently allow the user to more than double the output current of a given device–heat sink assembly over natural convection rating capability. Why buy more semiconductors than are really needed to do the job? A few fans and some baffling may provide a substantial savings in system design costs. Use forced air in place of natural convection air cooling when possible.

Water cooling is the most efficient type of cooling in general use today. To optimize the thermal efficiency of a water-cooled heat sink, pure copper heat sinks should be used. Copper alloys (bronze, aluminum, etc.) generally have much lower thermal conductivities and thus are not as efficient. The water-cooled assembly offers from 1.5 to 4 times more output current than the air-cooled assembly. It weighs only one-third as much, occupies about 40% of the space, and costs about the same. When heat sinks are compared, therefore, water cooling (if available) is usually the most economical choice; forced convection air is next; and natural convection air is last.

A final consideration when installing power semiconductors is to place them where the surrounding ambient temperature is as low as possible. Do not mount semiconductors near or above transformers or other heat-generating components in a cabinet, and avoid "dead air" pockets. Mount semiconductor heat sinks vertically to take advantage of the natural chimney-effect or updraft of air flow. When using forced air or water to cool components in a cabinet, make sure the cooling medium reaches the power semiconductors first. Good installation practices when mounting and cooling power semiconductors will ensure good operating performance and long life.

Critical Rate of Current Rise. Some data sheets specify a parameter called *critical rate of current rise*. Critical rate of current rise is expressed in amperes per microsecond. For example, the critical rate of current rise of the 2N1595 is about 25 A/μs. If this parameter is exceeded, hot spots develop within the device, which may destroy it.

One method used to prevent current damage is shown in Figure 6.32. The inductor L in the anode lead opposes any change in current, thus slowing down the rise of anode current.

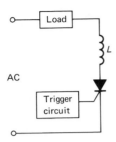

Figure 6.32   SCR Circuit with Inductor to Limit Current Rise

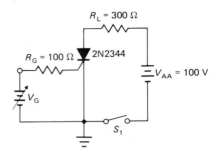

Figure 6.33   SCR in DC Circuit

Gate Triggering for the SCR.   Figure 6.33 shows a simple SCR circuit with a DC supply. Note that the SCR is forward-biased from anode to cathode.

Assume that gate voltage V_G is 0 V. If we close S_1, no current flows because of the absence of gate current I_G. What value of V_G will fire the SCR? Assume that gate turnon voltage V_{GT} is 0.6 V, and I_{GT} is 20 mA. Notice that the current I_G flows through R_G. If I_G flows through the 100 Ω resistor, it should drop about 2 V. Therefore, by Kirchhoff's law, V_G is as follows:

$$V_G = V_{GK} + V_{RG} = 0.6 \text{ V} + 2.0 \text{ V}$$
$$= 2.6 \text{ V}$$

where

V_{GK} = DC voltage from gate to cathode
V_{RG} = DC voltage drop across R_G

By this calculation, we find the gate voltage necessary to fire this device.

The next question we might ask is, "Will the SCR stay on after the gate voltage is removed?" Anode current must rise to a value called the latching current for the SCR to remain on. The *latching current* is defined as that amount of anode current necessary to keep the device on after switching from the off state and after the trigger has been removed. The latching value is normally two to three times the holding

current value. If we assume that the latching current is 300 mA, then the anode current must be more than this value if the SCR is to remain on.

Anode current (I_A) is calculated by assuming that the voltage drop across the SCR is zero (actually above 1 V). Thus, we assume that the entire supply voltage (V_{AA}) is dropped across the 300 Ω load (R_L). By Ohm's law, then, we have the following:

$$I_A = \frac{V_{AA}}{R_L} = \frac{100 \text{ V}}{300 \text{ Ω}} = 333 \text{ mA}$$

So, when the SCR fires, about 333 mA of current flows. Since this value exceeds the latching current value (300 mA), the device stays latched when the gate voltage is reduced to zero.

The gate trigger current and voltage in this example are assumed to be DC values. They may, however, be pulses with a certain pulse width. This pulse width must be long enough to allow the anode current to build up to the latching current value. The time it takes the anode current to build up to the latching current level is called *turnon time*.

In this example, we have also assumed that the load is resistive. Many times the load is inductive, such as in a relay or a motor. This inductance may keep the SCR from firing. As we know, an inductor opposes any change in current. With pulse triggering, the anode current may be kept below the latching current for the duration of the pulse. One way to overcome this problem is to make the pulse width longer, which is not always possible. Another solution is to bypass the inductive load with a resistor. This solution allows the anode current to rise quickly to the latching value.

Suppose the load in Figure 6.33 is inductive instead of resistive. How large does the resistor need to be? We can assume that 100 V is applied across the load when the SCR fires. If we divide this voltage by the latching current value, we have the minimum value of resistance needed to keep the SCR on. In this case the resistance bypassing the inductive loads needs to be at least 333 Ω.

When triggering SCRs with a gate signal, care must be taken that the maximum gate power dissipation is not exceeded. The actual gate power dissipated in a steady-state condition can be determined by multiplying the gate current drawn times the gate-to-cathode voltage. When the gate is triggered by repetitive pulses, the average gate power dissipation must be calculated. It can be found by first calculating or measuring the peak power of the triggering pulse. Once this is known, the average gate power can be determined by multiplying the peak power times the duty cycle of the pulse. You will recall that the duty cycle of a pulse is found by dividing the amount of time the pulse is on by the period of the repeating pulses.

Let us try an example. Suppose we are working with a triggering circuit for an SCR where repetitive pulses are used to trigger the SCR. The pulse repetition rate (PRR) is 200 Hz, with a pulse width of 1 ms. If the peak power in the pulse is 1 W, what is the average power dissipated by the gate? First, the period of the repeating pulses must be calculated. The pulse period (T_P) will be equal to the inverse of the PRR:

$$T_P = \frac{1}{PRR} = \frac{1}{200 \text{ Hz}} = 5 \text{ ms}$$

The duty cycle is equal to the time the pulse is on divided by the pulse period:

$$\text{duty cycle} = \frac{T_{on}}{T_P} = \frac{1 \text{ ms}}{5 \text{ ms}} = 0.2$$

The average gate power dissipation ($P_{G(av)}$) will then be the duty cycle times the peak pulse power ($P_{P(pk)}$) as follows:

$$P_{G(av)} = \text{duty cycle} \times P_{P(pk)}$$
$$= (0.2)(1 \text{ W}) = 200 \text{ mW}$$

Under these conditions, then, our gate would be dissipating 200 mW of power. This value could then be compared to the maximum average gate power dissipation of the SCR to see if it is excessive.

Turning Off the SCR. In most DC circuits the problem is not how to turn an SCR on; the problem is how to turn it off. The only way to turn an SCR off is to reduce anode current below the holding current value. In DC circuits this reduction usually is accomplished by adding a manual reset switch (S_1 in Figure 6.33). When the reset switch is opened, the blocking junction reestablishes itself.

Another common turnoff method for DC circuits is the commutation capacitor illustrated in Figure 6.34A. To simplify the schematic diagrams in this text, we represent the thyristor trigger circuits with a T at the gate. When SCR_2 is turned on by a gate pulse, anode current flows through the load and also charges C through R (SCR_1 is off). Because SCR_2 drops so little voltage in its on state, the capacitor C charges to about 100 V. SCR_2 may be turned off by triggering SCR_1 into conduction. When SCR_1 turns on, it reverse-biases SCR_2 by placing -100 V across it from anode to cathode.

Alternatively, the SCR may be turned off by turning transistor Q on, as shown in Figure 6.34B. When Q is forward-biased to saturation, it conducts current around the SCR. If the circuit is properly designed, the anode current

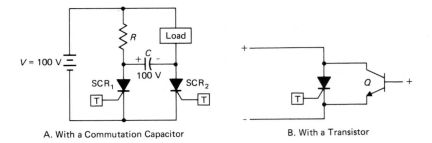

Figure 6.34 Turning Off SCR

A. With a Commutation Capacitor   B. With a Transistor

falls below the holding current, and the thyristor turns off. This method depends on the use of a power transistor with a low collector-to-emitter saturation voltage.

The problem of SCR turnoff does not occur in AC circuits. The SCR is automatically shut off during each cycle when the AC voltage across the SCR approaches zero. As zero voltage is approached, anode current falls below the holding current value. The SCR stays off throughout the entire negative AC cycle since the SCR is reverse-biased.

A simple AC application of the SCR is shown in Figure 6.35A. Here, the positive AC half-cycle is shown. As the supply goes positive, an increasing amount of gate current is drawn. The SCR blocks (between points A and B) until V_T rises to V_{GT} (point B). At point B the SCR quickly turns on and load current flows. The current through the SCR rapidly rises to a peak and decreases sinusoidally as the AC supply

voltage decreases toward zero. Finally, at point C the anode current falls below the holding current value. The SCR turns off.

Note that, as shown in Figure 6.35B, the SCR blocks current flow during the negative half-cycle. No current flows through the load at this time. Reverse-biased diode D_1 prevents potentially damaging current from flowing from the gate during the negative half-cycle. Resistor R_1 is placed in this circuit so that the total resistance cannot be reduced to a point where damaging gate current flows.

The distance from point A to B in Figure 6.35B represents the time in the positive half-cycle when the SCR is off. The distance between A and B, in degrees, is called the *firing angle* (θ_{fire}). Note that the SCR is conducting between points B and C. This angle is called the *conduction angle* (θ_{cond}). As this circuit is drawn, the firing angle can range between 10° and 90°.

Decreasing the resistance of R_2 in the cir-

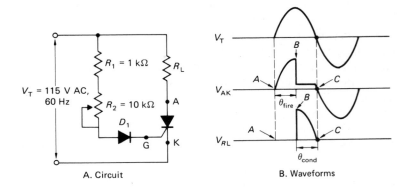

Figure 6.35 SCR in AC Circuit

A. Circuit   B. Waveforms

cuit allows more gate current to flow, thus firing the SCR earlier in the cycle. This technique applies a higher anode current to the load by increasing the average DC current and voltage to the load. In many cases, an ammeter is used to monitor the current through the load. Recall that most DC ammeters are calibrated to read average current. The following equation allows calculation of the average load voltage:

$$V_{L(av)} = \frac{V_{pk}}{2\pi}(1 + \cos\theta_{fire})$$

where

$V_{L(av)}$ = average load voltage in volts
V_{pk} = peak supply voltage in volts
θ_{fire} = firing angle in degrees

An example may help at this point. We are working with an SCR circuit with 120 V AC, 60 Hz applied, and the SCR is triggered at 45°. The load voltage is:

$$V_L = \frac{V_{pk}}{2\pi}(1 + \cos\theta_{fire})$$
$$= \frac{170}{(2)(3.14)}(1 + \cos 45°) = 46.2\ V$$

Alternatively, we can calculate the firing angle if we know R_1, R_2, I_{GT}, and V_{GT}. For I_{GT} = 10 mA, V_{GT} = 1 V, the forward-biased diode drop of D_1 at 0.6 V, and R_2 = 5 kΩ, we can calculate the supply voltage at which the SCR fires. At the firing point, I_{GT} is flowing through the gate circuit. This current drops 10 V across R_1, 50 V across R_2, 1 V across V_{GT}, and 0.6 V across D_1. These voltages total 61.6 V. In other words, the SCR will trigger when the supply reaches 61.6 V. Recall that any instantaneous voltage on a sine wave can be calculated from the following equation:

$$v = V_{pk}\sin(2\pi ft)$$

where

v = instantaneous voltage
V_{pk} = peak sine wave voltage
$2\pi ft$ = firing angle in degrees

Solving for the firing and using 162.6 V for V_{pk}, we get a firing angle of 22.3°.

An improvement to this circuit is shown in Figure 6.36. Here, a capacitor has been added between R_2 and the cathode of the SCR. The delayed gate voltage rise across the capacitor allows the SCR to fire past the 90° point. This circuit can control the SCR firing angle between about 15° and 170°. This type of AC control is called *phase control*.

Triac

After the SCR, the most commonly used thyristor is the triac. Like the SCR, the *triac* acts as a switch, with a gate that controls the switching state. Unlike the SCR, the triac can conduct in both directions. Thus, this device is a *bilateral, or bidirectional, device*. Recall that the SCR triggers only on a positive gate voltage (with respect to the cathode). The triac, on the other hand, fires on either positive or negative polarity gate voltages. The triac is superior to the SCR in that it can be used in full-wave AC control applications, such as control of AC mo-

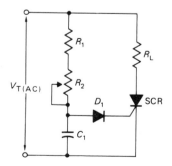

Figure 6.36   SCR Phase Control in AC Circuit

tors and AC heating systems. Triac current and voltage ratings do not yet approach those of the SCR, though.

The triac schematic symbol is shown in Figure 6.37A. Note that none of the terminals are labeled anode or cathode. The bidirectional nature of this device makes that labeling impossible. Instead, the two comparable terminals are called *main terminals* one (MT$_1$) and two (MT$_2$). The gate lead retains the same name and function as in the SCR.

The electrical behavior of the triac is depicted in Figure 6.37B. Quite simply, the triac can be thought of as two SCRs in inverse, parallel connection. When both high-current and full-wave operation are required, two inverse, parallel SCRs are used. Notice that the gate leads of both SCRs are connected together. When SCR$_1$ fires, current flows from MT$_1$ to MT$_2$. When SCR$_2$ fires, current flow is in the opposite direction. Thus, we have a bidirectional device. Its bidirectional nature is further shown in the current-voltage graph in Figure 6.38. Notice that the curve in quadrant III of the graph has the same shape as the one in quadrant I.

The triac is ideal for use in AC control circuits. Since the SCR is a unidirectional device, it has no control over half of each input cycle. The triac, however, can control current flow through a load driving both halves of the input cycle. A simple triac-controlled AC circuit is shown in Figure 6.39. The capacitor C_1 charges in either direction until $V_{(BO)}$ is reached. The triac then fires and stays on throughout the remainder of that half-cycle. The triac turns off when the AC input voltage nears zero. The main terminal current falls below the holding current.

Phase control is achieved by varying the resistance of R_2. Increasing R_2 resistance causes the capacitor to charge up more slowly, thus firing the triac later in the cycle. This increase thus delivers less energy to the load. Decreasing R_2 resistance causes the capacitor to charge up quickly, firing the thyristor earlier in the cycle. This decrease thus delivers more energy to the load.

Gate-Controlled Switch

As we have mentioned previously, the SCR and the triac can be turned off only by low-current dropout. In other words, anode (or main terminal) current must fall below the holding current. The *gate-controlled switch* (GCS), or *gate-turnoff SCR* (GTO), as it is sometimes called,

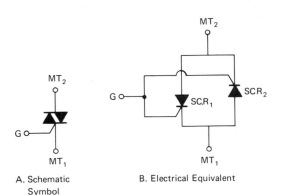

A. Schematic
Symbol

B. Electrical Equivalent

Figure 6.37   Triac

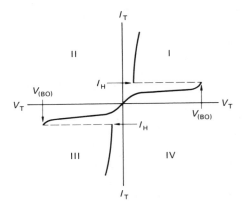

Figure 6.38   Triac Voltage-Current Characteristic Curves

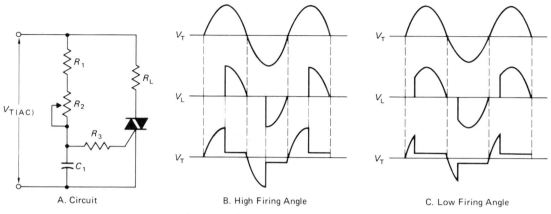

A. Circuit   B. High Firing Angle   C. Low Firing Angle

Figure 6.39 Triac in AC Circuit

turns on like a normal SCR, that is, with a gate signal. But unlike the SCR, it may be turned off by a negative trigger voltage. Positive and negative trigger spikes can be provided by differentiating a square wave and applying the signal to the gate. (Recall from Chapter 1 that a differentiated square wave is a series of alternating positive and negative voltage spikes.) The GCS schematic symbols are shown in Figure 6.40.

One problem with the GCS is that 10 to 20 times more gate current is needed to turn off the GCS than to turn it on. However, this disadvantage is more than outweighed when the GCS is used in DC circuits. At best, an additional SCR and commutation capacitor are required when SCRs are used in DC circuits (see the discussion on commutation in Chapter 3).

But the GCS may be used in DC circuits without these extra components.

The GCS finds applications in automobile ignition systems, hammer drivers in computer printers, and television horizontal deflection circuits.

Silicon-Controlled Switch

The final gate-controlled thyristor we will discuss is the *silicon-controlled switch* (SCS). Figure 6.41 gives a schematic diagram of the SCS. The SCS has two gates: an anode gate (AG) and a cathode gate (KG).

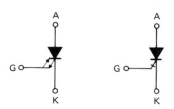

Figure 6.40 Schematic Diagrams for Gate-Controlled Switch

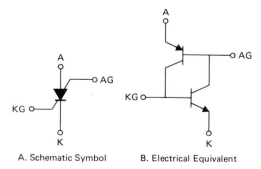

A. Schematic Symbol   B. Electrical Equivalent

Figure 6.41 Silicon-Controlled Switch

Figure 6.41B reveals an interesting fact. The electrical equivalent for Figure 6.41A is just the basic SCR circuit with both transistor bases accessible to trigger pulses. The device may be turned on with a positive gate pulse (with respect to the cathode), just as in the SCR. The anode gate exercises control over the thyristor when a negative gate voltage (with respect to the anode) is received. The SCS is used in logic applications and in counters and lamp drivers.

As a technician, you do not need to know how to design thyristors. However, you should know in a general sense how much current a thyristor is designed to carry. As an aid to understanding thyristors, we have included draw-ings, approximate case sizes, and approximate current-handling capability of the most popular case configurations in Figure 6.42.

THYRISTOR TRIGGERING

The thyristors discussed so far are used primarily as power control devices, high power in the case of the SCR and the triac, and somewhat lower power in the others. We also have seen that these devices are triggered by gate signals. The next thyristor devices to be discussed are breakover devices. In *breakover devices*, there are usually no gate structures, in contrast with

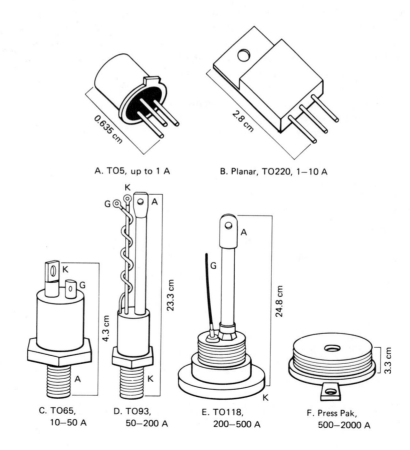

Figure 6.42 Thyristor Case Configurations

the SCR. Breakover devices are triggered by placing a voltage across the device high enough to break over the middle junction. Once the breakover occurs, the device quickly turns on by regeneration, just as in the SCR.

Shockley Diode

The *Shockley diode* (or *four-layer diode*) is a unidirectional thyristor used as a breakover device. Its structure is similar to that of the SCR without the gate, as shown in Figure 6.43A. The commonly used schematic symbol is shown in Figure 6.43B.

A circuit using the Shockley diode is illustrated in Figure 6.44. This circuit represents a relaxation oscillator. When the +15 V supply is connected, capacitor C_1 charges exponentially through the resistor R_1. When the breakover voltage is reached (5 V in this case), the capacitor discharges very quickly through the thyristor. At this time the voltage falls quickly toward zero. Near 0 V, the current through the device will fall below the holding current value. The output waveform approaches that of the sawtooth oscillator (see Figure 6.44B).

The oscillator period is calculated by using the following equation:

$$T = R_1 C_2 \ln \frac{1}{1 - (V_{(BO)}/V_{AA})}$$

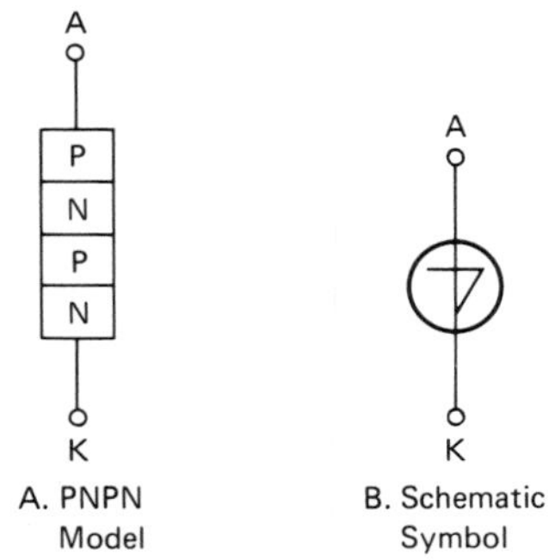

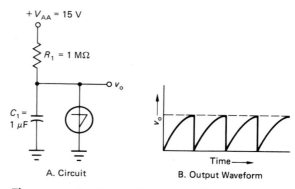

A. Circuit   B. Output Waveform

Figure 6.44 Relaxation Oscillator Using a Shockley Diode

where

T = time period in seconds
V_{AA} = voltage applied to circuit
$V_{(BO)}$ = breakover voltage of device
R_1 = resistance in megohms
C_1 = capacitance in microfarads

If we use the circuit in Figure 6.44 as an example, we have the following calculation:

$$T = R_1 C_1 \ln \frac{1}{1 - (V_{(BO)}/V_{AA})}$$
$$= (1 \text{ M}\Omega)(1 \text{ }\mu\text{F}) \ln \frac{1}{1 - (5 \text{ V}/15 \text{ V})}$$
$$= 0.405 \text{ s}$$

Since the frequency f of oscillation is the reciprocal of the period, we also have the following:

$$f = \frac{1}{T} = \frac{1}{0.405 \text{ s}} = 2.47 \text{ Hz}$$

This relaxation oscillator, then, is oscillating at 2.47 Hz.

A
P
N
P
N
K

A. PNPN
Model

A
K

B. Schematic
Symbol

Figure 6.43 Shockley Diode

Diac

We have seen that the thyristor family contains a unilateral breakover device called the Shockley diode. The *diac* is the bidirectional part of the family. Like the Shockley diode, the diac does not have a gate. The only way to get it to conduct is by exceeding the breakover voltage. The diac schematic symbol is shown in Figure 6.45. Notice in Figure 6.45B that, electrically, the diac is similar to the triac, with no gate leads available. The diac characteristic curve is shown in Figure 6.46.

The diac is used most often to trigger a triac into conduction, an application discussed in

Chapter 7. However, it may also be used in the same relaxation oscillator circuit as the Shockley diode.

Unijunction Transistor

Although the *unijunction transistor* (UJT) is not strictly a thyristor, we consider it here because it is often used to trigger an SCR or a triac. The UJT schematic symbol and construction are shown in Figure 6.47. The UJT's physical construction consists of an evenly doped block of N material with a portion of P material grown into its side.

Electrically, the UJT is illustrated by the circuit in Figure 6.48. The resistance between the PN junction and base B_1 is designated by r_{b1}; the resistance between the PN junction and

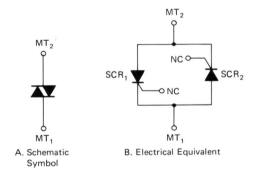

A. Schematic Symbol   B. Electrical Equivalent

Figure 6.45   Diac

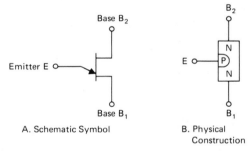

A. Schematic Symbol   B. Physical Construction

Figure 6.47   Unijunction Transistor (UJT)

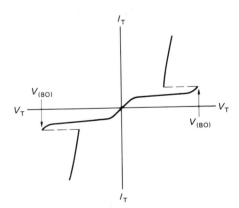

Figure 6.46   Diac Voltage-Current Characteristic Curve

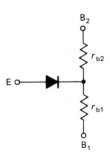

Figure 6.48   UJT Electrical Equivalent

base B_2 is called r_{b2}. The PN junction is represented by a diode symbol.

Since the UJT is used often as a relaxation oscillator, we will discuss that circuit, as shown in Figure 6.49. When supply voltage V_{BB} is applied to this circuit, the capacitor C_1 starts to charge at a rate determined by the time constant R_1C_1. Let us say that the UJT's P section is midway between base B_1 and base B_2. Since the block of N material is evenly doped, a voltage of about +10 V (with respect to ground) exists at the PN junction. The PN junction is reverse-biased until the capacitor voltage rises to about 0.6 V above the +10.6 V potential, or about +10.6 V. When this +10.6 V potential is reached across the capacitor, the PN junction becomes forward-biased. The capacitor discharges quickly through the forward-biased junction and R_3. When the capacitor voltage goes down near zero, the current flow goes below the holding current value, and the diode becomes reverse-biased again. The waveforms at the two outputs are shown in Figure 6.49B. Notice that the output taken across R_3 is a spiked waveform created by the rapid discharge of C_1 through R_3.

The period T of the oscillations (in seconds) is calculated from the following equation:

$$T = R_1C_1 \ln \frac{1}{1 - \eta}$$

where

 η = intrinsic standoff ratio

The *intrinsic standoff ratio,* electrically, is as follows:

$$\eta = \frac{r_{b1}}{r_{b1} + r_{b2}}$$

This parameter tells us basically how far the PN junction is from B_1. For instance, if the length of the N material is 10 mils, then an intrinsic standoff ratio of 0.6 means that the PN junction is 6 mils from B_1. UJTs have intrinsic standoff ratios that vary from 0.5 to 0.8.

The waveform's period (using $\eta = 0.5$) then is as follows:

$$
\begin{aligned}
T &= R_1C_1 \ln \frac{1}{1 - \eta} \\
&= (39 \text{ k}\Omega)(0.1 \ \mu\text{F}) \ln \frac{1}{1 - 0.5} \\
&= 2.7 \text{ ms}
\end{aligned}
$$

Taking the reciprocal of the period gives a frequency of 370 Hz.

The firing voltage may also be calculated. The intrinsic standoff ratio η times the supply voltage V_{BB} gives the voltage potential at the PN junction. Adding 0.6 V to this value yields the firing voltage, that is, the voltage at which

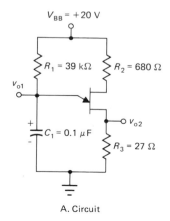

V_{BB} = +20 V

R_1 = 39 kΩ R_2 = 680 Ω

v_{o1}

C_1 = 0.1 μF v_{o2}

R_3 = 27 Ω

Figure 6.49 UJT Relaxation Oscillator

A. Circuit

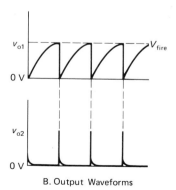

v_{o1} V_{fire}

0 V

v_{o2}

0 V

B. Output Waveforms

the PN junction is forward-biased. The equation for the firing voltage, V_{fire}, is as follows:

$$V_{fire} = \eta V_{BB} + 0.6 \text{ V}$$

In the example shown in Figure 6.49, where η is 0.5 and V_{BB} is 20 V, the PN junction is forward-biased when the capacitor charges to 10.6 V. Note also that the oscillation frequency of this circuit can be changed by varying the resistance of R_1.

Programmable Unijunction Transistor

The *programmable unijunction transistor* (PUT) is a thyristor that has a function similar to that of the UJT. It is commonly used in relaxation oscillators. Unlike the UJT, the PUT is a thyristor with a gate connected to the anode PN junction. The schematic symbol and electrical equivalent are shown in Figure 6.50. If you conclude from looking at the figure that the PUT is more like an SCR than a UJT, you are right. The PUT has an anode gate instead of a cathode gate. Thus, to trigger the PUT, we need a negative voltage on the gate with respect to the anode.

A relaxation oscillator using a PUT is shown in Figure 6.51. When supply voltage V_{BB} is applied to the circuit, capacitor C_1 charges through R_1 at a rate determined by the time constant. The R_2–R_3 voltage divider applies 10 V to the gate. When the voltage at the anode reaches 10.6 V, the PN junction is forward-biased and the PUT fires. The capacitor discharges quickly toward zero, eventually turning off the thyristor when anode current falls below holding current.

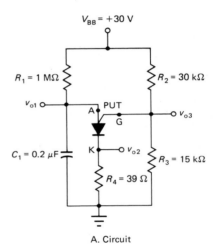

A. Circuit

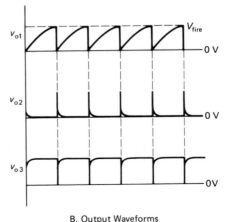

B. Output Waveforms

Figure 6.51 PUT Relaxation Oscillator

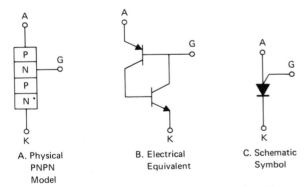

A. Physical PNPN Model

B. Electrical Equivalent

C. Schematic Symbol

Figure 6.50 Programmable Unijunction Transistor (PUT)

The functional difference between the PUT and UJT is that the PUT has an intrinsic standoff ratio that is variable or programmable, whereas the UJT does not. In Figure 6.51, the intrinsic standoff ratio is as follows:

$$\eta = \frac{R_3}{R_2 + R_3}$$

Calculations of peak firing voltage and frequency are identical to those for the UJT relaxation oscillator circuit discussed previously.

The circuit shown in Figure 6.52 is another example of a circuit using PUTs and SCRs. This circuit functions as a rear turn-signal flasher for an automobile. When S_1 is thrown, power is immediately applied to lamp 1, turning it on. At the same time power is applied to this lamp, power is applied to R_1 and C_1. C_1 will continue to charge until the anode of Q_1 is more positive than the cathode by about 0.6 V. When this potential is reached, Q_1 will fire, applying a positive voltage at the gate of SCR_1 and turning it on. When SCR_1 fires, lamp 2 will turn on, and a potential is applied to the Q_2 circuit. Capacitor C_2 charges until the anode of Q_2 is 0.6 V more positive than its anode. When the anode reaches this potential, Q_2 will fire, turning on

SCR_2, turning on the final lamp 3. After lamp 3 has been on for some time, the thermal-mechanical flasher (usually a bimetal strip) will disconnect all power from the circuit. The cycle will start all over again when the thermal-mechanical flasher unit closes the circuit.

Another circuit using SCRs and a PUT is shown in Figure 6.53. Here, the thyristors form a short-circuit-proof battery charger. This 12 V lead-acid battery charger provides an average charging current of 8 A. In addition to short circuit protection, the charger will shut down if the battery is connected incorrectly. With the battery correctly connected as shown, a small current is drawn through R_2, R_3, and R_4. A small current also charges C_1 through R_1. When the voltage on the anode of the PUT reaches the PUT firing potential, the PUT will conduct. The PUT conduction through T_2 will fire the SCR and start applying charge current to the battery. As the battery charges, its terminal voltage will increase, raising the firing potential of the PUT. The capacitor, therefore, must charge to a slightly higher voltage to fire the PUT. The firing potential on the capacitor will increase until the voltage on the capacitor starts to go above the zener diode voltage. The zener

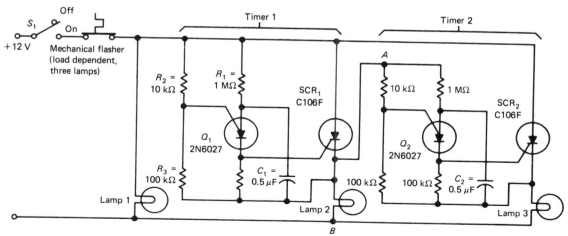

Figure 6.52 Turn-Signal Flasher

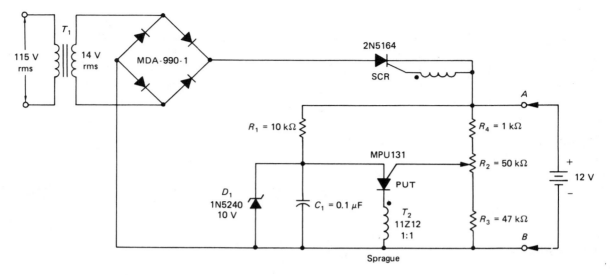

Figure 6.53 Battery Charger

diode will keep the voltage across the capacitor at 10 V, preventing the relaxation oscillator from oscillating, and charging stops. In this circuit the charging voltage can be set between 10 and 14 V, 14 V being the voltage of a fully charged, lead-acid battery. The lower charging voltage limit is set by the zener diode, while the 14 V upper limit is set by T_2 and is adjusted downward by setting R_2. This circuit is easily modified to charge lead-acid batteries with other voltages. The charging voltage may be lowered by replacing the zener diode with one that has a lower zener voltage. Charging current may be limited by adding a resistor in series with the SCR or by replacing transformer T_1 with one having a lower secondary voltage. Resistor R_4 prevents the PUT from being destroyed in the event that the wiper of R_2 is adjusted to the top.

Solid-State Relays

After performing switching tasks for several decades, the EMR is being replaced in some ap-

plications by a new type of relay, the solid-state relay (SSR). A *solid-state relay* differs from an EMR in that the SSR uses a thyristor to do the switching. An SSR has no moving parts.

A block diagram of an SSR is illustrated in Figure 6.54. A DC voltage turns on the LED. The diode turns on the phototransistor, which triggers the triac. The triggering circuit usually contains a zero-voltage switch that activates when the output voltage AC crosses zero. Zero-voltage switching is advantageous since it cuts down on electromagnetic interference (EMI), a common problem in EMRs. Well-designed SSRs have snubbers included across the output

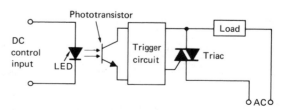

Figure 6.54 Functional Diagram of Optically Coupled SSR

terminals. You will recall from an earlier discussion that the snubber prevents false thyristor triggering caused by exceeding the critical rate of voltage rise.

SSRs are available in a great variety of shapes and sizes. Maximum currents of up to 50 A are possible at voltage ratings of 120, 240, and 480 V AC. The input voltages usually range from 3 to 32 V in the DC versions and 80 to 280 V in the AC versions.

The SSR has many advantages over the EMR. The SSR is more reliable and has a longer life because it has no moving parts. It is compatible with IC circuitry and does not generate as much EMI. The SSR is more resistant to shock and vibration, has a much faster response time, and does not exhibit contact bounce. As in the EMR, the SSR finds application in isolating a low-voltage control circuit from a high-power circuit. The degree of isolation between the control circuit and the power circuit depends on two factors: the dielectric characteristics and the distance between the source and sensor.

As shown in Figure 6.55, the SSR can be triggered by TTL or CMOS digital circuitry, including the output ports of microcomputers. In the application shown the logic gate is used in its sinking mode. The SSR triggers when the logic gate output goes low, drawing current through the LED in the SSR. The SSR parameters are slightly different when compared to the EMR parameters. The *must-operate voltage* is a parameter that specifies the minimum level of input control voltage required to change the

output from the off state to the on state. In EMR terms this would be equivalent to the pull-in voltage. The *must-release voltage* is the maximum input control voltage required to change the output from the on state to the off state.

As in every device SSRs do have some disadvantages. The SSR contains semiconductors that are susceptible to damage from transient voltage and current spikes. Such spikes are prevalent in the industrial AC line (especially on 480 V AC service), being generated by motors, solenoids, EMRs, and lightning. Some manufacturers add MOVs that add some protection to the delicate semiconductors within the SSR. (MOVs are discussed under the heading "Thyristor Protection" later in this chapter.) Unlike the EMR contacts, the thyristor has a significant on-state resistance, enough to cause a 1 V potential to be felt across the output thyristor when it is on. This may not seem like a significant voltage. Consider, however, that the power this device generates when drawing 10 A is 10 W. The heat generated by this 1 V potential must be dissipated or the output thyristor may be damaged. Other disadvantages include a significant off-state current, possibly as high as 1 mA, and damage to the device by nuclear radiation.

In most applications SSRs are used to interface between a low-voltage control circuit and a higher AC line voltage. These devices are used as switches in all kinds of industrial equipment and systems. Specifically, SSRs are used with incandescent lamps, motors, solenoids, transformers, motor starters, and industrial heaters. Recently, several manufacturers have offered I/O (input-output function) systems for computers made up of banks of SSRs. These systems usually consist of a mounting rack that contains from 4 to 32 separate plug-in SSR modules. These modules can be switched by input logic voltages of 5, 15, or 24 V DC. The outputs can be either AC or DC. A very important advantage of this type of system is that it

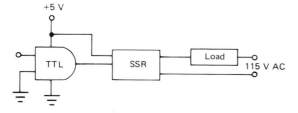

Figure 6.55 TTL Controlling an SSR

allows remote location of the SSR modules, keeping them away from the CPU. The newest generation of SSR I/O racks have an on-board microprocessor and a serial data link between the processor in the I/O rack and the processor in the computer. This system reduces wiring to a minimum, since only four wires are needed to control up to 512 relays. All relay actuation is controlled through coded messages between the two microprocessors.

Malfunctions in SSRs can be divided into three problem areas: the SSR will not turn on, it will not turn off, or it will not turn on and off reliably. Let us address each one of these situations to see how we can determine whether the problem is in the SSR or external to it. Refer to the basic functional diagram of the SSR in Figure 6.56.

If the relay fails to turn off on command, we might first disconnect the SSR and measure the control voltage across points *A* and *B*. The voltage may not be going low enough to turn off the device. Another way of saying this, using SSR terminology, is that the control voltage may be exceeding the must-release voltage mentioned previously. If this is the case, the malfunction is probably a control system malfunction, not a problem in the SSR. If the measured control voltage is lower than the SSR must-release voltage, the relay is probably damaged and should be replaced.

If the relay fails to turn on, a good starting place is to check all circuit connections. Connections should be checked for tightness and correct wiring, including proper polarity. If no problem is discovered, measurements must be taken. Again, the input voltage should be measured across points *A* and *B* on the diagram in Figure 6.56. If the input voltage is too low (under the must-operate voltage), the SSR will not turn on. If the input voltage is found to be too low, the malfunction is probably in the control system. This problem should be rectified before any further action takes place. The problem could be confirmed, if necessary, by simu-

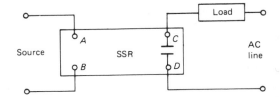

Figure 6.56   Basic SSR Functional Diagram

lating a control voltage with a special external power supply. If the SSR behaves properly, applying power to the load, the problem is definitely isolated to the input control circuit. A proper control circuit voltage, however, does not necessarily mean that the input circuit is operating properly. Since the LED at the input is basically a current device, current must be allowed to flow through the LED. Something in the input circuit may be keeping the proper current from flowing through the LED, even though proper voltage may be applied to the input. An input current measurement will confirm the proper amount of current through the LED. A lack of current indicates an open circuit in the relay input circuit and requires SSR replacement. A low current is more likely to be a control circuit problem, which should be diagnosed before continuing to use the SSR. If the current and the input voltage are normal, you should proceed to the output circuit.

With no input signal, measure across the relay terminals at points *C* and *D* (Figure 6.56). Full line voltage should be seen across the SSR output terminals. If no voltage is present, the load is probably open, preventing voltage from being felt at the output terminals of the SSR. The load should be thoroughly checked; it is possible that the load is shorted. In this case the SSR will probably be defective, due to an overcurrent situation caused by the defective load. If the load is defective, it may be checked by first removing power from the circuit. Then disconnect the output wires from the SSR and connect them together. Apply power to the circuit. The load, having the proper voltage applied to it, should behave normally. The SSR may be checked by replacing the output connections and applying the proper input signal and output power. The SSR should now control the load in a normal fashion.

TROUBLESHOOTING THYRISTORS

The most appropriate in-circuit test for thyristors is one performed with the oscilloscope. You should become familiar with the waveforms shown with each of the thyristors in this chapter. Bear in mind that most thyristors are open circuits when off and have an anode-to-cathode drop of about 1 V when on.

A simple static test for SCRs and triacs may be made with an ohmmeter connected as in Figure 6.57. If the anode to cathode is forward-biased with the ohmmeter's internal battery, the middle junction should block current flow. The ohmmeter should read infinite ohms. When lead L is connected to the gate, the middle junction will be forward-biased and the thyristor should turn on. The resistance should decrease and should stay low if the gate lead is removed.

Care is required with this method of checking SCRs and triacs. If the voltage applied by the ohmmeter to the gate-cathode junction is too high, the thyristor could be destroyed. Also, the resistance in the ohmmeter may be too

high. The thyristor may turn on but turn right off if the anode current is lower than the holding current value. The ohmmeter scale should be changed to lower the resistance.

Troubleshooting relaxation oscillators is made more difficult by the fact that circuit resistances can determine if the oscillator will work. For example, in Figure 6.49, the value of resistor R_3 is critical to circuit operation. If the UJT appears to be good, try checking the value of each of the resistors around the circuit.

Thyristor Protection

Thyristors are semiconductor devices and, as such, they are susceptible to damage by transient voltage and current spikes. Years ago the technology, being dominated by the more sturdy relay and vacuum tube, was relatively immune to this kind of damage. With the increased popularity of the more delicate semiconductor controls, transient voltage and current protection techniques are now more necessary.

Transient voltage and current surges come from many sources, but can be classified into two broad categories: internal and external. Internal sources of transients come from deenergizing a transformer primary. Opening a transformer secondary can generate extremely high transient voltage spikes as the magnetic field built up in the core collapses. The collapsing core field induces a high voltage spike in the secondary winding. In many cases spikes can be greater than 10 times the normal secondary voltage. Switches opening and closing are also known to produce high voltage transients, especially when switches close and contact bounce is present. Spikes resulting from contact bounce in relay contacts and switches can reach levels exceeding 3 kV in 120 V AC circuits.

Transients can enter a circuit externally from the commutation of motors and generators, from welding equipment, and from natu-

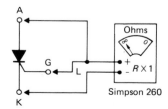

Figure 6.57 Checking SCR with Ohmmeter

ral sources such as lightning. These sources of transient voltages and currents usually enter circuits through the power supply.

Most semiconductors will be damaged or suffer a shortening of their useful life when operated with voltage and current above normal values. Relay contacts also may be damaged by the arcing across them during contact bounce. The insulation found on motors, generators, and transformers may be damaged as well by the application of transient voltage spikes. This damage may be prevented or at least reduced by the application of a special transient suppressing device called a *metal oxide varistor* (MOV).

Transient suppression is accomplished in two ways. First, transients can be suppressed by decreasing the size of the transient voltage or current. Low-pass filters placed in series with a load will attenuate high frequencies and pass the lower signal frequencies. The second method of dealing with transient voltage or current diverts the transient from delicate devices. This diversion is normally done with a voltage-clamping device. The MOV belongs under this classification. The MOV behaves in a nonlinear fashion; its impedance depends on the voltage across it. Simply put, the MOV has a high impedance to low voltage and a low impedance to high voltage. In normal circuit operation the MOV has a high impedance that does not affect circuit operation. In the presence of a high voltage spike, however, the MOV's high impedance changes to a low impedance, and the MOV clamps the voltage to a safe level.

The schematic symbols of the MOV shown in Figure 6.58 are similar in function to back-to-back zener diodes. The MOV, however, is more rugged than back-to-back zeners because it can absorb more energy. The MOV is made from zinc oxide with small amounts of cobalt, bismuth, manganese, and other metallic oxides added. The zinc oxide grains in the MOV form an interlocking matrix. The boundaries between the grains give the MOV a characteristic

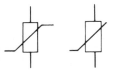

Figure 6.58 Metal Oxide Varistor (MOV) Schematic Symbols

similar to that of a PN junction. Since many boundaries exist in the material, any energy absorbed by the device is dissipated across the whole device. You will recall that in the zener diode, the energy absorbed is concentrated near the junction. The characteristics of the MOV are varied by the manufacturer by altering the size of the device. Generally, the higher the zinc oxide content, the more energy the MOV can absorb.

A good example of an application for a MOV is illustrated in Figure 6.59. During design and testing, the SCR and diodes in this motor control were destroyed when the power was removed from the primary. The manufacturer used components with voltage ratings up to 600 V without success. Placing a MOV across the secondary will clamp any transient to a safe level.

Another example of a use for a MOV is in the power supply shown in Figure 6.60. In this case the MOV is used to protect the rectifier diode from damage transmitted through the power line. The 100 mH (millihenry) choke and the 0.1 μF (microfarad) capacitor are used here as a filter to reduce RF interference. The filter will also cut in half the amplitude of any transient introduced. The MOV will reduce any remaining transients to a safe level.

CONCLUSION

In recent years more has been heard about the new IC technology than about power semicon-

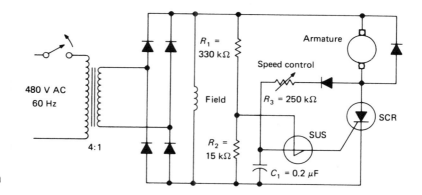

Figure 6.59 Circuit in
Need of MOV Protection

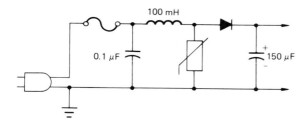

Figure 6.60 MOV Protection in Power Supply

ductor devices, such as the thyristor. However, the thyristor is still the workhorse of power control applications. Recent developments in power semiconductors have increased the voltage and current ratings to levels where vacuum tubes are virtually unnecessary. Even though the thyristor is far from an ideal switch, its usefulness can only increase in industrial applications. The entire thyristor family is summarized in tabular form in Figure 6.61.

QUESTIONS

1. The toggle switch, normally operated by hand, is an example of a _____ switch. Switches that are actuated automatically are called _____ _____ switches. A switch often used to start and stop motors by the switch's being contacted by a moving part of the system is called a _____ switch.
2. The relay, quite commonly used in industry, is an example of an _____ switch. In many applications, the use of relays has given way to the electronic switch, sometimes referred to as a _____.
3. When the thyristor is triggered, it turns on quickly, or latches, by means of _____ feedback. The SCR, the most popular thyristor, is usually triggered by a voltage on its _____.

4. The SCR is a thyristor that conducts in only one direction. Such a device is called a _____ device. The triac can conduct in both directions, so it is a _____ device.
5. The diac and Shockley diode are examples of _____ devices; they are used to turn on other thyristors.
6. The UJT has a _____ intrinsic standoff ratio that determines relaxation oscillator behavior. The PUT has an intrinsic standoff ratio that is variable or _____.
7. List the four classifications of control devices.
8. Give examples of each of the four switch classifications and their applications.
9. Describe each of the main parts of the armature type of EMR.

MAJOR GENERAL ELECTRIC SEMICONDUCTOR COMPONENTS

NAME OF DEVICE	CIRCUIT SYMBOL	COMMONLY USED JUNCTION SCHEMATIC	ELECTRICAL CHARACTERISTICS		MAJOR APPLICATIONS
GE-MOV® Varistor V series		WIRE LEAD, ELECTRODE, INTER-GRANULAR PHASE, ZINC OXIDE GRAINS, ELECTRODE, EPOXY ENCAPSULANT, WIRE LEAD		When exposed to high energy transients, the varistor impedance changes from a high standby value to a very low conducting value, thus clamping the transient voltage to a safe level.	Voltage transient protection High-voltage sensing Regulation
Diode or Rectifier 1N and A series	ANODE, CATHODE	ANODE, p, n, CATHODE	ANODE I, V_{ANODE} (-), V_{ANODE} (+)	Conducts easily in one direction, blocks in the other.	Rectification Blocking Detecting Steering
Tunnel Diode 1N3712-21	POSITIVE ELECTRODE, U, NEGATIVE ELECTRODE	POSITIVE ELECTRODE, p, n, NEGATIVE ELECTRODE	I_P, V_{ANODE} (+), I	Displays negative resistance when current exceeds peak point current I_P	UHF converter Logic circuits Microwave circuits level sensing
Back Diode BD2-7 400 Series	ANODE, CATHODE	ANODE, n, p, CATHODE	I, V_{ANODE} (-), V_{ANODE} (+)	Similar characteristics to conventional diode except very low forward voltage drop.	Microwave mixers and low-power oscillators
n-p-n Transistor	COLLECTOR, I_C, BASE, I_B, EMITTER	COLLECTOR, n, p, n, BASE, EMITTER	I_C, I_{B5}, I_{B4}, I_{B3}, I_{B2}, I_{B1}, 0, $V_{COLLECTOR}$ (+)	Constant collector current for given base drive	Amplification Switching Oscillation
p-n-p Transistor	COLLECTOR, I_C, BASE, I_B, EMITTER	COLLECTOR, p, n, p, BASE, EMITTER	$V_{COLLECTOR}$ (-), 0, I_{B1}, I_{B2}, I_{B3}, I_{B4}, I_{B5}, $I_{COLLECTOR}$ (-)	Complement to n-p-n transistor	Amplification Switching Oscillation
Unijunction Transistor (UJT) 2N2646,7 2N4871	EMITTER, BASE 2, BASE 1	BASE 2, EMITTER, p, n, I_E, BASE 1	VOLTAGE BETWEEN EMITTER & BASE 1, V_P, 0, EMITTER I_E	Unijunction emitter blocks until its voltage reaches V_P; then conducts	Interval timing Oscillation Level Detection SCR Trigger
Complementary Unijunction Transistor (CUJT) D5K1, 2	EMITTER, BASE 1, BASE 2	BASE 1, p, n, p, n, EMITTER, BASE 2	V_E, PEAK POINT, VALLEY POINT, I_E	Functional complement to UJT	High-stability timers Oscillators and level detectors
Programmable Unijunction Transistor (PUT) 2N6027, 8	ANODE, GATE, CATHODE	ANODE, p, n, p, n, GATE, CATHODE	I_A, A, G, C, VALLEY POINT, PEAK POINT, V_{AC}	Programmed by two resistors for V_P, I_P, I_V, Function equivalent to normal UJT.	Low-cost timers and Oscillators Long-period timers SCR Trigger Level Detector
Photo Transistor L14 series	COLLECTOR, BASE, I_B, EMITTER	COLLECTOR, n, p, n, BASE, EMITTER	I COLLECTOR, H4, H3, H2, H1, V_{CE}	Incident light acts as base current of the photo transistor	Tape readers Card readers Position sensor Tachometers

Figure 6.61 Thyristor Family Characteristics

MAJOR GENERAL ELECTRIC SEMICONDUCTOR COMPONENTS

NAME OF DEVICE	CIRCUIT SYMBOL	COMMONLY USED JUNCTION SCHEMATIC	ELECTRICAL CHARACTERISTICS	MAJOR APPLICATIONS
Opto Coupler (1) Transistor (H11A, H15A) (2) Darlington (H11B, H15B) Outputs			Output characteristics are identical to a normal transistor/Darlington except that the LED current (I_F) replaces base drive (I_B)	Isolated interfacing of logic systems with other logic systems. Power semiconductors and electro-mechanical devices. Solid-state relays.
Opto Coupler SCR Output (H11C)			With Anode voltage (+), the SCR can be triggered with a forward LED current. (Characteristics identical to a normal SCR except that LED current (I_F) replaces gate trigger current I_{GT}.	Isolated interfacing of logic systems with AC power switching functions. Replacement of relays, microswitches
AC Input Opto Coupler (H11AA)			Identical to a "standard" transistor cjoupler except that LED current can be of either polarity	Telecommunications – ring singnal detection, monitoring line usage. Polarity insensitive solid-state relay. Zero voltage detector.
Silicon Controlled Rectifier (SCR)	ANODE GATE CATHODE	ANODE CATHODE GATE	With anode voltage (+), SCR can be triggered by I_g, remaining in conduction until anode I is reduced to zero.	Power switching Phase control Inverters Choppers
Complementary Silicon Controlled Rectifier (CSCR)	ANODE GATE CATHODE	ANODE GATE CATHODE	Polarity complement to SCR	Ring counters Low-speed logic Lamp driver
Light-Activated SCR* L8, L9	ANODE GATE CATHODE	ANODE CATHODE GATE	Operates similar to SCR, except can also be triggered into conduction by light falling on junctions	Relay Replacement Position controls Photoelectric applications Slave flashes
Silicon Controlled Switch* (SCS) 3N83-6	ANODE CATHODE GATE ANODE GATE CATHODE	ANODE CATHODE GATE ANODE GATE CATHODE	Operates similar to SCR except can also be triggered on by a negative signal on anode-gate. Also, several other specialized modes of operation	Logic applications Counters Nixie drivers Lamp drivers
Silicon Unilateral Switch (SUS) 2N4983-90	ANODE GATE CATHODE	ANODE GATE CATHODE	Similar to SCS but zener added to anode gate to trigger device into conduction at ~8 volts. Can also be triggered by negative pulse at gate lead.	Switching Circuits Counters SCR Trigger Oscillator
Silicon Bilateral Switch (SBS) 2N4991,2	ANODE 2 GATE ANODE 1	GATE ANODE 2 ANODE 1	Symmetrical bilateral version of the SUS. Breaks down in both directions as SUS does in forward	Switching Circuits Counters TRIAC Phase Control
Triac SC92-265	ANODE 2 GATE ANODE 1	ANODE 2 GATE ANODE 1	Operates similar to SCR except can be triggered into conduction in either direction by (+) or (−) gate signal	AC switching Phase control Relay replacement
Diac Trigger ST2			When voltage reaches trigger level (about 35 volts), abruptly switches down about 10 volts.	Triac and SCR trigger Oscillator

Figure 6.61 *(Continued)*

10. List the advantages of thyristors over EMRs.
11. List the advantages of reed relays over EMRs.
12. Describe the regenerative switching action that makes a thyristor latch.
13. Draw the schematic symbols for the thyristors listed, and describe the usual method of triggering each: (a) SCR, (b) triac, (c) diac, (d) UJT, (e) PUT.
14. Describe how an LASCR is triggered.
15. Why is it necessary to have a snubber across a thyristor? How does a snubber work?
16. Distinguish between the PUT and UJT in terms of function and construction.

17. Define the intrinsic standoff ratio for a UJT.
18. List four advantages of a solid-state relay over the electromagnetic version.
19. Define critical rate of voltage and current rise. How do these parameters affect the operation of thyristors?
20. Describe how temperature affects the following parameters: (a) I_{GT}, (b) V_{GT}, and (c) $V_{(BO)}$.
21. Draw the schematic symbol of the following devices and describe how each one is triggered: (a) GCS, (b) SCS, and (c) Shockley diode.

PROBLEMS

1. Assume you have the following voltages and resistances for the circuit shown in Figure 6.33: R_L = 2 kΩ, R_G = 40 Ω, V_{AA} = 50 V, and $V_{F(ON)}$ = 1 V. The SCR used is a 2N881 with I_{GT} = 200 mA, V_{GT} = 0.8 V, and I_H = 5 mA.
   a. Will a V_G of 10 V fire the SCR?
   b. Calculate the power dissipated by the SCR, assuming that it turns on.
2. Given $R_{\theta JA}$ = 345°C/W, $R_{\theta JC}$ = 124°C/W, an ambient temperature of 40°C, and $T_{J(max)}$ = 200°C.
   a. Calculate the junction temperature when the thyristor is dissipating 300 mW without a heat sink.
   b. Calculate $R_{\theta CA}$ of the heat sink needed to dissipate 1.2 W at the junction. Assume $R_{\theta CS}$ = 1°C/W.
   c. Calculate the power dissipated at the junction with a sink having an $R_{\theta CA}$ of 30°C/W and a junction temperature of 100°C.
   d. Calculate the case temperature when the device is dissipating 1.2 W at the junction.

3. Suppose the load in Figure 6.33 is resistive. What size bypass resistor is necessary to ensure that the SCR will turn on? Assume I_L = 15 mA, $V_{F(ON)}$ = 1 V, and V_{AA} = 50 V.
4. If R_L in Figure 6.35A is 10 Ω, what is the firing angle with R_2 set at 3 kΩ? Assume I_{GT} = 200 μA and I_H = 10 mA.
5. At the firing angle calculated in Problem 4, what is the average current flowing through the load?
6. For the following values for the circuit shown in Figure 6.44A, calculate the frequency and the period of the relaxation oscillator: R_1 = 25 kΩ, C_1 = 0.1 μF, $V_{(BO)}$ = 30 V, and $+V_{AA}$ = 50 V.
7. Calculate the peak firing voltage and the oscillator frequency for the circuit shown in Figure 6.51A. Use the following values: R_1 = 220 kΩ, C_1 = 0.01 μF, R_2 = 16 kΩ, R_3 = 27 kΩ, R_4 = 22 Ω, and V_{BB} = 25 V.
8. A triggering circuit for an SCR has a pulse repetition rate of 100 Hz with a pulse width of 1 ms, and the peak power is 2 W.
   a. Calculate the average power dissipated by the device.
   b. Calculate the duty cycle.

9. An SCR circuit has 230 V AC and 60 Hz applied to the SCR. The SCR circuit has a firing angle of 25°.
   a. Calculate the average voltage at the load.
   b. What is the conduction angle?
   c. Find the average current in the load if the load resistance is 100 Ω.

10. In Figure 6.49, R_1 = 50 kΩ and η = 0.6.
   a. Calculate the frequency and period of the oscillator.
   b. Calculate the potential across the capacitor when the UJT fires.

11. In Figure 6.49, what will the value of R_1 need to be for the oscillator to oscillate at 100 Hz?

12. What will the value of R_1 need to be in Figure 6.51 to give an output frequency of 200 Hz?

13. If R_3 in Figure 6.51 is increased to 200 kΩ, what will the new output frequency be?

14. In Figure 6.52, how long a delay will occur between the time power is applied to Q_1 and the time Q_2 is triggered?

CHAPTER

7

POWER
CONTROL CIRCUITS

OBJECTIVES

On completion of this chapter, you should be able to:

- Describe how hysteresis produces the snap-on effect in phase control;
- Describe the operation and advantages of the ramp-and-pedestal circuit, zero-voltage switching, and chopper motor control;
- Describe open-loop and closed-loop motor control;
- Explain how pulse-width modulation can accomplish motor speed control;
- List and describe the methods by which an AC motor's speed may be varied;
- Describe how different levels of power are applied to a load with phase control circuitry;
- Define a converter, and describe the conversion that takes place;
- Describe how the PLL is used to control the speed of a DC motor;
- List the advantages of transistors and power MOSFETs as power control devices.

INTRODUCTION

In the preceding chapter we discussed the basic building blocks of semiconductor power control circuits, the thyristors. Now that we have provided a background in the operation of these devices, we are ready to discuss applications. Applications will fall into two basic categories: (1) general power control circuit principles, such as phase control and zero-voltage

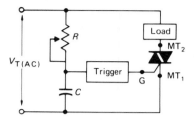

Figure 7.1 Triac Phase Control Circuit

Hysteresis in Phase Control

One major problem with phase control circuits is one that designers call the *hysteresis effect*. The hysteresis effect can be explained through the use of the triac phase control circuit in Figure 7.1. From examining this circuit, we would expect that the load (say, a lamp) would have AC power applied to it gradually. Such is not the case. In reality the lamp will turn on suddenly with moderate brightness. After the lamp comes on, its intensity can be varied from low to high levels. This *snap-on effect* is due to the hysteresis inherent in this design.

During the time when the triac is off, the capacitor charges through the variable resistor *R*. Since the capacitor is a reactive component, the capacitor voltage lags the line voltage by about 90°. When the triac fires, the capacitor discharges through the gate–main terminal 1 junction of the triac and the trigger device. During the next half cycle, the capacitor voltage will now exceed the trigger device's breakover voltage sooner, because it will start to charge from a lower voltage.

This undesirable hysteresis effect can be reduced or eliminated in several ways. Two ways of accomplishing the reduction are shown in Figure 7.2. The addition of the extra *RC* network in Figure 7.2A allows C_1 to partially recharge C_2 after the diac has fired, thus reducing hysteresis. This addition also adds a greater range of phase shift across C_2.

The second method of reducing hysteresis, illustrated in Figure 7.2B, uses a device called an asymmetrical trigger diode. This diode has a

switching, and (2) motor control. Motor control will be broken down further into AC and DC motor control. Since motor control is one of the major applications of thyristors in industry, we will use most of this chapter to discuss this important subject.

As a technician, you will need to be able to analyze power control circuits. Consequently, in our presentation we not only show simplified schematic diagrams, but we also give an idea of what the voltage waveforms should look like around the circuit.

PHASE CONTROL

In the previous chapter, we discussed AC phase control circuits using SCRs and triacs as power control devices. Examples of these circuits are shown in Figures 6.36 and 6.39. These circuits work well but do have some problems. In this section we will show circuits that overcome some of the problems inherent in the circuits presented in Chapter 6.

Figure 7.2 Reducing Hysteresis

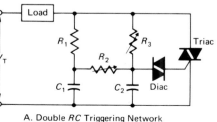

A. Double *RC* Triggering Network

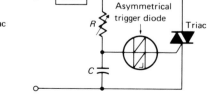

B. Asymmetrical Trigger Diode

greater breakover voltage in one direction than in the other. It is so made that when it triggers for the first time, the capacitor discharges into the triac gate. During the next half cycle, however, the triggering voltage is equal to the original breakover voltage plus the capacitor's voltage decrease. This result allows the capacitor voltage to have the same time relationship to the applied voltage. Thus hysteresis is reduced.

The circuits shown in Figure 7.2 can be used in virtually any AC low-power control circuits, such as universal and induction motors, lamps, and heaters. In fact, one designer calls this circuit the universal power controller.

UJT Phase Control

As discussed in Chapter 6, when thyristors are used in power control circuits, the thyristor is fired at some variable phase angle. But if we want maximum power delivered to a load, the thyristor must fire as soon as the AC voltage across it goes positive (and/or negative in the triac). The thyristor must conduct for 180° after firing for maximum power. If less than full power is needed, the thyristor firing must be delayed past 0° or 180°. A control circuit is needed, therefore, to control the firing angle between 0° and 180° for the SCR and also between 180° and 360° for the triac.

We have seen that the RC lag network can accomplish this control. These circuits, however, are affected by loading when connected to thyristor gate leads. They also are affected by supply voltage variations. The UJT oscillator, on the other hand, has none of the disadvantages of the RC networks. Thus, the UJT can be used as a phase control device in control circuits using either triacs or SCRs.

Figures 7.3A and 7.3B show a UJT oscillator controlling both an SCR circuit and a triac circuit. Both circuits function in essentially the same way. The thyristor is in the off state until the UJT fires. In Figure 7.3A, when the UJT fires, the capacitor will discharge quickly through the resistor R_4. This positive pulse will turn on the SCR. The SCR will remain on until the line voltage approaches zero. At this point, the SCR will turn off and remain off throughout the entire negative half cycle.

The triac circuit, Figure 7.3B, works in the same way. The positive spike is coupled to the triac's gate by the pulse transformer. When the triac turns on, the voltage to the control section of the circuit is taken away. No more

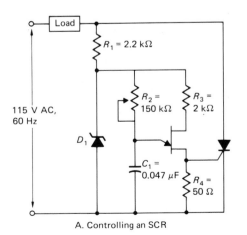

A. Controlling an SCR

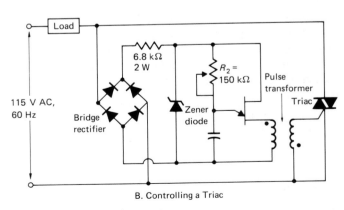

B. Controlling a Triac

Figure 7.3 UJT Oscillator Control Circuits

pulses can occur during that half cycle of the input voltage. Since the line voltage is rectified by the bridge, the operation is the same for both positive and negative half cycles. This circuit can control up to 97% of the power available to the load.

Ramp-and-Pedestal Phase Control

Most of the preceding phase control circuits work well in manually controlled applications. But signals in automatic control situations are generally not large enough to produce adequate control. We might say that these circuits have relatively low gain. That is, in the RC networks used in phase control, a relatively large change in resistance is needed.

An alternative to the RC network is the *ramp-and-pedestal control circuit,* which provides much higher gain. Such a control circuit is illustrated in Figure 7.4A. When the line voltage goes positive, the zener diode keeps point A at a positive 20 V. The capacitor C_1 charges quickly through diode D_2, R_2, and R_1. Let us say that resistor R_3 is adjusted to 3.3 kΩ. With this resistance, R_3 will drop about 10 V. The capacitor will continue to charge until the voltage at point C reverse-biases the diode D_1, which will occur at about 10 V. The capacitor, which was charging through about 8.3 kΩ, will now charge

much more slowly through the 5 MΩ resistance of R_4. Eventually, the voltage at point C will reach the firing potential of Q_1. The UJT will then turn on and fire the SCR.

The voltage rise at point C is diagramed in Figure 7.4B. Note that the voltage rises very quickly between points 1 and 2 (the pedestal). At point 2, diode D_1 becomes reverse-biased and charges more slowly from point 2 to 3 (the ramp). At point 3 the UJT will fire, turning on the SCR. If we increase the resistance of R_3, the capacitor will charge quickly to a higher voltage before D_1 becomes reverse-biased. This higher voltage is represented by point 4. Note that since the capacitor started charging slowly from a higher voltage, it will reach the firing potential sooner in the input cycle. Thus, more power will be delivered to the load.

IC Ramp-and-Pedestal Phase Control

The simple ramp-and-pedestal circuit we have just discussed can be improved by adding more components. Of course, these extra components add more cost as well as improved performance. Since one of the advantages of ICs is their ability to replace large and sometimes costly collections of discrete devices, ICs are well suited to this application.

Several manufacturers make an IC that

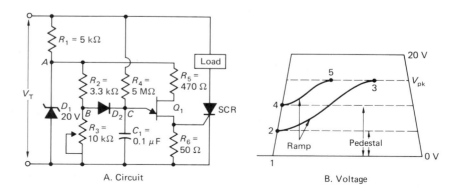

Figure 7.4 Discrete-Component, Ramp-and-Pedestal Control

A. Circuit

B. Voltage

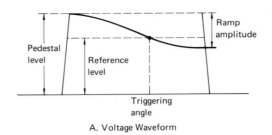

A. Voltage Waveform

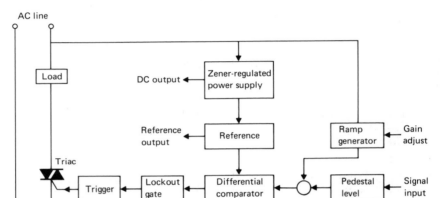

Figure 7.5 IC Ramp-and-Pedestal Control

B. Block Diagram

does the phase control task we discussed. One such IC is the PA436 phase controller manufactured by General Electric. Figure 7.5A shows the ramp-and-pedestal waveform used by the PA436. Note the difference between this waveform and the UJT waveform (Figure 7.4B). The UJT ramp-and-pedestal has a positive-going ramp, while the PA436 has a negative-going cosine ramp and a positive pedestal and reference. (A *cosine ramp* is a waveform whose average DC level is changing at an inverse sine rate.) In the UJT circuit the reference, or firing, voltage remains constant and the pedestal changes. In the PA436 system the pedestal remains constant and the reference changes.

The block diagram of the PA436 is illustrated in Figure 7.5B. The input signal is a DC voltage that establishes the pedestal level. The cosine ramp is developed from the supply voltage and is adjustable externally by the gain ad-

just. The ramp-and-pedestal waveform is then compared to a reference waveform by the differential comparator. The differential comparator produces an output waveform only when the ramp is below the reference level. The lockout gate keeps the differential comparator signal from reaching the trigger circuit until the AC supply voltage passes through zero.

ZERO-VOLTAGE SWITCHING

Several problems arise in the design of thyristor power control circuits such as the ones we have described previously. Opening and closing switches with applied voltages cause large variations in line current during short periods of time. These rapid current variations produce unwanted RF interference and potentially dam-

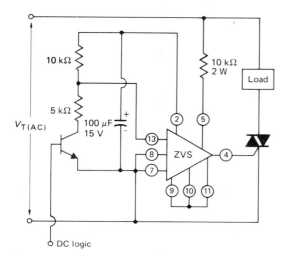

Figure 7.6 IC Zero-Voltage Switch (ZVS) Controlling a Triac

One such circuit is the CA3058 manufactured by RCA. This device is a monolithic IC designed to act as a triggering circuit for a thyristor. Internally, the CA3058 contains a threshold detector, a diode limiter, and a Darlington output driver. Together, these circuits provide the basic switching action. The DC operating voltages for the device are provided by an internal power supply. This power supply is large enough to supply external components, such as transistors and other ICs.

An example of a circuit using this IC zero-voltage switch (ZVS) is shown in Figure 7.6. Note that the ZVS output is connected directly to the gate of the thyristor. Whenever the DC logic level is high, the output to the gate of the triac is disabled. A low-level input enables the gate pulses, thus turning on the triac and providing current to the load.

Since the ZVS turns on only at or near the zero-voltage point of the applied voltage, only complete half or full cycles are applied to the load. This point is illustrated in Figure 7.7. The amount of power delivered to the load depends on how many full or half cycles are applied during a particular time. The average power delivered to the load, then, depends on the ratio of how long the thyristor was on to how long the thyristor was off.

Zero-voltage switching, while suitable for controlling heating elements, is not very useful

aging inductive-kick effects. (Recall that an inductor with current flowing through it will produce a large magnetic field around it. The field will collapse when current stops, generating a very high voltage.)

One solution to this problem is to make sure that the thyristor fires only when the supply is at or near 0 V. With no voltage across the switch, no current should flow. Several manufacturers provide ICs that accomplish this type of switching, called *zero-voltage switching*.

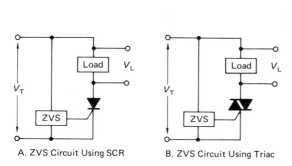

A. ZVS Circuit Using SCR   B. ZVS Circuit Using Triac

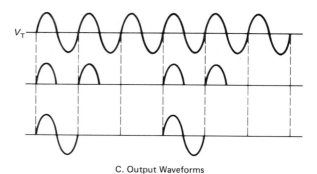

C. Output Waveforms

Figure 7.7 ZVS Applying Voltage to Load

in controlling the speed of motors. Motors have a tendency to slow down during the time that power is applied to the circuit, especially under heavy loads. This factor is not bothersome in heating elements because of their long time constant.

The IC ZVS is a very versatile device. It is used in industry in relay and valve control applications as well as in heating and lighting controls.

ELECTRICAL MOTOR CONTROLS

As we saw in the previous chapter, motors can be controlled manually by special manual switches and magnetically by special contactors. When manual switches are used to start a motor, the device is called a manual motor starter. Manual motor starters usually have overload protection for the motor as well as the switch to apply power to the motor. When motors cannot be operated by hand or when they must be operated by other switches, a special contactor called a magnetic motor starter is used. Circuits using manual motor starters generally are not complicated or difficult to understand. Magnetic motor starters, on the other hand, can be quite complex. In this section we will discuss several common circuits that use magnetic motor starters.

Before we examine these circuits, it is necessary to understand the special types of diagrams we will be using in the text. Indeed, these same types of diagrams are widely used in industry. Two of these types of diagrams are illustrated in Figure 7.8: the wiring diagram and the simplified diagram.

Wiring diagrams, like the one shown in Figure 7.8A, include all of the devices in the system and show their physical relation to each other. All poles, terminals, coils, and so on, are shown in their proper place on each device. If

you are involved in wiring a motor starter, you will find this type of diagram helpful, since connections can be made exactly as they are shown on the diagram. The wiring diagram, however, is not very useful in explaining how the system works. The connections are not made in a way that is easy to follow.

If the wiring diagram is rearranged and simplified, the result is the line diagram (Figure 7.8B). The *line diagram,* also referred to as a *simplified,* or *ladder,* diagram, is a representation of the system showing everything in the simplest way. No attempt is made to show the various devices in their actual positions. All the control devices are shown between vertical lines. The lines represent the source of power, either AC or DC. Circuits are shown connected as directly as possible from one of these lines to the other. All connections are made in such a way that the functions of the various devices can be easily traced. Note that the same terminal identification letters and numbers are used in both the wiring and line diagrams. They designate the control and power connections. For troubleshooting, it is much easier to work with a line diagram than a wiring diagram.

The line diagram in Figure 7.8B illustrates a circuit called a three-wire circuit with low-voltage protection. When the start button is pushed, current flows through the start switch, through the motor starter coil labelled *M*, through the overload relay contacts, and back to the other side of the line. All relay contacts associated with the M coil change state. This means that all NO switches close and all NC switches open. Note that the three NO contacts that supply power to the motor close. Power is then applied to the motor. When the start button is pushed and current flows through the M coil, the contacts around the start switch will close. Current will then flow through the circuit even when the start switch is released. This configuration is called a *latching relay circuit.* This circuit is protected against a low-voltage or a voltage loss condition. If the supply voltage

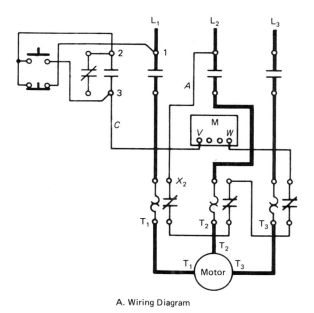

A. Wiring Diagram

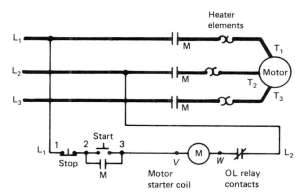

B. Simplified Diagram

Figure 7.8 Motor Starter Diagrams

goes low or is lost, the contactor will open. When proper line voltage is restored, the operator must press the start button to energize the contactor. The contactor can be de-energized by pressing the stop button. The contacts will then go back to their normal state, removing power from the motor.

As mentioned in our discussion of AC motors, the three-phase, squirrel-cage AC motor is well suited to reversing rotation. This reversal is accomplished by switching any two of the three line conductors. Reversing starters usually use two separate contactors, one for forward rotations and the other for reverse. The reversing starter is electrically and mechanically interlocked so both contactors cannot be energized at the same time and cause a short circuit.

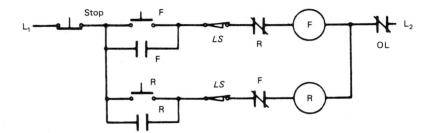

Figure 7.9 Reversing
Starter Circuit

A *reversing circuit* is shown in Figure 7.9. In this circuit the motor is brought to a stop before the direction of rotation is changed. Pressing the forward push button will energize the forward contactor coil (F), causing the motor to rotate in the forward direction. At the same time, the F coil opens the normally closed F contact in the reverse contactor coil circuit and closes the normally open F contact around the forward push button. As long as the F contactor is energized, pressing the reverse push button will have no effect, because the F contact is open in the reverse coil circuit. The operator must press the stop button before the direction of rotation can be changed. When the forward contactor drops out, the NC contact F in the reverse coil circuit is reclosed. The rotation of the motor now can be started in the reverse direction. The limit switches (*LS*) are shown in this circuit since they are sometimes used for equipment such as overhead doors, which are stopped with a limit switch at the end of the door travel.

Some motors must have the capability to inch or jog. A motor is said to be *jogged* when power is applied briefly in controlled bursts. Jogging motors are necessary when motor loads need to be precisely positioned prior to starting the motor. A jogging circuit applies power to the motor only when a push button is held down. The motor should stop when the push button is released. The circuit shown in Figure 7.10 is such a circuit. When the jog button is pressed, current flows through the M coil, energizing it. Although not shown, the M coil will have NO contacts that, when closed, will apply power to the motor. Note that when the jog push button is released, power is removed from the M coil and any NO contacts will go back open. The M coil is not in a latching relay configuration. When the start button is pushed, current flows through the control relay (CR) coil. Both CR contacts close, one latching the CR coil and the other applying power to the M contactor coil. The M contactor coil will remain energized as long as current flows through the CR coil.

These three motor starter applications have been based on AC motors. DC motors use similar circuits, such as the one in Figure 7.11. This circuit uses a resistor as its basic control mechanism. As you will recall from Chapter 3 on DC motors, a large inrush of current occurs at starting, due to the low cemf. If a resistor is placed in series with the armature, the voltage and current will be decreased at starting. The resistor, however, must be removed after the motor has picked up speed. M and A_1 in Figure 7.11 are

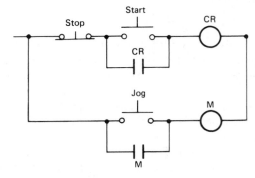

Figure 7.10 Jogging Circuit

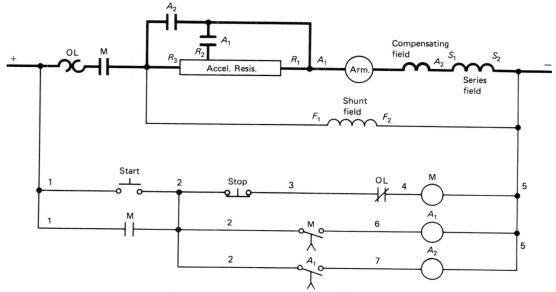

Figure 7.11 DC Motor Starter Circuit

TDRs. When the start push button is pressed, the M contacts close immediately, applying supply voltage across the full acceleration resistance and the field. After a certain delay the A_1 contacts close, shorting out part of the acceleration resistance. Of course, by the time that M times out, the motor has accelerated to a speed short of its full speed. After another time delay the motor will have reached its final speed. At this time A_1 times out, and A_1 contacts close, applying power to A_2. Since A_2 is a normal control relay, A_2 contacts close immediately, shorting out the remainder of the acceleration resistance. The motor has accelerated to full speed with an adjustable resistance, which lowers the armature current and torque at starting.

DC MOTOR CONTROL

In the process of converting factories to automatic controls, engineers are faced with a bewildering variety of new, high-tech motor controls. In some cases the choice is between old, reliable, well-understood techniques and newer, less-understood, and possibly less reliable techniques. For many years the choice was simple. If you wanted a variable-speed motor drive, you chose a DC motor and a Ward Leonard drive control system. Invented in the 1890s and patented by H. Ward Leonard, the Ward Leonard system held sway in motor control until the late 1950s, when other types of control became prominent.

Even after the introduction of thyristor-controlled motors, however, motor controls were still constructed component by component to control field current and armature voltage. Today, everything needed to control the speed of DC motors is in one easy-to-install package. All rectification and control functions are self-contained in the DC drive. There is no need to assemble a drive component by component. The only connections necessary are to an input source of AC power, either single- or three-phase, and an output to field and armature leads. There can be little justification today for installing anything other than a modern DC drive.

DC motors are widely used in modern industry for many reasons. Speed is relatively easily controlled with precision in DC motors. Furthermore, speed can be controlled continuously over a wide range, both above and below the motor's rated speed. Many industrial applications call for a machine that has very high starting torque, such as those found in traction systems. The DC motor satisfies most of these requirements adequately. Compared with AC drives, DC drives are generally less expensive, an important advantage in keeping initial equipment costs down.

As we have seen in this chapter, the thyristor is the major power control element in power controllers. And the SCR is the thyristor most often used in DC motor control applications. SCRs are used in two types of DC motor control systems: phase control circuits and chopper circuits. The phase control circuits, as we have discussed, control power by changing the firing angle (α). When the SCR fires early in the cycle, corresponding to a small α, more average power is delivered to the load. As firing angle increases, the SCR fires later in the cycle, delivering less power to the load.

Some DC motors have power supplies that are also DC. A good example of this application is the traction motor found in industrial delivery vehicles and forklifts. In the past, variable-resistance control was used in these situations. But such control is very inefficient. Today, the electronic chopper is used to control machines with DC power sources. A *chopper* is a thyristor switching system in which an SCR is turned on and off at variable intervals, producing a pulsating DC. The longer the SCR is on, or the higher the switching frequency, the more power is delivered to the load.

√ Phase Control

Many small industrial plants, farms, and homes are supplied with only single-phase power. DC motor speed control systems in these cases depend on the rectifying capabilities of the SCR to change the AC into the DC needed to develop torque in the motor. Because the reverse-blocking thyristor converts AC to DC, this circuit is often referred to as a *converter.*

Since in these cases the thyristors are supplied with AC voltage, commutation of the thyristor is not a problem. The SCR naturally commutates when the line voltage passes near 0 V. In addition to natural commutation with AC line voltages, reverse-blocking, thyristor converter circuits are very efficient. Because of the low voltage drop across the device, as much as 95% of available power can be delivered to the load.

Basic Converter. The simplest of the converters, the *half-wave converter,* is illustrated in Figure 7.12A. During the time that the supply goes negative, SCR$_1$ blocks current flow. As a result, no power is applied to the motor (M). When the input goes positive, SCR$_1$ is forward-biased. It will conduct when a positive trigger pulse is applied to the gate (T represents the trigger circuit in the figure). The SCR will remain in conduction until the supply goes negative.

When SCR$_1$ conducts, it applies power to the motor. Diode D_1 is a flywheel, or freewheeling, diode. After current flows through the armature and the SCR turns off, the magnetic field built up around the armature will collapse. The field reverses polarity and conducts through the diode. The collapsing field supplies energy to the load during the time the SCR is off.

As shown in the armature voltage (V_A) waveform (Figure 7.12C), the half-wave converter has a large amount of ripple. As a result, this form of control is not used very much in DC motor speed control.

Half-Controlled Bridge Converter. Figure 7.12B shows a more useful converter, the *half-controlled bridge converter.* When the supply

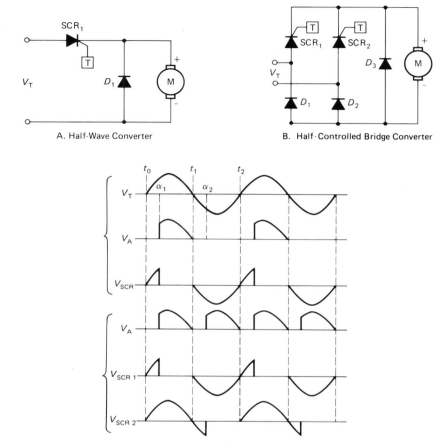

Figure 7.12 Half-Wave and Half-Controlled Bridge Converters Driving DC Motors

goes positive, SCR$_1$ and D_2 are forward-biased. Current will flow (when SCR$_1$ is triggered) through D_2, the armature, and SCR$_1$. Near 0 V, SCR$_1$ commutates (at time t_1 in Figure 7.12C) and the armature field collapses, causing current to flow in the flywheel diode. At α_2, SCR$_2$ is triggered, causing current to flow through D_1 and the motor armature. Current will continue to flow until time t_2, when SCR$_2$ commutates. From that point, the cycle starts over again.

If you compare the armature waveforms of these two systems, you will note that the half-controlled bridge has far less ripple than the half-wave system. As in any phase control system, decreasing the firing angle will cause more

power to be delivered to the load, increasing motor speed.

Full-Bridge Converter. A *full-bridge converter* is illustrated in Figure 7.13A. When the line voltage goes positive, SCR$_1$ and SCR$_4$ are forward-biased. They will conduct when triggered. Note in Figure 7.13C that the armature voltage goes negative between time t_1 and α_2. At time t_1 the SCRs have commutated. Beyond this point the motor armature field collapses and reverses polarity, supplying power back to the supply.

If the firing angle goes past 90°, the average armature voltage will be negative. That is, the

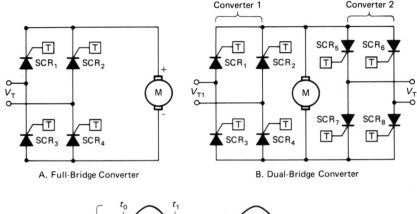

A. Full-Bridge Converter   B. Dual-Bridge Converter

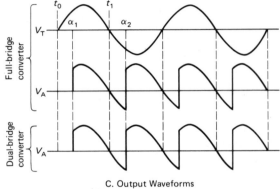

C. Output Waveforms

Figure 7.13 Full-Bridge and Dual-Bridge Converters Driving DC Motors

motor will be returning more power than it is using. Any rotating system that returns more power than it uses will decrease rotational speed. In fact, some motors use this system as a brake. Such a system is called a *regenerative braking system.*

The average DC voltage applied across the armature can be easily calculated if a flywheel diode is applied across the motor. The flywheel diode will change the armature voltage (V_A) waveform by clamping the negative voltage excursions to zero. When the line goes through zero volts, the flywheel diode will conduct, rather than force current to flow through the SCRs. The average armature voltage can be found using the following equation:

$$V_A = 0.637 V_{pk} \cos \alpha$$

where

V_A = average armature voltage

V_{pk} = maximum value of the line voltage

α = firing angle

Taking an example, what is the average armature voltage if a 120 V AC line is applied to this circuit with a firing angle of 30°?

$$V_A = 0.637 V_{pk} \cos \alpha$$
$$= (0.637)(170 \text{ V})(\cos 30°) = 93.8 \text{ V}$$

A change to a firing angle of 45° would give a new average voltage:

$$V_A = 0.637 V_{pk} \cos \alpha$$
$$= (0.637)(170 \text{ V})(\cos 45°) = 76.6 \text{ V}$$

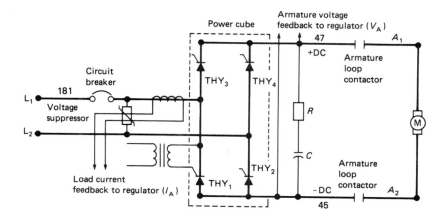

Figure 7.14 Reliance Electric Company DC Motor Drive

Note that the average voltage decreases as the firing angle increases, as we would expect.

Recall from Chapter 3 that a DC motor can reverse direction by reversing the direction of current in the armature. One way to accomplish this reversal is to have a switch that, when energized, will reverse the armature connections. In a shunt or a separately excited motor, we could reverse the direction of the current in the field winding by the same method. But mechanical or electromechanical switching is often too slow for some applications. However, almost instantaneous reversals of current flow can be achieved by using the circuit shown in Figure 7.13B, the *dual-bridge converter.* In this circuit, changing from converter 1 to converter 2 will reverse the direction of current flow in the motor.

We can see an example of a full-bridge (or full-wave bridge, as it is sometimes called) in Figure 7.14. This illustration was taken from an instruction manual on a MiniPak Plus DC motor drive, manufactured by Reliance Electric Company. The drive controls a wound-field, shunt-connected DC motor. It allows adjustable speed and gives constant torque output from 50% of base speed to base speed, provided the current to the shunt field is held constant. The controller changes the speed by adjusting the average armature voltage. Armature speed is closely proportional to the armature voltage. If used with PM DC motors, no field excitation is required. The amount of armature current drawn is proportional to the load torque on the motor shaft. The MiniPak Plus system contains an adjustable-voltage armature supply, a fixed-voltage field supply, and a regulator to control and adjust armature voltage under varying torque and speed demands.

The four SCRs (labelled THY in the figure) are contained in a package called a power cube. The armature voltage is regulated by adjusting the firing angle of these thyristors. Each of the SCRs receives a gate pulse at a point on the incoming line voltage where the anode is positive with respect to its cathode. Each SCR turns on when a low-power gate pulse is applied for a duration of 350 μs. This pulse will turn the device on and will keep it on as long as the cathode is positive with respect to the anode. When the anode goes negative with respect to the cathode, this device will turn off, or commutate.

As we have noted earlier, the gate pulse timing controls the firing angle. If the SCR fires early in the cycle, the SCR will conduct for a relatively long time. The longer the SCR remains on, the more average power is delivered to the load. Firing the thyristor late in the positive line alternation allows only short periods of conduction through the load. The less time the SCR conducts, the less average power delivered to the load.

When L_1 is positive with respect to L_2, current flows through thyristor 3, through the motor armature, through thyristor 2, and back to the supply. When L_2 is positive with respect to L_1, current flows through thyristor 4, through the motor armature, through thyristor 1, and back to the supply. Reliance has also supplied special circuitry to prevent thyristors 3 and 2 and thyristors 4 and 1 from turning on at the same time. Energizing these pairs would cause a short circuit, possibly damaging the motor and controller.

Note the transient protection in the MOV connected across the line. The MOV is part of a module for easy replacement when it fails.

Three-Phase Systems.   Large DC motors in industry use a three-phase power source. One reason three-phase power is used to supply DC drive systems is ease of filtering. The three-phase system has significantly less ripple voltage than a single-phase system and therefore takes less filtering to achieve the same result.

We will not discuss three-phase DC motor drive systems, because they do not differ significantly in theory of operation from single-phase systems. Of course, there will be two additional phases to contend with, each displaced 120° from the other. Because of the additional number of phases, circuit complexity increases. Also, waveform analysis of such a system becomes very complex, but little added understanding is gained through such analysis.

Chopper Control

In those applications where the power source is DC and the load is a motor, the DC chopper motor speed control is one of the most efficient methods of control. Chopper controls are used to control the speed of electric vehicles such as forklifts, delivery cars, and electric trains and trolleys. Chopper control is used not only because it is the most efficient, but also because it provides smooth acceleration characteristics.

Basic Chopper Control.   The basic chopper control circuit is illustrated in Figure 7.15A. Note that the SCR provides DC power to the motor by switching on and off. The average DC voltage presented to the motor is controlled by keeping the frequency constant and increasing and decreasing the amount of time that the SCR is on. When the SCR is on 50% of the time and off by the same amount (Figure 7.15B), an average of 50% of the voltage is delivered to the load. With a duty cycle of 25% (the SCR on 25% of the time), only a quarter of the applied voltage is delivered to the load (Figure 7.15C). Changing from a duty cycle of 50% to one of 25% will be accompanied by a decrease in the motor's speed. You may recognize this technique as pulse-width modulation.

Another method of chopper control, which we will not discuss, keeps the pulses the same width and increases or decreases the frequency. This method is not often used in controlling DC motors because it requires too great a frequency change to be practical.

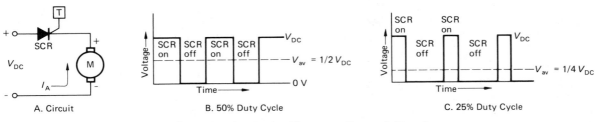

| A. Circuit | B. 50% Duty Cycle | C. 25% Duty Cycle |

Figure 7.15   Basic Chopper Control Circuit

Jones Circuit. Figure 7.16 shows a circuit often used in chopper control of DC motors, the *Jones circuit.* In this circuit when SCR_1 is fired, current flows through the motor, L_1, and SCR_1. Because of the mutual inductance between L_1 and L_2, current flow in L_1 induces current flow in L_2. Current flows down through D_1 and C_1, resulting in the charge shown. As long as SCR_1 is conducting, power is applied to the motor.

SCR_1 is commutated as follows: SCR_2 is triggered into conduction, causing C_1 to discharge through SCR_1. This discharge path is opposite the path of the load current flowing through SCR_1. Total anode current is reduced in SCR_1 to the point where it falls below the holding current level. SCR_1 then turns off.

External Commutation. Chopper circuits may be commutated externally. *External commutation* is shown in Figure 7.17. When SCR_1 is fired, current flows through the motor from the supply, developing torque to turn the motor. The thyristor is commutated by turning on the transistor Q_1. The transistor applies a reverse-biased voltage V_{com} across the thyristor, turning it off. Power to the motor is regulated by proper choice of the transistor turnon point.

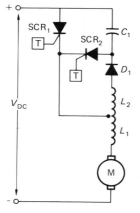

Figure 7.16 Jones Circuit

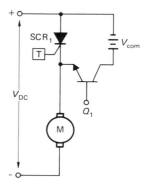

Figure 7.17 External Commutation Circuit

Closed-Loop Speed Control

Up to this point we have assumed that the control of the motor's speed was adjusted in an open-loop situation. *Open-loop control,* as we will see in a later chapter, is control without feedback. *Feedback control* of DC motors is an excellent method to regulate the motor's speed precisely. There are many methods used to control a motor by using feedback. The one we will discuss uses a device called a phase-locked loop.

Phase-Locked Loop (PLL). The PLL, discussed in detail in Chapter 2, is an electronic feedback control system having a phase detector, a low-pass filter, and a voltage-controlled oscillator (VCO). A PLL can also be considered a frequency-to-voltage converter (FVC).

Every system using negative feedback has the same characteristics. Negative (180° out of phase) feedback is fed from the output back to the input. The feedback signal or voltage is then compared with a reference. If the feedback is not equal to the reference, an error voltage or signal is generated. This error component is proportional to how different the two values are. This error component is then used to control the system and reduce the error to zero.

In the PLL system shown in Figure 7.18, the reference is a VCO. The VCO, as discussed

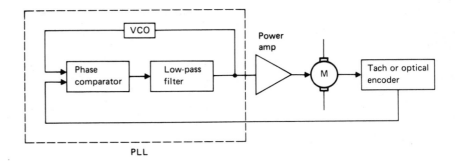

Figure 7.18 PLL Motor Control

in Chapter 2, is nothing more than a free-running multivibrator whose frequency of oscillation is controlled by an *RC* network. This frequency is compared with a frequency fed back from the motor. These two frequencies are compared in the phase comparator. If they are different in frequency, the phase comparator will produce an output signal proportional to the difference. This signal is then rectified by the low-pass filter and used to control a DC amplifier. The DC amplifier adjusts the motor's speed until there is little or no difference between the reference signal and the feedback signal.

This type of control can regulate motor speed to within 0.001% of the desired speed. Other methods of analog feedback are used to control DC motor speed, but none is as efficient as the PLL.

Pulse-Width Modulation. Figure 7.19 illustrates *pulse-width modulation* for DC motor control. Here pulse-width modulation is generated by the LM3524 IC. The pulse-modulated output of the LM3524 drives a transistor, which, in turn, controls the speed of the DC motor. A variable-reluctance sensor acts as a tachometer, converting the motor's mechanical motion to a variable frequency. This frequency is converted to a voltage by the LM2907 FVC. Both the voltage produced by the LM2907 and the speed-adjust potentiometer are connected to the input of the LM3524, where they are compared. Any error voltage generated by this comparison is used to adjust the amount of pulse-width modulation applied to the DC motor.

Stepper Motor Control

You will recall from Chapter 4 that the stepper motor is the only motor with a truly digital output. The stepper motor converts electrical pulses applied to the stator into discrete rotor movements called steps. Some low-power stepper motors are directly driven by TTL circuits; usually, however, TTL circuits do not have enough power to drive stepper motors. In this section we present a brief discussion of stepper motor drive circuits.

A basic stepper motor drive circuit is illustrated in Figure 7.20. In this circuit the switch-

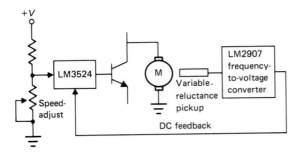

Figure 7.19 Pulse-Width Modulation Motor Control

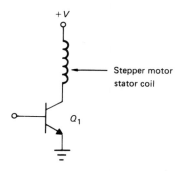

Figure 7.20 Transistor Driving Stepper Motor

ing transistor energizes the stator coil. Each coil in the stator would have a circuit identical to this one. Although not shown, the transistor is usually driven by the output of a TTL gate or

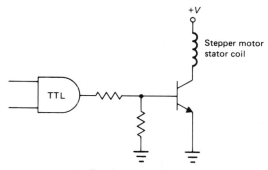

A. Transistor Driven by TTL

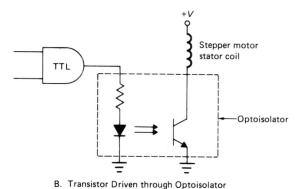

B. Transistor Driven through Optoisolator

Figure 7.21 Methods of Driving Transistor in Stepper Control Circuit

through an optoisolator, if isolation is required. The schematic in Figure 7.21A shows a TTL circuit driving the transistor. The optoisolator in Figure 7.21B provides the isolation needed by some circuitry.

The stepper motor stator is an inductive load. As such, when the transistor interrupts current flow, the stator coil collapses, producing a transient voltage spike. This spike can destroy the switching transistor if it is unprotected. A transistor protection circuit is shown in Figure 7.22. The diode D is reverse-biased when Q is conducting and current is flowing through the stepper motor stator. When the transistor turns off, the field built up around the coil collapses and forward-biases the diode. The pulse is then shorted out. The resistor R, connected in series with the diode, shortens the time constant. You will recall that, unlike the RC circuit, an LR circuit has a time constant inversely proportional to resistance. The time constant is L divided by R (L/R). If R were not included, the time constant would be long. The diode would be less effective as it shorts the transient pulse. Adding R decreases the time constant. By the same token many stepper motor drive circuits include a resistance in series with the stepper motor stator coil. The re-

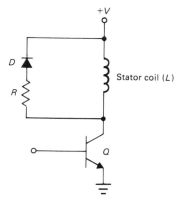

Figure 7.22 Transistor Protection Methods in Stepper Motor Control

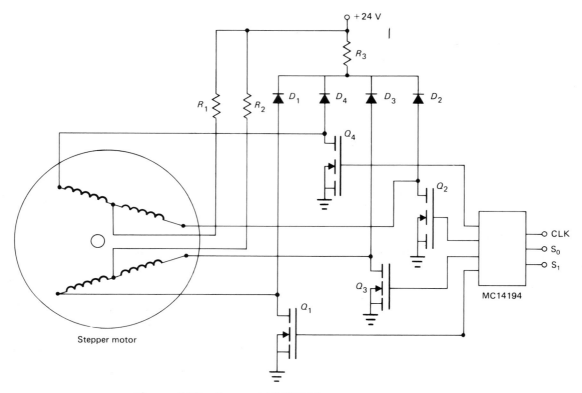

Figure 7.23   Power MOSFET Stepper Motor Driver

sistor in series with the stator coil reduces the time constant and increases the rotor torque. When resistance is added to the stator, the supply voltage must be increased to make up for the voltage dropped across the resistor.

The power MOSFET (metal-oxide semiconductor field effect transistor) has been used recently in stepper drive circuits instead of the transistor. The power MOSFET, described later in this chapter, is a special high-current FET (field-effect transistor). The current through the device can be controlled by nanoamperes of current. Power MOSFETs can be driven directly from TTL or CMOS circuitry. The use of power MOSFETs and CMOS circuitry in stepper motor drives simplifies control circuitry and allows considerable flexibility in control.

A circuit using power MOSFETs and CMOS control is illustrated in Figure 7.23. A full-step, center-tapped stepper motor is controlled by four power MOSFETs. The power MOSFETs are controlled by an MC14194 CMOS four-bit presettable shift register. Tapping the appropriate shift register outputs gives the required phasings to turn the rotor.

The motor rotates clockwise by right-shifting the MC14194. Left-shifting causes the rotor to turn counterclockwise. Control signals on the S_0 and S_1 inputs plus the clock (CLK) input control the stepping. When power is applied to the device, the MC14194 needs to be preset by putting highs at S_0 and S_1, and by receiving a leading clock pulse. These three requirements, if met, will put the logic in a known state. The

Table 7.1 Control Functions for MC14194

S_0, S_1	Result
0, 0	Hold
0, 1	Shift right
1, 0	Shift left
1, 1	Preset

other control functions are shown in Table 7.1. Stepping will occur when a leading edge clock pulse is received. The diodes serve the same purpose as in the transistor circuit previously described. They protect the power MOSFET from damage due to spikes. Power MOSFETs switch very quickly. As a result, the diodes may not be able to turn on fast enough to prevent damage to the power MOSFETs. To prevent this damage, a small capacitor (0.01–0.1 μF) is sometimes placed across the motor windings.

Several manufacturers make drives using custom ICs. These chips develop the required pulse trains and require a minimum of external components. A stepper driver IC made by Airpax Corporation is shown in Figure 7.24. This IC (the SAA1027) controls a four-coil, two-phase stepper motor. It can drive stator coils with a maximum of 350 mA per phase. The three inputs—trigger, set, and direction—are

controlled by applying high and low voltage levels to them.

The voltage on T, pin 15, is normally high when not being pulsed or stepped. A change from low to high and back again to high will trigger the IC. The motor connected to the output steps on the positive-going edge, low to high, of the pulse. When the T input is held high and a low is applied to S, the motor is set to the following condition: Q_1 low, Q_2 high, Q_3 low, and Q_4 high. When a high is placed at input R, the rotor will rotate counterclockwise. A low at input R will cause the motor to step clockwise. The R input should always be held either high or low for best immunity to noise.

Stepper Motor Control Example. As we have discussed previously, a stepper motor easily lends itself to microcomputer control. Figure 7.25 shows an example of an interface circuit that can be used to drive a small stepper motor. For convenience the microcomputer used, but not shown in the diagram, was an Intel 8085 based system. Any single-board microcomputer could be used, but the control lines may be different from the ones shown in the diagram.

The three inputs to the system are: (1) the address line, or decoded port, to select this particular port, (2) the control line IO/$\overline{\text{M}}$, and (3)

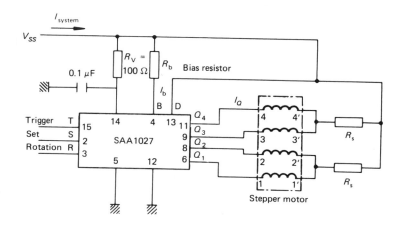

Figure 7.24 IC Stepper Motor Control

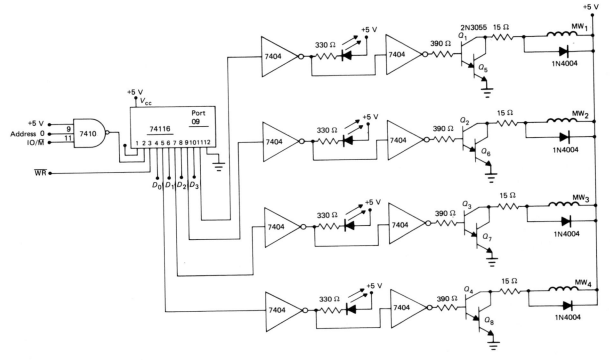

Figure 7.25   Microprocessor Controlling Stepper Motor

the control line $\overline{WR}$, used in conjunction with the OUT port instruction to enable the 74116 four-bit latch. The 74116 latches the data that were presented to it during the OUT port instruction and uses these data to drive the 7404 inverters. The inverters in turn drive other inverters and LEDs that display the data code that was presented to the 74116 latch. The second set of 7404 inverters drive the 2N3055 transistors that drive the stepper motor coils. The motor coils have flywheel diodes that prevent inductively generated voltages from destroying the transistors. The data codes and the time between changes in these codes determine the direction and speed of the stepper rotation.

Figure 7.26 shows various combinations of codes that could be used to rotate the stepper motor.

AC MOTOR SPEED CONTROL

In recent years the AC motor has become increasingly popular in certain areas of industry. Most of the interest in AC motors comes, at least in part, from the advantages it has over DC motors. The AC motor is smaller, and therefore less expensive, than the DC motor of equivalent horsepower rating. Generally, the AC motor is less costly to maintain than the DC motor. In the induction motor reduced maintenance is due to the absence of a mechanical connection between rotor and stator. Not only does this feature mean lower maintenance costs, but it also means increased safety. Machines with commutators and brushes generate sparks, which could ignite in an explosive atmosphere. It is true that large synchronous mo-

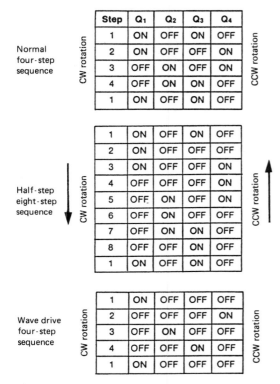

Step	Q_1	Q_2	Q_3	Q_4
1	ON	OFF	ON	OFF
2	ON	OFF	OFF	ON
3	OFF	ON	OFF	ON
4	OFF	ON	ON	OFF
1	ON	OFF	ON	OFF

Normal four-step sequence — CW rotation / CCW rotation

	Q_1	Q_2	Q_3	Q_4
1	ON	OFF	ON	OFF
2	ON	OFF	OFF	OFF
3	ON	OFF	OFF	ON
4	OFF	OFF	OFF	ON
5	OFF	ON	OFF	ON
6	OFF	ON	OFF	OFF
7	OFF	ON	ON	OFF
8	OFF	OFF	ON	OFF
1	ON	OFF	ON	OFF

Half-step eight-step sequence — CW rotation / CCW rotation

	Q_1	Q_2	Q_3	Q_4
1	ON	OFF	OFF	OFF
2	OFF	OFF	OFF	ON
3	OFF	ON	OFF	OFF
4	OFF	OFF	ON	OFF
1	ON	OFF	OFF	OFF

Wave drive four-step sequence — CW rotation / CCW rotation

Figure 7.26 Changing Stepper Motor Direction

tors have slip rings and, therefore, mechanical contacts. However, motors with slip rings produce less sparking and are easier to maintain than mechanically commutated motors. AC motors also can be run at generally higher speeds. Most general-purpose, wound-field DC motors have a top speed of about 2500 r/min. A comparable AC induction motor typically has a top speed twice that.

Just a few years ago the SCR was the leading component in power control in AC and DC drives. Today, large, high-voltage (>1000 V) transistors are used in some AC drive applications. The use of transistors also eliminates the need for bulky and expensive commutation cir-

cuits in DC drives. The GTO (gate-turnoff SCR), a thyristor discussed in Chapter 6, has been used as a replacement for SCRs, due to recent advances in GTO power handling capability. The Westinghouse Accutrol 300 uses GTOs exclusively as the power driver.

Several years ago the AC motor was used predominantly in fixed-speed applications. Recall from Chapter 5 on AC motors that the synchronous stator speed is directly proportional to the line frequency and inversely proportional to the number of magnetic poles. It is relatively easy to change the stator speed by varying the number of poles per phase. This type of speed control, however, results in speed control by steps only. There is no way to adjust the speed proportionally by altering the number of poles.

Speed may also be changed in an AC motor by varying the frequency of the voltage applied to the stator field. Although this type of speed control was known for a number of years, it was not economically sound until the advent of the thyristor in the 1950s. Research and development was soon undertaken to apply the thyristor to adjustable-frequency AC drives.

Universal Motor Speed Control

One circuit that will control the power applied to a universal motor is illustrated in Figure 7.27A. In this circuit, an open-loop control circuit, the capacitor charges up to the firing voltage of the diac in either direction. Once fired, the diac will apply a voltage to the gate of the triac. The triac will conduct and apply power to the motor. Note that the triac will conduct in either direction. Since this device is basically a series DC motor, current flowing in either direction will tend to cause rotation in only one direction. Speed may be changed by varying the resistance of the potentiometer, as discussed previously under phase control.

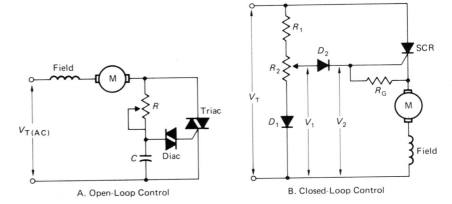

Figure 7.27 Universal Motor Control

A. Open-Loop Control   B. Closed-Loop Control

Closed-loop control for the universal motor is shown in Figure 7.27B. When line voltage is applied, the voltage at R_2 will rise proportionally to the voltage divider network. When the motor is turning at the desired speed, it develops a voltage V_2, which is mostly the cemf of the motor. If V_1 and V_2 are equal, no voltage is applied to the gate of the SCR. In this condition the reverse-biased diode D_2 blocks the voltage. If the motor's speed decreases, however, the cemf will decrease, causing voltage V_2 to decrease. When V_1 is greater than V_2, a gate voltage is applied to the SCR. This voltage generates torque in the motor. The motor will then pick up speed, returning to near-normal speed.

The firing angle of the SCR is inversely proportional to the load on the motor. As the load on the motor increases, the speed of the motor will decrease. The more the motor speed decreases, the earlier the thyristor fires in the cycle. The earlier the thyristor fires, the more power is applied to the motor. Thus, the universal motor speed is kept relatively constant under varying load conditions.

Adjustable Frequency AC Drives

With the exception of the universal motor, the remainder of the AC motor drives covered in this chapter operate on the principle of adjustable frequency control (AFC). As we have seen, in this type of control the rotor speed is directly proportional to the stator frequency. The AC motor can, therefore, be run at any speed, from a few r/min to the top speed of the motor.

We can see the effect of a change in stator frequency on the induction motor by looking at the speed-torque curve in Figure 7.28. Curve A is a typical curve for a four-pole machine with an applied stator frequency of 60 Hz. The motor will operate where the curve crosses the line labelled 100% load torque. If the stator frequency is decreased to 40 Hz, the motor will shift its operating point to curve B. Further reduction in frequency to 30 Hz will take the motor to curve C. This reduction in frequency could continue until the rotor is at a standstill. Although we moved the stator frequency in steps, the actual control is infinitely variable up to the maximum motor speed.

The induction motor is named for the way the rotor gets its power, that is, electromagnetic induction. Like the transformer, the AC induction motor needs to have a constant flux in the rotor and stator. Unless this requirement is met, the motor will not be able to generate full torque. If the frequency applied to the stator is decreased, the voltage applied to the stator must be decreased by the same amount. An-

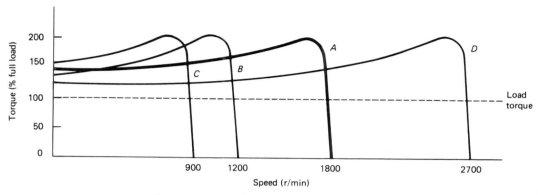

Figure 7.28 Adjusting AC Motor Speed

other way to state this requirement is that the ratio between the voltage applied to the stator and the frequency of the voltage applied to the stator must be constant. This ratio is called the *volts/hertz ratio,* or the *constant volts/hertz characteristic,* of an AC induction motor. To find the volts/hertz (V/Hz) ratio for an induction motor, simply divide the rated voltage on the motor nameplate by the rated frequency. For example, a motor nameplate rates an AC induction motor for 230 V AC, 60 Hz operation. The V/Hz ratio is:

$$\frac{\text{volts}}{\text{hertz}} = \frac{230 \text{ V AC}}{60 \text{ Hz}} = 3.83 \text{ V/Hz}$$

Above a frequency of about 15 Hz the amount of voltage needed to keep the flux constant is a constant value, as shown in Figure 7.29. To get the voltage needed above 15 Hz, multiply the V/Hz constant by the applied frequency. Below 15 Hz the voltage applied to the stator must be boosted above the value predicted by the constant V/Hz curve. The extra voltage is needed to make up for the I^2R losses in the motor at low speeds. The amount of voltage boosting will depend on the motor. Taking an example, we can find the stator voltage needed at 30 Hz by multiplying the 30 Hz by the 3.83 V/Hz ratio as follows:

$$V_{st} = f(\text{V/Hz}) = (30 \text{ Hz})(3.83 \text{ V/Hz})$$
$$= 115 \text{ V AC}$$

where

V_{st} = stator voltage required
f = frequency applied to the stator
V/Hz = volts/hertz ratio

Another consequence of operating in the constant V/Hz ratio relates to the power out of the machine. Since the torque remains constant, the power out of the machine will be directly proportional to the speed of the machine. As we learned in Chapter 3, the power out of a rotating machine is related to speed and torque as follows:

$$P_o = \frac{Tn_{rt}}{5252}$$

As an example, let us consider a 1 hp machine with a base speed of 1750 r/min at 230 V AC and 60 Hz putting out 3 ft-lb of torque. At 875 r/min (one-half of rated speed) the power out would be:

$$P_o = \frac{(3 \text{ ft-lb})(875 \text{ r/min})}{5252} = 0.5 \text{ hp}$$

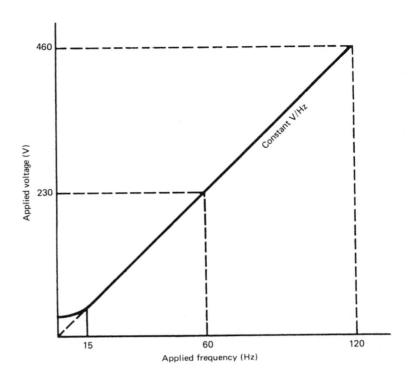

Figure 7.29 AC Induction Motor Applied Frequency-Voltage Curve

At one-third the rated speed the maximum power available would be 0.33 hp. You should also bear in mind that the motor may need extra cooling when operated below rated speed with full-load torque.

We can now calculate the voltage and frequency needed to operate a motor at any given rotor speed. First, we determine the speed at which the rotor would be turning the load. Next, we add the value of the slip at that load to get the synchronous speed. Recall that the induction motor must rotate at a slower speed compared to the stator for induction to take place. We next solve for the frequency applied to the stator. Finally, we multiply the stator frequency by the V/Hz ratio.

Consider an example. We have a four-pole, 5 hp motor with a rated voltage of 230 V AC at 60 Hz. We wish to operate the motor at 850 r/min at full-load torque. If slip is 5.55% at full-load torque, then the stator speed is 900 r/min. Rearranging the formula for finding stator speed (see Chapter 5) and solving for frequency we get:

$$f = \frac{n_{st}p}{120} = \frac{(900 \text{ r/min})(4)}{120} = 30 \text{ Hz}$$

The V/Hz ratio is found by dividing the rated voltage by the rated stator frequency:

$$\frac{\text{volts}}{\text{hertz}} = \frac{230 \text{ V AC}}{60 \text{ Hz}} = 3.83 \text{ V/Hz}$$

Finally, we find the stator voltage needed with a V/Hz ratio of 3.83 V/Hz:

$$V_{st} = f(\text{V/Hz}) = (30 \text{ Hz})(3.83 \text{ V/Hz})$$
$$= 115 \text{ V AC}$$

We can see an interesting relationship in these calculations. The synchronous speed at 30 Hz is exactly half of the speed at 60 Hz. The voltage needed with a 30 Hz stator frequency is also exactly half of the voltage needed for operation at 60 Hz. We should expect these relationships to hold true because of the proportionality between applied stator frequency and stator voltage.

An induction motor's rated speed may be exceeded in one of two ways. First, the frequency and voltage may be increased according to the V/Hz ratio to about 125% of the full-load speed. Higher speeds are not recommended without specially constructed motors. Induction motors may also be operated above rated speed in the constant-voltage mode, as shown in Figure 7.30. As the graph shows, the line voltage is held at the rated level and the applied stator frequency is increased past 60 Hz. In this area of operation the motor does not have enough voltage to produce full-load torque as speed increases. As speed increases, therefore, the torque decreases. In this constant-voltage range both the rated torque and the breakdown torque will decrease. Breakdown and rated torque decrease as the square of the increase in speed past 60 Hz. The new torque can be found by using the following equation:

$$T = \left(\frac{60 \text{ Hz}}{f}\right)^2$$

where

T = torque available as a percentage of rated torque

f = stator frequency

Taking an example, what percentage of rated torque would be available at 75 Hz? Calculations follow:

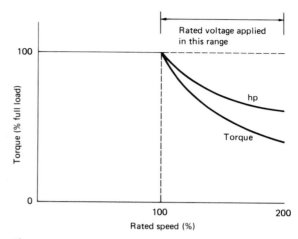

Figure 7.30 Operating an AC Induction Motor above Rated Speed

$$T = \left(\frac{60 \text{ Hz}}{75 \text{ Hz}}\right)^2 = 0.64, \text{ or } 64\%$$

The motor power output decreases proportionately as speed increases:

$$P = \frac{60 \text{ Hz}}{f}$$

where

P = percentage of output power at rated speed and frequency

f = new frequency in Hz

In the example just given, the percentage of output power is:

$$P = \frac{60 \text{ Hz}}{75 \text{ Hz}} = 0.80, \text{ or } 80\%$$

Changing the frequency of the voltage applied to the stator field will give smooth linear control of motor speed. Frequency control of synchronous and induction motors can be bro-

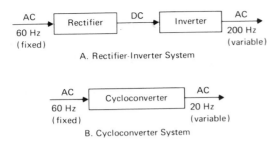

A. Rectifier-Inverter System

B. Cycloconverter System

Figure 7.31 ·Frequency Control Methods for AC Motors

ken down into two methods: rectifier-inverter systems and cycloconverter systems. Figure 7.31A is a block diagram of a *rectifier-inverter system*. Note that the frequency conversion takes place in two steps. First, the fixed 60 Hz line frequency is rectified to DC. Second, the DC is converted to variable-frequency AC by the inverter. Thyristor inverters can supply an output frequency of up to 1 kHz over a range of about 15:1. The *cycloconverter system* is shown in block diagram form in Figure 7.31B. Cycloconverters give output frequencies of up to 25 Hz.

Rectifier-Inverter System. Three types of adjustable-frequency controls are in common use today, all using the rectifier-inverter as a basic building block. These three types are the variable-voltage input (VVI), the current-source inverter (CSI), and the pulse-width modulation (PWM).

The VVI type of inverter, shown in Figure 7.32, receives DC input power from the phase-controlled bridge rectifier. The phase-controlled bridge varies the amount of DC input voltage to the inverter. The inverter switches the output circuit between the plus and minus bus. Each of the three phase outputs is displaced from one another by 120 electrical degrees. These waveforms (A to B, B to C, and C to A) are created by the algebraic summing of each output with respect to the DC bus. During

time periods T_1 and T_2 the voltage from outputs A and B are equal to V_{adj}, since B is at the $-V_{adj}$ bus potential. The entire V_{adj} voltage would be seen across whatever motor winding is connected across A to B. During the time period T_3, the A to B voltage is zero, since both A and B are at the same potential. Note that the AC voltages applied to A, B, and C are displaced by 120 electrical degrees and, therefore, will give the rotating field effect when applied to the stator windings. Direction of rotation may be changed by firing the inverter thyristor diodes in different sequences. The stator frequency is changed by firing and commutating each thyristor for different time periods. For example, if we wanted to increase the frequency, we would shorten time periods T_1 to T_7 by decreasing the time the appropriate thyristors are gated on. The voltage applied to the stator is increased by increasing V_{adj}. V_{adj} is increased, in turn, by gating on the rectifier thyristors earlier in the cycle. The VVI drives are the most popular general-purpose AFC drive today. They are used in applications up to 400 hp in pumps, fans, grinders, and saws.

The CSI drive uses the same basic components as the VVI drive. The CSI drive, shown in Figure 7.33, has a variable DC bus controlled by a thyristor regulator, a large DC filter inductor, and an inverter. The rectifier and filter provide a controlled, regulated DC current to the inverter. The source of power to the inverter is a controlled current source, rather than a controlled voltage source, as in the VVI. Current is usually detected by a sensor after the inductor. The amount of current sensed is used to provide feedback for current adjustment. The CSI drive system is most often found in applications above 10 hp.

The final AFC drive method is achieved by pulse-width modulating the input to the inverter. As you can see from the diagram, the PWM system converts AC to DC using a diode rectifier system, not a thyristor rectifier, as in the CSI and VVI drives. The thyristors are not

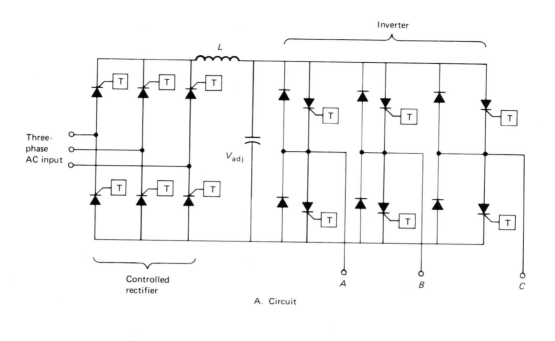

A. Circuit

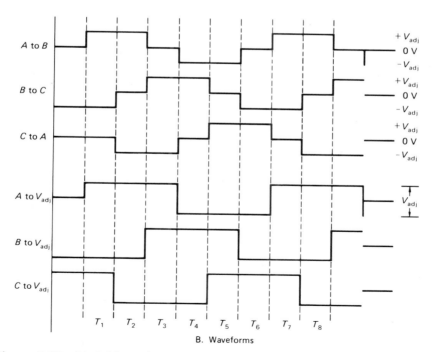

B. Waveforms

Figure 7.32   Variable-Voltage Inverter (VVI) Three-Phase AC Motor Drive

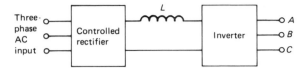

Figure 7.33 Current Source Inverter (CSI) Three-Phase AC Motor Drive

necessary because the PWM system uses a fixed DC voltage input to the inverter. A pulsed output is constructed by varying the gating and commutation times, as shown in Figure 7.34. Instead of turning on only once in a cycle, the SCRs trigger and commutate many times. Pulse-width modulation also is used to reduce the amount of undesired harmonic frequencies generated by other methods of inversion. You may have noticed that we have used the pulse-width modulation system in DC motor control earlier in this chapter. PWM is used frequently in both AC and DC applications. As in DC applications, PWM AFC drives are very responsive to speed changes. They find applications in conveyors, fans, and pumps.

Cycloconverter System. The cycloconverter method is the second way to provide variable stator frequency control. The cycloconverter changes the fixed-frequency input voltage to a variable-frequency output voltage. The cycloconverter commutates naturally from the input line voltage. Thus, the cycloconverter is less complicated when compared with systems requiring forced commutation, such as the pulse-width modulation inverter just discussed. The cycloconverter has another advantage compared with the inverter. The inverter requires a DC input, which necessitates rectification before it can be used. The cycloconverter, in contrast, operates from the AC input directly.

A simplified circuit diagram of a cycloconverter is shown in Figure 7.35A. During the first input half cycle, SCR_1 and SCR_2 are gated on, producing the first two positive pulses shown in the output of Figure 7.35B. SCR_3 and SCR_4 are gated on during the second input half cycle, producing the next set of positive pulses. The process then repeats. The output voltage, after it is filtered, is sinusoidal in shape. The cycloconverter shown is actually a 7:1 frequency divider.

The cycloconverter finds applications in low-speed synchronous motors where high torques are demanded. Previous to the advent of thyristors, low speeds were achieved by gearing, especially when the application demanded an induction motor. The cycloconverter allows the induction motor to run at a low speed while developing maximum torque.

Before moving on to another topic, we should mention another method of controlling induction motor speed. Varying the voltage applied to the stator winding will give a limited amount of speed control while keeping the frequency constant. This type of speed control can be achieved by phase control techniques such as those discussed earlier in the chapter. For example, the circuit illustrated in Figure 7.27A can be used for this type of control. Control is achieved in this case by changing the torque produced by the motor. Recall that the torque produced by an induction motor is directly proportional to the square of the stator voltage. Thus, changing this voltage produces a change in torque. This method of control is not used very much in industry, for several reasons. First, it gives only limited speed control. Sec-

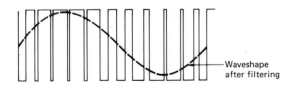

Figure 7.34 Pulse-Width Modulation of Voltage

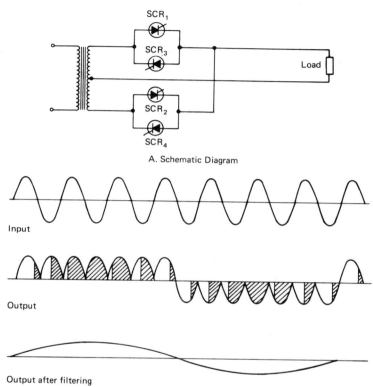

Figure 7.35 Cyclo-converter

ond, it changes the breakdown torque of the motor. Third, decreasing the speed by this method can result in the motor overheating.

THYRISTORS IN MOTOR CONTROLS

Two semiconductors were used extensively prior to 1978 in power control applications: the SCR and the transistor. The SCR was used in AC drive applications greater than 20 kVA, with transistors in the lower ratings. In the years since 1978 we have been seeing a greater variety of semiconductors in power control applications, including power FETs, GTOs, and higher-powered transistors and Darlingtons.

The power GTOs and transistors are used extensively in PWM AC drives. The power-handling capabilities of GTOs, transistors, and Darlingtons are a result of improvements in the methods of doping, device geometry, and packaging. Gains in power Darlingtons have increased to 500 with an increase in reliability. Better and more efficient packaging of semiconductors improves not only the reliability of the devices but the cost as well.

Semiconductors in Power Control

Special switching transistors can turn on very quickly and have low collector-to-emitter resis-

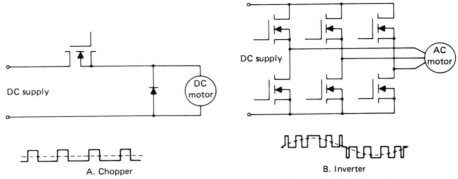

Figure 7.36 Power MOSFET Motor Control Applications

tance. They are, therefore, very efficient. Also, transistors do not have the disadvantage that SCRs have in the ability to turn off in DC circuits. In DC circuits transistors turn off easily by a removal of base current or voltage.

One of the major problems in using transistors in switching circuits involves the base current requirements. Transistors require that base current flow to provide collector current. Normally, power transistors have low current gain (β), some as low as 10. Such a low β means that considerable base current may be needed to keep the device on. This current is not doing useful work. Thus, because of this situation the transistor is inefficient in power applications.

The power MOSFET (metal-oxide semiconductor FET) has taken a firm hold on the power control market for power levels up to 30 hp (about 22 kW). Recall that, unlike the transistor, the MOSFET is basically a voltage-controlled device, like the vacuum tube. The power MOSFET, which can presently control up to 40 A, draws only nanoamperes of gate current. Power MOSFETs are presently being used in motor controls, especially choppers and inverters. Chopper and inverter applications are illustrated in Figure 7.36. The power MOSFET has advantages over transistors in the areas of ruggedness and much lower drive requirements. Circuit requirements are simplified, thereby reducing costs and improving reliability and re-

sponse time. Another advantage of the power MOSFET lies in its ability to increase power-handling capability by connecting power MOSFETs in parallel. Connecting transistors in parallel to get higher current ratings provides designers with an adventure in fusing and current-sharing techniques. Power MOSFET choppers with currents in the hundreds of amperes are not beyond the capabilities of the power MOSFET. Power MOSFETs also offer significant advantages in gain and switching times over standard power transistors and Darlingtons.

Applications for power MOSFETs are increasing every day. These hardy devices are used in switching power supplies, audio amplifiers, AM transmitters, induction heaters, high-frequency welding machines, and fluorescent lighting controls. The hammer driver application shown in Figure 7.37 is an interesting application demonstrating the strength of the power MOSFET. The power MOSFET in this application drives the coils that move the hammers in a computer dot matrix printer. You have no doubt seen the output of this type of printer. Each letter is made up of usually five to seven dots, providing the typical printed computer output. In this application the drive coil is energized by a power MOSFET when it gets a triggering voltage on its gate. Removing the triggering voltage stops the current through the

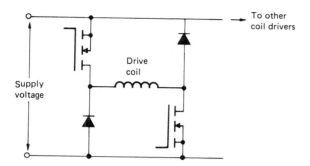

Figure 7.37 Printer Hammer Drive Application for Power MOSFET

MOSFET, turning it off. Before the advent of the power MOSFET, this switching was done by Darlingtons. The Darlington solution to this application has some problems. Darlingtons are slow and have high losses, especially when pulse-width modulated. The collector-to-emitter voltage drop of the Darlington may be as high as 2.5 V, increasing the heat dissipated by the TO-220 package usually used. The TO-220 package can dissipate the heat but must be hand inserted, making the manufacturing of printed circuit boards containing TO-220 packages more expensive. The power MOSFET has a voltage drop ten times less than that found across the Darlington, improving efficiency and lessening heat buildup. The power MOSFET switches faster and can be automatically inserted in printed circuit boards in a dual in-line package (DIP), thus reducing assembly costs.

Operating Power Semiconductor Equipment

Thyristor Identification. Technicians are often called on to replace semiconductors that have failed. Semiconductor failure may be caused by external overloads or by an internal semiconductor fault. In either case, the correct replacement part may not be available. The competent technician should be able to suggest a replacement part that will perform as well as the part being replaced. Suggesting a substitute part is not always an easy task. Every technician should be aware of the proper techniques for substituting semiconductors. This section will give some suggestions on how to do this effectively.

Many semiconductors are identified by a JEDEC (Joint Electronic Device Engineering Council) number. JEDEC numbers for rectifiers begin with a "1N" prefix (for example, 1N4001, 1N1206A), whereas SCRs and transistors begin with a "2N" prefix (for example, 2N681, 2N3055). A JEDEC number is an attempt by the industry to standardize electrical and mechanical parameters so that products of one manufacturer will be interchangeable with those of another manufacturer. The technician should be careful, however, when making a direct JEDEC substitution of another manufacturer because the JEDEC registered parameters do not always cover all of the critical parameters. As a result, manufacturers using completely different manufacturing processes often sell devices to meet the same JEDEC number. While this may not pose a problem in most general-purpose and phase control applications, the user must recognize that various manufacturers' devices marked with the same JEDEC number can exhibit completely different secondary characteristics and safety margins.

An engineer may unknowingly design a circuit around a particular JEDEC type number that was manufactured using process A, which built in a large safety factor. For example, our engineer chooses a 6 A rated device that really has a 12 A actual capability. Everything works fine until the user chooses a replacement JEDEC from another manufacturer. Then the new devices keep failing in the circuit. The circuit works properly with one manufacturer's device but not with another's. No one knows why, because the replacement part was not carefully researched. To be safe, all JEDEC substitutions should be evaluated thoroughly before use.

Most power semiconductors (especially those rated over the 40–100 A range) are marketed under their respective manufacturer's part number. Even though each manufacturer uses its own device nomenclature, the mechanical packages and electrical ratings are reasonably well standardized throughout the industry. Device substitution is not difficult, therefore, providing the user has a good cross-reference guide and the technical data sheets for the devices being evaluated. As with any cross-reference, the technician must determine an acceptable substitution. This is done by reviewing the detailed electrical and mechanical characteristics of the devices being considered. This comparison will ensure that the manufacturer's suggested replacement will perform properly in the given application.

Thyristor Replacement. Be sure to order the exact part number of the device you are replacing. If in doubt, include the entire device marking on your order. The most likely place to get off-the-shelf delivery of an exact replacement is directly from the equipment manufacturer. The original equipment manufacturer (OEM) normally carries a good supply of spare parts to service its equipment market. If you cannot get a replacement from the OEM, contact a local industrial or electronic distributor that handles power semiconductors. A user should buy a semiconductor type number only from an OEM authorized distributor outlet or directly from the OEM factory. The user could receive inferior or counterfeit devices if purchased from an unknown source. Unless otherwise directed, deal only with a manufacturer or an authorized distributor.

The technician must be thoroughly familiar with equipment service manuals, with the power semiconductors and their technical data sheets, and with the various tools and instruments available for testing and replacing semiconductors. It is wise to carry spare semiconductor components in stock in case of equipment breakdown, especially for critical pieces of equipment.

Many phase control applications—for example, welders, battery chargers, DC motor controls, and general-purpose power supplies—can use rectifiers, SCRs, and transistors with fairly broad parameters. As long as the device selected has an adequate current and voltage rating and fits mechanically, it should work. The technician must be extremely careful in selecting a replacement semiconductor when general-purpose or phase control devices are used in series and/or parallel combination, when fast-recovery rectifiers and fast-switching SCRs are used in inverters, choppers, and the like, or when semiconductor devices are marked with special (noncatalog) part numbers. Specially selected, tested, and matched units may be needed for the semiconductor to operate properly in the equipment. Failure to use the correct semiconductor device could result in device failures, equipment damage, and plant downtime.

When selecting a device for replacement, the technician can always use one with a higher voltage rating, provided all other device ratings are equal or better. Likewise, a higher current rated unit can be selected as long as all of the other ratings are equal or better and the mechanical package is the same.

By following these guidelines, the user often can locate a suitable replacement faster. Also, spare parts inventories may be reduced by standardizing a smaller number of replacement semiconductors, and gain increased reliability with greater voltage and/or current safety factors. Of course, the additional cost for a higher-rated semiconductor must be weighed against the savings in inventory reduction, fewer failures, and reduced downtime.

In the absence of any other information, a good rule of thumb for specifying the proper device voltage rating (based upon the supply voltage to the semiconductor equipment) is as follows: 110 V line—use a 300 V device, 220 V line—use a 600 V device, and a 440 V line—use a 1200 V device. If in doubt about what device to use, call the semiconductor manufacturer and ask for a recommendation.

The user must realize that power semiconductors, like any other components, fail for a reason. It is acceptable merely to replace a suspect semiconductor if the semiconductor is the problem. Frequently, however, a semiconductor fails because of a current or voltage overload elsewhere in the circuit. Simply replacing the suspected semiconductor in these instances will result only in destruction of more semiconductors. Therefore, always look for the cause of failure before replacing any devices.

There is still a tendency to consider the state-of-the-art in the design, manufacture, and application of power semiconductors as something new, mysterious, and glamorous. This is not correct. Industry has long been using silicon power semiconductors, and a vast knowledge in the use of these devices has evolved. Along with this knowledge has come a new state-of-the-art for using and maintaining semiconductor equipment. The emphasis placed on selecting, purchasing, and installing devices is part of this new knowledge. Of equal importance are the proper use and maintenance of equipment using power semiconductors, such as we have discussed in these last two chapters.

Power Semiconductor Heat Protection. Semiconductors are made from materials that have different rates of thermal expansion. Therefore, both short- and long-term temperature changes not only impair the electrical characteristics of these devices, but also set up internal mechanical stresses at each of the material interfaces. These combined effects of high temperature changes can lead ultimately to device malfunction and destruction. The maximum temperature limits of various materials used in semiconductors vary from 150°C for soft solder to 1300°C for silicon. Once these materials are combined into a fabricated device, however, electrical deterioration begins at lower temperatures.

For example, forward and reverse blocking capability of a junction begins to decrease and leakage currents increase when allowable maximum junction temperatures are exceeded. These temperatures are generally between 125°C and 200°C, far below the maximum temperature limit of silicon itself.

Periodic checks of ambient temperatures, of cooling fan operation, of cooling water temperatures, and of case temperatures of the semiconductors themselves will pinpoint many malfunctions before they cause catastrophic failures. Because of the importance of preventing high temperatures from developing, this monitoring must be emphasized in the design, testing, and operation of electronic equipment.

Equipment Overload Protection. Avoid overloading the semiconductor, which, in effect, means protect against overloading the equipment that the semiconductors are feeding and/or controlling. Overloading a motor operated with semiconductor controls puts a direct overload on the semiconductors themselves. Even if an overload occurs only for a short time, a semiconductor's useful life can be shortened. Controls are designed to include motor overloads, and the maximum values of overload should never be exceeded. Even short-term overload should be avoided. The damage caused by overload accumulates and is not detectable. The result is random, often unexplainable semiconductor failures.

Operators may cause overload conditions that could result in excessive semiconductor currents. Operator overloads are caused by high on/off duty cycle or jogging for high production needs, rapid reversing, motor jam, and long acceleration time. Some mechanical problems that can cause motor overloads are high equipment temperature due to lack of cooling, wiring or bearing failure due to improper installation or maintenance, phase failure due to blown fuses or loose connections, phase imbalance, overvoltage, transient overvoltage, contaminants, and dirty environment.

Table 7.2 shows various motor stress conditions and the common protection method employed. This table can be used to determine adverse load conditions and the type of protection that should be used. In a properly designed system, overload conditions are anticipated and ways are found to deal with the overload. The user or operator has an obligation to operate the equipment within the design ratings. The table shows the equipment user and designer methods to prevent overloading a motor. Following these guidelines will help avoid situations that cause semiconductor failure or reduce semiconductor life. A good maintenance program should include a check of all the protective devices mentioned in Table 7.2.

Protection against many of the adverse conditions shown in this table are often used in the design of semiconductor controls. They include water flow thermal relays, thermal overloads on the semiconductors and motors, phase failure relays, overvoltage relays, transient suppressors such as metal-oxide varistors (MOVs), capacitor-resistor networks, air filters, and fuses with blown fuse indicators.

In a properly designed system, overload conditions are considered and provisions are made to accommodate the overloads. However, the user and the operator have an obligation to operate the equipment within the design ratings. Table 7.2 gives the equipment user and the designer methods by which to prevent the overloading of a motor and to avoid the situation that would cause semiconductor failure or reduce semiconductor life.

Power semiconductors must be protected from transient overloads. Silicon semiconductors will have a long, useful life if they are ad-equately protected against surge currents and transient voltage spikes. Protection against failure due to current and voltage transients is relatively simple. Normally this protection is designed into the circuit. In industrial plant environments most voltage transients in semiconductor circuits arise from three major causes: switching, commutation transients, and regenerative surges.

Switching is the most common source of voltage transients. Whenever current is switched on or off in an inductive circuit, a transient voltage is generated at the switch terminals. Transformers and motor windings are highly inductive components, and the circuit wiring itself is inductive. Surges caused by lightning also add to switching transients. These transients can originate in remote circuits and still feed back to the semiconductor circuit through the power supply line. Protecting semiconductors from such transients is primarily a matter of reducing the surge to a level the semiconductor can tolerate.

Commutation transients are associated with the reverse recovery characteristic of a rectifier junction. In normal use a semiconductor is continually switching from a conducting state to a nonconducting state. This switching process causes rapid changes of current (high di/dt). Especially where fast switching rectifiers are used, keep circuit inductance at a minimum. Here again, suitable suppression networks should be designed into the circuit.

Regenerative surges in inductive or dynamic loads is the third major source of voltage transients. Such loads include motors, lifting magnets, solenoids, relays, and many other devices involving stored inductive energy in the form of high induced voltage. Reducing or eliminating these transient voltages usually requires protective devices with high energy storage capacity.

Some of the common devices used for transient protection are capacitors, zener diodes,

Table 7.2 Motor Overloads and Their Effect on Semiconductors

Problem	Cause	Motor Current Variance	Semiconductor Current or Voltage Variance	Common Semiconductor Protection	Comments
Overload	Operators choice	Increases three phase current approaching 200%	Increases semiconductor current, eventual motor burnout can cause surge on semiconductors.	Thermal overload on semiconductors and motors. Design for semiconductor current overload.	Operation at thermal overload conditions can reduce semiconductor life.
Motor Jam	Load blockage	High operating time at locked rotor current to 600%.	Increases semiconductor current, eventual motor burnout can cause surge on semiconductors.	Thermal overload on semiconductors and motors. Design for semiconductor current overload.	Depending on the design up to 600% increases in semiconductor currents are possible resulting in reduced life or shorted semiconductors.
High On/Off Duty Cycle	Jogging for high production needs.	High operating time at locked rotor current to 600%.	Increases semiconductor current, eventual motor burnout can cause surge on semiconductors.	Thermal overload on semiconductors and motors. Design for semiconductor current overload.	Surge currents can also reduce semiconductor lifetime and cause semiconductor failures.
Rapid Reversing	Production needs	High operating time at locked rotor current to 600%.	Increases semiconductor current, eventual motor burnout can cause surge on semiconductors.	Thermal overload on semiconductors and motors. Design for semiconductor current overload.	High ambient temperature results the same results as high currents i.e. reduced life and failed semiconductors.
Long Acceleration Time	High inertia slow starting loads	High operating time at locked rotor current to 600%.	Increases semiconductor current, eventual motor burnout can cause surge on semiconductors.	Thermal overload on semiconductors and motors. Design for semiconductor current overload.	Never operate semiconductors without proper cooling.
High Equipment Temperature.	High ambient temperatures, Lack of cooling	No increase but can cause wiring and insulation failure.	Increases semiconductor current, eventual motor burnout can cause surge on semiconductors.	Thermal overload on semiconductors and motors. Design for semiconductor current overload.	
Motor Wiring or Bearing Failure.	Overcurrent, Improper installation or Maintenance	Locked rotor current to 600%.	Increases semiconductor current, eventual motor burnout can cause surge on semiconductors.	Thermal overload on semiconductors and motors. Design for semiconductor current overload.	
Phase Failure	Blown Fuse Loose Connection	Decrease in current until motor core reaches saturation and current increases.	Increases semiconductor current, eventual motor burnout can cause surge on semiconductors.	Thermal overload Phase failure relay.	
Phase Unbalance	Unbalanced single phase loads on same line, poorly regulated service	Decrease in one phase and increase in the other two phases of current.	Increases semiconductor current, eventual motor burnout can cause surge on semiconductors.	Thermal overload Phase failure relay.	
Overvoltage	High source	Slight voltage increase decreases current. Large increase may saturate core and increase current.	Decrease or increase semiconductor current.	Overvoltage relay.	Increase of currents can cause semiconductor failure and reduced life.
Transient Overvoltage	Lightning, opening inductive switches etc.	Slight average current increases.	High transient voltages across semiconductor when semiconductor is in the off position.	Capacitor-Resistor Networks, Voltraps MOV, etc.	Short Duration high transients can cause semiconductor failure.
Underload	Operators choice	Decrease in motor current.	Decrease in semiconductor current.	None	Light load increases semiconductor life.
Contaminants	Dirty Environment	Causes Corona	Causes Corona	Clean Room Filters	Clean equipment periodically even with filters. Corona carbonizes dirt and causes eventual semiconductor shorts.

free-wheeling diodes, and varistors (MOVs). A discussion of the MOV is included at the end of the previous chapter.

Power semiconductor circuits must be protected from current overloads caused by short circuits or other component breakdowns. The following methods are available for this type of protection.

1. Semiconductor fuses protect against overloads, and when properly applied remove the semiconductor from the power source when an overload occurs. Fuses, however, cause more downtime and higher operating costs. Therefore, fusing is generally limited to applications where the power source can damage components before the slower circuit breakers remove the power.

2. Magnetically operated breakers provide an inexpensive means of limiting current. This type of mechanical breaker operates best under short circuit conditions and offers speedy, low-cost restarting and relatively fast circuit interruption.

3. Thermal breakers employ heating elements to operate bimetallic contact actuators. This type of breaker offers reasonable protection for wires and good protection for components with high thermal capacity. Their ability to protect semiconductors is limited, however.

4. Overrated semiconductors, semiconductors with current ratings high enough to accommodate anticipated current overloads, offer another method of protection. This approach allows the cost of fuses, wiring, and mounting to be put into the semiconductor. It also gives only minimal protection. Adequate safety measures require a conventional circuit breaker to disable the circuit in extreme malfunctions.

5. Combinations of circuit breakers with fuses, oversized semiconductors, or current feedback circuits are frequently applied protection techniques. Another approach is using special branch protection. For example, each leg of a single-phase bridge might be separately protected instead of using a single fuse or breaker to fuse the mains. This arrangement prevents overstressing a motor while the semiconductor devices are operating at high current peaks that are within fuse limits.

6. Semiconductors and semiconductor equipment must be kept clean. Why is this so important? Because the glass or ceramic seal on a high-power semiconductor package can be a source of trouble when dirt accumulates. Good engineering practice calls for the positioning of components and support material to reduce or eliminate excess voltage stress and voltage gradients. These gradients can change with moisture, dirt, pressure, temperature, and aging. In a clean system high transients can initiate corona that will then be extinguished upon return to normal voltage. However, with any buildup of moisture or dirt on the insulating glass or ceramic surface, the corona will remain when the voltage returns to normal. This residue can cause extensive damage to the circuit. Even semiconductors without a visible layer of contaminants may be covered with conductive particles that can cause corona. Therefore, periodic cleaning is advisable. The time between cleanings can be lengthened by the proper use of filters, but filters will not eliminate the need for cleaning. A good maintenance program is essential to ensure proper operation of semiconductors.

Preventive Maintenance

The key to successful and efficient operation of high-power semiconductors is a good, well-planned maintenance program. Routine maintenance of equipment using semiconductors must include good operating procedures. Emphasis should be placed on detection and prevention of (1) temperature buildup, (2) dirt accumulation, and (3) loose mountings and connections. A regular maintenance schedule should be followed for checking and eliminating these conditions.

Temperature Buildup.   Temperature buildup may be caused by excess ambient temperature, poor or blocked air circulation, failure of cooling devices such as fans or water circulating equipment, dirty heat sinks and air filters, and equipment overloading. Since temperature buildup usually is gradual, it should be constantly monitored with strategically placed thermometers. Thermocouples often are used to measure semiconductor temperatures. One method of attaching a thermocouple to a semiconductor base or heat sink is first to drill a small shallow hole in the device. Then fit the thermocouple into it and peen the surface around the hole to secure it. Equipment designed for forced air cooling or water cooling should never be energized without proper air or water flow.

Dirt Accumulation.   Dirt accumulation on the glass or ceramic surfaces of semiconductor packages must be regularly and thoroughly removed. The safest method of cleaning semiconductors is the one that gives the best results with the fewest disadvantages. Solvents can be hazardous and should be used with care. Even a solvent that is considered nontoxic can kill or injure when used with improper ventilation. One of the safest methods of cleaning is simply by wiping with clean cloths. Often the areas that need to be cleaned cannot be reached properly, however, and a liquid cleaner must be used. Table 7.3 lists solvents that can be used as cleaners.

The safest method of cleaning semiconductors is to use a detergent wash and then flush with distilled water. The system must be completely dry before the reapplication of power. Regular tap water may contain many conductive impurities and should not be used as a final rinse.

Methyl, ethyl, and isopropyl alcohol can be used for cleaning, but they can attack sleevings and insulations. Before using alcohol or any solvent, test the cleaner on a small sample of sleeving and insulation to determine if they are attacked. The amount of time a material is exposed to a solvent determines adverse solvent effects. The time of exposure during the test, therefore, should be the same as the time of exposure expected during actual cleaning. If you observe any stretching, softening, or deterioration of the material being cleaned, do not use the solvent, although a temporary softening may not necessarily be detrimental. Remember that alcohol is toxic and should be used with care.

Follow all safety precautions when using any solvents. Read the labels on all the chemicals. Do not mix solvents (or any chemicals) unless recommended. The information in Table 7.3 regarding the fire, explosion, and toxicity hazards of solvents is believed to be accurate, but cannot be guaranteed. This information provides very general guidelines only; always carefully follow any precautions and directions for use printed on a solvent label.

Loose Mountings and Connections.   Along with scheduled periodic cleaning, a good maintenance program should include regular checks for mounting and terminal connection tightness. If a loosely mounted device is discovered, it should be removed. Next, both mounting surfaces should be cleaned thoroughly, thermal compound reapplied, and the device remounted to the specified torque.

Safety

Safety in operating and servicing solid-state equipment is especially important. Unlike mechanical equipment that has rotating parts, moving contactors, and so on, there is nothing

Table 7.3   Solvents for Cleaning Semiconductors

Solvent	Fire Hazard	Explosion Hazard	Toxicity Hazard	Electrically Conductive	Attacks Rubber Insulation
Water, tap	None	None	None	Yes	None
Precautions: Rinse with distilled water and dry thoroughly.					
Water, distilled	None	None	None	No	None
Precautions: Dry thoroughly.					
Water and detergent	None	None	None	Yes	None
Precautions: Rinse with distilled water and dry thoroughly.					
Methyl alcohol	High	High	High	No	Slight
Precautions: Avoid skin contact and use ventilation.					
Ethyl alcohol	High	High	High	No	Slight
Precautions: Can cause internal damage; ingestion can cause blindness.					
Isopropyl alcohol	High	High	High	No	Slight
Precautions: Mix with distilled water to reduce hazards.					
Paint thinner (mineral spirits)	Low	Low	Low	No	Slight
Precautions: Use proper ventilation and limit exposure to rubber.					
Acetone	High	Moderate	Low	No	Slight
Precautions: Limit exposure to prevent rubber degeneration.					
Perchloroethylene (dry-cleaning solvent)	Low	Low	Moderate	No	Slight
Precautions: Use short exposure to prevent rubber damage; irritates eyes and causes headaches; heating causes fumes.					
Trichloroethylene	Low	Low	Moderate	No	Slight
Precautions: Use adequate ventilation and limit exposure time to prevent rubber damage.					
Freon	Low	Low	Low	No	Slight
Precautions: Use adequate ventilation and limit exposure time.					
Maltier XL-100	Low	Low	Low	No	Slight
Precautions: Use adequate ventilation and limit exposure time.					
Miller Stephenson MS-180	Low	Low	Low	No	Slight
Precautions: Use adequate ventilation and limit exposure time.					

to warn the unwary technician that semiconductor equipment is energized. There also is a tendency to believe that voltage and current levels associated with semiconductor electronics are too low to be hazardous. This is a dangerous assumption and not true. Thyristor voltage can be in the thousands of volts. Obvious safety measures must be carefully observed. Most high-power equipment includes built-in safety interlocks. These interlocks ensure that the equipment is turned off before anyone can gain access to the high-voltage areas.

Maintenance personnel, however, often defeat these interlocks to simplify their service work. This action is both careless and foolish. In cases when circuits must be checked while energized, there are a few simple rules that could prevent injury and equipment damage.

First, if possible, always work with another person present—one who knows how to shut down the equipment quickly. Next, make sure that the floor is covered with a rubber mat, and follow the good practice of working with one hand in your pocket. This sounds simple, but it could prevent you from completing an electrical circuit. Always use insulated tools to avoid short circuits that could further damage electronic components and circuits. The following common shop practices for safety are too often overlooked or ignored:

- Always wear safety glasses. Hot metal from a short or a soldering iron can ruin an eye as quickly as a metal chip.
- Keep long hair contained with a cap or net.
- Lock out the disconnect or breaker for the circuit you are about to service.
- Check with a voltmeter to be sure the circuit is completely dead.
- Maintain off-limit areas for high-voltage equipment. Access should be allowed only to authorized personnel.
- When measuring high voltage, use the following procedure:

1. De-energize equipment and tag breakers so power will not be turned on accidentally.
2. Discharge any capacitors that may have a dangerous voltage present.
3. Attach meter probes.
4. Energize equipment and take a reading.
5. De-energize equipment, discharge capacitors, and remove probes.

Too many accidents are caused by carelessness or laziness. Bear in mind that accidents can happen to anyone, so be careful.

CONCLUSION

We have seen that the thyristor is a versatile and powerful control device. Its most prominent disadvantage derives from its use in DC circuits. Despite the problems associated with its use, it is very popular in industry. Because of its widespread applications, a thorough understanding of the thyristor is a necessity for technicians.

QUESTIONS

1. The ramp-and-pedestal control circuit is classified as a _____ control. The ramp-and-pedestal has a greater control over the _____ _____ than other types of phase control.
2. The snap-on effect in some phase controls is due to _____. This effect can be reduced by adding a special diode or by adding another _____ network.
3. The type of switching that occurs only when the supply is at or near 0 V is called _____ switching. This type of switching is appropriate for heating controls but not for _____ controls.
4. The chopper motor control converts pure DC to _____. When the thyristor's on time increases compared with its off time, the _____ voltage at the load increases.
5. Closed-loop, or feedback, control is used to control a motor's speed by using an IC called a _____. This type of control can regulate a motor's speed to within _____% of desired speed.
6. The speed of AC motors is controlled by

varying the number of poles and by varying the _____ _____. In the rectifier-inverter speed control, the supply voltage is converted to DC, and the DC is converted to variable-frequency _____.

7. Explain how phase control varies the power delivered to a load.
8. Describe hysteresis, and explain how it affects some phase control circuits.
9. Why is the UJT used as an oscillator to trigger thyristor circuits?
10. Explain how the ramp-and-pedestal circuit works.
11. Explain the advantages and disadvantages of using zero-voltage switching in control circuits. Define zero-voltage switching.
12. Describe basic converter action, and explain how this action changes the power delivered to a load.
13. Define chopper. Describe its operation. How are choppers commutated?
14. Describe how pulse-width modulation can be used to control a DC motor's speed.
15. List the methods by which an AC motor's speed may be controlled for (a) the universal motor and (b) the synchronous and the induction motors.
16. Explain how a closed-loop speed control system works.
17. Explain the operation of the following circuits: (a) rectifier-inverter and (b) cyclo-converter.
18. Match the types of conversion with the circuits that perform them.
    a. AC to AC   1. inverter
    b. DC to AC   2. rectifier
    c. AC to DC   3. chopper
    d. DC to DC   4. converter
                   5. cycloconverter
19. List the applications of the cycloconverter. What are its main advantages?
20. What are the advantages of using transistors as power switches?
21. What problems occur if transistors are used as power switches?
22. Describe how the power MOSFET overcomes one of the transistor problems you listed in answer to Question 21.
23. Explain how a PLL can be used as a motor speed controller.

PROBLEMS

1. In the full-bridge converter with a flywheel diode, the applied voltage is 230 V AC. With a firing angle of 60°, what is the average armature voltage?
2. A four-pole, 100 hp, 440 V AC motor is designed to operate at 60 Hz and has a base speed of 2000 r/min.
   a. Calculate the V/Hz ratio.
   b. Determine the amount of voltage needed to keep the flux constant when operating the motor at 20 Hz.
   c. Calculate the rotor speed with a slip of 5%.
   d. Find the power output in watts when the motor produces a torque of 100 Nm.
3. A six-pole, 1000 W motor has a rated voltage of 230 V AC at 60 Hz. We would like to operate the motor at 1000 r/min at full-load torque. Slip is 3% at full-load torque.
   a. Calculate the V/Hz ratio.
   b. Calculate the stator voltage needed to operate the motor at 1000 r/min.
   c. Determine the stator frequency needed to operate the motor at 1000 r/min.
   d. Find the power output of this machine when producing 1000 Nm of torque.
   e. Calculate the percentage that full-load torque and power will decrease at the operating frequency.

CHAPTER

8

TRANSDUCERS

OBJECTIVES

On completion of this chapter, you should be able to:

- Define a transducer in terms of input, conversion, and output;
- Explain the transduction principle behind individual transducers;
- Identify the schematic symbols of the optoelectronic transducers;
- Classify individual transducers into one of eight basic areas;
- Calculate the unknown resistance in and the sensitivity of a bridge circuit.

INTRODUCTION

Control and regulation of industrial systems and processes depend on accurate measurement. It would not be going too far to state that a variable must be measured accurately to be controlled. In industrial systems the device that does this measurement is the transducer, or sensor. *Transduction* is the process of converting energy from one form to another. We can define a *transducer* as a device that converts a *measurand* (that which is to be measured) into an output that facilitates measurement. In other words, a transducer converts a variable (fluid flow, temperature, humidity) into an analog of the variable.

Let's use an example: the mercury thermometer. A thermometer is a transducer. The measurand is temperature. The transducer converts temperature into an analog of temperature: fluid level in a glass tube. The level of mercury is directly proportional to (an analog of) ambient temperature.

In this chapter we describe several transducers and the principles by which they operate. Our treatment of this topic is in no way exhaustive. Rather, we intend to describe the major sensors in use in industry today. The sensors covered are primarily electrical sensors, that is, sensors that give an electrical output.

Sensors giving other outputs are discussed when appropriate. In our treatments we have classified sensors according to the variable measured. In this chapter we also briefly discuss bridges, which are common devices for detecting sensor changes.

TEMPERATURE

Temperature is undoubtedly the most measured dynamic variable in industry today. Many industrial processes require accurate temperature measurement, because temperature cannot be precisely controlled unless it can be accurately measured.

Before we discuss temperature sensors, we must define temperature. Very simply, *temperature* is the ability of a body to communicate or transfer heat energy. Alternatively, we can define temperature as the potential for heat to flow. Recall that heat will flow from a hotter body to a cooler one.

Temperature generally is measured on three arbitrary scales: Fahrenheit, Celsius, and Kelvin. The *Fahrenheit scale* is referenced to the boiling point of water at 212°F and the freezing point of water at 32°F. The *Celsius scale* references the boiling point at 100°C and the freezing point at 0°C. The *Kelvin scale* is based on the divisions of the Celsius scale, with absolute zero at 0°. To convert from Celsius to Kelvin, you simply add 273°.

Temperature can be measured in many different ways. For purposes of simplicity, we divide temperature sensing into two areas: mechanical and electrical.

Mechanical Temperature Sensing

Mechanical temperature sensing depends on the physical principle that gases, liquids, and solids change their volume when heated. Furthermore, different substances change volume in differing amounts. For example, liquid mercury expands about 0.01% per degree Fahrenheit. Methyl alcohol, on the other hand, expands at about 0.07% per degree Fahrenheit (0.1% per degree Celsius). In this section we discuss mechanical temperature sensors—devices that convert temperature to position or motion.

Glass-Stem Thermometers. The *glass-stem thermometer* is one of the oldest types of thermometers. Its invention is credited to Galileo in about 1590. As shown in Figure 8.1, it consists of a bulb filled with liquid, a capillary tube, and a supporting glass stem. The capillary tube is quite thin, which helps increase the sensitivity of the device. Some type of magnifying lens is usually included in the glass stem to aid in reading temperature measurement. Linearity of this device is improved by evacuating the remaining air from the tube.

The glass stem has several advantages. It is inexpensive to manufacture and has excellent linearity and accuracy characteristics. It also has several disadvantages. It is fragile and difficult to read, it is not easily used for remote measurement or control, and it allows considerable time lag in temperature measurement. This last disadvantage arises because of the poor thermal conductivity of glass. In spite of these disadvantages, the glass-stem thermometer is still popular in industry today.

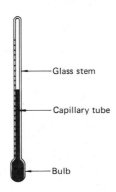

Figure 8.1   Glass-Stem Thermometer

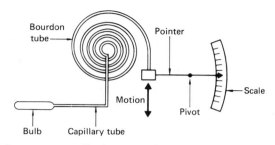

Figure 8.2 Filled-System Thermometer

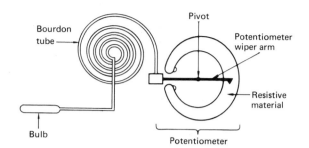

Figure 8.3 Filled System Connected to Potentiometer

Filled-System Thermometers. The *filled-system thermometer* works on the same basic principle as the glass-stem thermometer. The bulb, shown in Figure 8.2, is filled with a gas or a liquid. As the bulb is heated, the gas or liquid expands, exerting pressure on the Bourdon tube (to be described later in the chapter) through the capillary tubing. The Bourdon tube uncoils, moving the pointer. Since this system exerts a pressure, it can be used for chart-recording displays of temperature.

Although this system is mechanical, it can be converted into an electrical transducer by using a potentiometer, as shown in Figure 8.3, or a linear variable differential transformer (LVDT). Potentiometers and LVDTs convert position or displacement into an electrical parameter. These devices are discussed later in the chapter.

Filled systems react very quickly to temperature changes, can be as accurate as 0.5%, and can be used for remote measurement up to 300 ft (100 m). The major disadvantage with filled systems is the need for temperature compensation. Although the sensing element in the filled system is the bulb, the Bourdon tube and capillary tube are also temperature-sensitive. Thus, ambient temperature changes around the Bourdon and capillary tubes can give a false temperature reading. Filled systems can be temperature-compensated by use of bimetal strips or systems using dual elements.

Bimetallic Thermometers. The *bimetallic thermometer* operates on the principle of differential expansion of metals; that is, metals increase their volume when heated and different metals expand by different amounts. How much a metal expands when heated is indicated by a parameter called the *linear expansion coefficient*. This parameter shows volume expansion in millionths per degree Celsius. A list of the parameter values for a few metals is given in Table 8.1.

Table 8.1 Linear Expansion Coefficient of Various Metals

Substance	Linear Expansion Coefficient (millionths/°C)
Aluminum	23.5
Brass	20.3
Copper	16.5
Invar (copper-nickel alloy)	1.2
Kovar (copper-nickel-cobalt alloy)	5.9

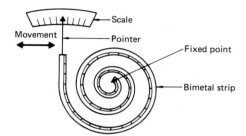

Figure 8.4   Bimetallic Thermometer

When two metals are joined together and heated, physical displacement occurs. In designing these devices engineers try to ensure maximum movement with a given temperature change. They do so by joining a metal with a low expansion coefficient to one with a high expansion coefficient. A popular combination is brass and invar (a copper-nickel alloy). From Table 8.1, we see that this combination meets the design criteria.

A bimetallic thermometer is shown in Figure 8.4. The bimetal strips usually are wound in a spiral or helix (coil). The helical type is more sensitive.

The bimetallic thermometer is one of the most popular thermometers in industry. It is relatively inexpensive, has a wide range (800°F, or 400°C), is rugged, and is easily installed and read. Its major disadvantages are that it cannot be used in remote measurement or in analog process control. Bimetallic sensors are used in simple on-off control systems. For this use a mercury switch is placed on the bimetal strip, as in home thermostats.

An important calculation when using mercury switches concerns how much the strip moves with a given temperature change. The amount of deflection of a straight bimetallic strip is given by the following equation:

$$y = \frac{3(c_A - c_B)(T_2 - T_1)l^2}{4d} \qquad \textbf{(8.1)}$$

where

y = amount of deflection

c_A = the linear expansion coefficient of metal A

c_B = the linear expansion coefficient of metal B

T_1 = lower temperature

T_2 = higher temperature

l = length of strip

d = thickness of strip

This formula tells us that the deflection of a straight bimetal strip is directly proportional to the change in temperature and inversely proportional to the thickness of the strip. The amount of deflection is directly proportional to the square of the length of the strip.

For example, suppose we have a 20 cm long, 0.05 cm thick, brass-invar strip that is straight at room temperature (25°C). How much will the strip deflect if the temperature changes 5°C? From Table 8.1 we see that invar and brass have coefficients of 1.2 and 20.3, respectively. Using Equation 8.1, we find that the deflection is 0.144 cm:

$$y = \frac{3[(2.03 \times 10^{-5}) - (1.2 \times 10^{-6})](5°C)(20 \text{ cm})^2}{4(0.05 \text{ cm})}$$

$$= 0.573 \text{ cm}$$

Note that the amount of deflection is directly proportional to the differences between the temperatures and the coefficients of expansion and proportional to the square of the length. Also, the deflection is inversely proportional to the strip's thickness.

The equation for finding the deflection of a spiral strip is similar to the equation for a straight strip:

$$y = \frac{9(c_A - c_B)(T_2 - T_1)rl}{4d} \qquad \textbf{(8.2)}$$

where

r = spiral's radius

l = length of strip if it were extended

We can see the value of the spiral strip if we use the data from the previous example, but convert the straight strip to a spiral. Let us say we wind the 20 cm strip into a spiral with a radius of 4 cm. Using Equation 8.2 and substituting values, we get

$$y = \frac{9(c_A - c_B)(T_2 - T_1)rl}{4d}$$

$$= \frac{9[(2.03 \times 10^{-5}) - (1.2 \times 10^{-6})](5°C)(4)(20)}{4(0.05 \text{ cm})}$$

$$= 0.340 \text{ cm}$$

You will notice that, all things being equal, the spiral bimetallic strip is more sensitive than the straight strip. Spiral winding makes the response of this device linear over a 200°C range.

We see in the bimetallic strip a good example of a practical application of the principle of the thermal expansion of materials. These sensors are used in single-temperature control circuits to make or break an electrical contact that produces corresponding changes in temperature.

Electrical Temperature Sensing

We have considered several methods of mechanical sensing. Because mechanical rather than electrical principles are used, these sensors are not suitable for use in analog process control systems. In this section, therefore, we discuss four commonly used electrical sensors: the thermocouple, the thermistor, the resistance temperature detector, and the semiconductor temperature sensor.

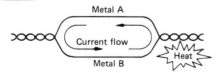

Figure 8.5 Seebeck Effect

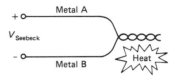

Figure 8.6 Potential Difference across Heated Junction

Thermocouple. Of the electrical temperature-sensing devices, the *thermocouple* enjoys the widest use in industry. Its discovery dates back to 1821, when Thomas Seebeck, a German physicist, joined two wires made of different metals. He found that when he heated one end, electric current flowed in the loop formed by the wires. This effect is called the *Seebeck effect* and is illustrated in Figure 8.5.

When the circuit was broken, as shown in Figure 8.6, Seebeck observed that there was a voltage between the two terminals. He discovered that the size of the voltage varied with heat. An increase in heat caused an increase in voltage. He also found that different combinations of metals produced different voltages.

Thermocouples come in many different forms, as shown in Figure 8.7. The thermocouple junction may be exposed, grounded, fashioned into washers, or placed in a well for protection. Each of these methods has its own advantages.

Generally speaking, the thermocouple is simple, rugged, inexpensive, and capable of the widest temperature measurement range (about 4500°F, or 2500°C) of any electrical temperature transducer. On the other hand, it is the least sensitive and stable of the electrical temperature transducers. Perhaps its biggest disadvantage is the need for a reference junction.

Consider the schematic diagram in Figure 8.8. As shown in the figure, we must connect a readout device to a thermocouple to read the voltage produced. But in doing so, we create two more thermoelectric junctions, as indicated in Figure 8.9. These extra junctions inter-

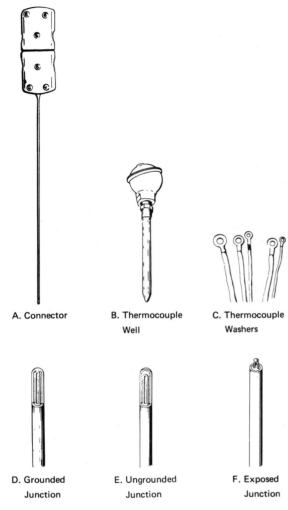

A. Connector

B. Thermocouple
Well

C. Thermocouple
Washers

D. Grounded
Junction

E. Ungrounded
Junction

F. Exposed
Junction

Figure 8.7 Thermocouples

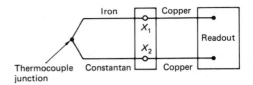

Figure 8.8 Indicating Device Connected to Thermocouple

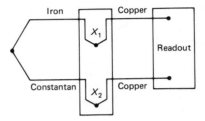

Figure 8.9 Two Additional Thermocouple Junctions Created

fere with the voltage produced by the iron-constantan thermocouple. (Constantan is an alloy of copper and nickel.)

This problem can be overcome by keeping the X_1 and X_2 junctions at a constant reference temperature. If these junctions are held at a constant temperature, it is a simple matter of converting the total voltage produced at the readout device to a temperature. As a matter of

fact, most conversion tables are based on a reference junction temperature of 32°F (0°C). This reference temperature has been chosen because for many years the reference junction was immersed in an ice bath to keep its temperature constant.

Because ice baths are often inconvenient to use and therefore impractical, several other methods are used to achieve the same result. One such method is shown in Figure 8.10. Both reference junctions and the bridge resistor R_2 are thermally integrated on a substrate. The resistor R_2 is a temperature-sensitive resistor. This resistor will change its resistance with ambient temperature changes, producing a voltage that is opposite to the reference voltage change. The remaining resistors in the bridge are not sensitive to temperature changes. As the temperature of the cold junction varies, the bridge produces a voltage to cancel out the cold junction potential produced. The only voltage that varies is the one that is produced by the thermocouple.

Another method for compensating ther-

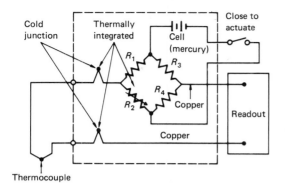

Figure 8.10 Thermocouple System with Thermally Integrated Cold Junctions and Resistors

mocouples, popular in recent years, uses a computer. The reference junction temperature is measured accurately with an electrical transducer. An ADC changes the analog temperature to a digital representation of the temperature and sends it into the input port of a computer. A computer program uses the reference junction measurement to calculate the required correction voltage and makes an estimate of the temperature of the measuring junction.

Thermocouple temperature versus emf curves are shown in Figure 8.11. The different thermocouples are represented by different letter designations.

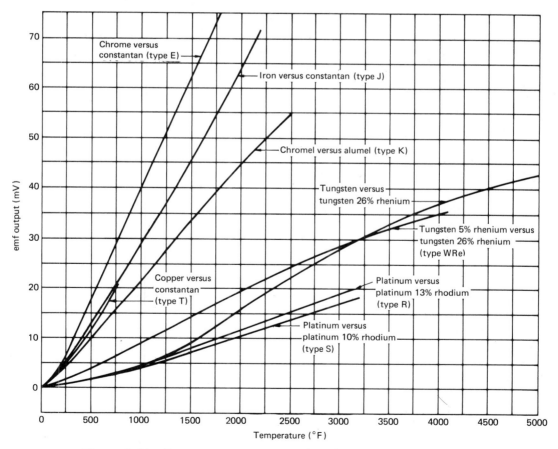

Figure 8.11 Temperature versus emf (Reference Junction at 32°F)

Tables are available that give the output voltage of a particular type of thermocouple at various temperatures. These charts usually use a reference junction of 0°C, as mentioned previously. Table 8.2 is an example of a table for a type J thermocouple. Notice that the type J device puts out 5.431 mV at 103°C.

Keep in mind that the voltage shown is a function of the temperature difference between the two junctions. Knowing this fact we can use this table for references other than 0°C. Using type J as an example, we can find the voltage when our measuring junction is 250°C with the reference at 25°C. At 250°C, the output from the measuring junction is 13.553 mV, while the reference junction's output is 1.277 mV. The thermocouple's output voltage is the difference between the two voltages—that is, 13.553 mV − 1.277 mV = 12.176 mV. Reversing this process allows us to find the temperature given the voltage output of the thermocouple.

As an example, we observe an output of 2.69 mV from a type J thermocouple (iron-constantan) with a reference junction at 0°C. What is the temperature at the measuring junction? Looking down through the table of voltages in Table 8.2, we see that 2.69 mV corresponds to a temperature of 52°C. A similar procedure is used when a measurement is taken with a reference junction at another temperature. If we observe an output voltage of 16.05 mV from a type J thermocouple with a reference junction at a room temperature of 27°C, what is the measuring junction temperature? We cannot use the 16.05 mV voltage in Table 8.2, since the table assumes a reference junction of 0°C. We must first get a corrected voltage by subtracting from the output voltage the voltage produced by a type J thermocouple at 27°C. The corrected voltage is 16.05 mV − 1.277 mV, which equals 14.773 mV. Looking in our type J table, we see that this value corresponds to a measuring junction temperature of 272°C.

At this point, we should note that the temperature of a thermocouple (or any sensor) takes a certain amount of time to react to a temperature change. The thermocouple's response time depends on the mass of the thermocouple, the specific heat of the thermocouple, the coefficient of heat transfer from one boundary to another, and the area of contact between the thermocouple and the material to be measured. These factors are related in the following equation:

$$t_c = \frac{mc}{kA} \tag{8.3}$$

where

t_c = thermal time constant

m = mass of sensor in g

c = specific heat in cal/g°C (calories per gram-degree Celsius)

k = heat transfer coefficient in cal/cm-s°C (calories per centimeter-second-degree Celsius)

A = area of contact between sensor and sample in cm^2

Just as in a capacitor, the rate of change of a thermocouple's temperature follows the time constant curve. Using Equation 8.3 we can calculate the time constant for any thermocouple. For example, let us suppose a thermocouple has a mass of 0.002 g, a specific heat of 0.08 cal/g°C, a heat transfer coefficient of 0.01 cal/cm-s°C, and an area of 0.02 cm^2. The time constant, based on this information, is 0.8 s, as shown in the following calculation:

$$t_c = \frac{(0.002 \text{ g})(0.08 \text{ cal/g°C})}{(0.01 \text{ cal/cm-s°C})(0.02 \text{ cm}^2)} = 0.8 \text{ s}$$

When temperature changes quickly, the temperature of a thermocouple at any time is calculated by using the following equation:

$$T - T_2 = (T_1 - T_2)e^{-t/t_c} \tag{8.4}$$

Table 8.2 Output Voltage for Type J Thermocouple

DEG C	0	1	2	3	4	5	6	7	8	9	10	DEG C
				THERMOELECTRIC VOLTAGE IN ABSOLUTE MILLIVOLTS								
-210	-8.096											-210
-200	-7.890	-7.912	-7.934	-7.955	-7.976	-7.996	-8.017	-8.037	-8.057	-8.076	-8.096	-200
-190	-7.659	-7.683	-7.707	-7.731	-7.755	-7.778	-7.801	-7.824	-7.846	-7.868	-7.890	-190
-180	-7.402	-7.429	-7.455	-7.482	-7.508	-7.533	-7.559	-7.584	-7.609	-7.634	-7.659	-180
-170	-7.122	-7.151	-7.180	-7.209	-7.237	-7.265	-7.293	-7.321	-7.348	-7.375	-7.402	-170
-160	-6.821	-6.852	-6.883	-6.914	-6.944	-6.974	-7.004	-7.034	-7.064	-7.093	-7.122	-160
-150	-6.499	-6.532	-6.565	-6.598	-6.630	-6.663	-6.695	-6.727	-6.758	-6.790	-6.821	-150
-140	-6.159	-6.194	-6.228	-6.263	-6.297	-6.331	-6.365	-6.399	-6.433	-6.466	-6.499	-140
-130	-5.801	-5.837	-5.874	-5.910	-5.946	-5.982	-6.018	-6.053	-6.089	-6.124	-6.159	-130
-120	-5.426	-5.464	-5.502	-5.540	-5.578	-5.615	-5.653	-5.690	-5.727	-5.764	-5.801	-120
-110	-5.036	-5.076	-5.115	-5.155	-5.194	-5.233	-5.272	-5.311	-5.349	-5.388	-5.426	-110
-100	-4.632	-4.673	-4.714	-4.755	-4.795	-4.836	-4.876	-4.916	-4.956	-4.996	-5.036	-100
-90	-4.215	-4.257	-4.299	-4.341	-4.383	-4.425	-4.467	-4.508	-4.550	-4.591	-4.632	-90
-80	-3.785	-3.829	-3.872	-3.915	-3.958	-4.001	-4.044	-4.087	-4.130	-4.172	-4.215	-80
-70	-3.344	-3.389	-3.433	-3.478	-3.522	-3.566	-3.610	-3.654	-3.698	-3.742	-3.785	-70
-60	-2.892	-2.938	-2.984	-3.029	-3.074	-3.120	-3.165	-3.210	-3.255	-3.299	-3.344	-60
-50	-2.431	-2.478	-2.524	-2.570	-2.617	-2.663	-2.709	-2.755	-2.801	-2.847	-2.892	-50
-40	-1.960	-2.008	-2.055	-2.102	-2.150	-2.197	-2.244	-2.291	-2.338	-2.384	-2.431	-40
-30	-1.481	-1.530	-1.578	-1.626	-1.674	-1.722	-1.770	-1.818	-1.865	-1.913	-1.960	-30
-20	-0.995	-1.044	-1.093	-1.141	-1.190	-1.239	-1.288	-1.336	-1.385	-1.433	-1.481	-20
-10	-0.501	-0.550	-0.600	-0.650	-0.699	-0.748	-0.798	-0.847	-0.896	-0.945	-0.995	-10
0	0.000	-0.050	-0.101	-0.151	-0.201	-0.251	-0.301	-0.351	-0.401	-0.451	-0.501	0

DEG C	0	1	2	3	4	5	6	7	8	9	10	DEG C
0	0.000	0.050	0.101	0.151	0.202	0.253	0.303	0.354	0.405	0.456	0.507	0
10	0.507	0.558	0.609	0.660	0.711	0.762	0.813	0.865	0.916	0.967	1.019	10
20	1.019	1.070	1.122	1.174	1.225	1.277	1.329	1.381	1.432	1.484	1.536	20
30	1.536	1.588	1.640	1.693	1.745	1.797	1.849	1.901	1.954	2.006	2.058	30
40	2.058	2.111	2.163	2.216	2.268	2.321	2.374	2.426	2.479	2.532	2.585	40
50	2.585	2.638	2.691	2.743	2.796	2.849	2.902	2.956	3.009	3.062	3.115	50
60	3.115	3.168	3.221	3.275	3.328	3.381	3.435	3.488	3.542	3.595	3.649	60
70	3.649	3.702	3.756	3.809	3.863	3.917	3.971	4.024	4.078	4.132	4.186	70
80	4.186	4.239	4.293	4.347	4.401	4.455	4.509	4.563	4.617	4.671	4.725	80
90	4.725	4.780	4.834	4.888	4.942	4.996	5.050	5.105	5.159	5.213	5.268	90
100	5.268	5.322	5.376	5.431	5.485	5.540	5.594	5.649	5.703	5.758	5.812	100
110	5.812	5.867	5.921	5.976	6.031	6.085	6.140	6.195	6.249	6.304	6.359	110
120	6.359	6.414	6.468	6.523	6.578	6.633	6.688	6.742	6.797	6.852	6.907	120
130	6.907	6.962	7.017	7.072	7.127	7.182	7.237	7.292	7.347	7.402	7.457	130
140	7.457	7.512	7.567	7.622	7.677	7.732	7.787	7.843	7.898	7.953	8.008	140
150	8.008	8.063	8.118	8.174	8.229	8.284	8.339	8.394	8.450	8.505	8.560	150
160	8.560	8.616	8.671	8.726	8.781	8.837	8.892	8.947	9.003	9.058	9.113	160
170	9.113	9.169	9.224	9.279	9.335	9.390	9.446	9.501	9.556	9.612	9.667	170
180	9.667	9.723	9.778	9.834	9.889	9.944	10.000	10.055	10.111	10.166	10.222	180
190	10.222	10.277	10.333	10.388	10.444	10.499	10.555	10.610	10.666	10.721	10.777	190
200	10.777	10.832	10.888	10.943	10.999	11.054	11.110	11.165	11.221	11.276	11.332	200
210	11.332	11.387	11.443	11.498	11.554	11.609	11.665	11.720	11.776	11.831	11.887	210
220	11.887	11.943	11.998	12.054	12.109	12.165	12.220	12.276	12.331	12.387	12.442	220
230	12.442	12.498	12.553	12.609	12.664	12.720	12.776	12.831	12.887	12.942	12.998	230
240	12.998	13.053	13.109	13.164	13.220	13.275	13.331	13.386	13.442	13.497	13.553	240
250	13.553	13.608	13.664	13.719	13.775	13.830	13.886	13.941	13.997	14.052	14.108	250
260	14.108	14.163	14.219	14.274	14.330	14.385	14.441	14.496	14.552	14.607	14.663	260
270	14.663	14.718	14.774	14.829	14.885	14.940	14.995	15.051	15.106	15.162	15.217	270
280	15.217	15.273	15.328	15.383	15.439	15.494	15.550	15.605	15.661	15.716	15.771	280
290	15.771	15.827	15.882	15.938	15.993	16.048	16.104	16.159	16.214	16.270	16.325	290

Table 8.2 (*continued*)

DEG C	0	1	2	3	4	5	6	7	8	9	10	DEG C
300	16.325	16.380	16.436	16.491	16.547	16.602	16.657	16.713	16.768	16.823	16.879	300
310	16.879	16.934	16.989	17.044	17.100	17.155	17.210	17.266	17.321	17.376	17.432	310
320	17.432	17.487	17.542	17.597	17.653	17.708	17.763	17.818	17.874	17.929	17.984	320
330	17.984	18.039	18.095	18.150	18.205	18.260	18.316	18.371	18.426	18.481	18.537	330
340	18.537	18.592	18.647	18.702	18.757	18.813	18.868	18.923	18.978	19.033	19.089	340
350	19.089	19.144	19.199	19.254	19.309	19.364	19.420	19.475	19.530	19.585	19.640	350
360	19.640	19.695	19.751	19.806	19.861	19.916	19.971	20.026	20.081	20.137	20.192	360
370	20.192	20.247	20.302	20.357	20.412	20.467	20.523	20.578	20.633	20.688	20.743	370
380	20.743	20.798	20.853	20.909	20.964	21.019	21.074	21.129	21.184	21.239	21.295	380
390	21.295	21.350	21.405	21.460	21.515	21.570	21.625	21.680	21.736	21.791	21.846	390
400	21.846	21.901	21.956	22.011	22.066	22.122	22.177	22.232	22.287	22.342	22.397	400
410	22.397	22.453	22.508	22.563	22.618	22.673	22.728	22.784	22.839	22.894	22.949	410
420	22.949	23.004	23.060	23.115	23.170	23.225	23.280	23.336	23.391	23.446	23.501	420
430	23.501	23.556	23.612	23.667	23.722	23.777	23.833	23.888	23.943	23.999	24.054	430
440	24.054	24.109	24.164	24.220	24.275	24.330	24.386	24.441	24.496	24.552	24.607	440
450	24.607	24.662	24.718	24.773	24.829	24.884	24.939	24.995	25.050	25.106	25.161	450
460	25.161	25.217	25.272	25.327	25.383	25.438	25.494	25.549	25.605	25.661	25.716	460
470	25.716	25.772	25.827	25.883	25.938	25.994	26.050	26.105	26.161	26.216	26.272	470
480	26.272	26.328	26.383	26.439	26.495	26.551	26.606	26.662	26.718	26.774	26.829	480
490	26.829	26.885	26.941	26.997	27.053	27.109	27.165	27.220	27.276	27.332	27.388	490
500	27.388	27.444	27.500	27.556	27.612	27.668	27.724	27.780	27.836	27.893	27.949	500
510	27.949	28.005	28.061	28.117	28.173	28.230	28.286	28.342	28.398	28.455	28.511	510
520	28.511	28.567	28.624	28.680	28.736	28.793	28.849	28.906	28.962	29.019	29.075	520
530	29.075	29.132	29.188	29.245	29.301	29.358	29.415	29.471	29.528	29.585	29.642	530
540	29.642	29.698	29.755	29.812	29.869	29.926	29.983	30.039	30.096	30.153	30.210	540
550	30.210	30.267	30.324	30.381	30.439	30.496	30.553	30.610	30.667	30.724	30.782	550
560	30.782	30.839	30.896	30.954	31.011	31.068	31.126	31.183	31.241	31.298	31.356	560
570	31.356	31.413	31.471	31.528	31.586	31.644	31.702	31.759	31.817	31.875	31.933	570
580	31.933	31.991	32.048	32.106	32.164	32.222	32.280	32.338	32.396	32.455	32.513	580
590	32.513	32.571	32.629	32.687	32.746	32.804	32.862	32.921	32.979	33.038	33.096	590
600	33.096	33.155	33.213	33.272	33.330	33.389	33.448	33.506	33.565	33.624	33.683	600
610	33.683	33.742	33.800	33.859	33.918	33.977	34.036	34.095	34.155	34.214	34.273	610
620	34.273	34.332	34.391	34.451	34.510	34.569	34.629	34.688	34.748	34.807	34.867	620
630	34.867	34.926	34.986	35.046	35.105	35.165	35.225	35.285	35.344	35.404	35.464	630
640	35.464	35.524	35.584	35.644	35.704	35.764	35.825	35.885	35.945	36.005	36.066	640
650	36.066	36.126	36.186	36.247	36.307	36.368	36.428	36.489	36.549	36.610	36.671	650
660	36.671	36.732	36.792	36.853	36.914	36.975	37.036	37.097	37.158	37.219	37.280	660
670	37.280	37.341	37.402	37.463	37.525	37.586	37.647	37.709	37.770	37.831	37.893	670
680	37.893	37.954	38.016	38.078	38.139	38.201	38.262	38.324	38.386	38.448	38.510	680
690	38.510	38.572	38.633	38.695	38.757	38.819	38.882	38.944	39.006	39.068	39.130	690
700	39.130	39.192	39.255	39.317	39.379	39.442	39.504	39.567	39.629	39.692	39.754	700
710	39.754	39.817	39.880	39.942	40.005	40.068	40.131	40.193	40.256	40.319	40.382	710
720	40.382	40.445	40.508	40.571	40.634	40.697	40.760	40.823	40.886	40.950	41.013	720
730	41.013	41.076	41.139	41.203	41.266	41.329	41.393	41.456	41.520	41.583	41.647	730
740	41.647	41.710	41.774	41.837	41.901	41.965	42.028	42.092	42.156	42.219	42.283	740
750	42.283	42.347	42.411	42.475	42.538	42.602	42.666	42.730	42.794	42.858	42.922	750
760	42.922											760

where

T = thermocouple temperature

T_1 = starting temperature

T_2 = final temperature

t_c = thermocouple time constant

t = time from start

For example, suppose a thermocouple at 20°C is placed into a 100°C water bath. With the time constant calculated previously, the temperature after 1 s is about 77°C:

$$T - 100° = (20° - 100°)e^{-1/0.8} = 77°C$$

We can see from these calculations that the thermocouple does not react instantly to a temperature change. All sensors take some time to respond to a change in the measurand.

A short time constant is a desirable feature for a thermocouple then, since the thermocouple can monitor changing temperatures without a large time delay. Equation 8.3 indicates four areas where the design of the thermocouple and the system can keep the time constant small. First, the mass of the thermocouple should be kept small. Second, materials with a low specific heat should be used, if possible. This choice often conflicts with the need for a large thermoelectric voltage, a design trade-off. Third, the area of contact between the sensor and the object being measured should be as large as possible. Fourth and last, the thermocouple and the measurand should have a good thermal bond between them.

In some applications a compromise must be made between the thermal time constant and the ruggedness of the sensor. For example, in an application where there is a large amount of vibration, a larger, more rugged device would be chosen. A device with a larger mass to withstand shock and vibration will necessarily have a larger mass and, therefore, a larger time constant.

A recent development in thermocouple technology is the *thin film* thermocouple. The advantage of the thin film thermocouple is its fast speed of response. A look at Equation 8.3 will tell us why this device responds more quickly to temperature changes. As we have just discussed, in the conventional two-wire thermocouple two metal wires are joined together, usually by welding or soldering. The two-wire thermocouple is then attached to the sample, usually with a thermal grease to im-

prove heat transfer. In the thin film thermocouple, two metals are vapor-deposited directly on the material in layers about 10^{-5} cm thick. The thermal mass of this kind of thermocouple is much lower than that of the conventional two-wire thermocouple. Also, since the metals are directly deposited on the material, there is no need to use thermal grease. Both of these factors improve the heat transfer from the sample to the thermocouple. Thin films also may be spread over a larger area than conventional thermocouples.

Thermal time constants on thin film thermocouples have been measured to be as low as 1 μs. These low-mass devices offer significant improvements in response time over conventional thermocouples. Conventional thermocouples with small wires have response times between 0.1 and 0.4 s, considerably longer than the times of thin film thermocouples. The cost of thin film thermocouples is greater than that of conventional thermocouples, however, and thus these devices may not be cost-effective for all applications.

Many of the inaccuracies that result from using thermocouple sensors come from improper applications and inadequate system design criteria. One such system consideration not often taken into account is lead length. Thermocouples must be connected to measuring instrumentation by lead wires. As a general rule of thumb, lead lengths between measuring junction and measuring instrument should be kept as short as possible. If the total resistance of the thermocouple and leads exceeds 100 Ω, you should check to see what resistance your measuring instrument requires. Some instruments will not work properly with resistances over 100 Ω.

Thermocouples are used in industry to measure oven and furnace temperatures. Sometimes they are used as part of a process control system where temperature is not merely measured, but also carefully controlled. Thermocouples also are used to measure the temperature of flowing fluids, especially when their temperature fluctuates widely. A more critical application involves measuring the temperatures in nuclear reactor cores. Sensors need to respond quickly and accurately to allow the system to keep temperature within acceptable limits. Thermocouples, because of their ruggedness and wide temperature range, are also used to measure the temperatures of exhaust gases of nose cones in missile and rocket applications.

Thermistor. A *thermistor* is a thermally sensitive resistor, usually having a negative temperature coefficient. As temperature increases, the thermistor's resistance decreases, and vice versa.

We have talked about the fact that the thermistor has a negative temperature coefficient. The temperature coefficient of resistance of a material, signified by α (the Greek letter alpha), is represented by the following equation:

$$\alpha = \left(\frac{\Delta R}{R_s}\right)\left(\frac{1}{\Delta T}\right)$$

where

ΔR = change in resistance caused by a given temperature change

R_s = resistance of the material at a reference temperature

ΔT = change in temperature above or below the reference temperature

Let us compare the temperature coefficient of iron, which is 50×10^{-4}, with that of nichrome (a nickel-chromium alloy), which is 2×10^{-4}. We can see from these coefficients that iron increases its resistance about 25 times more than nichrome per degree Celsius of temperature rise. Contrast these coefficients with the coefficient of the thermistor, which can be as high as 600×10^{-4} per degree Celsius, corresponding to a resistance change of 6%.

Thermistors are made out of oxides of nickel, manganese, cobalt, copper, and other metals. Figure 8.12 shows the almost bewildering variety of packages in which thermistors come. Modern thermistors have many of the requirements system designers want for tem-

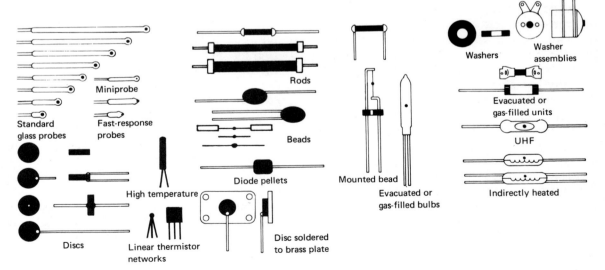

Figure 8.12 Thermistors

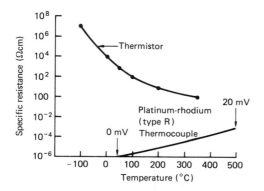

Figure 8.13 Temperature-Resistance Curves for Thermistor and Thermocouple

perature sensors. They have excellent sensitivity, a wide range of impedance levels, small and varied sizes, high precision, and, recently, much better long-term stability.

A typical thermistor's temperature-resistance curve is shown in Figure 8.13. As we can see, there are several differences between this curve and that for the thermocouple, which is also shown in Figure 8.13. First, the thermocouple curve is relatively linear. The thermistor's resistance varies exponentially. As a result, the thermistor's thermal sensitivity is very high, as much as 5% resistance change per degree Celsius. Thus, the thermistor is the most sensitive temperature sensor in common use. The thermistor is therefore reasonably linear over a relatively narrow range of temperatures.

Thermistors often are used in conjunction with a parallel resistor to improve linearity, as shown in Figure 8.14A. Resistor R_s (a shunt resistance) is chosen to equal the resistance of the thermistor at the median temperature expected to be measured. Figure 8.14D shows the change in the response curve when a resistor is added.

There are many ways to use thermistors. Figure 8.14 shows several circuits using thermistors. Note the schematic representation of the thermistor as a resistor with a T inside a circle. The circuit shown in Figure 8.14A represents a potentiometric circuit. As temperature increases, the resistance of the thermistor decreases, increasing the current through the ammeter A. Figure 8.14B depicts a more sensitive bridge circuit, where changes in temperature unbalance the bridge, causing a meter indication. Differential temperature is measured with the circuit shown in Figure 8.14C. Bridge unbalance comes only when there is a difference in temperature between the two sensors. Notice that none of these circuits use an amplifier. Because of the large voltage output produced by a typical thermistor bridge, amplification is normally unnecessary. A 4000 Ω thermistor in a bridge network at 25°C produces approximately 18 mV/°C.

The thermistor is the most sensitive of the electrical temperature sensors. Some thermistors actually double their resistance with a temperature change of 1°C. Thermistors are relatively inexpensive, react quickly to temperature

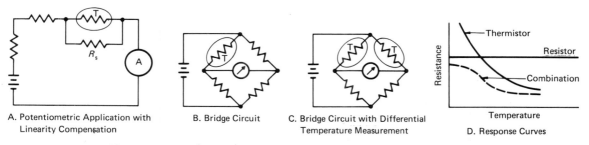

A. Potentiometric Application with Linearity Compensation

B. Bridge Circuit

C. Bridge Circuit with Differential Temperature Measurement

D. Response Curves

Figure 8.14   Thermistors in Potentiometric and Bridge Circuits

changes, and require only very simple circuitry. Resistances range from 0.5 Ω to 80 MΩ. Thermistors are not without disadvantages, however. They are extremely nonlinear and fragile, and they have a limited temperature range.

Not all thermistors have a negative temperature coefficient. Some specialized thermistors have a positive temperature coefficient (PTC) of resistance. These PTC thermistors have a very nonlinear resistance-temperature curve. Their application is usually found in motors to indicate an overload condition.

Most of the applications we have been discussing up to this point have used the thermistor in what is called the *externally heated mode.* In this mode of operation the thermistor changes its resistance due to heating from an external source. Another mode of thermistor operation, frequently used in time-delay circuitry, is called the *self-heating mode.* In this mode the thermistor uses the heating effect of its own current flow. If a thermistor is connected in series with a variable resistor, a source, and a relay, a variable time-delay circuit can be constructed. When current is allowed to flow in the circuit, the amount allowed to flow is small, limited by the high resistance of the cold thermistor. As current flows in the circuit, the thermistor gradually heats up, allowing more current to flow, until the relay is triggered into conduction. Increasing the variable resistance will increase the time lag. With careful circuit design, a wide range of time delays are possible, due to the lag in the response of the self-heated thermistor.

Over the past several years thermistors have become more popular as sensors in industry. They are used as primary sensors; that is, they are used to measure temperature as a variable. Many applications in the chemical industry require temperature control to 1°C. For example, the chemical activity of some enzymes may change as much as 4% with each degree Celsius change in temperature. Thermistors are ideal for this kind of application because of their extremely high sensitivity. Thermistors are also used in compensation circuits for thermocouples, making the thermocouples more accurate and linear.

Thermistors find applications in more than temperature measurement and control. They also are used as secondary sensors to measure liquid level or fluid flow. If a thermistor is placed in a tank, its resistance changes, depending on whether or not the device is immersed in a liquid.

Resistance Temperature Detector (RTD). All metals change resistance when subjected to a temperature change. Pure metals, such as platinum, nickel, tungsten, and copper, have positive temperature coefficients. Thus, for pure metals, temperature and resistance are directly proportional. As temperature increases, a pure metal's resistance will increase. This result is the idea behind the *resistance temperature detector* (RTD).

The resistance of the RTD can be calculated as temperatures above 0°C using the following formula:

$$R_T = R_0(1 + \alpha T)$$

where

R_T = resistance at temperature T

R_0 = resistance at 0°C

α = temperature coefficient of resistance

Let us take an example of an RTD with a resistance of 100 Ω at 0°C and a temperature coefficient of resistance of 0.00392 and calculate the resistance of the RTD at 50°C:

$$\begin{aligned} R_T &= R_0(1 + \alpha T) \\ &= (100\ \Omega)[1 + (0.00392\ \Omega/\Omega/°C)(50°C)] \\ &= 119.6\ \Omega \end{aligned}$$

At 50°C, this RTD would be expected to have a resistance of 119.6 Ω.

RTDs come in many packages; the most popular are shown in Figure 8.15. The helical

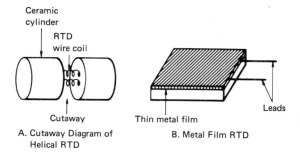

Figure 8.15 Two Popular RTDs

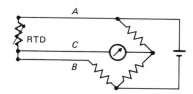

Figure 8.16 Three-Wire Bridge Configuration

RTD (Figure 8.15A) consists of a platinum wire wound in a tight spiral (helix) threaded through a ceramic cylinder. A newer construction technique is found in the metal film RTD (Figure 8.15B). Platinum is usually deposited, or screened, on a small flat ceramic substrate. It is then etched with a laser trimming system and sealed. The film RTD offers a substantial reduction in cost over other forms of RTDs and reacts more quickly to temperature changes.

Common resistance values for RTDs range from 10 Ω (for platinum) to several thousand ohms. The most common value of resistance is 100 Ω. Because of their low resistance, long lead wires can produce inaccurate temperature measurements. This problem can be solved by using a three-wire bridge configuration, as shown in Figure 8.16. If wires A and B are identical in resistance, their resistive effects cancel out, because each is in an opposite leg of the bridge. The third wire (C) carries no current; it is only a sense lead.

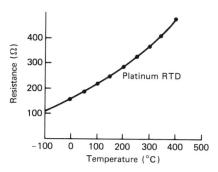

Figure 8.17 Temperature-Resistance Curve for Platinum RTD

Of the temperature sensors we have presented, the RTD is the most accurate and stable. Furthermore, as can be seen from the curves in Figure 8.17, it is very linear and covers a high temperature range. Disadvantages include slowness of response, small resistance changes (usually requiring amplification), and high cost.

A circuit using an RTD is shown in Figure 8.18. It measures temperature between 0° and 266°C, the output ranging between 0 and 1.8 V. The 2.5 V reference is amplified to 6.25 V by the first op amp. The span-adjust sets the output to 1.8 V at 266°C, while the offset is adjusted to 0 V at 0°C.

The three most common metals that make up the sensing element of the RTD are platinum, copper, and nickel. Platinum has the greatest temperature range and stability with moderately good linearity, at about ±0.4°C over the 0°–100°C range. Copper provides nearly perfect linearity, and nickel offers low cost, high resistance, and sensitivity.

Semiconductor Temperature Sensors. In the semiconductor area, evenly doped crystals of germanium can be used to sense temperature near absolute zero. The temperature-resistance response curve of the device resembles that of the thermistor. It has a negative temperature coefficient and is very nonlinear.

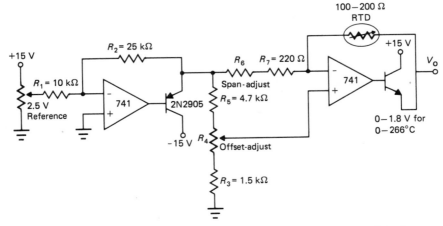

Figure 8.18 Low-Cost Temperature Sensor Using RTD with Fixed Reference and Op Amp

Silicon crystals are also employed as temperature sensors. They usually are shaped in the form of discs or wafers for measuring surface temperature. Their useful range extends from −67° to 275°F. Unlike germanium crystals and thermistors, silicon crystals have a positive linear temperature coefficient in this range. Below this range the temperature-resistance response curve becomes negative and very nonlinear.

Recall that a reverse-biased PN junction conducts a small amount of leakage, or minority carrier, current flow. The amount of current flow depends on temperature. As temperature increases, leakage current increases exponentially. This effect is used to measure temperature with transistors and reverse-biased diodes. Normally, germanium semiconductors are used, because their leakage is much greater than that of silicon.

The diode may be used in a CDA, as illustrated in Figure 8.19. Note that the germanium diode is reverse-biased by the positive supply. As temperature increases, the leakage current will increase. The increase in current will be mirrored in the inverting terminal, increasing

the output voltage of the CDA. The CDA output will then be directly proportional to temperature; as temperature increases, the output voltage will increase, and vice versa.

The bijunction transistor, shown in Figure 8.20, is sometimes used in thermometers. The transistor base-emitter voltage V_{BE} depends on temperature, varying −2.25 mV per degree Celsius. For every degree Celsius increase in temperature, the base-emitter voltage will decrease by 2.25 mV.

One of the most recent innovations in thermometry is the IC temperature sensor. The output of the IC (either voltage or current) is directly proportional to temperature. Although limited in temperature range (below +300°F),

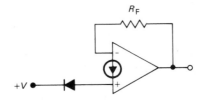

Figure 8.19 CDA with Diode as Temperature Sensor

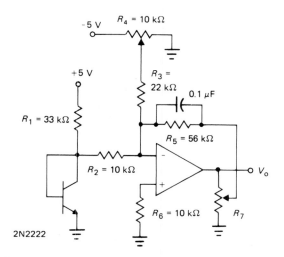

Figure 8.20 Op Amp Thermometer with Transistor as Temperature Sensor

it produces a very linear output over the operating range. Linearity exceeds that of the RTD. A typical sensitivity value is 1 μA/°C or 10 mV/°C. Figures 8.21A and 8.21B show the voltage and current modes of the AD590 manufactured by Analog Devices.

Generally speaking, semiconductor temperature sensors produce good linearity, are small and low in cost, and produce a high impedance output.

Tables 8.3 and 8.4 list the temperature

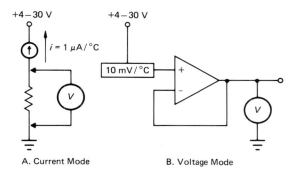

A. Current Mode B. Voltage Mode

Figure 8.21 IC Temperature Sensor

ranges of the thermometers we have discussed in this section and give a summary of information about them.

HUMIDITY

Many industrial processes depend on accurate assessment and control of humidity. *Humidity* is defined as the amount of water vapor in the air. How much water vapor air can hold depends on several factors, the most important of which is the air temperature. Warm air can hold more water vapor than cold air can. The most common measure of water vapor in air is called relative humidity. *Relative humidity* is a ratio of the amount of moisture air holds at a certain temperature compared with how much it could hold at that temperature.

In this section we discuss humidity sensors in two classifications: direct methods through hygrometers and indirect methods by comparing changes in temperature.

Generally, the measurement of the amount of moisture in a gas is accomplished by one of three techniques. First, the water can be extracted from the sample and either weighed or detected in some other manner. Second, some property of the water may be measured to estimate how much water is in a sample. The properties of water sometimes used are conductivity, dielectric constant, and infrared, ultraviolet, or microwave absorption. Third, water content may be analyzed by a change it makes in another material. This principle is used in the hair hygrometer, discussed a little later in this chapter. When you are reading through this section on humidity sensors, try to identify each sensor with one of these three techniques.

Some areas of industry use other methods to measure the amount of water in a material. The mass or volume ratio is often used. This ratio is specified in a percentage or in parts per

Table 8.3 Temperature Ranges of Mechanical and Electrical Temperature Transducers

SENSOR RANGES

°F	−450	−300	−100	0	100	200	500	1000	2000	5000	10000
°C	−450	−200 −300	−100 −100	0 0	100 100	200 200	500 500	500 1000	1000 2000	2000 5000 5000	10000
°F	−459.67	−297.332		32.02	212		787.244		1763.474 1947.974		

FIXED CALIBRATION POINTS	ABSOLUTE ZERO	BOILING POINT OF OXYGEN	TRIPLE POINT OF WATER	BOILING POINT OF WATER	FREEZING POINT OF ZINC	FREEZING POINT OF SILVER	FREEZING POINT OF GOLD

| °C | | −273.15 | −182.962 | 0.01 | 100 | 419.58 | 961.93 | 1064.43 |
| °F | −450 | −300 | −100 | 0 | 100 | 200 | 500 | 1000 | 2000 | 5000 | 10000 |

Sensor	Range
Glass Stem	−321; −200 to 600; 1100
Bimetalic	−100; −80 to 810; 1000
Filled	−450; −300 to 1200; −1400
Thermistors	−150 to 600
Resistance Thermometers	
Platinum	−450; −364 to 1382; 1800
Nickel	−150; −76 to 356; 600
Copper	−100 to 300
Balco	32 to 400
Tungsten	−100 to 5000
Semiconductors	
Silicon	−100 to 300
Thermocouples	
J(Fe/Const)	32 to 1400
K(NiCr/NiAl)	−418; 32 to 2300; −2500
T(Cu/Const)	−450; −300 to 700
E(NiCr/Const)	−450; −300 to 1600
R(Pt87-Rh13/Pt)	32 to 2700
S(Pt90-Rh10/Pt)	32 to 2700
B(Pt70-Rh30/Pt94-Rh)	1600 to 3100
WRe(W95-Re5/W74-Re26)	32 to 4200
KPAuFe(NiCr/AuFe-.07)	−454 to 45
Pyrometers	
Radiation	0 to 7000
Optical	1400 to 6300
Infrared	0 to 6000

million (PPM) by weight or volume of a sample. This measure is also called the *specific humidity*. Volumetric specific humidity is usually expressed as parts per million by volume (PPM$_v$). Specific humidity also may be expressed on a weight basis. Like the volumetric measure, the weight, or gravimetric, measure is expressed in parts per million by weight (PPM$_w$). Another measure involves the dew, or frost, point. The *dew point* is the temperature at which water vapor condenses and deposits on a solid surface. It is measured in degrees Celsius or Fahrenheit.

Psychrometers

The *psychrometer*, shown in Figure 8.22, is a common means of measuring relative humidity. Note that its construction is based around two bulbs, one wet and one dry. Air must be passed over the wet bulb, causing the water to evaporate.

Both bulbs contain temperature sensors. If the air temperature remains constant, the dry bulb sensor remains at a constant temperature. However, the wet bulb's temperature varies with the relative humidity. More evaporation

Table 8.4 Summary of Temperature Sensors

Type/Name	Temperature Range	Linearity	Advantages	Disadvantages
Mechanical				
Glass stem	−100°–500°C	Good	Inexpensive Linear Accurate (0.05°C)	Local measurement only Fragile Hard to read Time lag
Bimetallic	−100°–600°C	Good	Inexpensive Wide range Rugged Easy to read and install	Local measurement only
Filled	−200°–650°C	Fair	Reacts quickly Accurate (0.5%) Remote measurement (300 ft) Can be used with chart recorders	Often needs temperature compensation
Electrical				
Thermocouple	−273°–2000°C	Good	Simple, low cost Rugged Wide range Self-powered	Low sensitivity Reference needed Poor stability
RTD	−200°–800°C	Good	Very stable Very accurate Linear Wide range	Slow response Low sensitivity Expensive Self-heating Limited range Subject to thermal runaway
Semiconductor	−50°–150°C	Poor	Inexpensive Small Output impedance high	
IC	−50°–150°C	Excellent	Excellent linearity Inexpensive Very sensitive	Slow response
Thermistor	−100°–300°C	Very poor	Small size Low cost High sensitivity Fast response	Very nonlinear Poor stability at high temperatures Limited range

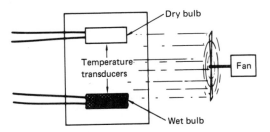

Figure 8.22 Psychrometer Humidity Sensor

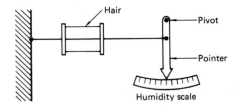

Figure 8.23 Hair Hygrometer

takes place as relative humidity decreases. Thus, as humidity drops, the wet bulb becomes cooler. The difference between the temperature of the two bulbs, then, reflects the relative humidity.

Conversion tables are usually necessary to estimate relative humidity with the psychrometer. However, relative humidity conversion may be accomplished automatically with appropriate hardware or software.

Hygrometers

A *hygrometer* is a device used to measure humidity. These devices contain a material whose properties change when moisture is adsorbed. As we have seen, the psychrometer indirectly measures humidity by comparing the temperatures of two bulbs. The hygrometer, on the other hand, measures humidity directly. In this section we will describe several common hygrometers in use today.

Hair Hygrometer. The simplest and oldest hygrometer is made from human hair or an animal membrane. A simple *hair hygrometer* is shown in Figure 8.23. Human hair lengthens about 3% over a range from 0% to 100% humidity. This change in length is detected by the pointer, giving a readout in relative humidity.

The hair hygrometer is accurate to within 3%. It is used for measuring relative humidity only between 15% and 90% over a temperature range from 1° to 40°C. Although the hair hy-

grometer is strictly mechanical, it can be converted to an electrical sensor by attaching the hair to the core of, say, an LVDT or any appropriate position-to-voltage transducer.

Impedance Hygrometer. Several hygrometers use a change in impedance to detect levels of humidity. One, the *resistance hygrometer,* varies its resistance (and, hence, impedance) as humidity changes. The resistance hygrometer, as shown in Figure 8.24, is composed of two electrodes separated by a thin layer of lithium chloride. The lithium chloride film is *hygroscopic,* which means that it adsorbs moisture from the air. As relative humidity increases, the device's resistance decreases. The change is then sensed by potentiometric or bridge methods.

A similar principle is used in the *electrolytic hygrometer.* In this device two wires, usually made of platinum or rhodium, are wound in a

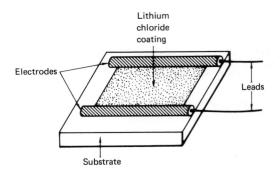

Figure 8.24 Resistance Hygrometer

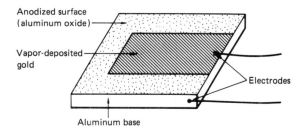

Figure 8.25   Impedance Hygrometer

spiral and encased in a tube. The gas under measurement is continually circulated through the tube. The metal wires are coated with phosphorous pentoxide, a hygroscopic material with properties similar to those of lithium chloride. A high fixed voltage is applied across the two wires, which electrolyzes the water adsorbed by the coating. The amount of current flow is directly proportional to the concentration of water vapor in the gas.

Another hygrometer that uses an impedance change is made out of aluminum oxide. This device, sometimes called a *resistance-capacitance hygrometer,* is shown in Figure 8.25. The surface of the aluminum base is anodized to form a layer of aluminum oxide. Gold is then vapor-deposited on top of the oxide coating. Note that one electrode is attached to the gold film and another is attached to the aluminum base. Since the gold is very thin, water vapor diffuses through to the oxide layer. An increase in relative humidity causes the impedance (resistance and capacitance) of the oxide coating to decrease. The change in impedance is detected by an impedance bridge calibrated to read percent relative humidity.

The aluminum oxide element operates over the range from 0% to 100% relative humidity. The element is very sensitive, small in size, and very linear. Condensation of moisture on the surface of the element does not affect it.

Sorption Hygrometer. The *sorption hygrometer* uses the principle of an oscillating

crystal to measure humidity. The moisture increases the mass of the crystal and decreases the frequency of oscillation. Frequencies used are normally around 9 MHz. Commercial units use two crystals: One is the sensor, and the other is exposed to a dry gas and acts as a reference. This device is also called a *piezoelectric hygrometer,* and can measure moisture in gases from 1 to 25,000 PPM.

Capacitive Hygrometer. The *capacitive hygrometer* works on the principle that the dielectric constant of a capacitor changes in the presence of water. Recall that changes in the dielectric constant vary the capacitance of a capacitor. Measurements of capacitance are made through impedance bridges or oscillator frequency changes.

Microwave Absorption. The absorption of microwave radiation by water vapor is also used in humidity measurements. Microwave radiation is part of the electromagnetic spectrum and has values between 1 and 100 GHz. Water absorbs thousands of times more microwave radiation than a dry gas does. More water vapor in the air increases microwave energy absorption and decreases energy transmitted to a sensor. The level of radiation transmitted is inversely proportional to the relative humidity.

A summary of the humidity sensors discussed in this section is presented in Table 8.5.

LIGHT

In 1887 Heinrich Hertz discovered the photoelectric effect. Albert Einstein received the Nobel Prize in 1921 for his explanation of this effect. Since that time optoelectronic devices and systems have become an integral part of industrial control. Popularity of optoelectronic devices stems from the fact that they are noncontact sensors. No parts wear out from metal

Table 8.5 Summary of Humidity Sensors

Name	Operating Principle	Relative Humidity Range	Temperature Range	Accuracy
Psychrometer	Evaporation rate of water	2%–98%	32°– 212°F	± 5%
Hair hygrometer	Hair changes length with humidity	15%–90%	0°– 160°F	± 5%
Resistance hygrometer	Lithium chloride changes resistance with humidity	1.5%–99%	−40°– 160°F	± 1.5%
Resistance-capacitance hygrometer	Aluminum oxide changes impedance with humidity	0%–100%	0°– 300°F	± 2%
Capacitive hygrometer	Dielectric constant changes with humidity	0%–100%	0°– 200°F	± 0.5%

fatigue caused by continuous use. As a technician you should be familiar with the various devices used to sense light in industry.

Our discussion of optoelectronic sensors is broken down into three areas: photoconductive devices, photovoltaic devices, and photoemissive devices.

Photoconductive Transducers

A *photoconductive transducer* is a transducer that converts a change in light intensity to a change in conductivity. The method used to produce this conductivity change differs with the photoconductive device. We will study two basic types of photoconductive devices: bulk photoconductors, such as the photoresistor, and PN junction photoconductors, such as the photodiode, phototransistor, and photo Darlington.

Photoresistors. For several years before Hertz discovered the photoelectric effect, Willoughby Smith noticed that the resistance of a block of selenium decreased when exposed to light. This same effect is used in the *photoresistor*. The photoresistor, shown in Figure 8.26, is made of either cadmium sulfide (CdS) or cad-

mium selenide (CdSe). These substances are vapor-deposited on a glass or ceramic substrate and then hermetically sealed in plastic or glass. When light strikes the photoresistive material, it liberates electrons. These electrons are then available as current flow, and the resistance decreases. More light falling on the photoresistor causes resistance to decrease further.

The spectral response (how the device reacts to different light wavelengths) depends on the type of material used. Notice in Figure 8.27 that the spectral response of the CdS photoresistor closely matches the response of the human eye. The CdSe spectral response is shifted toward the infrared.

The specific material used determines the

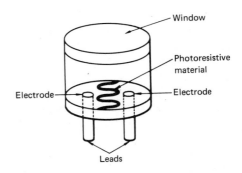

Figure 8.26 Cutaway View of Photoresistor

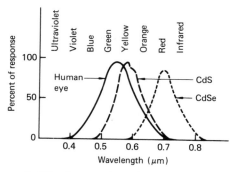

Figure 8.27 Spectral Response of CdS and CdSe Photoresistors Compared with Response of Human Eye

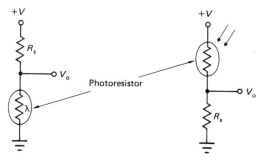

Figure 8.28 Photoresistors Used in Voltage Divider Circuits

ratio of dark-to-light resistance, which can range from 100:1 to 10,000:1. The photoresistor, then, is very sensitive to changes in light. It is generally easy to use and inexpensive. Disadvantages include narrow spectral response, poor temperature stability, and light history effect. The light history (hysteresis) effect is particularly annoying, since it causes the resistance to depend on past light intensity. In addition, photoresistors are slow to react to light intensity changes. The CdS takes about 100 ms to respond, while the faster CdSe takes about 10 ms.

Because the photoresistor is so sensitive to light changes, it does not normally need amplification. A simple potentiometric circuit, like the ones shown in Figure 8.28, is sufficient. Note the different schematic symbols in common use. The Greek letter λ (lambda), as indicated in Figure 8.28, is often used to symbolize light.

Photodiodes. Light levels are also detected by PN junctions, like those found in photodiodes and phototransistors. *Photodiodes* are simply PN diodes with their junctions exposed to light.

Early experimenters with PN junctions discovered that minority carrier current flow was influenced by temperature and light levels. In-

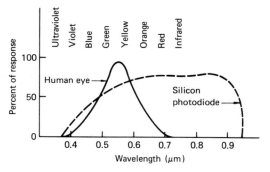

Figure 8.29 Spectral Response of Silicon Photodiode Compared with Response of Human Eye

creasing light falling on the junction creates more electron-hole pairs, which increase minority carrier current flow. Since we are dealing solely with minority carrier current, the photodiode is operated with reverse bias.

Note the spectral response curve for the photodiode shown in Figure 8.29. This curve is much wider than that of the photoresistor, covering the visible spectrum as well as the infrared.

Photodiode construction is illustrated in Figure 8.30. The simple PN device (Figure 8.30A) is nothing more than a PN junction whose structure is optimized for light reception. The PIN device (Figure 8.30B) adds a layer of undoped semiconductor material (I,

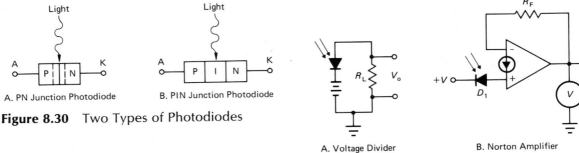

Figure 8.30 Two Types of Photodiodes

A. PN Junction Photodiode

B. PIN Junction Photodiode

A. Voltage Divider

B. Norton Amplifier

Figure 8.32 Applications of Photodiode

which stands for "intrinsic"). The layer of un-doped material increases the size of the deple-tion region. Widening the depletion region ef-fectively decreases junction capacitance. This decrease makes the PIN diode respond much faster to light-level changes than the PN pho-todiode does. PIN photodiode dark currents are much smaller, also.

Photodiodes produce relatively small cur-rents even under fully illuminated conditions. Some manufacturers add an amplifier in the same package to increase sensitivity. Packages for photodiodes without amplifiers are shown in Figure 8.31, along with common schematic symbols for photodiodes.

Detecting the change in current can be ac-complished in several ways. Figure 8.32A shows the photocurrent developing a voltage

across an external load R_L. Normally, this method is not sensitive enough to develop small voltages. For better development of small voltages, a differential amplifier may be used, or the Norton amplifier, shown in Figure 8.32B, may be used. In the Norton amplifier, as light level increases, diode D_1 increases its conduc-tion. The increase in current is mirrored in the current flowing through the feedback resistor R_F. Hence, more voltage is produced at the output.

Photodiodes, with their extremely fast re-sponse, are ideal for laser detection. They also find applications in ultrahigh-speed demodula-tion, switching, and decoding.

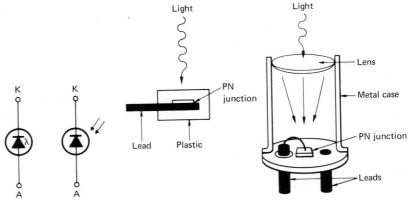

Figure 8.31 Photodiodes   A. Schematic Symbols   B. Packaged Photodiodes

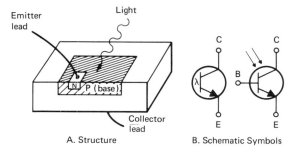

Figure 8.33 Phototransistor

Phototransistors. As we might expect, if diodes can be made photosensitive, transistors can, too. The *phototransistor* is electrically similar to a small-signal silicon transistor. Structurally, the only difference is that in the phototransistor, the collector-base junction is larger and is exposed to light.

A diagram of the structure of the phototransistor is shown in Figure 8.33. The base material is thin so that impinging light can strike the collector-base (CB) junction. When light strikes the collector-base junction, electron-hole pairs are created, as in the diode. In this way base current is created and amplified by the current gain of the transistor. The CB junction then acts like a current source, as indicated in Figure 8.34. Although the photodiode needs amplification, the phototransistor does not: The current generated in the phototransistor is automatically amplified in the collector.

The phototransistor is much more sensitive than the photodiode. However, it has higher junction capacitances, which give it poor frequency response compared with the photodiode.

Phototransistors can be either two-lead or three-lead devices (see Figure 8.33B). In the two-lead package, only the emitter and collector are connected; the base is not electrically available for biasing. The only drive for the two-lead device is the light falling on the CB junction. Two-lead packages remain the most common form of phototransistor. The three-lead form allows electrical connection to the base for purposes of biasing and decreasing device sensitivity.

Spectral response characteristics of the phototransistor are shown in Figure 8.35. Notice that the response characteristics of the phototransistor extend well into the infrared. For this reason tungsten lamps, which emit radiation in this area, are often used to illuminate phototransistors.

Because its response is slower than that of the photodiode and because it is nonlinear, the phototransistor most often is used for computer punch and paper tape readers. When the phototransistor is employed in a digital application, linearity considerations are not important.

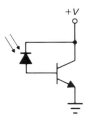

Figure 8.34 Equivalent Circuit for Phototransistor

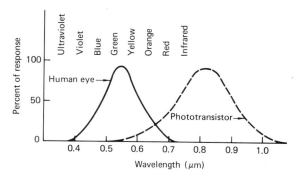

Figure 8.35 Spectral Response of Phototransistor Compared with Response of Human Eye

Photo Darlington transistors have been developed by several manufacturers. The photo Darlington is characterized by higher sensitivity compared with the phototransistor, but it has slower response time.

Photovoltaic Transducers

A *photovoltaic transducer* is a device that generates a voltage when irradiated by light. This phenomenon, called the *photovoltaic effect*, is exhibited by all semiconductors. Recall from semiconductor theory that the area around the PN junction becomes depleted of carriers. The P type region receives an excess of electrons, while the N type region receives an excess of holes. An equilibrium condition finally is reached where no more migration of carriers occurs. The PN junction then blocks any more movement of majority carriers.

The photovoltaic cell is so constructed that when light falls on it, electron-hole pairs are formed, as shown in Figure 8.36. These pairs become minority carriers. Unlike majority carriers, minority carriers can and do move across the junction. Thus, a negative potential is developed at the terminal attached to the N type material. If a load is attached to the two terminals, current flows; the current is proportional to the amount of carriers created by light.

Two basic types of photovoltaic devices are in general use today: silicon cells and selenium cells. Pictorial representations of both are

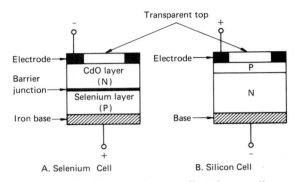

Figure 8.37   Photovoltaic Cells (Photocells)

shown in Figure 8.37. In the *selenium cell* (Figure 8.37A), a thin layer of selenium is deposited on a metal base. Cadmium is then deposited on top of the selenium. A PN junction forms between the cadmium and selenium. In the *silicon cell* (Figure 8.37B), a thin layer is doped with P type impurities, the remainder with N type.

The photovoltaic cell (photocell) has low internal resistance in order to prevent the voltage generated from being lost within the device under load conditions. This low internal resistance is achieved by heavy doping. A depletion region then forms around the PN junction. The operation of both types of cells is the same: Light falls on the PN junction, creating electron-hole pairs.

Spectral responses of these two types of cells differ, as shown in Figure 8.38. Under high light intensities, silicon cells produce as much as 0.5 V between their terminals and supply as much as 100 mA of current. Selenium cells, although possessing better spectral qualities, are much less efficient as power converters. For this reason silicon cells are more useful as solar cells. The purpose of a solar cell is to convert radiant energy from the sun into electrical energy. Solar cells are often found in series, parallel, or series-parallel combinations to increase current and voltage output.

Selenium cells are often used in light me-

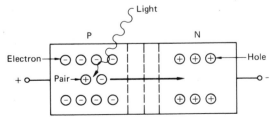

Figure 8.36   Construction of Photovoltaic Cell

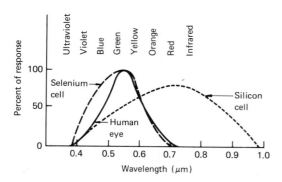

Figure 8.38 Spectral Response of Selenium and Silicon Photovoltaic Cells Compared with Response of Human Eye

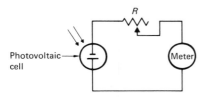

Figure 8.39 Light Meter Using Photovoltaic Cell

ters, an application illustrated in Figure 8.39. Such circuits are very simple, requiring only a sensitive meter movement, a variable resistor, and the photovoltaic cell.

A more sensitive meter may be obtained by using an op amp, as shown in Figure 8.40. The op amp amplifies the small voltage change from the photovoltaic cell. Note the different schematic symbols used for photovoltaic cells in Figures 8.39 and 8.40. Both symbols are used interchangeably.

Photoemissive Transducers

Photoemissive transducers emit electrons when struck by light. Most of the transducers using this principle are vacuum tubes. Although most vacuum tube devices have been replaced in in-

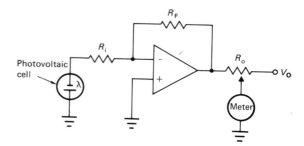

Figure 8.40 Op Amp Combined with Photo-voltaic Cell

dustry, two are still being used as photosensors: the phototube and the photomultiplier.

Phototubes. The *phototube,* shown in Figure 8.41, has a peculiar construction. The rod-shaped device is the anode. The cathode, shaped as an open cylinder, is coated with a *photoemissive material.* When light strikes this material, it emits electrons. Increasing light levels increase the anode current.

Although the phototube has been replaced by the photodiode in many applications, it is still used. Its spectral response is very linear, and it has excellent frequency response charac-

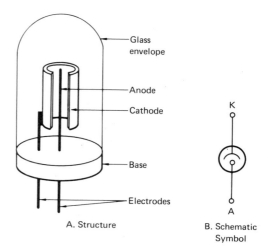

Figure 8.41 Phototube

teristics. Because current flows are only a few microamperes, the phototube usually needs amplification.

Photomultipliers. The *photomultiplier* operates on a principle similar to that of the phototube. It has a photosensitive cathode that emits electrons when light strikes it. In addition to one anode, the photomultiplier has between 10 and 15 electrodes, which are called *dynodes*. The structure of the photomultiplier is shown in Figure 8.42.

Starting with the dynode closest to the cathode, each dynode has an increasingly high positive potential applied to it. When a photon strikes the cathode, electrons are emitted. The emitted electrons are attracted to dynode 1, striking it. Secondary emission occurs from dynode 1, and more electrons are emitted. This cascading process occurs until a great many electrons are emitted from dynode 8. Photomultipliers have gain factors up to 10^9, which makes them one of the most sensitive photodectors available.

Photosensors have gained a wide following in industry in recent years. Their popularity is due in large measure to their versatility. Figure 8.43 illustrates the photosensor's wide area of application.

Web breaks, that is, the breaks that some-times occur in a roll of paper or cloth, can be detected with the arrangement shown in Figure 8.43C. Any break will let the light from a source pass through to a detector, thus setting off an alarm or turning off the drive motor. In Figure 8.43E boxes on a conveyor may be merged without colliding. The box on the right will be stopped until the box on the left passes by. In Figure 8.43F information from registration marks (reflected back to a sensor) can initiate related operations, such as printing, cutting, or folding. Gluing can be accomplished, as shown in Figure 8.43G, by using the photodetector to detect the presence of a product. Obviously, the manufacturer would not want glue applied when there is no product present to receive it. Guillotine blades can be controlled by photosensors in cutting operations, as shown in Figure 8.43H. A common and important industrial application of photosensors is in detecting fill levels, as shown in Figure 8.43L. The fill detectors (above the box) are turned on by the box detectors (on opposite sides of the box). Without this circuit the fill inspection pair might mistake the space between the boxes as an improper fill.

A summary of optoelectronic sensors is given in Table 8.6. A summary of their schematic symbols is given in Figure 8.44.

DISPLACEMENT, STRESS, AND STRAIN

In industrial process control it is often necessary to have accurate information about an object's position. *Displacement transducers* provide this kind of information. In measurement systems displacement and the force that produces the displacement are inseparably linked. We can, then, measure either the displacement or the force (stress) producing the displace-

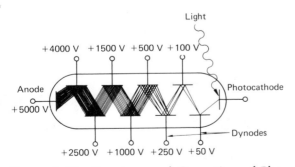

Figure 8.42 Structure and Operation of Photomultiplier Tube

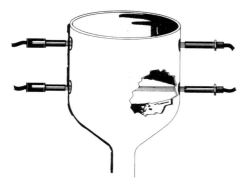

A. Keeping Hopper Fill Level between High and Low Limits

B. Counting Products

C. Detecting Web Break

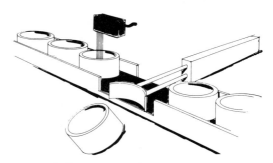

D. Checking Dark Caps for White Liners

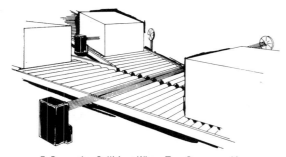

E. Preventing Collisions Where Two Conveyors Merge

F. Detecting Registration Marks

Figure 8.43 Applications of Optoelectronic Sensors

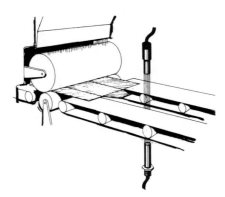

G. Gluing, Buffing, or Flattening

H. Detecting Products and Operating Guillotine Blade

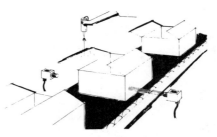

I. Detecting Thread Breaks

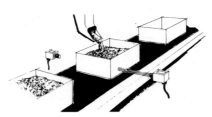

J. Slowing Conveyor and Filling Carton

K. Controlling Size of Paper or Fabric Roll

L. Checking Fill Level

M. Turning Glue Nozzle On and Off

Figure 8.43   (*continued*)

Table 8.6 Summary of Optoelectronic Sensors

Type	Name	Spectral Response (μm)	Response Time	Advantages	Disadvantages
Photoconductive	Photoresistor			Small	Slow
	CdS	0.5– 0.7	100 ms	High sensitivity	Hysteresis
	CdSe	0.6– 0.9	10 ms	Low cost	Temperature range limited
				Visual range	
	Photodiode	0.4– 0.9	1 ns	Very fast	Low-level output (current)
				Good linearity	
				Low noise	
				Wide spectral range	
	Phototransistor	0.25–1.1	1 μs	High current gain	Low frequency response (500 kHz)
				Can drive TTL	Nonlinear
				Small size	
Photovoltaic	Photovoltaic (solar) cell	0.35–0.75	20 μs	Linear	Slow
				Self-powered	Low-level output (voltage)
				Visual range	
Photoemissive	Phototube	0.15–1.2	1 μs	Good stability	Fragile
				Fast response	
				High impedance	
	Photomultiplier	0.2– 1.0	1–10 ns	Fast response	Fragile
				High sensitivity	Need high-voltage power supply
				Good linearity	

ment. In some applications we are interested in measuring the deformation of an object caused by force or stress. This deformation is called strain. In this section we deal with transducers that can be used to measure each of these concepts.

Displacement Transducers

Displacement is defined as an object's physical position with respect to a reference point. Dis-

placement breaks down into two categories: linear displacement and angular displacement. *Linear displacement* is defined as the position of a body in a straight line with respect to a reference point. *Angular displacement* is the angular position of an object with reference to a fixed point. Industry uses transducers for both types of displacement.

Angular-Displacement Transducers. One of the most common angular-displacement transducers is the potentiometer. You are prob-

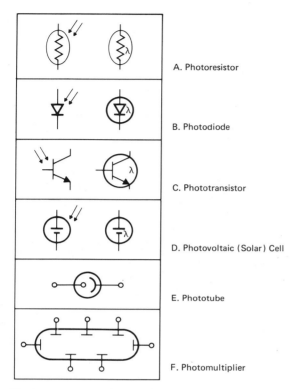

Figure 8.44 Schematic Symbols for Opto-electronic Sensors

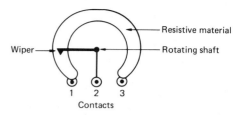

Figure 8.45 Potentiometer as Indicator of Angular Displacement

ably familiar with the basic structure of a potentiometer. It is composed of a resistor shaped in a circle with a wiper sliding on it, as shown in Figure 8.45. If the shaft is rotated clockwise, the resistance between contacts 1 and 2 increases. This increasing resistance can be used to indicate the position of a rotating shaft, for example, on a motor. The resistive element may be carbon, conductive plastic, or thin wire wrapped around a nonconductive form.

Linear-Displacement Transducers. Linear displacement can be measured in many ways. For example, if the resistive element in the angular-displacement transducer is straightened

out, we have a linear-displacement transducer. The position of the tap or wiper is directly controlled by linear displacement.

Variations of capacitance and inductance are also used to indicate linear displacement. Inductance normally is changed by withdrawing the core from the windings, as shown in Figure 8.46A. Capacitance is varied by moving the plates or by withdrawing or inserting the dielectric, as shown in Figure 8.46B.

By far the most common displacement transducer is the *linear variable differential transformer* (LVDT). The LVDT is basically a transformer with two secondaries, as illustrated in Figure 8.47. A movable core is connected to the shaft. An object attached to the shaft moves the core. The primary winding is excited by an AC frequency between 50 Hz and 15 kHz with an amplitude of up to 10 V. The secondary

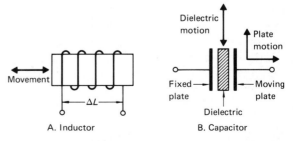

Figure 8.46 Capacitors and Inductors as Indicators of Linear Displacement

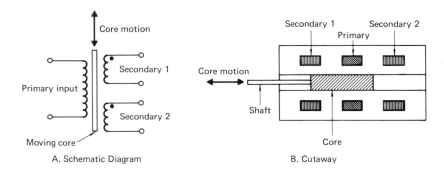

Figure 8.47 LVDT as Indicator of Linear Displacement

A. Schematic Diagram

B. Cutaway

windings usually are connected series opposing so that when the core is centered, there is no output from the secondary. If the core material moves in either direction, an output voltage is produced, because the mutual inductance changes.

The LVDT gives an output voltage that is linear with changes in the core position. The diagram in Figure 8.48A shows a plot of output voltage and the core position. Note how straight the line is, within the ± 0.06 core range. Outside this range the LVDT output is less linear. The diagram in Figure 8.48B shows the LVDT output voltage as a function of core position with reference to the phase of the output voltage. When the core is in position A, the phase relationship is shown below the horizontal axis. As the core is moved to position B, the phase changes 180°.

The LVDT is a passive transducer—that is, it does not produce an output voltage without some form of excitation. The excitation usually is provided by an oscillator that is called the *carrier generator*. This AC voltage used by the LVDT is at a frequency and amplitude not normally generated by other electronic circuits and, therefore, must be generated by special signal-conditioning circuits. Since most readout and control devices operate on DC, a circuit called a *demodulator* converts the AC output of the LVDT to DC. This small DC potential may

need to be amplified or otherwise conditioned to make an appropriate input for other circuitry.

A block diagram of a typical LVDT circuit is shown in Figure 8.49. The LVDT is supplied with an AC voltage from the carrier generator. When the core is moved within the form, an AC potential is produced by the LVDT and rectified and filtered by the demodulator. The DC voltage produced by the demodulator is amplified by the DC amplifier. An example of a passive demodulator is illustrated in Figure 8.50. In this circuit the difference between the two DC voltages is sensed by the differential amplifier. Since both the secondary winding outputs, S_1 and S_2, are positive when rectified, the differential amplifier reverses the polarity as the core passes through null. Therefore, if more of the core is between primary winding P and S_1, the DC produced by the rectifier associated with it will produce more DC voltage than the bottom circuit. Since the S_1 DC input is connected to the noninverting terminal of a differential amplifier, the output will be a positive voltage. If, however, the core position is such that the signal from S_2 is larger than that from S_1, then the output of the differential amplifier will be negative.

LVDTs are used extensively in industry as displacement transducers. Displacements of as little as 50 μin. are detected by LVDTs. In ad-

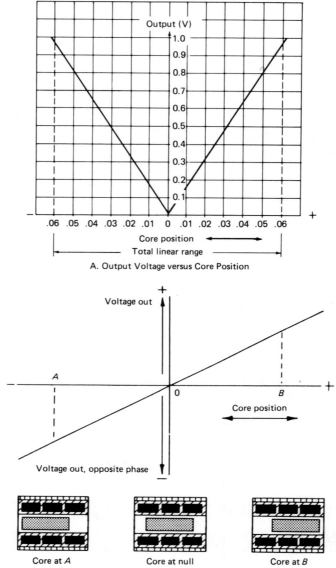

A. Output Voltage versus Core Position

Figure 8.48 Relationship of LVDT Output Voltage and Core Position

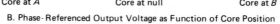

Core at *A*   Core at null   Core at *B*

B. Phase-Referenced Output Voltage as Function of Core Position

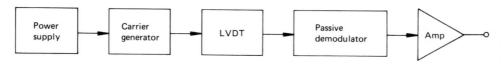

Figure 8.49 LVDT Driver Functional Block Diagram

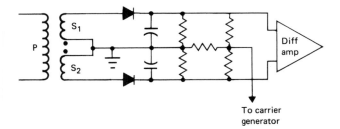

Figure 8.50 LVDT Driver
Circuit

dition to displacement, LVDTs are used to measure weight and pressure in combination with mechanical transducers.

We can see in Figure 8.51 an application where an LVDT is used in a Bourdon tube and a diaphragm to sense pressure. The LVDT is mounted on a block to which the fixed end of the Bourdon tube (covered later in this chapter) is attached also. A spring attached to the LVDT core keeps the core centered in the form. When pressure increases, the Bourdon tube tries to uncoil, pulling the core of the LVDT upward. With small changes in pressure, the output of the LVDT output voltage is linear with respect

to changes in pressure. The diaphragm (also covered later in this chapter) also is used in conjunction with the LVDT to change pressure to a voltage. An increase in pressure felt on the diaphragm will cause the diaphragm to expand, moving the LVDT core.

The LVDT has many useful features that are helpful when a transducer is used in automatic control applications. The LVDT exhibits little frictional drag since the core is not mechanically connected to the coil. Because the coil and core do not touch, friction and mechanical wear are reduced, thus reducing maintenance and improving reliability. The LVDT

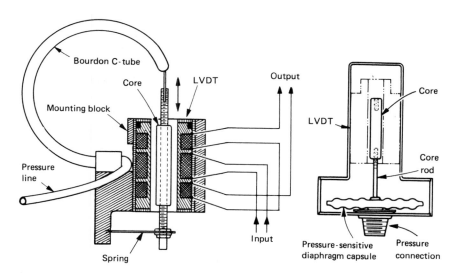

Figure 8.51 LVDT Being Driven by a Bourdon Tube and a Diaphragm

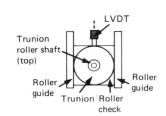

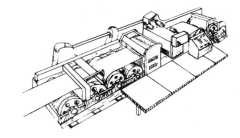

Figure 8.52 LVDT in a Rolling Mill Application

output has infinite resolution, unlike displacement transducers like the wire-wound potentiometer. The wire-wound potentiometer gives an output resistance curve made up of many small steps. As the wiper slides over a wire, it produces a small resistive change. The LVDT, on the other hand, responds to the smallest change in motion with essentially stepless resolution. Like any transformer the LVDT has good isolation between the excitation circuitry and the output.

No device is without some disadvantages and limitations. The LVDT is susceptible to interference from strong electromagnetic fields. Such fields are created by induction welding and brazing machinery. LVDTs are not cost-effective for linear displacements of greater than one foot and require a high-frequency excitation source.

We can see the versatility of the LVDT by examining its industrial applications. One use of the LVDT in a rolling mill is shown in Figure 8.52. The rolling mill is an essential part of any factory that makes metal in sheets. The metal is melted, then pressed and rolled into sheets of uniform thickness. Part of the quality control of such a factory is the measurement of the thickness of the sheets to ensure that the thickness is within allowable limits. A key element in rolling mill production is the LVDT, used as a position transducer. The LVDT core is attached to a chromium-steel-tipped probe. The LVDT probe senses the displacement caused by the metal as it fills up the roll. A computer usually

makes a running computation of how thick the metal is by estimating what the displacement should be at that point with metal of a uniform known thickness.

Figure 8.53 illustrates another LVDT application where ceramic tiles are sorted according to thickness. The LVDT probe is mounted above a moving conveyor belt. Thickness of the tiles is measured when the LVDT core is pressed down on the tile The core is spring-loaded, moving upward as it touches the tile. The greater upward movement of the core, the greater the tile thickness. Once the tiles are accurately measured, they can be sorted appropriately according to thickness. LVDTs allow sorting to thousandths of an inch in applications like this. This same type of system presently is being used to sort semiconductor chips as small as 2.5 mm^2.

If you have ever adjusted the valves on an automobile, you known the great care required to make the difficult adjustments and readjustments with feeler gauges. Some automobile manufacturers have automated this process on

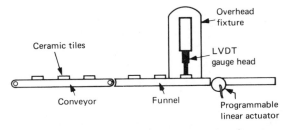

Figure 8.53 LVDT in a Conveyor Application

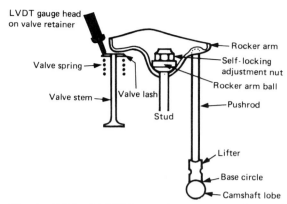

Figure 8.54 LVDT in an Automatic Valve Adjustment Application

assembly lines with LVDTs, as shown in Figure 8.54. A valve clearance may need to be set at 0.33 mm with a tolerance of 0.025 mm. In this application the engine camshaft is rotated until the lifter is riding on the lowest part of the camshaft lobe. The adjusting nut is loosened under computer control until the valve is fully seated, a position indicated by the LVDT. The locking nut is then tightened until the LVDT indicates that the valve is starting to move downward. This, on some engines, is the correct valve setting.

Stress and Strain Transducers

Stress, strain, and force are included in this section on displacement because displacement is the primary means of measuring these quantities. *Force* is defined as the quantity that changes the motion in a body. *Stress,* a similar concept, is the force acting on a solid's unit area. *Strain,* on the other hand, is the change in shape or form resulting from stress. The change in shape may be a change in length or width. Thus, the force applied to a solid material is the stress, and the deformation of that material is the strain.

Strain is defined as the change in length of an object per unit of length of the object. The unit of strain in the English system is microinches per inch (μin./in.). In the SI system strain can be given in microcentimeters per centimeter (μcm/cm) or micromillimeters per millimeter (μmm/mm). The unit of strain in either case is called *microstrain.* You will notice that strain is actually a dimensionless number, like voltage or current gain. Like gain, strain is a ratio.

When you are working with semiconductor materials, temperature and power dissipation are critical, due to the temperature dependencies of semiconductor parameters. Bridge excitation voltages, therefore, must be chosen with care. The maximum excitation voltage of the piezoresistive semiconductor strain gauge is limited by its maximum power dissipation. Factors that affect heat dissipation are gauge size, amount of glue that fixes the gauge to the object, size of the object being measured, ambient temperature, and amount of air circulation. For gauges with lengths of 0.15 to 0.25 in., power dissipations of from 20 to 50 mW are acceptable.

In many scientific and industrial applications it is necessary to measure the strain capabilities of new alloys. Measurements may need to be made of mechanical parts under different operating conditions. How much stress and strain a material can endure before permanent distortion occurs is valuable knowledge when designing mechanical components. Companies manufacturing automobiles, for example, test the stress and strain on axles of new models.

In this section we discuss two popular types of strain gauges: the wire gauge and the semiconductor gauge.

Bonded-Wire Strain Gauge. Probably the most popular strain-sensing device is the *bonded-wire strain gauge.* Strain gauges are made of metal, in the form of either wire (about

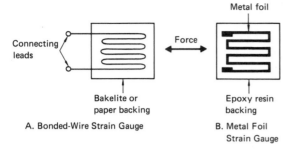

Figure 8.55 Pictorial Representations of Wire and Foil Strain Gauges

0.001 in. in diameter) or foil (about 3 μm thick). Strain gauges usually are shaped in a serpentine fashion, as shown in Figure 8.55.

The wire strain gauge is cemented firmly to a paper or Bakelite backing. The metal foil type normally is photoetched on an epoxy resin backing, as a printed circuit board is. The backing serves two purposes. First, it supports the delicate strain gauge and protects it from damage. Second, it insulates the gauge from the object being measured.

Both wire and foil gauges react to force with a change in resistance. Recall that the resistance of a conductor depends on length and cross-sectional area, among other factors. As length increases and diameter decreases, resistance increases. This principle is the one on which the wire and foil strain gauges work.

The strain gauge is bonded firmly to the surface of the device under test, usually with a cement developed specifically for this purpose. As force is applied to the object under test, the surface deforms and so does the gauge. The dimensions of the gauge change, and this deformation is what causes the change in electrical resistance.

As mentioned in the introduction to this section, the change in an object's length compared to its original length is called strain ($\Delta l/l_0$). Strain is often represented by ϵ (Greek letter epsilon). How much the resistance of a strain gauge changes depends on several things.

Any resistance change is directly proportional to the original resistance, the elasticity of the material, and the strain, or deformation. These factors are related in the following equation:

$$\Delta R = R_0 G \epsilon \qquad (8.5)$$

where

ΔR = change in resistance

R_0 = original resistance

G = gauge factor

ϵ = strain

The gauge factor contains a constant called Poisson's ratio, which is an index of elasticity.

Dividing both sides of this equation by the original resistance and the strain, we see that the *gauge factor G* is a measure of the fractional change in resistance per unit of strain:

$$G = \frac{\Delta R/R}{\epsilon} \qquad (8.6)$$

Or we can say that the gauge factor tells us the percent change in resistance of the gauge compared to its percentage change in length. Most metals have a gauge factor between 1.8 and 5.0.

Stress and strain are related through a law called *Hooke's law*. Hooke pointed out that for many materials, there is a constant ratio between stress and strain. This constant of proportionality between stress and strain is called *Young's modulus*. This relationship is expressed mathematically as follows:

$$E = \frac{\sigma}{\epsilon} \qquad (8.7)$$

where

E = Young's modulus in force per unit area

σ = stress in force per unit area

ϵ = strain

As an example let us say we have a gauge with a modulus of elasticity (Young's modulus) of 1.03×10^8 kPa undergoing a stress of 206

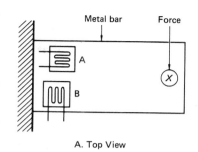 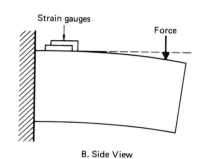

Metal bar   Force

A

B

A. Top View

Strain gauges   Force

B. Side View

Figure 8.56 Strain Gauges Mounted on Bar and Subjected to Stress

kPa. Solving Equation 8.7 for the strain ϵ, we get

$$\epsilon = \frac{\sigma}{E} = \frac{206 \text{ kPa}}{1.03 \times 10^8 \text{ kPa}} = 2 \text{ } \mu\text{cm/cm}$$

Using the preceding equations we can find how much stress a material is receiving. We will use these equations for the system shown in Figure 8.56.

Strain gauge A in Figure 8.56 is the sensor. As a force is applied down at point X, the metal bar bends. So does strain gauge A. In bending, the strain gauge wire increases its length and decreases its cross-sectional area. Thus, its resistance increases. (Strain gauge unstrained resistances range from 10 to 10,000 Ω.) This change in resistance is related to strain and the stress (or force) that produced it. From the gauge factor equation (8.6) we can see that the important quantity is the change in the gauge resistance, not the actual resistance.

One problem with this type of strain gauge

is its inherent lack of sensitivity. Let us see how great a resistance change we would expect in the previous example with a gauge factor of 2 and a sensor with an unstrained resistance of 120 Ω. Solving Equation 8.5 for the change in resistance we get:

$$\Delta R = G\epsilon R_0 = (2)(1.0 \times 10^{-6})(120 \text{ }\Omega)$$
$$= 0.000240 \text{ }\Omega$$

Detecting this very small change in resistance means we must use instrumentation with microohm sensitivity. Measuring so small a resistance change usually requires a bridge, as shown in Figure 8.57. The unbalance in the bridge caused by the strain gauge resistance change is detected by a differential amplifier. Resistances R_A and R_B are designed to be greater than ten times the strain gauge resistance in order to make the circuit a constant-current source for the strain gauge. The design current is generally 1 mA in order to keep I^2R heat to a minimum and still have the strain

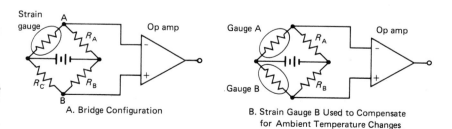

Strain gauge   A R_A Op amp

R_C R_B

B

A. Bridge Configuration

Gauge A R_A Op amp

Gauge B R_B

B. Strain Gauge B Used to Compensate for Ambient Temperature Changes

Figure 8.57 Using Bridge to Measure Change in Strain Gauge Resistance

gauge voltage high enough to be above the background noise level.

Let us say that strain gauge A in Figure 8.57B has a resistance of 1000 Ω. We have measured this resistance and found it to have increased by 2 Ω under stress. Let us further specify that Young's modulus E is 10^{12} dyn/cm² and the gauge factor G is 2. Solving Equation 8.6 for strain, we find it to be 0.001:

$$\epsilon = \frac{\Delta R/R}{G} = \frac{2\,\Omega/1000\,\Omega}{2} = 0.001$$

To find the stress we solve Equation 8.5 for stress. Substituting the values for Young's modulus and the strain, we find stress to be 1.0×10^7 N/cm²:

$$\sigma = E\epsilon = (1.0 \times 10^7 \text{ N/cm}^2)(0.001)$$
$$= 1.0 \times 10^4 \text{ N/cm}^2 \text{ or } 10 \text{ kPa}$$

Now that we have a basic understanding of the bridge, let us turn our attention to the bridge's voltage behavior. Using the voltage divider equation, we can write an equation that tells us the output voltage of a bridge if the bridge supply voltage and resistances are known:

$$V_o = V_i\left(\frac{R_A}{R_A + R_g} - \frac{R_B}{R_C + R_B}\right) \quad \textbf{(8.8)}$$

where

R_g = resistance of the strain gauge

The configuration in Figure 8.57A is called a quarter bridge, since only one of the resistances is an active gauge. The other three are fixed resistors. When the ratio $R_B/R_C = R_A/R_g$, the voltage difference between the two legs of the bridge is zero. The output voltage is then 0 V, which can be seen by examining Equation 8.8. At this time the bridge is said to be in balance.

Referring to Equation 8.6, we see that the quantity we want to measure is the gauge resistance change when a strain is applied. If we adjust resistor R_B with no strain applied to the

gauge and then apply strain, the output voltage will be some value other than zero. The output voltage is produced by the unbalance caused by the change in the gauge resistance. With the strain applied, we again can adjust the bridge to a balanced or null condition. The amount of resistance change in R_B will equal the change in R_g caused by the strain. Some static strain indicators calibrate R_B and directly read out strain on a dial with a pointer fixed to the shaft of a variable resistor in the position of R_B.

In our previous example we used a gauge with an unstrained resistance of 1000 Ω that changed to 1002 Ω under strain. If $R_A = R_B = 2000$ Ω, $R_C = 1000$ Ω, and $V_i = 10$ V, what is the output voltage? Using Equation 8.8, we can solve for the output voltage of the bridge:

$$V_o = V_i\left(\frac{R_A}{R_A + R_g} - \frac{R_B}{R_C + R_B}\right)$$
$$= 10 \text{ V}\left(\frac{2000}{2000 + 1002} - \frac{2000}{1000 + 2000}\right)$$
$$= 4.44 \text{ mV}$$

It is not always convenient to balance the bridge in a strain measurement, because it would need to be done manually. Fortunately, we do not need to balance the bridge to get a strain measurement, as we did in the preceding example. By rearranging Equation 8.8, we get the following equation:

$$\frac{V_o}{V_i} = \frac{R_A}{R_A + R_g} - \frac{R_B}{R_C + R_B}$$

This equation is true for both strained and unstrained measurements; it also holds whether the bridge is balanced or not. At this point it may be helpful to define a new term, V_r:

$$V_r = \underset{\text{(strained)}}{\frac{V_o}{V_i}} - \underset{\text{(unstrained)}}{\frac{V_o}{V_i}} \quad \textbf{(8.9)}$$

This term, V_r, is the difference between the ratios of the input and output voltages of the bridge under strained and unstrained conditions. By substitution, we can derive a formula relating strain, gauge factor, and this new variable, V_r:

$$\epsilon = \frac{-4V_r}{G(1 + 2V_r)} \qquad (8.10)$$

By measuring the output voltage V_o under strained and unstrained conditions, and having a knowledge of the gauge factor, we can calculate the strain an object undergoes. An example at this point will help you understand this concept.

Let us say we are working with an unbalanced bridge with an input voltage that remains constant at 1.980 V. In the unstrained condition the output voltage is 0.000128 V. The unstrained output-to-input ratio then is:

$$\frac{V_o}{V_i} = \frac{0.000128 \text{ V}}{1.98 \text{ V}} = 0.000065$$
(unstrained)

In the strained condition we measure an output voltage of 0.000418 V. The strained output-to-input ratio is:

$$\frac{V_o}{V_i} = \frac{0.000418 \text{ V}}{1.98 \text{ V}} = 0.000211$$
(strained)

We can now compute V_r, the difference between the two ratios:

$$V_r = \frac{V_o}{V_i} - \frac{V_o}{V_i}$$
$$\qquad \text{(strained)} \quad \text{(unstrained)}$$
$$= 0.000211 - 0.000065 = 0.000146$$

Using Equation 8.10, we can now calculate the strain, assuming a gauge with $G = 2$:

$$\epsilon = \frac{-4V_r}{G(1 + 2V_r)} = \frac{-4(0.000146)}{2[1 + 2(0.000146)]}$$
$$= 291 \ \mu\text{in./in.}$$

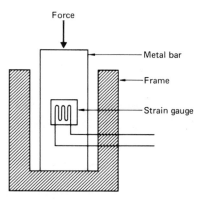

Figure 8.58   Load Cell

The strain on this object then is 291 μin./in. or 291 μcm/cm.

Up to this point we have not talked about the other strain gauge in Figure 8.56, strain gauge B. Another problem in strain gauge measurements results from resistance changes due to ambient temperature fluctuations. Strain gauge B (called the dummy) is used for temperature compensation. Figure 8.57B shows both strain gauges in a balanced bridge. If temperature rises, both strain gauges change resistance by the same amount, creating no imbalance. Only the change in strain gauge A is proportional to strain.

Strain gauges are also found in load cells, which are commonly used to measure force. The *load cell* consists of one or more strain gauges mounted on some form of metal beam or bar, as shown in Figure 8.58. The gauges usually are wired in a bridge configuration. The imbalance is then taken from the load cell directly.

Semiconductor Strain Gauge. A more sensitive device is the *silicon semiconductor strain gauge.* Silicon is a *piezoresistive substance;* that is, it changes its resistance under pressure. The semiconductor strain gauge typically is used in a bridge (with temperature compensation), as wire or foil strain gauges are. It is 30 to 50 times more sensitive than the comparable bonded-wire or foil strain gauge. Over

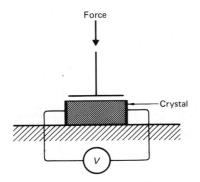

Figure 8.59 Piezoelectric Crystal Generating Voltage

wide ranges linearity sometimes is a problem. So, these devices commonly are used over narrow ranges. Where the application calls for wide ranges and high sensitivity, these devices are combined with linearity-compensating devices or circuits.

Another device used for strain or stress measurement is the *piezoelectric crystal*. Such crystals (commonly quartz, Rochelle salts, or tourmaline) exhibit a potential difference when compressed, as illustrated in Figure 8.59. The voltage that appears at the edges of the crystal can be high in voltage value but is normally low in the amount of current it provides. Also, the charge leaks off very quickly under static conditions. Therefore, this device usually is used to measure forces or stresses that change rapidly, such as vibrations.

Many natural and man-made crystals exhibit a property called the *piezoelectric effect*. When force is applied to a piezoelectric crystal, the crystal lattice structure is strained or deformed, as shown in Figure 8.59. This deformation produces an electrical charge on the edges of the crystal. When the force causing the deformation is taken away, the charge dissipates quickly. The transduction principle involved here is pressure to voltage. Generally, the more pressure applied to the crystal, the larger the voltage. The piezoelectric effect is not

exhibited in every crystal, nor is every crystal equally efficient at producing voltage. The amount of voltage produced depends on the crystal type, its thickness, and the amount of force exerted on it. Table 8.7 lists some crystals that exhibit the piezoelectric effect and the amount of voltage they produce.

We can use this parameter to calculate the output voltage of a crystal by multiplying the appropriate constant in Table 8.7 times the force (pressure) applied to the crystal times the thickness of the crystal. An example will help at this point. We would like to know how much voltage to expect from a 1 mm thick Rochelle salt crystal with a pressure of 10^6 N/m^2 applied to it. We will substitute values in the following formula:

$$
\begin{aligned}
V &= KFd \\
&= \left(\frac{0.098 \text{ V} \cdot \text{m}}{\text{N}}\right)\left(\frac{10^6 \text{ N}}{\text{m}^2}\right)(0.001 \text{ m}) \\
&= 98 \text{ V}
\end{aligned}
$$

Semiconductor strain gauges have gauge factors as high as 200. Some manufacturers of semiconductor strain gauges state that the linearity of these gauges is as good as that of the best foil or wire gauges. When assessing linearity we must consider the linearity of the circuit as well as that of the gauge itself. Semiconductor gauge linearities of $\pm 1\%$ are not unusual.

Table 8.7 Crystals Exhibiting Piezoelectric Effect

Crystal	$K = \dfrac{\text{V} \cdot \text{m}}{\text{N}}$ (Voltage per unit thickness per unit applied pressure)
Quartz	0.055
Rochelle salt	0.098
Lithium sulfate	0.165
Tourmaline	0.0275

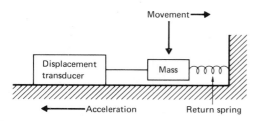

Figure 8.60   Basic Accelerometer

Gauge resistance values range from under 100 to 10,000 Ω. Gauge resistance is determined by length, cross-sectional area, and resistivity of the crystal.

Acceleration Transducers

We have seen how displacement transducers can be used to measure stress and strain. The displacement transducer can also be used to measure acceleration. Such a device is called an *accelerometer*.

Most accelerometers use the same principle, which is illustrated in Figure 8.60. As acceleration starts, the mass, which is free to move, goes backward. The mass moves backward because all mass has inertia, or resistance to motion. The higher the acceleration, the more backward movement that results. Any displacement, force, strain, or pressure transducer can be used to detect the motion or backward force.

Many accelerometers use the piezoelectric crystal to convert the force to electricity. Measuring the charge on the edges of a piezoelectric crystal gives the most accurate estimate of acceleration. Because measuring charge is both difficult and costly, however, voltage is measured instead. Acceleration in these cases is calculated by using the following equation:

$$a = \frac{V}{s}$$

where

a = acceleration

V = peak voltage in mV

s = sensitivity of the sensor in mV/g

Recall that 1 g is the acceleration due to the earth's gravity, which is about 981 cm/s². Using this equation we can calculate the acceleration of a system with a sensitivity of 20 mV/g and an output of 45 mV peak. We find the acceleration to be 2.25 g:

$$a = \frac{V}{s} = \frac{45 \text{ mV}}{20 \text{ mV}/g} = 2.25 \ g$$

MAGNETISM

Recall that magnetic fields are distributed spatially around an object as lines of force. Sensors that detect the strength of magnetic fields are used in measuring devices called gauss meters. Other common applications of magnetic field sensors are in ammeters. Two devices in particular are used in industry to detect the presence of magnetic fields: Hall effect devices and magnetoresistors. We will describe each type in turn.

Hall Effect Devices

The most popular device for detecting magnetic fields is the *Hall effect sensor*. The transduction principle used in the Hall effect device was discovered in 1879 by Edward H. Hall at Johns Hopkins University. He found that when a magnetic field was brought close to a gold strip in which a current was flowing, a voltage was produced. This effect is named the *Hall effect*.

Today, semiconductors are used in Hall devices instead of the gold with which Hall

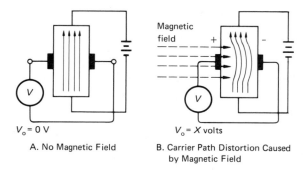

$V_o = 0$ V

A. No Magnetic Field

$V_o = X$ volts

B. Carrier Path Distortion Caused by Magnetic Field

Figure 8.61   Hall Effect

worked. Semiconductors produce the highest Hall voltages of any solid material.

The Hall effect is illustrated in Figure 8.61. Note that there is no distortion of current carrier paths in Figure 8.61A. When a magnet is brought close to the semiconductor material, as in Figure 8.61B, electrons are forced to the right side of the material. A differential voltage is then felt across the sides of the device.

Hall effect devices are typically four-terminal devices. Two terminals are used for excitation and two for output voltage. Hall effect devices come in two basic functional classifications: linear and digital.

Linear Hall effect devices are used for gauss meters, with a sensitivity of 1.5 mV/g, and for DC current probes. Normally, to measure DC current we must break the circuit and insert an ammeter. But breaking the circuit is not necessary with the Hall probe. The Hall effect DC current probe senses the magnetic field strength around a wire, which is proportional to current flow. Other applications include cover interlocks and ribbon speed monitoring in computer printers. Hall effect devices also find uses in brushless DC motors in computer disc and tape drives. When a PM rotor and a stator of coil windings are used, Hall sensors can replace brushes.

Two further applications of linear Hall effect devices are illustrated in Figures 8.62A and

8.62B. The diagram in Figure 8.62A shows a means of keeping tabs on a conveyor operation. The Hall effect sensor is mounted on the frame. A magnet is mounted on the drum. The Hall effect sensor actuates every time the magnet passes by. This system is used for speed control or for informing the operator that the conveyor is still in operation. Figure 8.62B shows a beverage dispenser. Magnets are attached to the

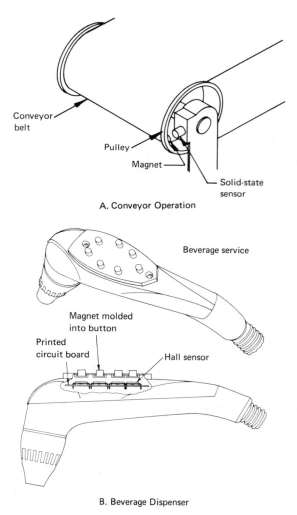

A. Conveyor Operation

B. Beverage Dispenser

Figure 8.62   Applications of Hall Effect Sensors

buttons. When the buttons are pushed, the magnets actuate the sensors, dispensing the correct drink. The unit can be completely sealed for ease of cleaning.

Digital Hall effect devices are used in counting, switching, and proximity sensors. Proximity is sensed by placing a magnet on the object whose proximity is to be detected.

Hall effect devices typically produce about 10 mV potentials at the output terminals. Therefore, many manufacturers include an amplifier in the package with a Hall effect device.

Magnetoresistors

Magnetoresistors, although not as popular as Hall effect devices, are also used to sense the presence of a magnetic field. Magnetoresistors look like photoresistors in structure. In these devices indium antimonide is deposited in a serpentine pattern on a substrate, as shown in Figure 8.63. When a magnetic field impinges on the semiconductor, the carrier paths are distorted, as in the Hall effect device. However, in the magnetoresistor, carrier path distortion effectively narrows the cross-sectional area of the conductor, thereby increasing resistance.

Note that the magnetoresistor is a two-terminal device whose resistance varies with magnetic field proximity. As field strength increases, magnetoresistor resistance increases.

Magnetoresistors, being two-terminal devices, are used to replace resistors in low-voltage circuits, which Hall effect devices cannot be

used for. Magnetoresistors also are used to sense mechanical position (proximity), current, and magentic fields, as Hall effect devices are. An advantage of the magnetoresistor is its sensitivity. In a bridge circuit magnetoresistors produce up to 1 V output, whereas the Hall effect devices typically produce less than 10 mV.

PRESSURE

Pneumatic and hydraulic systems abound in industry, even in the electronic age. There are many advantages to such systems, advantages that are discussed in Chapter 9. Suffice it to say here that these systems are still used; and it is often necessary to measure the pressures exerted at different points in these systems.

Pressure can be defined as the action of one force acting against an opposing force. Pressure normally is measured as a force per unit area, such as pounds per square inch (commonly abbreviated as psi rather the $lb/in.^2$ in industry). In the English system if the pressure is referenced to ambient pressure, the measurement is called gauge pressure (symbolized by psig). If the pressure measurement is referenced to a vacuum, the measurement is called absolute pressure (symbolized by psia). The metric system unit for absolute pressure is the pascal (Pa) and is equal to $1 N/m^2$. Also, $1 Pa = 1.45 \times 10^{-4}$ psi. Because the pascal is so small, the kilopascal (kPa) is often used.

Pressure transducers are classified into two basic types: gauge and absolute pressure transducers. The *gauge transducer* measures pressure with respect to the local atmospheric pressure, which, you will recall from physics, is 14.7 psi or 75 cm of mercury. This transducer is vented to the outside atmosphere. When the pressure-measuring port is exposed to the atmospheric pressure, the gauge should read 0 psig. The *absolute pressure transducer* measures pressure with reference to an internal chamber

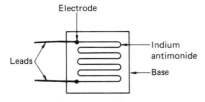

Figure 8.63 Magnetoresistor Construction

that is evacuated of air. When the pressure-measuring port is exposed to atmospheric pressure, the transducer indicates atmospheric pressure of 14.7 psia, since the reference chamber exerts no pressure, being evacuated of all gas. Another type of pressure transducer you may see is called the *differential pressure transducer*. This device has two input ports and measures the difference (psid) between the pressures applied to the two ports.

Pressure-measuring devices will be broken down into two basic classifications: pressure-to-position transducers, such as the manometer, Bourdon tube, bellows, and diaphragm; and pressure-to-electrical transducers, such as the piezoresistive transducer.

Although the transducers discussed in this section primarily are pressure-to-position transducers, they can be converted easily to produce an electric output. As we proceed within this section we will point out methods to achieve this conversion.

Manometers

The U tube manometer, shown in Figure 8.64, is representative of the liquid-column pressure transducers. The *U tube manometer* is a differential pressure-measuring device constructed

from a glass or plastic tube bent in a U shape. The tube contains an amount of liquid, whose makeup depends on the application. Note that pressure p_1 is greater than pressure p_2 in the figure. The difference in pressure causes the liquid to move into the left leg until all pressures are in equilibrium. The difference in the height ($\triangle h$) of the liquid is proportional to the difference in pressures.

The major disadvantage of manometers is that they are strictly mechanical in nature. They can, however, be converted to electrical systems with a float device attached to a displacement transducer, such as the LVDT.

Elastic Deformation Transducers

All *elastic deformation pressure transducers* use the same transduction principle. Pressure causes a bending, or deformation, of the transducer material, usually a metal. This deformation results in a deflection or displacement. We might say, then, that these transducers are pressure-to-position converters. As we will see, the indicators can be purely mechanical, or they can use electrical transducers to convert the displacement to an electrical parameter. We will discuss three types of elastic deformation transducers, all based on the same principle but each different in physical shape and size. The transducers we will consider are the Bourdon tube, the bellows, and the diaphragm.

Bourdon Tube. The *Bourdon tube* is one of the oldest and most popular (even today) pressure transducers. It was invented in 1851 by Eugene Bourdon and has been used in industry ever since. Bourdon tubes come in three basic shapes: C tube, spiral tube, and helical tube, as illustrated in Figures 8.65A, 8.65B, and 8.65C.

All three devices work in essentially the same fashion. They are all composed of a flat-

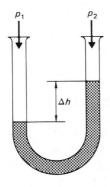

Figure 8.64   U Tube Manometer

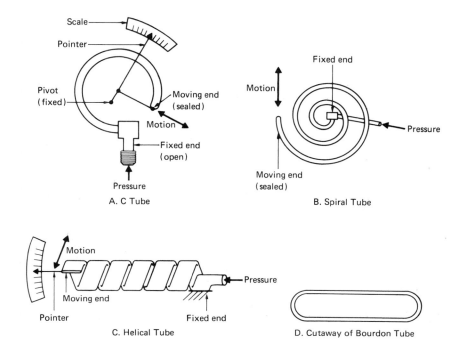

Figure 8.65 Bourdon Tubes

A. C Tube

B. Spiral Tube

C. Helical Tube

D. Cutaway of Bourdon Tube

tened metal tube, usually made of brass, phosphor bronze, or steel, as shown in cross section in Figure 8.65D. The pressure at the input is communicated to the inside of the device. Since there is more area on the outside of the tube than on the inside, the tube unwinds, much like a paper-coil party toy. The unwinding of the Bourdon tube is not linear. Therefore mechanical pointers fastened to the moving tips have displacement applied through a gearing system. This gearing system compensates for the nonlinearity of movement. Nonlinearity may also be compensated for by using a nonlinear pointer scale.

The Bourdon tube is simple to manufacture, inexpensive, and accurate, and it measures pressures up to 100,000 psi (700,000 kPa). The C-shaped Bourdon tube is the least sensitive to pressure changes, and the helical tube is highest in sensitivity. These devices do not work well with pressures under 50 psi (750

kPa). Bourdon tubes are converted very simply to electrical transducers by using electrical displacement transducers such as potentiometers and LVDTs.

Bellows. The *pressure bellows* is a cylindrical-shaped device with corrugations along the edges, as shown in Figure 8.66A. As pressure increases, the bellows expands, moving the shaft upwards. The shaft may be attached to a pointer or an electrical displacement transducer like an LVDT.

In Figure 8.66B, the bellows is connected to indicate differential pressure. If p_1 is greater than p_2, the pointer moves up. The amount of displacement is proportional to the difference in pressures.

Bellows are more sensitive to pressures in the 0–30 psi (0–210 kPa) range than are Bourdon tubes. Thus, they are more useful at lower pressures.

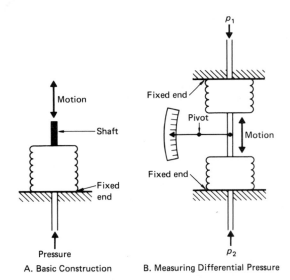

Figure 8.66   Pressure Bellows

$$p = \frac{F}{A}$$

Diaphragm. The *diaphragm* is nothing more than a flexible plate, as illustrated in Figure 8.67. The plates are usually made of metal or rubber. The flat (or curved) metal plates generally are less flexible (and less sensitive) than the corrugated type. Increases in pressure will cause the shafts to move up. Diaphragms are used in the 0–15 psi (0–105 kPa) pressure range.

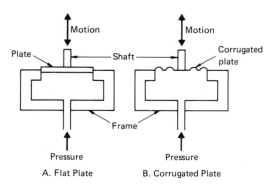

Figure 8.67   Pressure Diaphragms

Piezoresistive Transducers

Both National Semiconductor and the Micro Switch Division of Honeywell manufacture very sensitive and linear piezoresistive pressure transducers. These sensors use the piezoresistive effect and have their own built-in amplifiers and temperature compensation networks. National's line of products can sense pressures ranging from 10–5000 psi (70–35,000 kPa) with linearity less than 1%. Figure 8.68 shows one of National's transducers.

FLUID FLOW

Along with temperature, fluid flow is a very important measure. Many industrial processes, especially biomedical applications, depend on accurate assessment of fluid flow. Fluid flow measurements are an integral part of chemical, steam, gas, and water treatment plants. The usual approach to measuring the rate of a flowing fluid is to convert the kinetic energy that the fluid has to some other measurable form. This process is carried out by the fluid flow transducer.

There are many different transducers available to measure fluid flow. As with the other transducers we have discussed, we limit our treatment to only the more common fluid flow measuring devices: differential pressure flowmeters, variable-area flowmeters, positive-displacement flowmeters, velocity flowmeters, and thermal heat mass flowmeters.

Differential Pressure Flowmeters

The most common industrial flowmeters come from this group. Most *differential pressure flowmeters* use a pressure difference caused by a restriction as a transduction principle. Each flow-

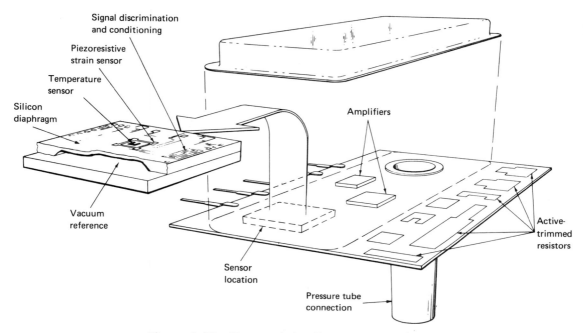

Signal discrimination
and conditioning

Piezoresistive
strain sensor

Temperature
sensor

Silicon
diaphragm

Amplifiers

Vacuum
reference

Active-
trimmed
resistors

Sensor
location

Pressure tube
connection

Figure 8.68 Piezoresistive Pressure Transducer

meter has a different method of producing this pressure differential. The principle involved here is called the Bernoulli effect. The *Bernoulli effect* states that as the velocity of a fluid (liquid or gas) increases, the pressure decreases, and vice versa. We say that an inverse relationship exists between the pressure and velocity of a fluid.

If the fluid is incompressible, the same volume of fluid must pass through a given area in the same time interval. According to the laws of conservation of mass and energy, the mass entering any protion of the pipe must equal the mass leaving that portion of the pipe during some time interval. The same is true for the kinetic energy of the fluid. The fluid flow velocity is, therefore, increased in the smaller section of the pipe, causing an increase in kinetic energy per unit of mass.

In this section we consider five of the commonly used differential pressure flowmeters:

orifice plate, Venturi tube, flow nozzle, Dall tube, and elbow.

Orifice Plate. We will use the most common of the differential pressure flowmeters, the *orifice plate flowmeter*, as an example. It is shown in Figure 8.69. This flowmeter consists of a plate with a hole in it. If we place liquid columns all along the tube to indicate pressure, we see a *pressure gradient*, or pressure distribution, as pictured in Figure 8.69. Pressure increases slightly just before the restriction and falls off quickly after the restriction. The pressure decreases sharply because of the increase in velocity at and just after the restriction. A manometer or other differential pressure indicator can be calibrated to read the flow rate. As fluid flow increases, the pressure difference increases also.

The orifice plate flowmeter is the most widely used of all the flowmeters. It is simple in

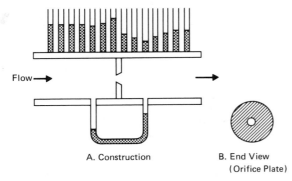

A. Construction   B. End View
                          (Orifice Plate)

Figure 8.69   Static Pressure Gradient in Orifice Plate

structure, low in cost, and easy to install. It suffers from the disadvantages of poor accuracy, limited range, and inability to be used with slurries (liquids with suspended particles of solid matter).

Venturi Tube.   Another popular flowmeter is the *Venturi tube*, shown in Figure 8.70. The Venturi tube employs the same principle as the orifice: differential pressure. The pressure difference is caused by the restriction in the tube.

The Venturi has two major advantages over the orifice. First, it creates less turbulence than the orifice because of the gently sloping sides. Thus, the final pressure loss is less than that in the orifice. This loss is called *insertion loss.* Up to 90% of the initial pressure can be recovered with the Venturi. Second, the Venturi can be used for slurries because there are no sharp edges or projections for solids to catch

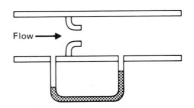

Figure 8.71   Flow Nozzle Flowmeter

on. Its major disadvantage is cost. It is expensive to buy and install.

Flow Nozzle.   The *flow nozzle*, shown in Figure 8.71, falls somewhere between the Venturi and the orifice plate in cost and pressure loss. In the flow nozzle the edges of the restriction are somewhat rounded. Thus, the flow nozzle permits up to 50% more flow rate than the orifice. An additional advantage is its ability to be welded into place, which cannot be done with the other differential pressure flowmeters.

Dall Tube.   Of all the differential pressure flowmeters, the *Dall tube* has the least insertion loss. The Dall tube, shown in Figure 8.72, consists of two cones, the shorter one upstream of the restriction, the larger cone downstream. It is generally cheaper to purchase and install compared with the Venturi tube. However, it cannot be used with solids and slurries, as the Venturi can.

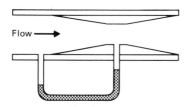

Figure 8.70   Venturi Tube Flowmeter

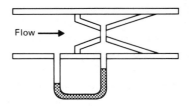

Figure 8.72   Dall Tube Flowmeter

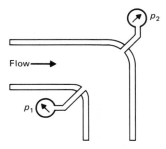

Figure 8.73   Elbow Tube Flowmeter

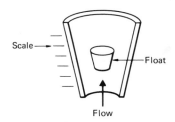

Figure 8.74   Rotameter

Elbow.   The *elbow tube flowmeter* is a differential pressure flowmeter that does not use the Bernoulli effect. It is shown in Figure 8.73. The differential pressure is developed by the centrifugal force exerted by the fluid flowing around the corner. In this case, p_2 is greater than p_1. Since elbows (corners) usually already exist in pipes, installation of elbows poses no problems. There is no additional pressure loss because there is no restriction. However, the elbow flowmeter has only $\pm 5\%$ accuracy at best.

The devices we have discussed thus far are called *primary elements.* They are so named because, in a primary sense, they convert fluid flow to pressure differential. *Secondary elements* are needed to convert this pressure difference into a usable parameter. This conversion can be accomplished with any of the pressure transducers discussed previously. In addition, National Semiconductor manfactures a differential pressure transducer that is used as a secondary element to measure flow.

Variable-Area Flowmeters

The *variable-area flowmeter* is a flowmeter that controls the effective flow area and, therefore, the flow rate within the meter. An example of the variable-area flowmeter is the rotameter, also called the float or rising ball. Shown in Figure 8.74, the *rotameter* consists of a float that is

free to move up and down in a tapered tube. The tube is wider at the top and narrows toward the bottom. Fluid flows from the bottom to the top, carrying the float with it. The float rises until the area between it and the tube wall is just large enough to pass the amount of fluid flowing. The higher the float rises, the higher is the fluid flow rate.

Normally, the rotameter is used for visual flow measurement. It cannot be used for automatic process control.

Positive-Displacement Flowmeters

Positive-displacement flowmeters divide the flowing stream into individually known volumes. Each volume is then counted. Adding each volume increment gives an accurate measurement of how much fluid volume is passing through the meter. Positive-displacement meters are accurate to less than 1%, simple, inexpensive, and reliable. Furthermore, pressure losses generally are lower compared to those of other flowmeters.

The *nutating-disc flowmeter,* shown in Figure 8.75, is a common example of a positive-displacement flowmeter. The nutating-disc meter is used in residential water metering as well as in industrial metering. As the liquid enters the left chamber, it fills quickly. Because the disc is off center, it has unequal pressures

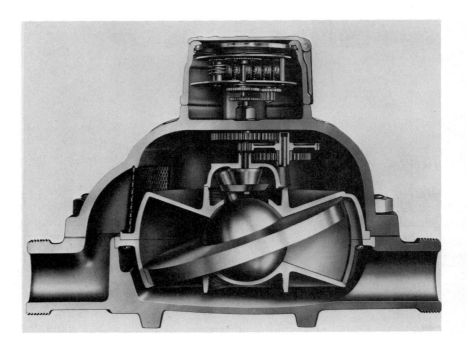

Figure 8.75   Nutating-Disc Meter

on it, causing it to wobble, or nutate. This action empties the volume of liquid from the left chamber into the right. The process repeats itself, driven by the force of the flow stream. Each nutation of the disc is counted by a mechanical or an electromagnetic counting mechanism.

Because of the mechanical structure of the nutating-disc flowmeter, it is not suitable for measuring the flow of slurries.

Velocity Flowmeters

Velocity flowmeters (sometimes called volumetric rate flowmeters) give an output that is proportional to the velocity of fluid flow. We will consider three of the most commonly used velocity flowmeters: turbine, electromagnetic, and target.

Turbine Flowmeter.   The *turbine flowmeter* is shown in Figure 8.76. The fluid flowing

past the turbine exerts a force on the blades. The blades then turn at a rate directly proportional to the fluid velocity. The turbine blades can be magnetized to produce pulse as voltage is induced in the coil. Turbine meters are useful because of their good linearity, range, and accuracy. They are among the most accurate flowmeters available.

Electromagnetic Flowmeter.   The *electromagnetic flowmeter* operates on the principle of eletromagnetic induction. The *principle of electromagnetic induction*, known as Faraday's law

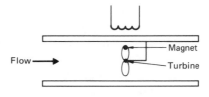

Figure 8.76   Turbine Flowmeter

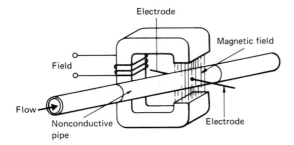

Figure 8.77 Electromagnetic Flowmeter

of electromagnetic induction, states that a conductor passing through a magnetic field (at right angles) produces a voltage. The voltage produced is proportional to the relative velocity of fluid flow.

The electromagnetic flowmeter, shown in Figure 8.77, generates a magnetic field through an electromagnet. The conductor is the flowing fluid. As we might expect, the fluid must be conductive. A potential is then produced between the electrodes, which are imbedded in a piece of nonconductive pipe. Since no obstructions exist, the pressure loss is small. Additionally, the lack of obstructions makes measuring slurries easy.

Target Flowmeter. The *target flowmeter*, shown in Figure 8.78, presents an obstruction to fluid flow. The fluid exerts a pressure on the target that is directly proportional to the fluid velocity. The force on the target is sensed by a strain gauge.

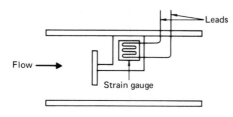

Figure 8.78 Target Flowmeter

Thermal Heat Mass Flowmeters

The *thermal heat mass flowmeter* is an accurate device for measuring fluid flow. In this device heat is applied to the flowing fluid. A temperature sensor is then used to measure the rate at which the fluid conducts heat away. This rate is directly proportional to the flow rate. One of the foremost advantages of the thermal heat mass system is its lack of obstructions to fluid flow.

Table 8.8 summarizes the fluid flow measuring devices discussed in this section.

LIQUID LEVEL

The methods of measuring liquid level are as numerous and varied as those for fluid flow. Also, there are as many reasons for wanting to know the level of a liquid (or solid) as there are reasons for wanting to know the rate of fluid flow. Liquid level is an important part of process control and accounting.

Liquid-level measurement is divided into two basic types: point and continuous. The *point level* is just what its name implies—measurement of a level at a point. *Continuous-level measurement* is analog in nature. The output of a continuous-level sensing instrument is directly proportional to the level of the liquid. Some sensors are more suited to one or the other of these types of measurement. In our description of the sensors we point out which application is more appropriate. For purposes of discussion we classify liquid-level sensors into sight, force, pressure, electric, and radiation sensors.

Sight Sensors

Sight level sensors are among the oldest and simplest. The *dipstick* is possibly the oldest,

Table 8.8 Summary of Flowmeters

Type	Name	Range	Operating Principle	Accuracy	Advantages	Disadvantages
Differential pressure	Orifice plate	3:1	Variable head	1%–2%	Lowest in cost Easy installation	Highest permanent-head loss Cannot be used with slurries
	Flow nozzle	3:1	Variable head	1%–2%	Can be used at high velocity Can be used with slurries	Expensive Difficult to remove Difficult to remove and install
	Venturi tube	3:1	Variable head	1%–2%	Most acccurate of these meter types Can be used with slurries Low permanent-head loss	Difficult to remove and install Costly
	Dall tube	3:1	Variable head	1%–2%	Lowest permanent-head loss	Costly
	Elbow tube	3:1	Centrifugal force	5%–10%	Saves space No head loss No obstructions,	Low accuracy Small differential pressure created
Positive displace-ment	Nutating disc	5:1	Measured volume	0.1%	High accuracy Simple to install	Costly to install and maintain Cannot be used with slurries
Velocity	Turbine	10:1	Spinning turbine or propeller	0.5%	High accuracy Low head loss Wide temperature range	Cannot be used with slurries Hard to install
	Electromagnetic	30:1	Electromagnetic voltage generation	1%	Linear No obstructions No head loss	Must have conductive fluid High cost
	Target	3:1	Fluid pressure on target	0.5%–3%	Compact Easy to install Low cost	Subject to temperature variations

Table 8.8   (*continued*)

Type	Name	Range	Operating Principle	Accuracy	Advantages	Disadvantages
Variable area	Rotameter	10:1	Balance between gravity and fluid pressure	1%–2%	Little head loss Linear response	Expensive Fragile
Mass flow	Thermal heat mass flowmeter	100:1	Temperature rise to mass flow rate	2%	Measures low flow rates Low head loss Fast response	Useful only for gases Costly and complex

and it is still used in some parts of industry. To make a measurement, you insert a stick (sometimes calibrated with lines, as in a car oil dipstick) into the container. You then withdraw it and observe the level of fluid on the dipstick.

Related to the dipstick method are the sight glass and the gauge glass. The *sight glass* is similar to the coffee urns found in cafeterias. The sight glass allows you to see the liquid level in the tube, as shown in Figure 8.79A. A *gauge glass* is similar (Figure 8.79B). It has windows in its sides to allow you to see the liquid level.

The *displacer* works on a different principle entirely. It uses Archimedes' principle. Archimedes found that a body immersed in water loses weight equal to the amount of water displaced. In Figure 8.79C, the weight of the displacer object is proportional to the liquid level. As the liquid rises, the displacer object has less weight. The weight measurement is made mechanically and displayed on a pointer, or electric force sensor. A strain gauge works nicely here.

The *tape float* (Figure 8.79D) works in a similar fashion. The float rides on top of the liquid. It is attached to a tape, at the end of which

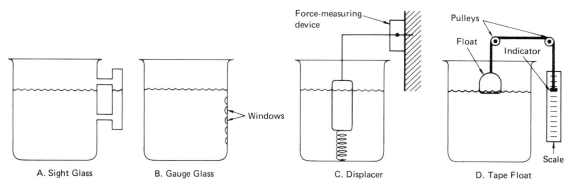

Figure 8.79   Liquid-Level Measurement by Sight

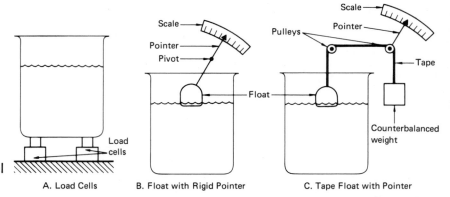

Figure 8.80 Liquid-Level
Measurement by Force

A. Load Cells   B. Float with Rigid Pointer   C. Tape Float with Pointer

is an indicator. As the level rises, the tape
moves down the scale.

All the sight methods have the same disadvantage: They require human operators to
see and record the levels. However, these methods are very inexpensive, and all are suited to
continuous measurement.

Force Sensors

The weight of a container (force) is often used
to indicate liquid levels. When more liquid (or
solid) enters the container, it weighs more. The
weight is detected by *force* or *strain sensors*, as
in the load cell of Figure 8.80A. Alternatively,
the buoyant force of a float moves a rigid rod
or flexible tape, cable, or chain, as illustrated in
Figures 8.80B and 8.80C.

All these methods are suited to continuous
measurement of liquid level. The float methods
shown may be converted to an electrical output
by placing a potentiometer at the pivot point.
Changes in the linear displacement of the float
are then changed to angular displacement by
the pulley or pivot. The potentiometer changes
angular displacement into an electrical resistance change.

Pressure Sensors

Pressure is another variable that is affected by
liquid level. You will recall from physics that
pressure in a liquid increases with the depth of
the liquid. For instance, if more water is added
to a tank, there will be more water above the
point at which the pressure is measured. The
increased weight of the water will, therefore,
create a greater pressure. The pressure at any
depth in a container filled with liquid depends
on the height of the liquid above that point, the
density of the liquid, and gravity. This relationship is shown in the following equation:

$$p = \rho g h$$

where

p = pressure in N/m^2

ρ = density in kg/m^3

h = depth of liquid above point where
measurement is taken

As an example, let us try to find the pressure at the bottom of a tank that contains water
at a depth of 30.5 m. Water has a density of
1000 kg/m^3 and the acceleration of gravity is 9.8
m/s^2. Using the preceding equation:

$p = \rho g h$

$\quad = (1000 \text{ kg/m}^3)(0.8 \text{ m/s}^2)(30.5 \text{ m})$

$\quad = 2.99 \times 10^5 \text{ kg/m-s}^2$

$\quad = 2.99 \times 10^5 \text{ N/m}^2$

The pressure exerted on the bottom of the tank is called the *hydrostatic head*, which is defined as the pressure a column of liquid exerts on the bottom of a container. As liquid level increases, then more pressure is exerted at the bottom of the tank. Devices that use this method of sensing level are called *hydrostatic-head devices*. In the *diaphragm type*, shown in Figure 8.81A, more pressure is exerted on the diaphragm as level increases. The air pressure inside the tube then increases and registers a pressure on the meter. Figure 8.81B shows a similar device mounted on the bottom of the tank.

Other hydrostatic-head devices use differential pressure, as shown in Figure 8.81C. The pressure difference between the bottom and top of the tank depends on liquid level. The difference is sensed by a differential pressure-measuring device, such as bellows or National Semiconductor's IC differential pressure transducers.

The *bubbler system*, shown in Figure 8.81D, is one of the oldest and simplest devices for measuring pressure. Air flows in through the tube placed in the tank. If enough pressure is exerted in the tube, bubbles are forced out of

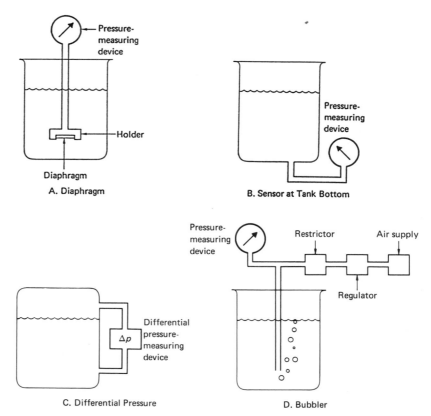

Figure 8.81 Liquid-Level Measurement by Pressure

A. Diaphragm

B. Sensor at Tank Bottom

C. Differential Pressure

D. Bubbler

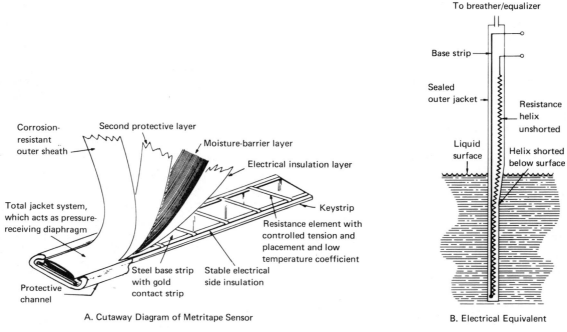

Figure 8.82   Resistance Changes with Changes in Liquid Level

Electric Sensors

the tube. The pressure required to force the bubbles out varies proportionally with the liquid level. This pressure is indicated by the pressure-measuring device.

Electric Sensors

Beyond the mechanical-to-electrical conversions mentioned previously, certain sensors give direct electric outputs with changing liquid levels. One interesting device, manufactured by Metritape, changes the resistance of a wire helix as pressure is applied. This device, illustrated in Figure 8.82, is suited to continuous-level measurement.

Electrodes are also used to sense electrical changes. In Figures 8.83A and 8.83B, the capacitance changes as the liquid level fluctuates.

The change in capacitance is caused by the changing dielectric constant between the liquid and the air. The changes in capacitance usually are sensed with an oscillator or an AC bridge.

Note the resistive electrode sensor in Figure 8.83C. With conductive liquids the electrodes in Figure 8.83C provide a path for current flow through the series circuit. This type of system is more suited to a point-sensing application.

Radiation Sensors

Most of the level measurement devices discussed previously have elements that are in contact with the liquid. In some industrial measurements, however, the level-sensing device must be used with corrosive liquids or liquids

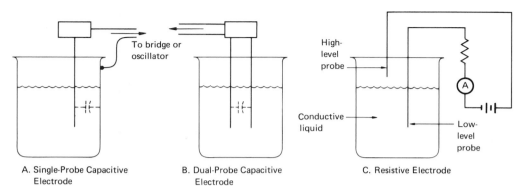

A. Single-Probe Capacitive Electrode

B. Dual-Probe Capacitive Electrode

C. Resistive Electrode

Figure 8.83   Liquid-Level Measurement by Electrical Means

under high pressures. Hence, contact with the liquid might destroy the device. In this section we will discuss one class of sensors that do not physically contact the liquid at all: radiation sensors. We divide these sensors into two classifications: sonic (including ultrasonic) and nuclear radiation sensing devices.

Sonic.   Sonic and ultrasonic beams (7.5–600 kHz) use the *echo principle* to measure liquid level. In Figure 8.84A the sonic beam is transmitted from the transducer to the liquid surface. It is then reflected back to the transducer from the surface. The time it takes for the beam to travel from the transducer to the surface and back to the transducer depends on the liquid level. The lower the level, the more time

it takes for the beam to travel the distance. The time interval then is inversely proportional to the liquid level.

Nuclear.   Nuclear radiation is sometimes used to make liquid-level measurement for point-sensing applications. In Figure 8.84B radiation is emitted from the transmitter in the form of gamma rays. The receiver is usually a Geiger-Müller tube. The radiation is absorbed by the liquid or solid to be measured. The output of the receiver, therefore, is dependent on whether or not the liquid is between the sensor and the transmitter. Nuclear sensors are expensive compared with other electrical methods of sensing level. They also require extensive in-plant safety precautions.

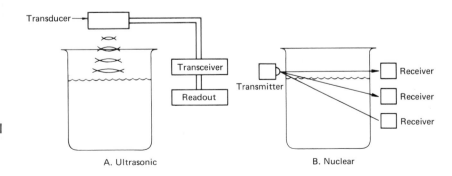

Figure 8.84   Liquid-Level Measurement by Radiation

A. Ultrasonic

B. Nuclear

Table 8.9 Summary of Liquid-Level Sensors

Type	Name	Use*	Temperature Range	Pressure (lb/in.²)	Accuracy	Range
Sight	Tape float	C	to 149°C	300	±⅛ in.	to 60 ft
	Displacer	P, C	to 260°C	300	±¼ in.	10 ft
	Sight glass	P, C	to 260°C	10,000	±½%–1%	6–8 ft
Pressure	Diaphragm	P, C	to 459°C	Atmospheric	±1 in.	Unlimited
	Differential pressure	P, C	to 649°C	to 6000	±½%–1%	—
	Bubbler	C	Dew point	Atmospheric	±1%–2%	Unlimited
Electrical	Resistive electrode	P	−26°–82°C	to 5000	±⅛ in.	to 66 ft
	Capacitance	P, C	−26°–982°C	to 5000	±1%	to 20 ft
Radiation	Sonic	P, C	−40°–149°C	to 150	±1%–2%	to 125 ft
	Ultrasonic	P, C	−26°–60°C	to 320	±1%–2%	to 12 ft
	Nuclear	P, C	to 1648°C	Unlimited	±1%–2%	Unlimited

*P = point; C = continuous.

Table 8.9 summarizes the different types of liquid-level sensors and some pertinent information about them.

MEASUREMENT WITH BRIDGES

Many of the transducers discussed previously use bridges to transform their resistance or impedance change to a voltage change. The *bridge* can transform a small change in resistance or impedance to a large change in voltage. The bridge circuit is capable of detecting changes on the order of 0.1%.

The basic four-arm DC bridge is shown in Figure 8.85. No current flows through the meter when the following resistance ratio is met:

$$\frac{R_1}{R_3} = \frac{R_2}{R_4} \tag{8.11}$$

When no current flows through the meter, the bridge is said to be in a *null state*.

In transducer applications one of the resistors of Figure 8.85 is replaced by the trans-

ducer. Since the three remaining resistors are constant, any change in the bridge null state is caused by the transducer. Furthermore, the resistance of the transducer can be calculated at null by solving Equation 8.11 for the unknown resistance.

If R_4 is the unknown resistance, it is related to the other resistances in the bridge by the following equation:

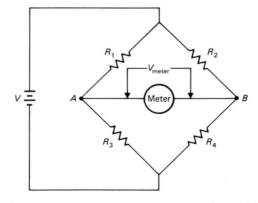

Figure 8.85 Four-Arm DC (Wheatstone) Bridge

$$R_4 = \frac{R_2 R_3}{R_1}$$

For example, if $R_1 = 75\ \Omega$, $R_2 = 100\ \Omega$, and $R_3 = 150\ \Omega$, then R_4 is $200\ \Omega$.

A resistance change causes a bridge to become unbalanced. The imbalance is in the form of a voltage between points A and B. This difference in potential causes current to flow through the meter. The potential difference (V_{meter}) between A and B can be approximated by the following equation:

$$V_{meter} = V\left[\frac{R_2\,\Delta R_4}{(R_2 + R_4)^2}\right]$$

where

V = supply voltage

ΔR_4 = change in resistance of R_4

R_4 = original resistance of R_4

Let us suppose that the bridge is balanced with $R_1 = 100\ \Omega$, $R_2 = 200\ \Omega$, $R_3 = 400\ \Omega$, and $R_4 = 800\ \Omega$. The supply voltage is 100 V. If R_4 changes by 10 Ω, then V_{meter}, previously at 0 V, will be at about 0.2 V, as shown in the following calculation:

$$\begin{aligned}
V_{meter} &= V\left[\frac{R_2\,\Delta R_4}{(R_2 + R_4)^2}\right]\\
&= 100\ V\left[\frac{(200\Omega)(10\ \Omega)}{(200\ \Omega + 800\ \Omega)^2}\right]\\
&= 0.2V
\end{aligned}$$

Another helpful relationship to know is the bridge's sensitivity. The *sensitivity* of a bridge is the change in output voltage compared with the transducer's change in resistance. That is, the sensitivity (s) of a bridge (in volts per ohm) is equal to the change in voltage between points A and B (V_{meter}) divided by the change in the transducer's resistance (ΔR_4):

$$s = \frac{\Delta V_{meter}}{\Delta R_4} = V\left[\frac{R_2}{(R_2 + R_4)^2}\right]$$

The sensitivity of the bridge in our example is 0.2 V/10 Ω, or 20 mV/Ω. In other words, for every ohm that R_4 changes, the voltage V_{meter} changes about 20 mV.

CONCLUSION

We have covered a great many sensors in this chapter. All sensors convert the variable to be measured to an analog of that variable. As we have seen, the most common conversion is to an electrical quantity, because electric control systems predominate in industry.

You should have a firm grasp of the transduction principles behind each sensor. Although you probably will not be asked to choose a sensor for a system, you will have to work with existing sensors in control systems. Thus, you will need a knowledge of the concept behind the sensor.

In the next chapter the sensor is assumed to be an integral part of a process control system.

QUESTIONS

1. A transducer is a device that converts a _____ into an output that facilitates measurement. The mercury thermometer converts temperature to _____.

2. The mechanical temperature transducers operate on the principle that gases and liquids change _____ when heated. The potential produced by the junction of two dissimilar metals is used in the _____ temperature sensor.

3. The psychrometer assesses relative humidity by comparing the _____ of a wet and dry bulb. The hair hygrometer element changes _____ with varying relative humidity.

4. The _____ photoresistor's spectral response closely approaches that of the human eye. Of the optoelectronic sensors discussed, the _____ is the most sensitive.

5. Because the phototransistor's response is so nonlinear, it is used more in _Digital_ systems than analog systems. The _photovoltaic_ cell increases its voltage output with increasing light levels.

6. The displacement sensor with a primary wound between two secondary windings is called a(n) _____. The bonded-wire strain gauge changes its _____ with strain.

7. Two devices used to measure magnetic field strength are the Hall device and the _____.

8. The _____ pressure transducer is more suited to measuring high pressures. A bellows converts pressure to _____.

9. The most commonly used differential pressure flowmeter is the _____ _____. Differential pressure flowmeters use the _____ effect to generate a pressure difference.

10. The two types of liquid-level measurement are called continous and _____. The electronic level probe is more suited to _____ measurement.

11. Define temperature.

12. Describe the function of a transducer.

13. Compare sensors that are active (give an output voltage with input energy) and those that are passive (give no output voltage).

Give an example of each of the two types of sensors from among the temperature transducers and optical transducers.

14. Name the parts of a filled-system, temperature-measuring device.

15. Describe the behavior of a thermistor when it is heated. What type of temperature coefficient does it have? Is its response linear?

16. Describe how a thermistor can be used in (a) a fluid flow system, (b) a liquid-level system, and (c) time delay. *Hint:* Does the thermistor change its resistance immediately when heated?

17. How can a thermistor be made more linear? Name two methods.

18. Of the temperature sensors discussed in this chapter, which is the most sensitive?

19. Describe how a thermocouple is used to measure temperature.

20. Describe the operational principle behind the bimetallic strip.

21. Describe how the following humidity sensors work: (a) hair hygrometer, (b) resistance hygrometer, and (c) microwave transceiver.

22. What is the relationship between gain and sensitivity in optoelectronic sensors?

23. Define the gauge factor for a strain gauge.

24. Describe two types of differential pressure transducers.

25. Describe measurement of liquid level and fluid flow using differential pressure transducers.

26. Compare the operation of the Bourdon tube with that of the bellows.

27. List the maximum ranges of the bellows, diaphragm, and Bourdon tube.

PROBLEMS

1. A type J (iron-constantan) thermocouple measures 6.907 mV across the device. Find the temperature of the test junction if the reference junction is at (a) 0°C and (b) 25°C.

2. The temperature of a type J thermocouple

is 400°C. Find the voltage across the thermocouple if the reference junction is at 0°C.

3. A straight bimetal strip is 20 cm long, 0.1 cm thick, and clamped at one end. How much would it move if it was made of brass-invar and heated to 15°C?

4. A spiral bimetal strip has a radius of 10 cm and a thickness of 0.2 cm, and it is made of metals with coefficients of 17 millionths/°C and 5 millionths/°C. The spiral is 50 cm long, and it deflects by 1 cm. What is the heat change that produced this deflection?

5. A thermocouple has a mass of 0.008 g, a specific heat of 0.08 cal/g°C, a heat transfer coefficient of 0.05 cal/cm-s°C, and an area of 0.05 cm^2. What is its thermal time constant?

6. A strain gauge has an initial resistance of 200 Ω. Under a strain of 10^{-4}, resistance increases to 402 Ω. What is the gauge factor?

7. A spiral bimetal strip has a radius of 6 cm, a length of 40 cm, and a diameter of 0.075 cm. If the metals are aluminum and invar, what will the deflection be for a 25°C change?

8. A thermistor at room temperature (25°C) is placed in an oven with a temperature of 100°C. With a time constant of 0.5 s, what will the thermistor's temperature be after 1.5 s?

9. An RTD has a resistance of 150 Ω at 0°C. If the RTD has a temperature coefficient of resistance of 0.00412, what is the resistance of the RTD at (a) 25°C and (b) 50°C?

10. If a transistor has a V_{BE} of 0.600 V at 25°C, what is the base-emitter voltage at 100°C?

11. The LM335 has a voltage out that rises 10 mV for every degree Celsius of temperature change. If the device is calibrated so that it has an output voltage of 2.98 V at 25°C, what will the output of the device be at (a) 55°C and (b) 75°C?

12. A strain gauge has an original, unstrained resistance of 150 Ω and a gauge factor of 1.8. If the gauge undergoes a strain of 2 μcm/cm, and assuming an increase in resistance occurs, what is the new resistance?

13. In Problem 12 if Young's modulus is 1.0×10^8 kPa, what is the stress in kPa?

14. What is the output voltage of the bridge in Figure 8.57A, if the gauge is the one referred to in Problems 12 and 13 and is strained with a 2 μcm/cm strain? Assume $R_A = R_B = 100$ Ω, $R_C = 150$ Ω, and the bridge starts out balanced.

15. A crystal made of lithium sulfate has a pressure of 5.0×10^5 N over a surface area of 0.005 m^2. What is the output voltage of the crystal?

16. A pressure sensor at the bottom of a tank filled with liquid is used to measure the depth of the fluid. If the tank contains water, what would the pressure be if the tank was filled to a level of 10 m? How deep would the tank be if the sensor measured a pressure of 1.0×10^5 N/m^2?

CHAPTER

9

INDUSTRIAL PROCESS CONTROL

OBJECTIVES

On completion of this chapter, you should be able to:

- Define process control;
- List the elements of a process control system;
- Define process load, process lag, and stability, and describe how they relate to a process control system;
- List and give examples of two types of process control systems;
- Distinguish between feedback and feedforward control;
- Describe the following controller modes: on-off, proportional, integral, and derivative;
- Contrast electrical controllers and pneumatic controllers in terms of the advantages and disadvantages of each.

INTRODUCTION

A *process control system* can be defined as the functions and operations necessary to change a material either physically or chemically. Process control normally refers to the manufacturing or processing of products in industry.

Every process has one or more controlled, or dynamic, variables. The *controlled (dynamic) variable* is a variable we wish to keep constant. Processes also have one or more manipulated variables, or control agents. A *ma-*

312

nipulated variable is a variable that we change to regulate the process. Specifically, the manipulated variable enables us to keep the controlled variable constant. Examples of controlled and manipulated variables are pressure, temperature, fluid flow, and liquid level. Finally, each process control system has one or more disturbances; *disturbances* tend to change the controlled variable. The *function* of the process control system is to regulate the value of the controlled variable when the disturbance changes it.

An example may help to identify the components of a process control system. Figure 9.1 illustrates a system in which milk is heated to pasteurizing temperature. The temperature of the milk is the controlled variable. Steam flows through the pipes, transferring its heat to the milk. The flow of steam controls the temperature of the milk. Therefore, the steam flow is the manipulated variable. The ambient temperature surrounding the tank can be considered a disturbance. For instance, if the ambient temperature decreases, the temperature of the milk will eventually decrease.

In an automatic control situation, the temperature of the milk would be monitored. If it decreased beyond allowable limits, the controller would make adjustments to bring the temperature under control. The *controller* is that part of a process control system that decides how much adjustment the system needs and implements the results of that system. For example, the controller might open the steam control valve, causing the milk to be heated and returning it to the required temperature. This valve is the final control element in a process control system.

In this chapter we will present a basic outline of process control systems. Since this subject is rather complex, we will present only a general overview of these systems. We will first discuss the characteristics that are found in most process control systems. Next, we will consider general types of process control and the ways in which controllers operate. Finally, we will give a short summary of those final control elements we have not discussed in previous chapters. Recall that we have already covered motors in Chapters 3, 4, and 5.

Well-designed process control systems can save an enormous amount of time and money, reduce error (which improves the quality of a product), and provide greater operator safety. Because of these advantages, sophisticated process control systems are frequently found in the manufacturing industry. The tasks of preventive and corrective maintenance vary directly with the complexity of the system. Therefore, it is wise for technicians to have a firm grasp of the concepts behind process control systems.

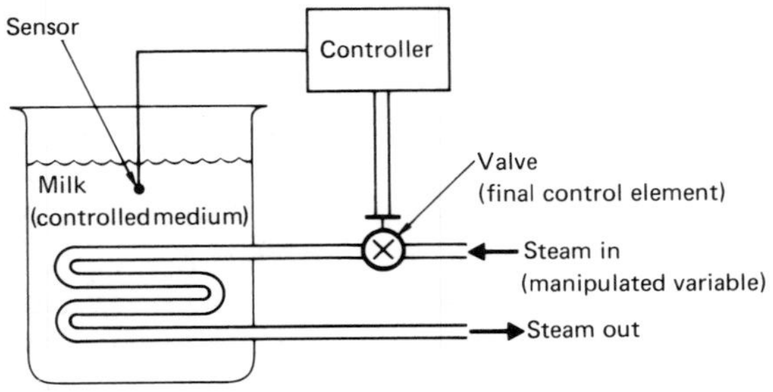

Figure 9.1 Process Control System for Pasteurizing Milk

CHARACTERISTICS OF PROCESS CONTROL SYSTEMS

All process control systems have three characteristics in common. First, each automatic process control system makes a *measurement* of the controlled variable. This measurement usually is made by a sensor or transducer. As we saw in Chapter 8, transducers change the controlled variable into another form, usually electrical. In our milk example a thermistor could have been used to change the temperature of the milk into a corresponding electrical resistance. As the temperature of the milk changed, so would the sensor's resistance. Since the sensor's output must be evaluated by the controller, the output must be in a suitable form. For example, if the controller is a computer, the sensor's output must be digital. Thus, the sensor's output may need to be converted, or conditioned, to an appropriate form for evaluation.

Second, as mentioned, the controller needs to *evaluate* the information from the sensor. It does so by comparing the sensor measurement with a reference called the *set point*. If the sensor measurement differs from the set point, an *error condition* occurs. The controller then decides whether or not this amount of difference is acceptable. If it is not acceptable, the controller initiates action to reduce the error.

Finally, each process control system must have a final control element. A *final control element* makes those adjustments that are necessary to bring the controlled variable back to the set point value.

PROCESS CHARACTERISTICS

Every process exhibits three basic characteristics that are important in the understanding of process control systems. These characteristics are process load, process lag, and stability.

Process Load

Process load can be defined as the total amount of control agent needed to keep the process in a balanced condition. In the milk pasteurization example we need a certain amount of steam (the control agent) to keep the milk at the correct temperature. Suppose the ambient temperature around the tank decreases. Then, we need a greater rate of steam flow to keep the milk at a constant temperature. Thus, a decrease in the ambient temperature constitutes a change in the process load.

The process load is directly related to the setting of the final control element. Any process load change will cause a change in the state of the final control element. The final control element's adjustment is what keeps the process balanced.

Process Lag

Process lag is the time it takes the controlled variable to reach a new value after a process load change. This time lag is a function of the process and not the control system. The control system has time lags of its own. Process lags are caused by three properties of the process: capacitance, resistance, and transportation time.

Capacitance can be defined as the ability of a system to store a quantity of energy or material per unit quantity of a reference. For example, a large volume of water has the ability to store heat energy. As a result we could say that the water has a large thermal capacitance.

A large capacitance in a process means that it takes more time for process load changes to occur. From one viewpoint large capacitance is a desirable characteristic, because it is then easier to keep the controlled variable constant. Small disturbances do not have very much effect on the process load. On the other hand, large capacitance also means that it is more difficult to change the controlled variable back to the desired point once it has been changed.

Resistance in a process can be defined as opposition to flow. In the pasteurization example we would notice thermal resistance in the walls of the steam pipes. In other words, the material of the pipes slows down the transfer of heat to the liquid. Sometimes gases and liquids surround the pipes in layers or films, increasing the thermal resistance compared with pipes without the layers or films. These layers have a slowing effect on the transfer of heat energy and thus have resistance. Large resistances will, therefore, increase the process lag by opposing the change in the controlled variable.

Often the capacitance and resistance of a system are combined into a factor called *RC delay,* or the *RC time constant.* Recall that in an electric system the time it takes to charge a capacitor to 63% of its final voltage is called one time constant. The rate at which the capacitor charges will depend on the capacitance of the capacitor and the resistance through which it must charge. The product of the resistance and the capacitance (*RC*) gives one time constant, in seconds.

Process control systems have the same *RC* lag. In the pasteurization example the product of the thermal capacitance, in joules per degree Celsius, and the thermal resistance, in degrees Celsius per joule per second, will give one time constant, in seconds. In practice one system may have many *RC* lags caused by many individual resistances and capacitances.

The third component of process lag, *transportation time,* or *dead time,* can be defined as the time it takes for a change to move from one place to another in a process. Or we can define transportation time as the time between the application of the disturbance and the changing of the process load. Dead time is most easily illustrated in applications where fluids or solids are moving, as in a flowing fluid or a conveyor belt application. In a fluid flow application, for example, let us say that a temperature change is made some distance upstream from the sensor. It will take some time for the temperature change to be transported downstream to the sensor. In the conveyor illustration the same principle applies. Any change in the condition of the product (weight, for example) will not be detected until that change is communicated to the sensor. This time delay is called dead time, or transportation time.

Note that any controller action is delayed by the amount of time delay present. To find the time delay, we use the equation $d = rt$, which can be rewritten as follows:

$$t = \frac{d}{r}$$

where

> t = transportation (delay) time in min
> d = distance in cm
> r = flow rate in cm/min

In the fluid flow example suppose that fluid is flowing at a rate of 100 cm/min, and say the temperature sensor is located 50 cm from the temperature change. Then, transportation (delay) time is as follows:

$$t = \frac{d}{r} = \frac{50 \text{ cm}}{100 \text{ cm/min}} = 0.5 \text{ min, or } 30 \text{ s}$$

In other words, it will take 30 s for the temperature change to reach the sensor. Thus, there will be a delay of 30 s *before* the controller can even start to react.

Lags occur in areas also: the sensor, the controller, and the final control element. We saw in the previous chapter on sensors that a temperature sensor takes some amount of time to respond to a temperature change. The sensor must be heated to the same temperature as the dynamic variable. In the case of the RTD, which may be enclosed in a protective sheath, some delay will occur while the temperature change is transmitted through the sheath to the sensor. You will also recall that the sensor itself will have a thermal mass, which will slow down the sensor temperature change. This delay is called the *transducer lag,* or *measurement delay.*

Sometimes the controller takes a significant amount of time to evaluate the input data and make a decision. As most controllers are chosen for their speed, this factor is not usually a problem unless the process conditions change quickly. Sometimes a *controller lag* is introduced into the system by the process engineer to meet certain system requirements.

Many final control elements have significant lags. Large motors, for example, may take a minute or so to come up to full speed after a load change. Electrically operated valves are notoriously slow in opening and closing. Some valves may take two minutes to fully open from a closed position.

Process engineers are especially interested in keeping dead time to a minimum. Long dead times make accurate process control difficult.

Stability

An important consideration when examining control systems using feedback is the stability factor. We say that a process control system is *stable* if it can return the controlled variable to a steady-state value. Typically, an unstable system will cause the controlled variable to oscillate above and below the desired value.

If the controlled variable oscillates above and below the desired value, three things will happen. First, the strength of the oscillations will increase in amplitude as time increases. This result occurs when the feedback is in phase and the loop gain is greater than 1. (Recall the Barkhausen criteria for oscillators discussed in Chapter 1.) Second, if the feedback is in phase with the oscillations and the loop gain is 1, the oscillations will have a constant amplitude. Third, if the loop gain is less than 1 and the feedback is out of phase with the oscillations, then the oscillations will gradually die out. The amount of time that it takes for the oscillations to die out, or damp, is called the *settling time*. A good process control system will reduce settling time to a minimum.

Important information can be gained from analysis of the system response to a load change. If too much control action is present, corresponding to a loop gain larger than 1, the system responds with an unstable and possibly dangerous response (Figure 9.2A). If the amount of control action or feedback is decreased to a loop gain of 1, we will see the system responding as in Figure 9.2B, with a stable, constant amplitude response. As the amount of control is decreased to a loop gain of less than 1, we see a type of response that is called underdamped. This system response is so named because it does not have enough damping or resistance to keep the system from oscillating, as shown in Figure 9.2C. Damping in a control system is a force that opposes the change in the manipulated variable. A bit less control (less positive feedback) will give a waveform known as a critically damped response (Figure 9.2D). Even less control action and more damping produce the waveform known as overdamped (Figure 9.2E).

Which system response is best? We can rule out the first response, the one with the increasing amplitude, as generally undesirable. The oscillating, constant amplitude waveform, typical of one particular type of controller, may not be particularly desirable, but may be acceptable if the system does not require close control of the dynamic variable. The choice usually is between the critically damped, underdamped, or overdamped responses. The overdamped response usually is chosen if close control is necessary and no oscillations are tolerable. The underdamped response is chosen where fast dynamic variable response is required. In general, however, the critically damped response is most desired, offering a compromise between speed of response and oscillations.

TYPES OF PROCESS CONTROL

Generally speaking, process control can be classified into two types: open-loop and closed-loop control. Closed-loop control can be further divided into feedback and feedforward control.

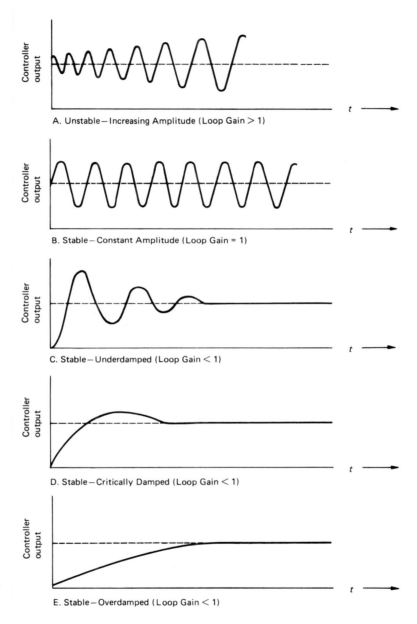

A. Unstable—Increasing Amplitude (Loop Gain > 1)

B. Stable—Constant Amplitude (Loop Gain = 1)

C. Stable—Underdamped (Loop Gain < 1)

D. Stable—Critically Damped (Loop Gain < 1)

E. Stable—Overdamped (Loop Gain < 1)

Figure 9.2 Process Control System Responses

Open-Loop Control

Open-loop control involves a prediction of how much action is necessary to accomplish a process. That is, in an open-loop system no check is made during the process to see whether corrective action is necessary to accomplish the end result.

Figure 9.3 shows a block diagram of a typical open-loop system. A washing machine is a

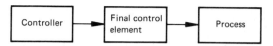

Figure 9.3 Block Diagram of Open-Loop Control System

good example of such a system. In this case the person operating the machine takes a measurement (the clothes are dirty), compares this information with a reference level (clean clothes), and makes a prediction (a cycle is chosen and soap added). The operator than starts the machine and attends to other business.

The operator assumes that the prediction made will accomplish the desired objective (clean clothes). If the prediction is absolutely correct in all aspects (amount of soap, amount of hot and cold water, and so forth), the clothes will be absolutely clean. Therefore, open-loop control is capable of perfect control. If the prediction was wrong for any reason, however, the objective will not be reached. Open-loop control does not always provide perfect control. Since the washing machine does not make any measurements or comparisons, any mistakes in the prediction will produce an undesirable outcome.

Closed-Loop Control

In a *closed-loop system* a measurement is made of the variable to be controlled. This measurement is then compared to a reference point called a set point. If a difference or error exists between the actual and desired levels, an automatic controller will take the necessary corrective action.

Feedback Control. A typical block diagram of a closed-loop control system is shown in Figure 9.4. A comparison of Figures 9.3 and 9.4 shows the difference between open-loop and closed-loop systems. In the open-loop system,

no actual measurement is made on the process. In contrast, in closed-loop control the variable to be controlled is measured and compared to a reference (the set point), and corrective action is taken.

The milk pasteurization example is a good illustration of a closed-loop system. This system is often called a *feedback control system,* since sensor information is fed from the output back to the input.

Figure 9.4 shows a block diagram for a closed-loop feedback control system. The measurement of the dynamic variable is taken by the sensor. The controller compares the information from the set point and the measurement. It then decides whether or not to take corrective action on the basis of its evaluation. The controller can be any circuit or system capable of evaluation and decision making, from a simple op amp comparator to a complex digital computer. In the block diagram the controller also performs any signal conditioning needed to make the sensor measurement compatible with the input to the evaluation circuit. The controller, if it decides action is to be taken, adjusts the final control element to bring the dynamic variable back under control.

We can relate this block diagram to the pasteurization example discussed at the beginning of the chapter. The measurement is taken by a temperature sensor. The controller, on the basis of this measurement, decides whether to adjust the final control element (a valve) or to leave it

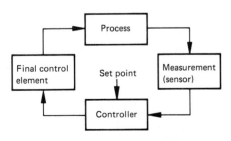

Figure 9.4 Block Diagram of Closed-Loop Feedback Control System

unchanged. Recall that the valve controlled the flow of steam in the pipes (the manipulated variable).

Feedforward Control. Closed-loop feedback control has disadvantages in two areas of process control systems. Feedback control is not satisfactory when there are large disturbances in the process load and when there are large process lags. These disadvantages can be dealt with by using a type of control called feedforward (or predictive) control. *Feedforward control* is defined as a closed-loop system that feeds a correction signal forward to the controller based on a measurement of the disturbance. A simple feedforward control system is shown in Figure 9.5

There is one basic difference between feedforward control and open-loop control: Open-loop control makes an assumption about the variables in the system. In contrast, feedforward control makes a measurement of a disturbance. From this measurement the controller estimates what action needs to be taken to keep the process from changing. In other words, the controller decides what change to make in the manipulated variable so that the change, when combined with the disturbance, will produce no change in the controlled variable. The controller anticipates the effect the disturbance will have on the process.

Feedforward control, like open-loop control, relies on a prediction. It differs from open-loop control in that it does not rely on a fixed program. Feedforward control is a type of closed-loop control.

Note the difference between feedback control and feedforward control. In feedback control a measurement of the dynamic variable is taken. In feedforward control a measurement of the disturbance is taken.

Like feedback control, feedforward control is not without its problems. The prediction made in feedforward control assumes that all significant disturbances are known. It also assumes that there are sensors present to measure these disturbances and that we know exactly how the disturbances will affect the process. However, the engineering and technical know-how needed to accomplish this type of control is immense. Thus, feedforward control is used only where the process load and the disturbances to it are well understood and can be accurately predicted.

BASIC CONTROL MODES

As we discussed previously, the controller performs evaluation and decision-making operations in a control system. It takes the difference or error signal from the comparator and determines the amount of change the manipulated variable needs to bring the controlled variable back to normal. The controller may contain the *error detector* (or comparator), a signal-conditioning device, a recording device, and a telemetering device. The entire system with all these possible combinations is shown in Figure 9.6.

Controllers may be classified in several different ways. For instance, they may be classified according to the type of power they use. Two common types in this category are electric and pneumatic controllers. *Pneumatic controllers* are decision-making devices that operate on air pressure. *Electric* (or *electronic*) *control-*

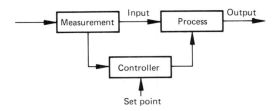

Figure 9.5 Block Diagram of Closed-Loop Feedforward Control System

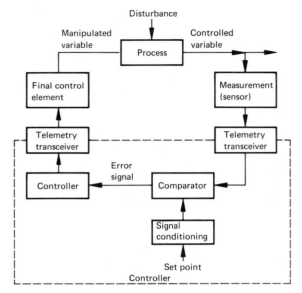

Figure 9.6 Expanded Block Diagram of Closed-Loop Feedback Control System

lers operate on electric signals. You will see both types used widely in industry. Pneumatic controllers are used in the chemical and petrochemical industries for reasons of safety. They are less expensive and simpler than comparable electric controllers. However, pneumatic controllers are difficult to interface with digital computers, a great disadvantage in light of the increasing popularity of computers in process control. Electric controllers suffer no such disadvantage.

Controllers are also classified according to the type of control they provide. In this section we will discuss five types of control in this category: on-off, proportional, proportional plus integral, proportional plus derivative, and proportional plus integral plus derivative.

On-Off Control

In *on-off control,* the final control element is either on or off. This controller is also called

bang-bang from the speed of the response of the on-off state. In the pasteurizing example if the controller were an on-off controller, the valve would be either open or closed. This type of control is sometimes also called two-position control, since the final control element is either in the open or the closed position. That is, the controller will never keep the final control element in an intermediate position.

On-off control is undoubtedly the most popular mode in industry today. On-off control also has domestic applications. For example, most home heating systems use the on-off control mode. If the room temperature goes below a predetermined point, the heater turns on. When the room temperature goes above that point, the heater shuts off. Thus, the control is on-off.

On-off control is illustrated in Figure 9.7. As soon as the measured variable goes above the set point, the final control element is turned off. It will stay off until the measured variable goes below the set point. Then, the final control

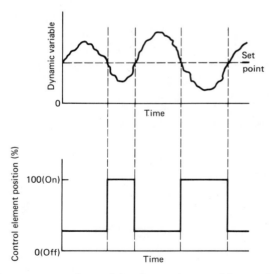

Figure 9.7 Plot of Measured Variable and Final Control Element Position versus Time in an On-Off Controller

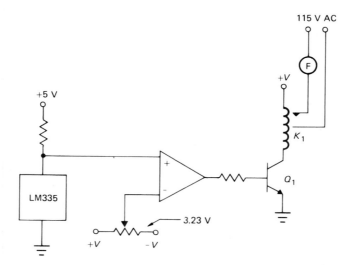

Figure 9.8 Op Amp On-Off Controller

element will turn on. The measured variable will oscillate around the set point at an amplitude and frequency that depend on the process's capacity and time response. This oscillatory response is typical of the on-off controller. For processes that require that the dynamic variable be held relatively constant, this type of controller is not a wise choice. The oscillations may be reduced in amplitude by increasing the sensitivity of the controller, but this would cause the controller to turn on and off more frequently, a possibly undesirable consequence.

An example of an on-off controller is shown in Figure 9.8. In this simple circuit the sensor is an LM335, a temperature-to-voltage converter. (A data sheet for this device is included at the back of this book.) The sensor is calibrated to give 2.98 V at 25°C and gives an output of 10 mV/K. Thus, if the temperature is 30°C, the output of the LM335 is:

$$V_o = \text{temperature in K} \left(\frac{10 \text{ mV}}{\text{K}} \right)$$

$$= 303 \text{ K} \left(\frac{10 \text{ mV}}{\text{K}} \right) = 3.03 \text{ V}$$

The output of the LM335 is applied to the comparator, which in this case is the controller. Note that the comparator has a reference voltage of 3.23 V. In process control terminology this voltage represents the set point. It is called the set point because it sets the point at which the controller switches from on to off or off to on. With a temperature of 30°C and an output of 3.03 V from the LM335, the comparator output will be low. Recall that in a noninverting comparator, the output will go high only when the input voltage goes above the reference voltage. In this case the comparator output will go high only when the input goes above the reference of 3.23 V. This will occur when the LM335 reaches a temeprature of 323 K, or 50°C. When the output of the comparator goes high, it turns on the transistor Q_1 and relay K_1 energizes. The relay in this case controls a fan, which brings the ambient temperature down. The fan will stay on until the ambient temperature goes below 50°C. When the temperature goes below this set point, the comparator output will go low, turning off the transistor and the relay. The relay contacts open, removing power from the fan.

This simple on-off controller has one main disadvantage. If the fan cools down the sensor too quickly, the comparator and the relay will turn on and off quickly, possibly damaging the relay. One way to stop this action is to add a dead zone or a differential gap, to be discussed.

Despite its disadvantages on-off control is useful. The on-off control mode is chosen by the process engineer under the following conditions:

1. Precise control must not be needed.
2. The process must have sufficient capacity to allow the final control element to keep up with the measurement cycle.
3. The energy coming in must be small compared with the energy already existing in the process.

On-off control is most often found in air-conditioning and refrigeration systems. It is also widely used in safety shutdown systems to protect equipment and people. In this application on-off control is called a shutdown or cutback alarm.

Differential-Gap Control. *Differential-gap control,* shown in Figure 9.9, is similar to on-off control. Notice that a band, or gap, exists around the control point. When the measured variable goes above the upper boundary of the gap, the final control element closes. It will stay closed until the measured variable drops below the lower boundary. Some home heating systems use this type of control mode rather than on-off.

Differential-gap control does not work as well as on-off, but it does save wear and tear on the final control element. In industry differential-gap control is often found in noncritical level-control applications, such as keeping a tank from running dry or from flooding.

The gap in differential-gap control is often called a *dead zone.* Engineers normally choose

a dead zone of about 0.5% to 2.0% of the range of the final control element.

Let us consider an example of a differential-gap controller, shown in Figure 9.10. This circuit uses the LM335 as a temperature sensor, and control circuitry is provided by the versatile LM3900 CDA. The first CDA (1IC) is used as a buffer amp, isolating the sensor from the controller. The second CDA (2IC) is the controller, a circuit you may recall from the first chapter of this text. 1IC is connected as a noninverting comparator, with a reference voltage of $+5$ V connected to the inverting input. The output of the comparator drives a transistor, which, in turn, drives a relay that controls 115 V AC power to a fan. When the comparator output goes high, the transistor is forward-biased, turning on relay K_1. The NO contacts of the relay then close, applying power to the fan.

As you know from studying this circuit in Chapter 1, the comparator output is either 0 V or $+5$ V, depending on the state of the input.

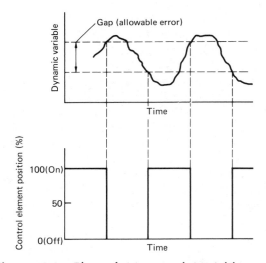

Figure 9.9 Plot of Measured Variable and Final Control Element Position versus Time in Differential-Gap Controller

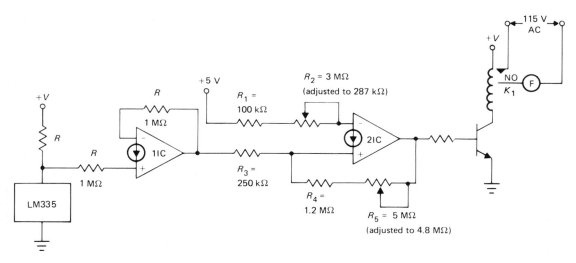

Figure 9.10   Op Amp Differential-Gap On-Off Controller

The output will go high when the input current from the buffer amp exceeds the current drawn in the reference circuit, which consists of the +5 V potential, R_1, and R_2. With the output at 0 V, the only current drawn from the noninverting terminal results from the input voltage. The current drawn from the reference circuit is:

$$I_{ref} = \frac{V_{ref}}{R_1 + R_2} = \frac{+5 \text{ V}}{100 \text{ k}\Omega + 287 \text{ k}\Omega}$$

$$= 12.92 \ \mu\text{A}$$

For the output voltage to go high, the input must draw an amount of current greater than 12.92 μA. This will occur at the following input voltage:

$$V_i = (12.92 \ \mu\text{A})(R_3) = (12.92 \ \mu\text{A})(250 \text{ k}\Omega)$$

$$= 3.23 \text{ V}$$

Combining these two equations we can find the input voltage that will make the output go high:

$$V_i = \left(\frac{V_{ref}}{R_1 + R_2} \right) R_3$$

If the LM335 has been properly calibrated, this voltage should be attained when the temperature of the sensor is 50°C. This set point can be changed by varying the size of R_2. Increasing R_2 will decrease the reference current. The output of the comparator goes high when a lower voltage is applied to 1IC by the LM335.

When the output goes high, current is drawn from the noninverting terminal to the output. This reduces the amount of current needed to trip the comparator to make it go low. Specifically, the input voltage needed to make the comparator go low is represented by the following equation:

$$V_i = \left(\frac{V_{ref}}{R_1 + R_2} - \frac{V_{sat}}{R_4 + R_5} \right) R_3$$

Using a value of 287 kΩ for R_2 and 4.8 MΩ for R_5, and solving the preceding equation, we have:

$$V_i = \left(\frac{V_{ref}}{R_1 + R_2} - \frac{V_{sat}}{R_4 + R_5} \right) R_3$$

$$= \left(\frac{5\text{ V}}{387\ \Omega} - \frac{5\text{ V}}{6\text{ M}\Omega} \right) 250\text{ k}\Omega$$

$$= 3.02\text{ V}$$

When the input voltage goes below 3.02 V, the output will go low again. This voltage (3.02 V) corresponds to a temperature of 29°C at the LM335. In this application the fan switches on when the temperature reaches about 50°C. This value, the set point, can be varied by changing R_2. The fan will not go off until the temperature is brought down to about 29°C. The difference between these two temperatures is called the dead zone, neutral zone, or differential gap. In this circuit the differential gap is changed by adjusting resistor R_5.

Another example of a differential-gap on-off controller, this time in a level-control application, is found in Figure 9.11. This application uses conductive probes as a transducer. The application requires the use of a liquid that is conductive but not flammable. When the circuit is energized, current flows through the primary and the NC relay contacts, applying power to the pump. Note that at this time no current is flowing in the secondary circuit. The tank starts to fill as the pump transfers fluid into it. When the fluid level reaches the high-level probe, current flows from the secondary into the fluid via the ground probe, through the high-level probe, R_4, R_1, and the diode 1CR. A negative potential is developed on Q_1 gate with respect to its cathode. Since this is a PUT (discussed in a previous chapter), the gate-to-anode voltage turns it on. The PUT conducts through resistor R_3,

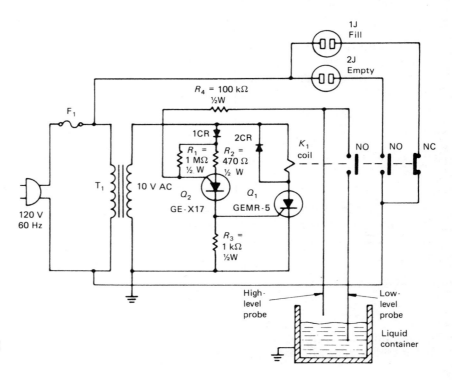

Figure 9.11 Liquid-Level
On-Off Controller

placing a positive potential on the gate of Q_1, turning it on. Relay K_1 turns on when Q_1 conducts, changing the state of the relay contacts. The NC contact opens, removing power from the pump. Note that during the process, when the liquid falls below the high-level probe, current can still flow in the control circuit through the low-level probe and the closed relay contacts. When the level falls below the low-level probe, however, the control circuit current flow is interrupted, since it can no longer flow through the fluid. When current can no longer flow through R_1, Q_2 will not turn on, keeping Q_1 and relay K_1 off. Power is then applied to the pump, filling the tank.

We can clearly see that this, too, is an on-off controller. Note that there is a neutral zone established between the two liquid-level probes. In the previous example the differential gap was temperature; in this case it is liquid level.

On-off control is almost always the simplest and least expensive controller mode. This type of control works equally well with electric systems and with pneumatic systems. However, many industrial processes need better, more sophisticated control. Proportional control was developed to meet this need.

Proportional Control

In *proportional control,* the final control element is purposely kept in some intermediate position between on and off. Proportional control is a term usually applied to any type of control system where the position of the final control element is determined by the relationship between the measured variable and the set point.

An example of a proportional control mode using a filled-system temperature sensor is shown in Figure 9.12. The bulb of the filled system senses the temperature of the fluid in the

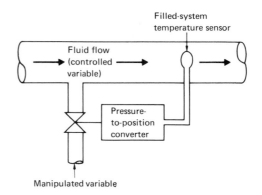

Figure 9.12 Proportional Control System Using Filled-System Temperature Sensor

line. Any change in the temperature causes the valve position to change.

The *gain* of the controller is determined by the relationship of change in temperature to change in valve position. If the valve moves a great deal for a given temperature change, the gain is high. If the valve does not move very much for a given temperature change, the controller gain is small. The valve normally is set so that when the temperature is in the center of its range, the valve is one-half open. In this system the set point is determined by the initial setting of the valve (50%). Once this system is installed, the controller gain is also fixed. More sophisticated proportional control systems have variable gain or set points.

The final control element setting when the measured variable equals the set point is called the *bias.* In our example the bias is 50%. As the measurement deviates from the set point, the final control element setting will change from 50%. The amount of change in output (final control element) for a given change in input is a function of the gain of the controller's amplifier, as shown in the following equation:

$$\text{controller gain} = \frac{\Delta \text{output}}{\text{set point} - \text{measurement}}$$

Most industrial controllers have a gain adjustment that is expressed in percent of the proportional band. *Proportional band* is the amount of change in the dynamic variable that causes a full range of controller output. Or, we can say that the proportional band is equal to the range of values of the dynamic variable that corresponds to a full or complete change in controller output. Proportional band is also called throttling range. Normally, the proportional band is expressed as a percentage:

$$\% \text{ proportional band} = \frac{1}{\text{gain}} \times 100$$

Then:

$$\text{gain} = \frac{100}{\% \text{ proportional band}}$$

Note the relationship between gain (or sensitivity) and proportional band. Systems with large proportional bands are less sensitive (have less gain) than narrow-band systems.

The equation for determining output for the proportional controller in Figure 9.12 is as follows:

$$
\begin{aligned}
\text{output} = \frac{100}{\% \text{ proportional band}} \\
\times (\text{set point} - \text{measurement}) \\
+ \text{bias} \quad\quad\quad \textbf{(9.1)}
\end{aligned}
$$

Thus, for a fixed gain (proportional band setting) and a fixed bias, we can calculate the output. To do so, we must know the measurement and the set point. For example, let us say the proportional band is 100% (gain = 1) and the bias is 50%. When the measurement equals the set point, the output (final control element setting) is 50%. When the measurement exceeds the set point by 10%, the output is 40%. When the measurement is below the set point by 10%, output is 60%.

The higher the gain, the more the output will move for a given change in either measurement or set point. Figure 9.13A shows the effect on the input-output relationship for a gain change.

Another way of showing the effect of changing the proportional band is illustrated in Figure 9.13B. Each position in the proportional band produces a controller output. The wider

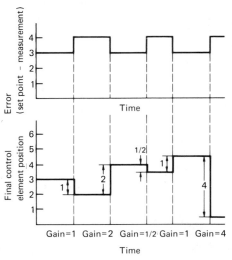

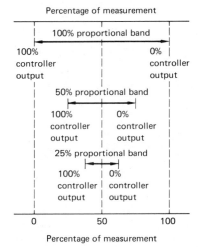

Figure 9.13 Plots of Controller Gain and Proportional Band Changes

A. Effect of Gain Change on Input-Output Relationships

B. Effect of Proportional Band Change on Controller Output and Measurement the band, the more the input signal (set point − measurement) must change to cause the output to swing from 0% to 100%. Changing the bias shifts the proportional band so that a given input signal will cause a different output level.

An example may help at this point. Let us go back to the example of the pasteurization process presented in the introduction to this chapter. Assume that the set point is 65°C. Since the pasteurization process requires that the milk stay between 60° and 70°C for one-half hour, we will assume that the extremes of our temperature range will be limited to these two points. If the input range is 10°C, the proportional band is 50% of the input range or 5% of total (±2.5°C). The controller will be at 100% when the sensor is measuring 62.5°C. If the sensor moves down to 60°, no further control action is possible. On the other hand, the controller will be operating at 0% when the temperature reaches 67.5°C. If the temperature goes above that value, say, to 70°C, no further control action will occur. When the input is between 62.5° and 67.5°C, the controller output will be proportional to the difference between the set point and the measurement. The gain of this system is:

$$\text{gain} = \frac{100}{\% \text{ proportional band}} = \frac{100}{50} = 2$$

Since a 50% change in the measurement produces a 100% change in the controller output, the gain is 2.

Rewriting Equation 9.1, we can calculate the controller output with the following formula:

$$C_o = Ke + C_b \qquad (9.2)$$

where

C_o = controller output in %

K = gain of controller (inverse of proportional band)

e = error term (% drift away from set point)

C_b = controller output bias in % when measurement equals set point or error equals 0

The error term in this equation is, as suggested in Equation 9.1, a function of the difference between set point and measurement but referenced to the full input error range, in this case 10°C. The error term may be calculated by the following formula:

$$e = \frac{M_{dv} - M_{sp}}{M_{max} - M_{min}} \times 100 \qquad (9.3)$$

where

M_{dv} = measurement of dynamic (controlled) variable

M_{sp} = set point

M_{max} = maximum possible value of dynamic variable

M_{min} = minimum possible value of dynamic variable

We can use Equations 9.2 and 9.3 to help us solve for the controller output. For example, if the measurement is 64°C, solving this formula for e, we get:

$$e = \frac{M_{dv} - M_{sp}}{M_{max} - M_{min}} \times 100$$
$$= \frac{64 - 65}{70 - 60} \times 100 = -10.0\%$$

Substituting into the controller output equation (9.2) and assuming a controller bias setting of 50%, we get:

$$C_o = Ke + C_b = [(-2)(-10\%)] + 50\%$$
$$= 80\%$$

From this example we can see that the controller output would be 80% with a dynamic variable measurement of 64°C.

Proportional control tries to return a measurement to the set point after a disturbance has occurred. However, it is impossible for a proportional controller to return the measurements so that it equals the set point. By definition, the output must equal the bias setting (normally 50%) when the measurement equals the set point. If the loading conditions require a different output, a difference between measurement and set point must exist for this output level. Proportional control may reduce the effect of a load change, but it can never eliminate it.

The resulting difference between measurement and set point, after a new balance level has been reached, is called *offset*. The amount of offset may be calculated from the following equation:

$$\Delta\text{offset} = \frac{\% \text{ proportional band}}{100} \\ \times \Delta\text{measurement}$$

The change in measurement is the change required by load upset.

From this equation we see that as the proportional band goes toward zero (gain approaches infinity), offset will approach zero. This result seems logical, because a controller with infinite gain is by definition an on-off controller. We know that we cannot have an offset

in an on-off controller. On the other hand, as proportional band increases (gain decreases), proportionately more offset will exist.

Many proportional control circuits use the op amp, like the one illustrated in Figure 9.14. This particular amplifier has a standard industrial control signal in the form of a current that varies between 4 and 20 mA. Resistor R_4 converts the 4–20 mA signal to a voltage between −1 and 5 V. If the resistor R_1 is properly adjusted, the output from 1IC and 2IC will be zero. Since 1IC is just a summing amp, an equal positive and negative voltage at the two inputs will cancel each other out. For example, if the system is to give a zero output voltage with a 4 mA input, the zero adjust must be set to +1 V. This voltage will cancel the −1 V potential developed across R_1. The span adjust sets the gain of the system. Let us further specify in our system that the output should go from 0 to 10 when the input goes from 4 to 20 mA. The zero adjust would be adjusted to produce +1 V to counteract the −1 V developed across the resistor R_4 by the 4 mA signal. A 20 mA current signal would develop a −5 V potential across R_4. The output of 1IC would be +4 V with a 20 mA signal. Adjusting R_7 to a resistance of 5 kΩ will give 2IC a gain of 2.5 and an output voltage of 10 V with an input of 20 mA. We can see, therefore, that our requirements have been met.

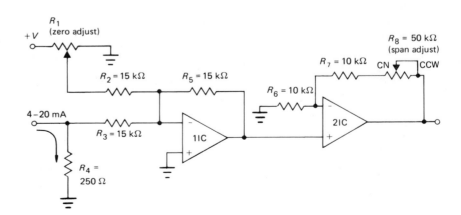

Figure 9.14 Op Amp Proportional Control System

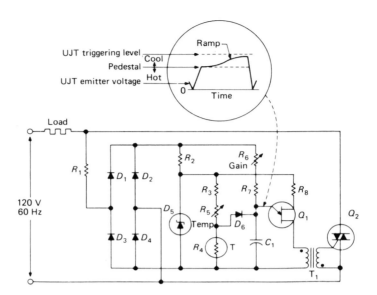

Figure 9.15 Proportional Controller—Precision Temperature Control

Using this system let us see how it responds to a change in the process. The full measurement range is 20 mA − 4 mA = 16 mA. Suppose that our set point is 12 mA and this corresponds to a 50% controller output (+5 V). What will the controller output be with an increase of 1 mA in the measurement? A 13 mA signal would develop a −3.25 V potential across R_4, giving an output of 5.625 V at the output of 2IC.

Note in this example that a full, 100% output swing of the controller was caused by the full, 100%, 4–20 mA control signal. The porportional band for this system would, therefore, be 100%. If we limited the input to 8–16 mA to produce 100% swing in output voltage (0–10 V), this would be a 50% proportional band. If the output was set to +5 V when the error was zero with a set point of 12 mA, an increase of 25% (4 mA) would cause the output to go to 100% (+10 V). Conversely, a decrease of the input by 25% (to 8 mA) would cause the output to go to 0% or 0 V. Note also that the inverse of the proportional band does not equal the gain of the op amp circuit. The relationship be-

tween gain and proportional band holds only when both measurement and output are in the same units.

Another example of a proportional control is shown in Figure 9.15. This circuit is a precision proportional temperature controller that uses a thermistor as a temperature sensor. The load is a heating element. You will recognize this circuit as a ramp-and-pedestal circuit presented in the previous chapter. Let us suppose that the temperature detected by the thermistor in the process decreases. A decreased temperature will increase the resistance of the thermistor R_4, increasing the voltage at the top of R_4. This action raises the pedestal. Since the capacitor charge now takes less time to reach the firing voltage of the UJT, the SCR will fire earlier in the cycle, thus applying more average power to the load. More power to the heating element will increase the amount of heat it gives off. The gain of this circuit can be adjusted over a wide range by changing the size of the charging resistor, R_6. When a ramp amplitude of 1 V is selected with a 20 V zener, a 22% change in the thermistor resistance will give a linear, full-

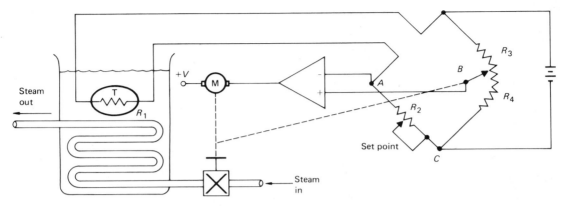

Figure 9.16 Proportional Control—Liquid Temperature

range change in the controller output. The controller in this case is the heater.

Recall the example presented at the beginning of this chapter concerning the pasteurization process. That example is one that uses proportional control. A simplified electrical circuit for such a control is presented in Figure 9.16. The sensor in this case is a thermistor, which, you will recall from the last chapter, has a negative temperature coefficient. The final control element is a motor-driven valve that controls the rate of steam flow through the pipes in the liquid. The motor in the valve is driven by an amplifier that is connected to a bridge. Please note the dotted line in the illustration. As we have noted, the line is connected to the valve, indicating that the motor drives the valve. But note also that the motor drives a potentiometer in the bridge. In the schematic diagram the upper half of the potentiometer is labelled R_3, while the lower half is labelled R_4. When the bridge is in balance no difference in potential exists between points A and B on the bridge. The output of the amplifier is 0 V and the motor is at rest. Let us suppose that at a temperature of 65°C each resistor, including the thermistor, is at a resistance of 2000 Ω. The voltage at each point A and B would then be +

5 V. Let us further say that the temperature of the liquid drifts downward. This change would be detected by the thermistor and reflected in an increase in resistance, say, to 3000 Ω. If R_2 stayed at 2000 Ω, the potential at point A with respect to C would decrease to +4.44 V. The potential at B would be +5 V. So, a difference in potential of about 0.56 V appears between A and B. This difference in potential is amplified by the amplifier and used to drive the motor on the valve. In this case the valve will open, letting in more steam, thus adding more heat to the system. The motor, which drives the valve, also drives the bridge. As the motor is driving the valve open, the motor is readjusting the bridge to a balanced condition again. When the null condition is reached on the bridge, the motor will stop because the voltage to the amplifier is zero again. How much the valve opens before the motor stops turning depends on the gain of the system.

An increase in the temperature of the liquid will have the opposite effect. The thermistor resistance will decrease, unbalancing the bridge in the opposite direction. The voltage at point A will be greater than that at point B. This voltage, amplified by the amplifier circuitry, will move the motor in the opposite direction, closing the valve slightly. Closing the valve will decrease the flow of steam to the system, reducing the amount of heat that the system receives. The bridge may be unbalanced manually by an operator who can adjust resistor R_2. Manipulating resistor R_2 adjusts the set point. The dial on the potentiometer R_2 would, then, be calibrated in degrees. The operator could adjust the temperature of the liquid by varying this potentiometer. The system would treat this as the new set point and try to maintain that temperature as disturbances tended to change it.

Proportional plus Integral Control

Often in industrial applications, the offset caused by proportional control cannot be tolerated. Process engineers solve this problem by adding another control mode. Recall from Chapter 1 that the op amp integrator integrates any voltage present at the input. *Integral control* (sometimes called reset action) in a process control system will integrate any difference between the measurement and the set point. The controller's output will change until the difference between the measurement and the set point is zero.

The response of a pure integral controller is shown in Figure 9.17A. Note that the controller output changes until it reaches 0% or 100% of scale (or until the measurement is returned to the set point). This figure also assumes that an open-loop condition exists where the controller's output is not connected to the process.

Figure 9.17B shows an open-loop, proportional plus integral controller's response to a step change. *Proportional plus integral* (PI) *control* combines the characteristics of both types of control. Reset time is the amount of time required to treat the amount of change caused by proportional action. In Figure 9.17B reset time t is the amount of time required to repeat the amount of output change y.

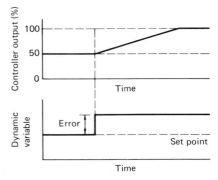

A. Pure Integral Controller (Open Loop)

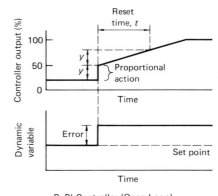

B. PI Controller (Open Loop)

Figure 9.17 Plots of Controller Output and Dynamic Variable Measurement versus Time

Another way to visualize PI action is shown in Figure 9.18. In this example a 50% proportional band is centered around the set point. If a disturbance comes into the system, the measurement will deviate from the set point. Proportional response will be seen immediately in the output, followed by integral action.

We may think of integral action as forcing the proportional band to shift. This shift causes a new controller output for a given difference between measurement and set point. Integral control will continue to shift the proportional

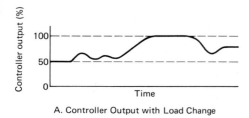

A. Controller Output with Load Change

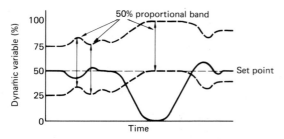

B. Dynamic Variable with Shift of Proportional Band

Figure 9.18 Shifting of Proportional Band in PI Controller When Error Changes

band as long as there is a difference between the measurement and the set point.

Integral control action does exactly what an operator would do by manually adjusting the bias in the proportional controller. The width of the proportional band stays constant. It is shifted in a direction opposite to that of the measurement change. Thus, an increasing measurement signal results in a decreasing output, and vice versa. Because integral control acts only on long-term or steady-state errors, it is seldom used alone.

The PI control mode is used in situations where changes in the process load do not happen very often; but when they do happen, changes are small. These small changes may occur for a long time before they finally go beyond the allowable error limits. In many industrial processes even small amounts of error that persist for long periods of time are undesirable. Consequently, PI control is a very popular mode of control in industry today.

Proportional plus Derivative Control

Some process control systems have errors that change very rapidly. This situation is especially true in processes that have a small capacitance. Neither proportional control nor PI control responds well to errors that change rapidly. Recall from Chapter 1 that the differentiator circuit responded to the input rate of change, or slope. The same concept applies to the type of controller mode called *derivative,* or *rate, control.* By adding derivative control to proportional control, we get a controller output that responds to the measurement's rate of change as well as to its size.

Proportional plus derivative (PD) *control* is illustrated in Figure 9.19. When a measurement changes, derivative action differentiates the change and maintains a level as long as the measurement continues to change at a given rate.

Derivative control is never used alone because it can react to measurements only when they are changing. It cannot react to steady-state errors.

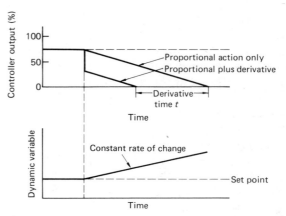

Figure 9.19 Plot of Measurement and Controller Output versus Time in PD Controller

Proportional plus Integral plus Derivative Control

All three controller modes we have discussed can be combined into one mode, *proportional plus integral plus derivative* (PID) *control.* PID control has all the advantages of the three types of control. The proportional mode portion produces an output proportional to the difference between the measurement and the set point. The integral control action produces an output proportional to the amount and the length of time the error is present. The derivative section produces an output proportional to the error rate of change. The PID mode of control applies to systems that have transient as well as steady-state errors.

CONTROLLERS

As we have seen, the controller is the part of a control system that compares the measurement of the controlled variable with the set point. The controller also directs the action of the final control element, which corrects or limits the deviation.

Controllers in industry may be classified in several ways. For example, they may be classified as either electric (electronic) or pneumatic. Also, electric and pneumatic controllers can both be broken down into two popular types: analog and digital. In this section we will discuss both analog and digital types of electric and pneumatic controllers.

Electric Controllers

Many electric (or electronic) analog controllers use the op amp as a basic building block. Recall that the op amp can perform the processes of comparison, summation, integration, and differentiation. These are precisely the building blocks we need to put together the types of controller modes we have just discussed. Process engineers choose the types of control to suit the particular process application. All the controller modes we have discussed can be accomplished with the op amp.

While each manufacturer of electronic analog process controllers produces a different system, many of these systems have common factors. For example, many electronic controllers have similar front panels, with metering systems similar to the one shown in Figure 9.20. This meter displays two types of information: the set point and the measured variable values. The set point normally is adjustable from the front panel by means of a potentiometer. Thus, it is simple to change the set point for different applications. The scale usually is calibrated to read from 0% to 100% of signal variation. In addition many controllers give operators the ability to switch back and forth from manual to automatic mode. The manual mode often is used when the process is started up and when it is shut down.

Digital electronic control is also found in industry. The computer, of course, can be clas-

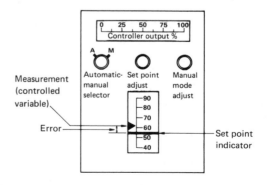

Figure 9.20 Front Panel of Typical Electronic Controller

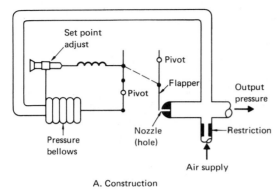

A. Construction

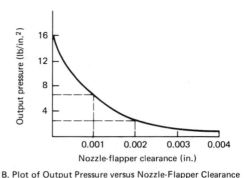

B. Plot of Output Pressure versus Nozzle-Flapper Clearance

Figure 9.21   Pneumatic Flapper-Nozzle Amplifier

sified as a digital controller. Because of the ability of the computer to make decisions and initiate action, it is ideally suited to process control applications. The most popular digital controller today is the programmable controller, which is a microprocessor-based system capable of a wide range of tasks. The programmable controller will be discussed in Chapter 13.

Pneumatic Controllers

Because of their low cost, low maintenance, and safety advantages, pneumatic controllers are still popular in industry. Pneumatic controllers, like electric controllers, come in both analog and digital versions.

The analog pneumatic controller uses the pressure bellows and the flapper-nozzle combination to effect its control. A simple *flapper-nozzle amplifier* is shown in Figure 9.21A. Constant air pressure is supplied to the nozzle through a restrictor whose diameter is about 0.010 in. (0.254 mm). The nozzle itself has a diameter of about 0.020 in. (0.5 mm). The flapper is positioned against the nozzle opening. The position of the flapper is determined by the transducer output and the set point. The nozzle's back pressure is inversely proportional to

the distance between the nozzle opening and the flapper. As shown in Figure 9.21B, a flapper motion of about 0.002 in. (0.05 mm) is sufficient to provide a full range of output. The sensitivity of this system can be adjusted by moving the flapper pivot point.

The flapper-nozzle amplifier shown in Figure 9.21A can be used as a proportional controller. When the controlled variable (output pressure) increases, the bellows expand, causing the flapper to move away from the nozzle. Air then escapes through the nozzle into the atmosphere. This action reduces the output pressure to the set point value. Conversely, a decrease in output pressure causes the output pressure to increases to the set point value.

A proportional plus derivative pneumatic controller is shown in Figure 9.22. The addition of a variable restrictor results in a delayed negative feedback. If the controlled variable increases or decreases suddenly, the transducer causes the flapper to open or close. This change in flapper position causes the output pressure to change suddenly, but the pressure in the bellows can only change slowly. The slower change in the bellows is due to the variable restriction and the capacity of the bellows. This slower change delays and reduces the amount of negative feedback. Since the feedback is negative, the output pressure is higher and leads the

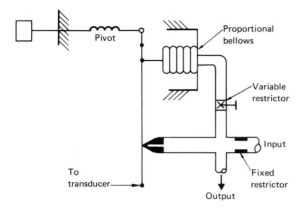

Figure 9.22 Proportional plus Derivative Pneumatic Controller

transducer signal. Therefore, the delayed negative feedback produces a derivative response.

These examples show that control can be accomplished by analog pneumatic means. However, digital pneumatic control also is available through fluidic devices. A *fluidic device* is a digital logic element that operates by air flow and pressure.

One of the most popular fluidic devices uses the *Coanda,* or *wall attachment, effect.* This effect is illustrated in Figure 9.23. A fluid

flowing past a wall has a tendency to attach itself to that wall. When air pressure flows from the supply, the stream of air will attach to one of the two walls. The stream will stay attached until a jet of air from the control port detaches it from the wall.

You may recognize this fluidic device, functionally, as a bistable flip-flop. By a rearrangement of the input ports and the addition of a restriction to one of the output legs, both AND/NAND and OR/NOR gates may be formed. The logic symbols are shown in Figure 9.24. With the three logic elements, a simple digital pneumatic computer may be made. It can be used for simple controlling and decision-making functions in the same way an electronic digital computer is used.

Fluidic systems have several advantages over electric systems. Since no electric potentials are present, there is less likelihood of explosive gases being ignited. This advantage is especially important in the chemical and petrochemical industries. Also, the fluidic system

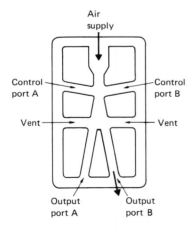

Figure 9.23 Wall Attachment Effect in Fluidic Device

A. OR/NOR Gate

B. AND/NAND Gate

C. Bistable Flip-Flop

Figure 9.24 Schematic Diagrams of Fluidic Logic

can be completely flushed out with cleaning solvents at regular intervals. It is difficult to do such cleaning in electric equipment. This advantage is particularly important in the food-processing industry, where bacterial contamination must be kept low.

There are some disadvantages to fluidics, however. Generally, fluidic systems are much slower than comparable electric systems. Also, fluidic systems take up much more space, since large-scale integration is not available with pneumatic systems.

FINAL CONTROL ELEMENTS

Previously we defined the final control element as a device that corrects the value of the controlled variable. For instance, in an assembly line the controlled variable might be conveyor belt speed. The conveyor belt is driven by a motor, which is the final control element. In a proportional control system an increase in the load on the belt would tend to slow the belt down. This speed change would be sensed and compared to a set point. The controller would increase the belt speed back to normal by varying the speed of the motor, the final control element. In the pasteurization example recall that the steam heated the milk to the proper temperature. Therefore, the valve controlling the flow of steam is the final control element. If the liquid were heated by a resistance element, that element would be the final control element.

In many process control applications, the final control element is a control valve. The *control valve* is a mechanism that regulates the flow rate of a fluid, which can be a gas, a liquid, or a vapor. The control valve can be classified on the basis of its flow characteristics or on the basis of its body style.

In electrical terms a valve can be considered analogous to a resistance. Flow through a valve is then proportional to two things: the area of the valve opening, and the pressure drop across the valve. The following formula expresses this relationship:

$$Q = KA\sqrt{\Delta p}$$

where

Q = quantity of fluid flow

K = constant of proportionality for conditions of flow

A = area of valve opening

Δp = pressure drop across valve

Figure 9.25 is a diagram of the percent of valve travel, or position, plotted against the percent of flow. Note that the *quick-opening valve* shows a large increase in flow rate with a small change in valve opening. This type of valve is used in on-off control situations. The *linear valve* has a linear response curve. That is, the valve position is linearly related to the flow rate.

The most commonly used valve type is the *equal-percentage valve*. Note that a change in the valve position produces an equal-percentage change in flow. If you were to plot this relationship on a logarithmic scale, the relationship would be a linear one. The equal-percentage valve is designed to be used between a

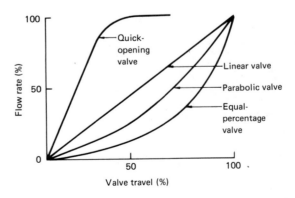

Figure 9.25 Plot of Valve Travel versus Flow Rate for Four Control Valve Designs

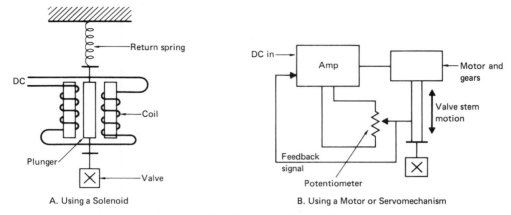

Figure 9.26 Electrical Valve Control

certain minimum and maximum flow rate. The maximum flow rate (Q_{max}) divided by the minimum flow rate (Q_{min}) is called the *rangibility* (*R*) of the valve. The following formula expressed the rangibility of the valve:

$$R = \frac{Q_{max}}{Q_{min}}$$

Most commercial valves have a rangibility between 30 and 50.

Generally speaking, valves are controlled in industry in two ways: electrically and pneumatically. Electrical actuation of valves may be done by a solenoid, such as the one illustrated in Figure 9.26A. When current flows through the coil, a magnetic field is generated, which moves the plunger. Such a device is used in conjunction with on-off controllers.

Another method of controlling valve actuation electrically is by motors or servomechanisms, as illustrated in Figure 9.26B. The DC signal from the controller is amplified and operates a geared motor, driving the valve stem. The valve position is converted to an electric signal that feeds back to the amplifier circuit. Here it is compared to the DC signal. Any difference is amplified to drive the valve to a po-

sition exactly proportional to the original DC signal.

Valves may also be actuated pneumatically. Pneumatic actuation is shown in Figure 9.27. In this case the diaphragm moves proportionally to the amount of air pressure at the inlet. Air pressure may be used to either open or close the valve.

Although control valves are the most common final control elements used in industry,

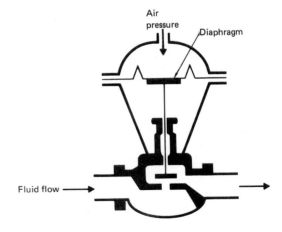

Figure 9.27 Pneumatic Valve Control

there are others in use. In Chapters 3, 4, and 5 we discussed DC and AC motors. These devices are used in industrial velocity and position control systems, such as conveyor belts and mixing operations. The speed of DC motors is easily controlled and is capable of reversible operation. Speed control is possible in AC motors as well, although not as conveniently.

CONCLUSION

At this point we have discussed most of the components of a process control system. These parts should be starting to fit together into a coherent whole for you. All the previous chapters have been building up to this point—a clear understanding of process control systems and their components.

QUESTIONS

1. The function of a process control is to keep the controlled variable constant when it tends to be changed by a _____.
2. The process of comparing the information from a sensor to a reference is called _____. The difference between the sensor and the set point reference is called the _____.
3. _____ control has no feedback from output to input. Feedback control is sometimes unsuited to control applications, since large _____ or _____ cannot be dealt with effectively by feedback control.
4. _____ control is sometimes called anticipatory control because it anticipates the effect of a disturbance and tries to correct it before it happens. _____ control is suited to systems that have large capacitances.
5. In proportional control the final control element action is directly related to the difference between the measurement and the _____. The amount of final control element change for a given input change is called _____.
6. Integral control provides controller action that changes as a function of _____ _____ the error is present. Integral control is also called _____ control.
7. Derivative control provides controller action that varies as a function of the _____ at which the error changes. Derivative control is also called _____ control.
8. Define the following terms: (a) dynamic variable and (b) manipulated variable.
9. Draw a block diagram of a closed-loop control system with blocks for measurement, evaluation, and control.
10. Define process load.
11. Define process lag, and list the three properties of a process control system that cause it.
12. Give an example of an open-loop control system.
13. Draw a block diagram of a feedforward control system, and describe the system's operation.
14. Is it necessary to have a sensor in an open-loop control system? Why or why not?
15. Is a sensor necessary in a closed-loop control system? Why or why not?
16. What is an error signal? How is it generated?
17. Describe and draw response curves for the following control modes (open loop): (a) on-off, (b) proportional, (c) integral, and (d) derivative.
18. As the controller gain increases, what happens to the controller output for a given change in the measurement?
19. Define offset. How may offset be calculated?
20. Explain the advantages and disadvantages of (a) proportional control, (b) integral control, and (c) derivative control.

21. Can integral and derivative control be used alone? Why or why not?

22. Explain how control can be achieved by pneumatic means. Describe digital and analog pneumatic controllers and how they work.

23. Define final control element.

24. What is the relationship between controller gain and proportional band? Is it ever possible to have a 0% proportional band? Why?

PROBLEMS

1. A process control system with a DC motor as a final control element has a proportional band of 100%. What is the gain of the controller? If the proportional band were 50%, what would the gain be?

2. In Problem 1 suppose a system had a gain of 3, a set point of 70°C, and a measurement of 50°C. How much would the controller output change over this difference?

3. Calculate the proportional band of a pneumatic controller that has a total range of 0–200 lb/in.^{2} with its set point at 50 lb/in.^{2}. A deviation of ± 10 lb/in.^{2} will cause the output of the controller to vary from 0 to 100 lb/in.^{2}.

4. A pneumatic-to-electric transducer provides the following information:

Pounds/in.^{2}	V_o
0.5	1.00
7.5	1.50
10.0	1.75

What is the gain of the transducer using the values in the first and second rows?

5. In the on-off controller in Figure 9.8 we would like the fan to turn on when the temperature goes to 60°C.

a. Calculate the output of the LM335 at 60°C.

b. Determine the reference voltage needed at the noninverting terminal.

6. The circuit in Figure 9.10 has the following resistor value changes: $R_2 = 250$ kΩ and $R_5 = 4.5$ MΩ. Calculate the temperatures at which the circuit will trip.

7. In the controller in Figure 9.10 calculate the resistance changes in R_2 and R_5 to cause switching at temperatures of 55° and 35°C.

8. The entire measurement range of a proportional control system is 100°C. The controller is a fan motor with an adjustable speed. Its bias point is 50%, which is equal to a speed of 1000 r/min. When temperature increases, the fan speed increases. The proportional band is 25%.

a. Calculate the gain or sensitivity of the system.

b. Calculate the fan speed if the measurement is 110°C.

9. In the circuit in Figure 9.14 what would the output voltage of the circuit be if the input were 18 mA?

10. In the proportional control system shown in Figure 9.14, what values of resistance would be needed to change this controller's proportional band to 25% with a bias point of 3.5 V at the output with a 13 mA signal in?

CHAPTER
10
PULSE MODULATION

OBJECTIVES

On completion of this chapter, you should be able to:

- Define pulse modulation;
- Determine the minimum sample rate for pulse modulation;
- Modulate and demodulate the four types of analog pulse modulation;
- List the advantages and disadvantages of the pulse modulation types;
- Generate and demodulate pulse-code modulation;
- Determine quantizing noise values for pulse-code modulation.

INTRODUCTION

Pulse modulation has been known to humanity for a long time. Modulated drum beats and smoke signals are two examples of this process. However, in this chapter we will be concerned with a more recent development: human-generated electric pulse modulation.

Pulse modulation is a process in which an analog signal is sampled at regular periods of time. Information contained in the analog signal is transmitted only at the sampling time

and may also have synchronizing and calibrating pulses. At the receiver the original waveform may be reconstructed from the information contained in the samples. If the original samples are taken frequently enough, the analog signal can be reproduced with minimal error or distortion.

This chapter will deal with the various types of pulse modulation and demodulation and some examples of circuits using these pulse

340

modulation methods. Pulse modulation is fundamental to the understanding of telemetry, the subject of Chapter 11.

ELECTRIC PULSE COMMUNICATION

Electric pulse communication had its origin in 1837 when Samuel Morse invented the telegraph. *Telegraphy* is the process of sending written messages from one point to another in the form of code. A few years later, in 1845, a Russian general, K. I. Konstantinov, and Dr. Poulié constructed a telemetry system. A *telemetry system* performs measurements on distant objects. The original system recorded and analyzed the flight of a cannonball. This system automatically reported the course of the cannonball as it passed through screens, and recorded the electric impulses and a timing impulse on a manually turned recording drum loaded with graph paper. This system was one of the first written records concerning telemetry.

More recently, with the development of television and radar, an additional type of pulse communication called data communication has come into widespread use. *Data communication* is the process of transmitting pulses, which are the output of some data source, from one point to another. Examples of data communication are computer-to-computer transmission, data collection, or telemetry and alarm systems. Other commerical uses are financial/credit information, travel and accommodation booking services, and inventory control for stores.

Facsimile is also considered to be a form of data communication. *Facsimile* is the process whereby fixed graphic material, such as pictures, drawings, or written material, is scanned and converted into electric pulses, transmitted, and, after reception, used to produce a likeness (facsimile) of the original. This process can be considered similar to the transmission of a single television picture, but in facsimile the picture is recorded on paper. Facsimile has been used by news services to transmit newspaper photos and by ships for reception of up-to-date weather charts and maps.

In recent years the volume of pulse communication has increased greatly. There are at least three factors contributing to this increase in pulse communication: (1) Much of the information to be transmitted is in pulse form to begin with, such as computer data or, to some extent, picture elements; (2) a more error-free transmission process is possible, for example, as in distant space probe picture transmission; and (3) the advent of large-scale integration has greatly simplified the necessary electronic circuitry. In addition, large-scale integration has permitted the use of complex coding systems that take the best advantage of channel capacities.

Telegraphy and telemetry can properly be considered subsets of data communication, but historically they have been kept separate, as indicated in Figure 10.1. Telegraphy has been

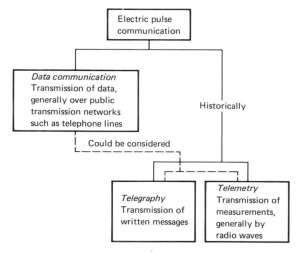

Figure 10.1   Subsets of Electric Pulse Communication

considered a separate branch because it is the transmission of messages, which are not judged to be data. Telemetry has been considered a separate branch because its first major use employed pulse modulation and radiotelemetry and did not use public transmission networks, such as the telephone line, for its transmission. Since the inception of data communication, however, these distinctions have become less clear.

It is instructive to understand the relationships among telegraphy, telemetry, and data communication. This background will be beneficial when you are studying the next chapter, which is concerned with telemetry.

PULSE MODULATION TYPES

Pulse modulation can be divided into two major categories: analog and digital. In *analog pulse modulation* a characteristic of the pulse, such as height or width, may be infinitely variable and varies proportionally to the amplitude of the original waveform. In *digital pulse modulation* a code is transmitted; the *code* indicates the sample amplitude to the nearest discrete level.

All modulation systems sample the information waveform to be telemetered, but they all have different ways of indicating the sampled amplitude. Each modulation system also has its advantages and disadvantages, as will be discussed in the following sections.

ANALOG PULSE MODULATION

There are four major types of analog pulse modulation: pulse-amplitude modulation, pulse-width (or duration) modulation, pulse-position modulation, and pulse-frequency modulation. We will discuss each type in detail in this section.

Pulse-Amplitude Modulation (PAM)

Pulse-amplitude modulation (PAM) is illustrated in Figure 10.2. PAM is a process in which the signal is sampled at regular intervals, and each sample is made proportional to the amplitude of the signal at the instant of sampling. As shown in Figure 10.2, there are two types of PAM. *Double-polarity* PAM can have pulse excursions both above and below the reference level (Figure 10.2B). *Single-polarity* PAM has a fixed DC level added to the pulses so that the pulse excursions are always positive (Figure 10.2C).

Noise. Since the amplitude of the pulse contains the information, it is very important that the amplitude be a true representation of the original level. However, in the transmission of the signal, noise can be added to the pulse,

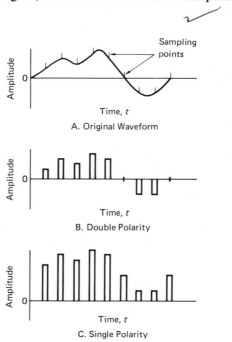

A. Original Waveform

B. Double Polarity

C. Single Polarity

Figure 10.2 Pulse-Amplitude Modulation (PAM)

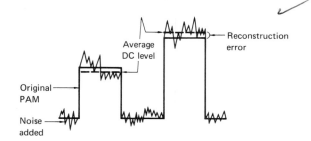

Figure 10.3   Effect of Noise on PAM

causing reconstruction errors. *Reconstruction errors* are errors that cause the reconstructed waveform to differ from the original waveform. These errors may be caused by noise or they may result from other phenomena.

Figure 10.3 illustrates reconstruction errors due to noise. Noise and the other reasons for error are discussed in more detail in Chapter 11. The added noise depends on many factors, such as the strength of the signal, the distance the signal must travel, and the environment the signal must pass through. Because PAM is greatly affected by noise, it is not used as often as other forms of pulse modulation.

Aliasing.   Another problem that can exist with any form of sampled telemetering is called aliasing. *Aliasing* is the reconstruction of an entirely different signal from the original signal. The different signal is a result of the sampling rate being low in comparison with the rate of change of the signal sampled. If, for example, the signal to be sampled varies at a rate of 10 Hz, the signal may be sampled ten times during its period (sampling rate of 100 Hz). This sampling rate will not result in severe aliasing. However, if the sampling rate were reduced to 9 Hz, or less than once each signal period, a certain amount of aliasing would occur. Figures 10.4 and 10.5 illustrate these ideas.

In Figure 10.4, the sampling rate is ten times the frequency of the waveform to be sampled. As shown, the resulting, or recovered, waveform is approximately equal to the original signal. In Figure 10.5, the original wave-

form frequency remains the same as in Figure 10.4, but the sampling rate is reduced by more than ten times the sample rate of Figure 10.4. The resulting, or recovered, waveform in Figure 10.5 no longer resembles the original waveform. Therefore, considerable aliasing is produced.

Sampling Theorem.   The aliasing problem associated with sample rate has been investigated by many researchers and has resulted in a theorem known as the *sampling theorem:*

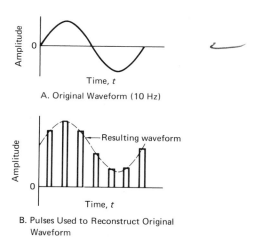

Figure 10.4   Sample Rate Resulting in Minimum Aliasing

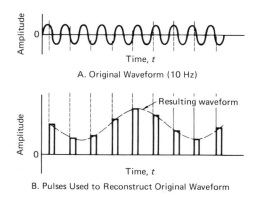

Figure 10.5   Sample Rate Resulting in Considerable Aliasing

If the sampling rate in any pulse modulation system exceeds twice the maximum signal frequency (or Nyquist rate), the original signal can be reconstructed in the receiver with vanishingly small distortion.

As a result of this theorem most pulse modulation sampling rates for speech over standard telephone channels, whose bandwidth is 300–3400 Hz, is standardized at 8000 samples per second. The 8000 samples per second is slightly more than twice 3400 Hz, and, therefore, the sampling theorem is satisfied..

To illustrate the theorem we will assume Figure 10.6A represents a small portion of a speech transmission. The portion of the waveform from point *A* to point *B* represents the steepest slope, or greatest rate of change, of the wave. Since it is telephonic speech, this rate of change corresponds to the frequency of 3400 Hz. A 3400 Hz sine wave (the dotted line) whose peaks are at points *A* and *B* is shown for comparison. The period of the 3400 Hz sine wave is also shown. According to the sampling theorem, the samples are taken 8000 times per second and therefore have a pulse period of 0.125 ms.

The waveform in Figure 10.6B is the same as the waveform in Figure 10.6A, but it shows how far apart the samples are taken to indicate 8000 pulses per second. This rate is the slowest pulse rate required to reproduce the original signal with "vanishingly small distortion."

Generation and Demodulation of PAM. In most electronic systems there are many different circuits that can be used to produce the same outcome. So it is here. The circuits in this chapter represent only a few of the circuits that can be used to produce pulse modulation. PAM can be produced by first generating a pulse, then varying the gain of the amplifier to which the pulse is going. A block diagram for producing PAM is shown in Figure 10.7. A National Semiconductor LM555 timer connected in the astable configuration or a CD4047 low-power, monostable/astable multivibrator in the astable mode can be used as the pulse generator. The CD4047 is CMOS (complementary metal-oxide semiconductor).

The variable-gain amplifier employed here is often used to generate analog amplitude modulation (AM). The basic circuitry is that of a differential amplifier, shown in Figure 10.8. The pulses are input into one of the differential transistors, while the modulating signal is applied to the normally constant current transistor. The modulating signal thus causes the am-

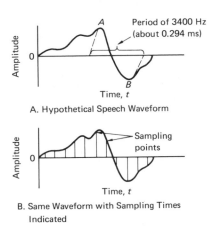

Figure 10.6 Waveform Illustrating Sampling Theorem

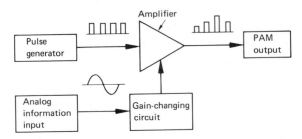

Figure 10.7 Block Diagram for Producing PAM

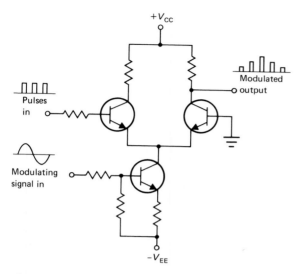

Figure 10.8 Variable-Gain Differential Amp

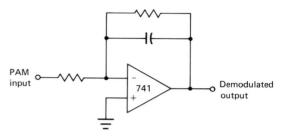

Figure 10.10 Low-Pass Filter for Demodulating PAM

plifier gain to vary. The output is now pulse-amplitude–modulated. This result is essentially what the transconductance amplifier, discussed in Chapter 1, produces.

Other classes of circuits that can be used for

the variable-gain amplifier are Motorola Semiconductor's MC1595L linear four-quadrant multiplier, National Semiconductor's LM1596 balanced modulator-demodulator, and RCA's CA3080 operational transconductance amplifier (OTA).

The complete circuit for PAM is shown in Figure 10.9.

Demodulation, or signal recovery, of PAM can be accomplished with a low-pass filter, as shown in Figure 10.10. Additional filtering or signal smoothing may be required.

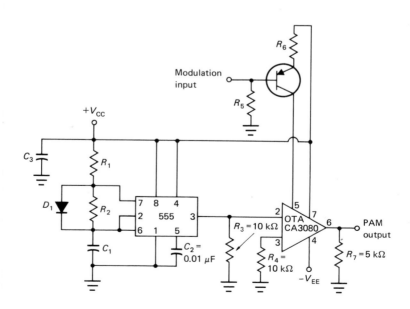

Figure 10.9 Circuit for Generating PAM

Pulse-Width Modulation (PWM)

Pulse-width modulation (PWM), sometimes referred to as *pulse-duration modulation* (PDM), is shown in Figure 10.11. In PWM the signal is sampled at regular intervals, but the pulse width is made proportional to the amplitude of the signal at the instant of sampling. As shown in Figure 10.11B the pulses are of equal amplitude and the leading edges of the pulses are the same time apart. However, the trailing edges of the pulses are not.

One concern in PWM is that the pulse-width variations do have practical limits. Obviously, the pulse width cannot exceed the spacing between pulses. If it did, there would be overlapping, and the proportional relationship between pulse width and signal amplitude

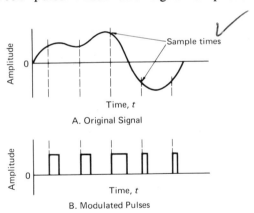

A. Original Signal

B. Modulated Pulses

Figure 10.11   Pulse-Width Modulation (PWM)

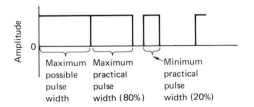

Figure 10.12   Practical Pulse-Width Limitations

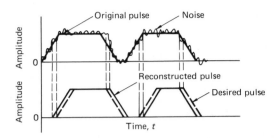

Figure 10.13   Exaggerated Effect of Noise on PWM

would no longer be true. And since a negative pulse width makes no sense, the most negative swing of the signal to be sampled can, at most, produce zero pulse width. In practice these limits are more restricted. As a rule of thumb the widest pulse width should not exceed 80% of maximum; the narrowest pulse width should not be less than 20% of maximum. Figure 10.12 illustrates these limitations.

PWM is less affected by noise than PAM. However, it is not completely immune. Figure 10.13 shows the possible error due to noise for PWM.

Generation of PWM can be produced by again employing the LM555. National Semiconductor's LM555 specification sheet contains

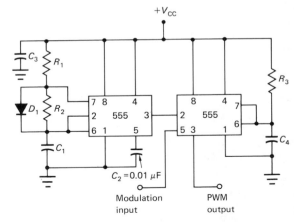

Figure 10.14   Circuit for Generating PWM

a circuit similar to the one shown in Figure 10.14. In this circuit two LM555s are used. However, the circuit can be made more compact by using an LM556, which is a dual LM555.

Demodulation of PWM can be accomplished in the same manner as for PAM (see Figure 10.10). A low-pass filter will demodulate pulses that vary in either amplitude or width.

Pulse-Position Modulation (PPM)

Pulse-position modulation (PPM) and a method of generating it are shown in Figure

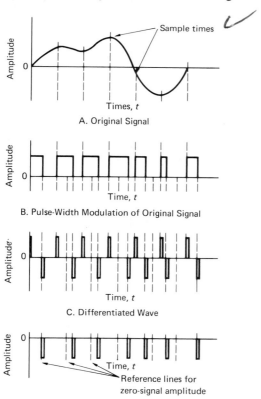

Figure 10.15 Generation of Pulse-Position Modulation (PPM)

10.15. The original signal is first pulse-width–modulated, then differentiated, and clipped. The resulting pulses vary their position about a reference line, which is established by the zero-signal pulse position. As shown in Figure 10.15D, the PPM pulses are on the reference line, or lag the reference, or lead it. The amount of lead or lag from the reference line is proportional to the amplitude of the original signal above or below the zero line. Note that the leftmost pulse of Figure 10.15D is on the reference line; the second, third, and fourth pulses from the left lag the reference lines; the fifth and sixth pulses lead the reference lines; and the last pulse is on the reference line.

PPM is as susceptible to noise on the pulse as PWM is, but it is not as bad as PAM. The same cause of error shown in Figure 10.13 can affect PPM. But, of course, for PPM the position of the pulse is affected; its width is not important.

As shown in Figure 10.15, in the generation of PPM the signal is first converted into PWM. The circuit of Figure 10.14 shows generation of PWM. The output from the circuit of Figure 10.14 is then input to the circuit of Figure 10.16. The PWM is thus differentiated, and the negative pulses are clipped. Because the differentiator is an inverter, the PPM output is now a positive-going pulse.

PPM can be demodulated in at least two different ways. In the first method, PPM is converted back to PWM, and then the output is fil-

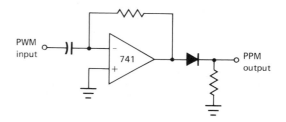

Figure 10.16 Circuit for Producing PPM from PWM

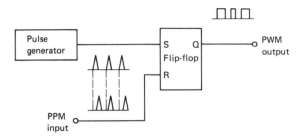

Figure 10.17   Converting PPM to PWM

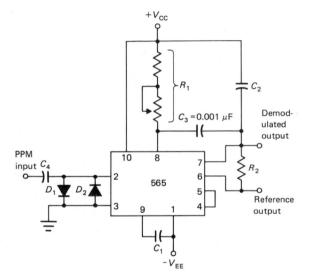

Figure 10.18   Demodulation of PPM Using a PLL

tered. Figure 10.17 illustrates the conversion to PWM. Here, a flip-flop is used for the conversion to PWM, which then can be filtered to the original signal. This circuit has a drawback, however: The pulse generator input must somehow be synchronized with the PPM input. This synchronization calls for more circuitry and is more difficult than necessary. Therefore, it will not be discussed further.

The second method for demodulating PPM does not require synchronization and is easier to do than the first. This method requires the

use of a PLL circuit, which can be used to demodulate any frequency-modulated signal. The circuit is shown in Figure 10.18. The two diodes, D_1 and D_2, are used to limit the input signal to a safe range in case it is large. The input of the LM565 responds to zero or reference crossings only, and therefore the input signal does not have to be a sine wave but can be any waveform.

The VCO free-running frequency f_o of the 565 is given by the following equation:

$$f_o = \frac{1}{4R_1C_1}$$

This free-running frequency should be adjusted to be near the center frequency of the input PPM frequency range. In the equation R_1 should be between 2 kΩ and 20 kΩ with 4 kΩ being the optimum value. The value of C_1 is not critical. Resistor R_2 in Figure 10.18 is not necessary but can be used to decrease the lock range of the 565. The output of the 565 (pin 7) will need filtering and amplification.

Pulse-Frequency Modulation (PFM)

Pulse-frequency modulation (PFM) may look very much like PPM, but it is generated in a different way. While PPM has only one pulse in each sample space, PFM can have more than one pulse in each sample space. (See Figure 10.21E.)

Of the four pulse modulation systems discussed up to this point, PFM is the least affected by noise. PFM depends not on pulse width or pulse location, but rather on the number of pulses per second. Thus, noise spikes would have to be very large in order to affect frequency.

One method for generating PFM uses a single LM555 connected as shown in Figure 10.19. This circuit is shown in National Semiconduc-

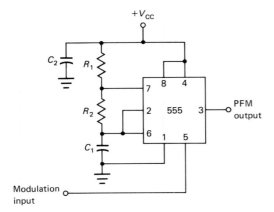

Figure 10.19 Generation of PFM Using the LM555

tor's specifications for the LM555 as PPM. However, according to the definitions in this chapter, Figure 10.19 is the circuit for PFM, not PPM.

Another circuit for generating PFM very simply is that of Figure 10.20. Here, the LM741 op amp is used as a relaxation oscillator. The R_1C_1 combination determines the rate of change of voltage at the inverting input to the op amp. The R_2R_3 combination provides positive feedback to the op amp and is the comparator-with-hysteresis portion of the circuit.

Some means of varying the value of R_1 would cause the output square wave to change its frequency. Therefore, if the modulation signal is applied such that R_1 varies its resistance at the modulation rate, then the output square wave will vary its frequency at the modulation rate. Figure 10.20 shows an LED (D_1) and a photoresistor (R_1) enclosed in a lightproof enclosure. The circuitry associated with the LED causes it to vary its intensity, which, in turn, varies the resistance of R_1. The output of the 741 is now PFM.

Demodulation of PFM can be done in the same way that demodulation of PPM is done, that is, with the PLL of Figure 10.18. The PLL *chip* (the semiconductor material that makes up the integrated circuit) is a very handy circuit. You should get to know it well.

Summary of Analog Pulse Modulation

Figure 10.21 gives a visual summary of the analog pulse modulation types. As we have seen, these pulse modulation types can be implemented easily by using the simple circuits illustrated in this chapter. These circuits have been chosen to facilitate understanding of the modulation types and also to illustrate the problems associated with each modulation type, such as modulation, demodulation, and noise. An industrial grade modulation/demodulation system would necessarily be more complex, but the principles involved would be the same.

DIGITAL PULSE MODULATION

Pulse-code modulation is the major type of digital pulse modulation in use today. In this section we will describe pulse-code modulation in

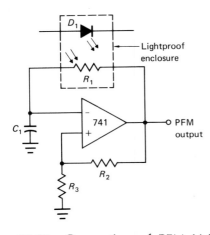

Figure 10.20 Generation of PFM Using the LM741

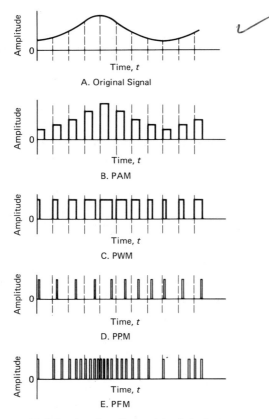

Figure 10.21 Analog Pulse Modulation Types

detail and briefly discuss some other digital systems found in industry.

Pulse-Code Modulation (PCM)

Pulse-code modulation (PCM), a form of digital pulse modulation, is very different from analog pulse modulation. Analog pulse modulation requires that some characteristic of the pulse, such as amplitude, width, position, or frequency, vary in accordance with the amplitude of the original signal. Pulse-code modulation, in contrast, transforms the amplitude of the

original signal into its binary equivalent. This binary equivalent represents the approximate amplitude of the signal sampled at that instant. The approximation can be made very close, but it still is an approximation.

To illustrate PCM let us suppose a signal could vary between 0 and 7 V. This voltage range is divided into a number of equally spaced levels, or *quanta,* which can be a representation of the integer values of voltages, as shown in Figure 10.22A.

When the signal is sampled (the vertical lines in Figure 10.22A), the digit produced depends on the level, or quantum, the signal is in at that instant. For instance, the leftmost sample time in Figure 10.22A finds the signal in the range between ½ and 1½ V. This range corresponds to the decimal number 1 and the binary number 001. The process of determining which level the signal is in is called *quantization.* This quantized digit is represented in binary code by

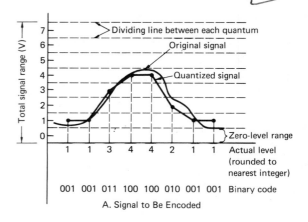

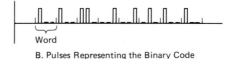

Figure 10.22 Encoding a Signal for Pulse-Code Modulation (PCM)

1s and 0s (Figure 10.22A) and sent as pulses (Figure 10.22B), generally in reverse order to make decoding easier.

As shown in Figure 10.22B, it may be difficult to determine when one group of pulses, called a *word*, ends and the next group or word starts. For this reason, a *supervisory*, or *synchronizing*, *bit* is generally added to each coded word to separate them. This bit is distinguishable from the other pulses in some way, such as having a longer pulse duration or a greater pulse amplitude. This bit allows easier decoding.

You probably immediately noticed in Figure 10.22 that the quantized signal does not look like the original signal. This difference is called *quantizing noise*. It is called noise because the errors are randomly produced. The largest quantizing error that can occur is equal to one-half the amplitude of the levels into which the signal is divided. In Figure 10.22 the largest quantizing error is ½ V.

For reduction of this quantizing noise, the obvious solution is to increase the number of quantizing levels used to encode the original signal. For example, ½ V intervals instead of 1

V intervals could have been used. In the case of Figure 10.22, then, four bits (16 voltage levels) instead of three bits (eight voltage levels) would have to be used to represent each level. The increased number of levels results in more pulses per second at the output of the PCM encoder and, therefore, an increase in the frequency of the signal to be transmitted. The advantage of increasing quantizing levels is that quantizing noise is reduced; the disadvantage is that a greater bandwidth is required to transmit the signal. In practical systems 128 levels for speech are considered adequate.

Generation and Demodulation of PCM. Generation and demodulation of PCM are much more complex than they are for analog pulse modulation. The block diagrams of Figures 10.23A and 10.23C give some idea of the complexity involved. Figure 10.23B shows the original input analog signal at the top, the analog signal transformed into PCM in the middle, and the demodulated PCM on the bottom line. The middle diagram of Figure 10.23B also shows when the sample-and-hold (S/H) circuit is sampling and holding (to the end of the S/H

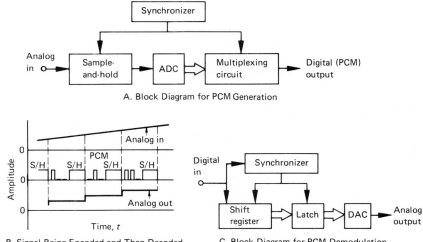

Figure 10.23 PCM Generation and Demodulation

A. Block Diagram for PCM Generation

B. Signal Being Encoded and Then Decoded

C. Block Diagram for PCM Demodulation

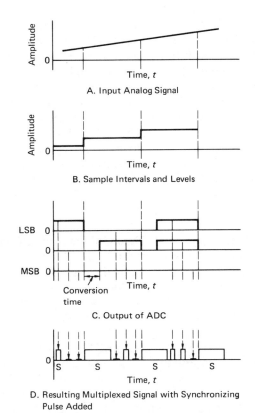

A. Input Analog Signal

B. Sample Intervals and Levels

C. Output of ADC

D. Resulting Multiplexed Signal with Synchronizing Pulse Added

Figure 10.24 Signal Process for Generating PCM

block). The dotted lines indicate the precise time the input signal is held to be converted to digital code, which follows the S/H block.

Figure 10.24 shows the encoding process for generating PCM. For encoding of the incoming analog signal into a serial pulse train, the signal is first sampled at regular intervals, as shown in Figure 10.24A. This sampled voltage is held at a constant level for a short period of time by an S/H circuit (Figure 10.23A). An example of an S/H circuit is shown in the bottom left corner of Figure 10.25. The components involved are the CD4016 bidirectional switch (the B part), the 1 μF capacitor, and the 741 voltage follower.

The next step in the process is to convert the sampled signal into a digital representation. This conversion is done by the ADC (or A/D) shown in Figure 10.23A. The ADC in Figure 10.25 is the ADC0804. The specifications and description of this device are available from the manufacturer. Because the ADC does not operate instantaneously, the input signal must be sampled and held at a steady-state voltage long enough for the digital conversion to occur. In addition the ADC converts the single input into a number of parallel outputs, each output representing a bit location. The thick arrow out of the ADC in Figure 10.23A represents this parallel output. Figure 10.24B illustrates the sampled intervals, and Figure 10.24C shows the output of the ADC.

Lastly, the outputs of the ADC are multiplexed onto the output line (Figure 10.23A), one at a time and each in turn. Multiplexing starts with the LSB (least significant bit) and ends with the MSB (most significant bit); see Figures 10.24C and 10.24D. At the same time a synchronizing pulse (S) is added in order to separate each word and allow for easier demodulation. In Figure 10.25 the multiplexing circuit is on the right side and is composed of the CD4017 and two CD4016s. There may or may not be a space between bits, depending on the system used. The resulting signal is shown in Figure 10.24D.

All these operations must occur repeatedly and within a given time frame. This job belongs to the synchronizer shown in Figure 10.23A. In Figure 10.25 the synchronizer is the pair of 555s at the top.

The encoded signal is now ready to be sent to the transmitter. The transmitter will convert the pulses into AM, FM, single sideband (SSB), or whatever type of modulation is desired. The signal does not necessarily have to be sent by radio, however. Light, sound, or telephone line transmission may be employed, depending on transmission distance and other factors.

At the receiver the transmitted signals are

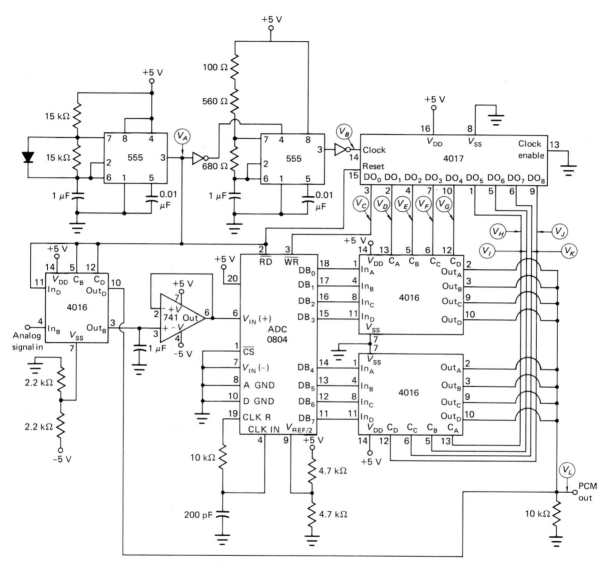

Figure 10.25 Schematic of PCM Modulator

converted back into pulses, then demodulated into the original analog signal (or a close approximation). Figures 10.23C and 10.26 show the PCM demodulation process.

The PCM signal is first input into a *shift register* (Figures 10.23C and 10.26B), which converts the serial pulses of each word into a parallel output. The first bit of the word is shifted into the portion of the register we will call the MSB location. As each bit, in turn, is shifted into the MSB location, the preceding bit in that location is shifted into the next bit lo-

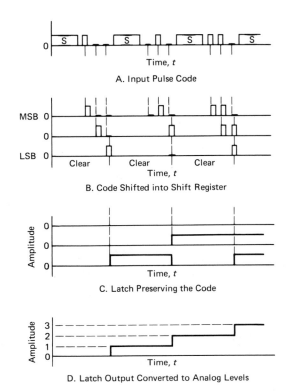

A. Input Pulse Code

B. Code Shifted into Shift Register

C. Latch Preserving the Code

D. Latch Output Converted to Analog Levels

Figure 10.26   Signal Process for Demodulating PCM

cation. Finally, when the last bit of the word is shifted in, the register contains the serial pulse code displayed in parallel fashion. This process is illustrated in Figures 10.26A and 10.26B. The shift register is shown in Figure 10.27; it is the 74164.

Also at this time the output of the shift register is latched into the latch (Figures 10.23C and 10.26C) to preserve the pulse code and present it to the DAC (or D to A, or D/A) shown in Figure 10.23C. *Latching* is the process of locking the input signal onto the output and holding it there regardless of what the input signal does following the latching process. The shift register is cleared during each synchronizing pulse time. The latch is shown in Figure 10.27; it is the 74LS377.

The DAC (Figure 10.23C) converts the code into the appropriate analog level, and the process is complete. Figure 10.26C shows the latched digital code, and Figure 10.26D shows the resulting analog output. The DAC in Figure 10.27 is the DAC0808. If this signal is now applied to a low-pass filter, the high-frequency components of the steps between voltage levels can be reduced, and the waveform will look more like the original analog signal.

If we look again at the block diagram in Figure 10.23C, we see that the PCM signal was input into the synchronizer as well as the shift register. The synchronizer detects the synchronizing pulse and synchronizes the shifting and latching process just described. In Figure 10.27 the entire top half of the circuit is the synchronizing pulse detector and synchronizer. This part of the circuit is a very critical part. If the synchronizer is off slightly, the latching process can occur too early or too late to give correct information. Most of the problems associated with this circuit will involve synchronizer timing.

We have gone through the process of generation and demodulation of PCM in detail. Certainly, there are many other ways to do the same thing. However, the ideas presented here are valid. Detailed schematics of a working PCM modulator and demodulator were presented in Figures 10.25 and 10.27. These circuits are designed not for efficiency but as teaching aids to illustrate PCM modulation and demodulation. With the ideas presented here, you can design your own system or improve on the circuits given.

Example of PCM. Figure 10.28 shows several *synchrograms,* which are diagrams showing waveforms synchronized in time. (Another name for synchrogram is timing diagram.) The diagrams in Figure 10.28 are synchrograms of the voltage waveforms for the voltages specified by the letters in the PCM modulator schematic of Figure 10.25. The data

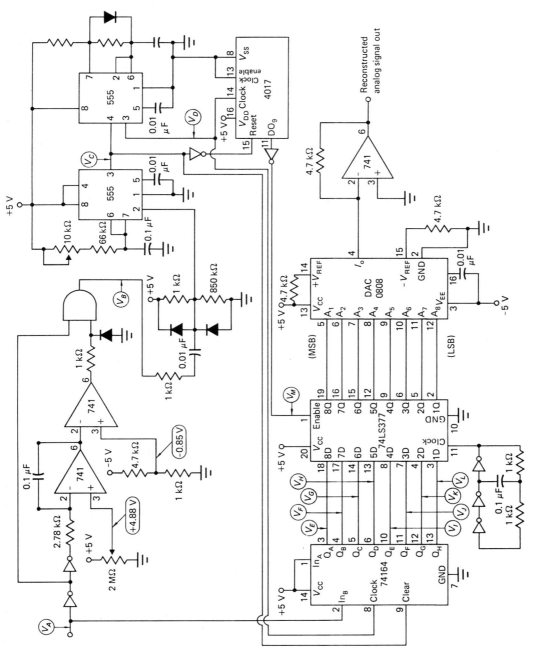

Figure 10.27 Schematic of PCM Demodulator

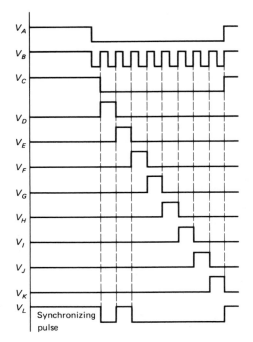

Figure 10.28 Synchrogram for PCM Modulation

word represented in Figure 10.28 is 00000010, which corresponds to an analog input of 0.05 V.

The photograph in Figure 10.29A presents the same information as the synchrogram in Figure 10.28. The voltage waveforms for V_A through V_L (the letters in the schematic in Figure 10.25) are consecutive in each photograph in Figure 10.29 from top to bottom.

Figure 10.29B shows one data word of 01010101, or an analog voltage input of 1.57 V. Notice the bottom line in each photograph in Figure 10.29 shows the data word as it comes from the PCM modulator. Figures 10.29C and 10.29D have an expanded time base in order to show more than one data word at a time; these photographs also show that the data words are varying with time.

Figure 10.30 shows synchrograms for the PCM demodulator whose schematic was given

in Figure 10.27. The voltage waveforms for V_A through V_M (the letters in the schematic) are consecutive in each photograph of Figure 10.30, from top to bottom. The last line of the photographs is unused.

Each photograph in Figure 10.31 shows the PCM signal (V_A at the top of each photograph) and the output of the integrator (bottom). (The integrator is also called the synchronizing pulse detector.) The bottom signal in each photograph is the input to the next 741 (the comparator) and triggers it when its input signal goes to -0.85 V. Notice the effect of the PCM code on the bottom trace in each photograph. The worst-case condition shown in Figure 10.31C has an input code of 11111111. However, the result of the integration at the end of the coded part is still not low enough to reach the -0.85 V level and trigger the comparator.

Go through the schematics and synchrograms to verify for yourself how the system operates.

The PCM system we have been discussing was designed to operate at very low frequencies. The photograph in Figure 10.32A shows a 1 Hz analog input and the resulting PCM demodulated output. The input is the top waveform; the output is the bottom waveform. As the input frequency increases (Figure 10.32B), the steps in the output become much more visible, and the output becomes increasingly phase-shifted. The output is being sampled approximately 44 times during one cycle of the input signal. This sampling rate is 22 times the Nyquist rate for this input signal. (Recall that the Nyquist sampling rate is twice the highest frequency expected. See Figure 10.32D.) The phase shift of Figure 10.32B is a direct result of the *frame time*—the time from synchronizing pulse to synchronizing pulse—compared with the input signal frequency.

In the circuit for Figure 10.32 the frame time is about 23 ms. The input analog signal is sampled at the beginning of the synchronizing pulse, but it is not converted back into an ana-

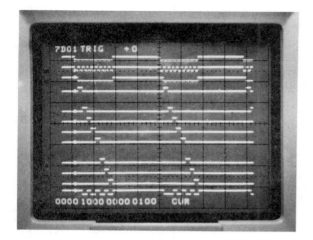

A. Data Word 00000010, Representing Analog Input of 0.05 V

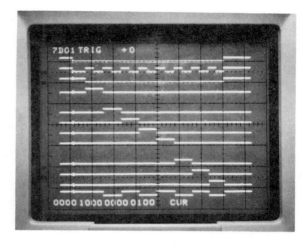

B. Data Word 01010101, Representing Analog Input of 1.57 V

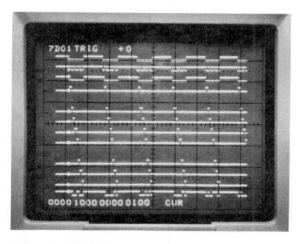

C. Two Data Words Representing Same Information as in Part B

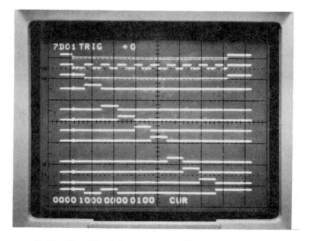

D. Five Data Words Representing a Time-Varying Signal

Figure 10.29   Synchrograms for PCM Modulator

log signal until the end of the frame in the demodulator. Therefore a phase shift results as the input analog period approaches the frame time. Figures 10.32A through 10.32D illustrate this progression.

Figure 10.32D shows the highest input frequency possible, resulting in two samples per period (Nyquist rate) of the input in order to be within the specifications set forth by the sampling theorem. In this case filtering would be necessary to recover the original signal.

Companding. Often we must reduce quantizing error for different-amplitude signals. To do so we employ the process called *companding*, which is the compression of large sig-

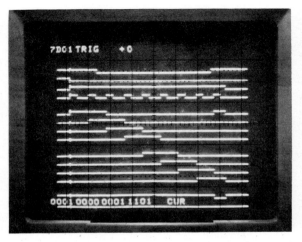

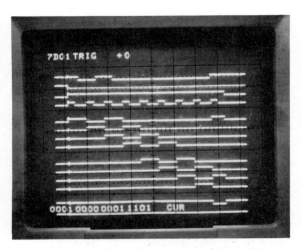

A. Data Word 00000001, Representing Analog Input of 0.03 V   **B. Data Word 00000010, Representing Analog Input of 0.05 V**

Figure 10.30   Synchrograms for PCM Demodulator

nals at the transmitter, followed by expansion of the signals at the receiver. This process is called *companding* because it *com*presses the large signals (at the transmitter) and then ex-*pands* those that were compressed (at the receiver). The following discussion demonstrates why this process is helpful.

As was previously shown, the maximum quantizing error for a PCM system is one-half the amplitude of a quantizing level. The PCM system of Figure 10.25 has a signal amplitude range of 5 V and uses eight bits for each word. Thus, the maximum quantizing error in this system is as follows:

$$\frac{1}{2^8} \times 5\,\text{V} = \frac{1}{256} \times 5\,\text{V} = 19.5\,\text{mV}$$

For an input analog signal of 5 $V_{(p-p)}$, 19.5 mV represents about a 0.4% error. However, for an input signal of 0.1 $V_{(p-p)}$, 19.5 mV is almost a 20% error. Thus, small-signal inputs can produce large errors at the output because of quantizing.

One way to reduce this problem is to have more quantizing levels for small signals, or, in other words, to *taper* the quantizing levels. Figure 10.33 illustrates a nonlinear quantizing-level arrangement. At the receiver the levels would be corrected by a corresponding amount so that the amplitude of the original signal is returned.

The circuitry necessary to produce nonlinear quantizing levels is much more complex than that required for linear levels. However, a suitable alternative can be employed. The large signals can be attenuated proportionally more than small signals before they are encoded. Then at the receiver the large signals can be amplified more than the small signals. The results here would be the same as those for the nonlinear circuitry, without greatly complicating the circuitry.

Integrated Circuit CODEC.   At this point we can appreciate the complexity of designing a PCM modulator and demodulator system. National Semiconductor now has two ICs that perform these functions. They are called PCM CODECs (pulse-code modulation coders/decoders). They are the TP3001 μ-law CODEC and the TP3002 A-law CODEC. The *μ-law* and

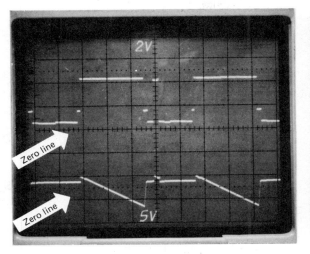

A. Zero-Level Input Signal and 00000010-Level Input

B. Time-Varying Input

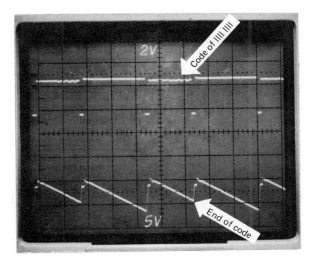

C. Worst-Case Condition

Figure 10.31   Results of Processing Input Signal in Order to Synchronize Demodulator (Figure 10.27)

A-law designations refer to the two different transfer functions used in companding. The μ-law function is used in North America and parts of the Far East. The A-law function is used in most of the remaining countries. For more applications and information, see National Semiconductor's application note 215. A reprint of the application note is given in National Semiconductor's *Special Functions Databook.*

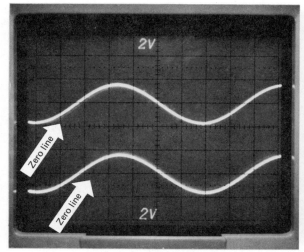

A. Well-Reconstructed Waveform

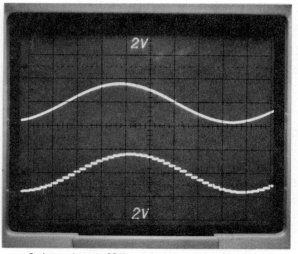

B. Approximately 22 Times the Nyquist Sampling Rate

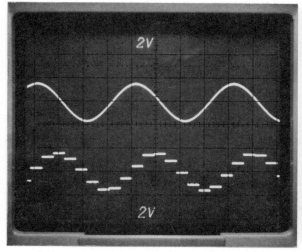

C. Approximately 5 Times the Nyquist Sampling Rate

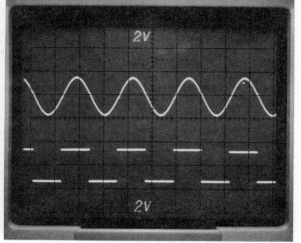

D. Highest Input Frequency (Nyquist Sampling Rate)

Figure 10.32   Waveforms of Input and Output Signals of Circuits in Figures 10.25 and 10.27

Advantages of PCM.   Unless the noise signal completely obliterates the PCM pulse or pulses, PCM is not affected by amplitude noise, as PAM is. Nor is PCM affected by rising or falling edge noise, as PWM, PPM, and PFM are. Thus, PCM is less affected by noise than any of the other pulse modulation methods so far presented. Insensitivity to noise is a great

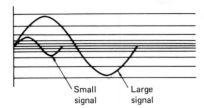

Figure 10.33 Tapered Quantizing Level

advantage, especially in systems where the signal must be relayed from point to point before it reaches its destination. Other signals are degraded slightly after each relay process; PCM is not.

If noise rejection is such an advantage, why is PCM not the only system in use today? Other systems are used because they were developed first, because PCM circuitry is much more complex, and because PCM requires a much larger bandwidth than the other systems.

PCM was patented in 1937 by Alex H. Reeves of Great Britain. Even though PCM was well documented, it did not come into its own until the early 1960s, when ICs could be used to implement the circuitry. As discussed, ICs make implementation of complex PCM much easier.

The bandwidth requirement for PCM will also become less of a problem as technology advances. With the steady increase in the use of lasers and fiber optics, noise-free transmission of wide-bandwidth signals can be realized.

PCM is becoming more popular as circuit complexity is solved. In 1965 the *Mariner IV* space probe transmitted pictures by using PCM. The system required 30 min to transmit each picture, but the transmitter was over 200,000,000 km away, and the transmitter power was only 10 W! Today PCM is being used to produce noise-free recordings on phonograph records (digital recordings) and on video recorder systems (laser discs). PCM will become more widespread as technology continues to develop.

Other Digital Systems

Other digital pulse modulation systems have been proposed, but none is as popular as PCM. We will discuss briefly two of the other digital modulation systems.

One digital system in use is called *differential PCM*. Differential PCM is very similar to regular PCM. The difference is that regular PCM quantizes the absolute amplitude of the signal, whereas differential PCM quantizes the difference in amplitude, positive or negative, between one sample and the previous sample. The idea here is that most signals do not change much from their previous level. Thus, it would take fewer pulses to represent a change in amplitude than it would take to represent an absolute amplitude. If this technique could be implemented, then a bandwidth reduction could be realized. Again, complex encoding and decoding circuits have kept differential PCM from becoming widely accepted.

The second and more popular digital system is called *delta modulation*. Delta modulation in its simplest form is similar to differential PCM. However, delta modulation changes only one bit, positive or negative, per sample. Figure 10.34 illustrates delta modulation.

One of the problems associated with delta modulation is *slope clipping*. Here the input

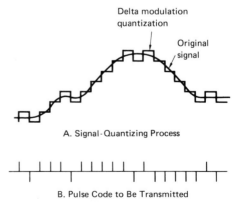

Figure 10.34 Delta Modulation

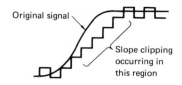

Figure 10.35 Slope Clipping in Delta Modulation

signal is changing so rapidly that the delta modulator cannot keep up with it. Slope clipping is illustrated in Figure 10.35.

CONCLUSION

This chapter has covered pulse modulation, the basis for telemetry. Of all the pulse modulation types, PCM is probably going to become the most widely used. PCM's advantage of excellent noise rejection is beginning to overshadow its disadvantage of complex circuitry, especially now that single ICs are acting as encoders and decoders. Already the industrial, consumer, and government markets have made extensive use of the process.

QUESTIONS

1. _____ is the process for performing measurements on distant objects.
2. News services transmit photographs electronically by a process known as _____.
3. The _____ measurement technique is easier to use over short distances. The _____ technique is more noise-free over long distances.
4. The four types of analog pulse modulation are _____, _____, _____, and _____.
5. The subclassification of PAM in which pulse excursions occur both above and below the reference level is called _____.
6. In PWM the widest pulse width should not exceed _____% of maximum, and the narrowest pulse width should not be less than _____%.
7. Pulse-_____ modulation is differentiated and clipped PWM.
8. Pulse-_____ modulation is a form of digital pulse modulation.
9. In PCM the analog signal voltage is divided into levels called _____.
10. Name one of the five modulation types that is most affected by noise.
11. Identify the reconstruction phenomenon that may occur when the sampling rate of the analog signal is too slow.
12. Name the type of modulation produced by an amplifier whose gain can be controlled by an analog signal.
13. Name the type of filter with which both PAM and PWM can be demodulated.
14. The PLL can be used to demodulate what type of modulation?
15. Does the input to the PLL have to be a sine wave for the PLL to work properly?
16. Name the term used for a group of pulses in PCM representing an analog voltage level.
17. Discuss why quantizing noise is random noise.
18. Discuss why the largest quantizing error is one-half the amplitude of the levels into which the analog signal is divided.
19. Discuss the use of a synchronizing pulse in PCM.
20. Define the term synchrogram.
21. Discuss the need for companding.
22. Explain the term CODEC.
23. Discuss two disadvantages that are keeping PCM from becoming more common.
24. Define slope clipping.

PROBLEMS

1. Observing the rules of the sampling theorem, determine the minimum sampling rate for high-fidelity voice (20 kHz).

2. Draw to scale the PPM waveform for a straight line that has a slope of 0.2 and a length of 20 cm, and goes through zero at the 10 cm length point. The pulse width (of PWM) at the lower end of the straight line is 20% of its maximum, and the pulse width (of PWM) at the upper end of the straight line is 80% of its maximum pulse width. Use eleven equally spaced sample points from one end of the 20 cm point to the other.

3. Determine the VCO free-running frequency of the 565 given in Figure 10.18. The value of R_1 is 4 kΩ, and the value of C_1 is 0.1 μF.

4. In a PCM system with six bits in each word, the maximum input signal swing is 10 V. For quanta that are equally spaced, determine the maximum quantizing error.

5. In Figure 10.20 suppose $C_1 = 0.01$ μF, $R_1 = 1$ kΩ, $R_2 = 4.6$ kΩ, and $R_3 = 5.4$ kΩ. What is the approximate output frequency of the circuit if the power connections on the 741 are ± 10 V?

6. If R_1 in Problem 5 changes to 1.5 kΩ, what is the new output frequency?

7. What are the digital bit values of the two data words shown in Figure 10.29C, given in the order of transmission (that is, given as in Figure 10.29A, but two data words instead of one)?

11

INDUSTRIAL TELEMETRY AND DATA COMMUNICATION

OBJECTIVES

On completion of this chapter, you should be able to:

- Describe frequency-division and time-division multiplexing;
- Determine the relationship among data bandwidth, frequency deviation, and modulation index;
- Relate pulse rise time to data bandwidth;
- List the sources of error in multiplexed transmission;
- Determine the classification of multiplexing systems;
- List methods used for data communication and recording on magnetic media;
- Identify the most important circuits in the RS-232C serial interface standard;
- Calculate frame time, frame rate, sample time, and commutation rate for pulse modulation systems.

INTRODUCTION

Telemetry was discussed briefly at the beginning of the preceding chapter because it relies heavily on pulse modulation. Now that pulse modulation has been presented, telemetry can be discussed in more detail.

Telemetry can be defined as the science in-

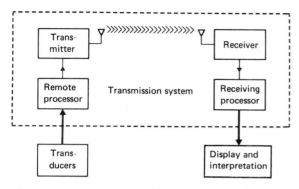

Figure 11.1 Functional Diagram of Telemetry System

volved in performing a measurement at a remote location and transmitting the data to a central location for processing or storage. The telemetry system can be divided functionally into three parts: transducers, transmission system, and display and interpretation system. See Figure 11.1.

The transducers are located at the remote station and transform the physical quantities to be monitored into electrical signals. The transmission system consists of a device for transforming the electric signals from the transducers into signals suitable for transmission to the receiving station; the transmission and receiving devices; and a device for transforming the output of the receiver to forms suitable for display and interpretation. The display and interpretation system consists of the devices that calculate the desired parameters and display them for final interpretation by automatic or human means.

Transducers were discussed in detail in Chapter 8 and that discussion need not be repeated here. Our attention in this chapter will focus on the portion of the telemetry system called the transmission system and, in particular, on the remote and receiving processor portions in Figure 11.1. These devices transform the signals for transmission and later transform them back to signals suitable for display.

APPLICATIONS OF TELEMETRY

Telemetry is considered a necessary topic in a text on industrial electronics because of the need in industry to transmit measurements over a distance. Typical applications, to name only a few, have been the monitoring and automatic adjustments of electrical power transmission grids or networks and other utility systems; the monitoring of meteorological (weather) conditions in atmospheric, subsurface, or severe climate locations; the monitoring of seismic (earthquake) conditions; the monitoring or control of satellite, space probe, missile, and ordnance systems; easier monitoring of vital signs of animals or humans who are ambulatory (able to move about); and the monitoring and control of rotating machines, such as the blades of a turbine.

Of the telemetry applications listed, most of the measurements to be telemetered are analog (or continuously variable) and may be transmitted by analog methods. A simple example of an analog method is a voltmeter (or ammeter) with long leads. Many factors must be considered when telemetering analog measurements, such as ground loops, line loss, and noise. Nevertheless, most of these applications are relatively easy to realize with telemetry when compared with pulse modulation, especially over short distances.

If analog transmission is easy and pulse modulation is more complicated, why is pulse modulation used so much? Some of the advantages have already been listed in Chapter 10. To summarize: (1) Some of the information is in pulse form to start with; (2) pulse transmission is a more error-free transmission process; and (3) large-scale integration has reduced circuit complexity. In addition to these advantages are two more: Transmitters can operate on very low duty cycles (percent of time the transmitter is on), and the time intervals between pulses can be filled with samples of other data. This last advantage illustrates how a number of dif-

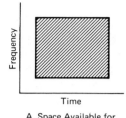

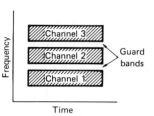

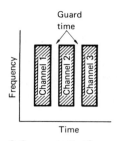

Figure 11.2 Frequency-Division and Time-Division Multiplexing

A. Space Available for Communication

B. Communication Space Divided into Frequency Channels for FDM

C. Communication Space Divided into Time Slots for TDM

ferent sets of data can be *multiplexed,* that is, transmitted over the same channel. We will discuss multiplexing in detail in this chapter.

MULTIPLEXING

In most telemetry applications it is desirable to perform a number of different measurements. Although a separate transmission line or link could be used for each meaurement, the problems of efficiency in terms of power, bandwidth, weight, size, and cost normally prevent us from doing so. Therefore, it is common to send many measurements over a single transmission channel. This process, as mentioned, is called multiplexing. Multiplexing can be used in systems for purposes other than telemetering, but it is so useful in telemetering that it will be the subject of much of this chapter.

There are two types of multiplexing in common use today: frequency-division multiplexing (FDM) and time-division multiplexing (TDM). These two types of multiplexing are illustrated in Figure 11.2.

Figure 11.2A is a representation of the space available for communication. This communication space can be divided into different frequency channels at the same time for *frequency-division multiplexing,* or into different time slots for the same frequency for *time-division multiplexing.* Of course, combinations

of these two methods could also be used. Notice that each type has a *guard space* to separate channels. Generally, the guard space is used to reduce cross talk (interference between channels) and therefore reduce circuit cost and complexity.

FREQUENCY-DIVISION MULTIPLEXING (FDM)

A familiar example of FDM is radio broadcasting. The signals received by a radio antenna contain many different programs traveling together but occupying different frequencies on the radio spectrum. The tuning circuits in the radio receiver allow the selection of one station from all the others.

The use of the radio frequency spectrum for telemetry and other applications is regulated by various government agencies. The Inter-Range Instrumentation Group (IRIG) publishes a set of telemetry standards that are revised periodically and that specify all pertinent details of telemetry applications.* Table 11.1 is an excerpt from these telemetry standards.

*A copy of the standards may be requested from Secretariat, Inter-Range Instrumentation Group, White Sands Missile Range, New Mexico 88002.

Table 11.1 Proportional Bandwidth FM Subcarrier Channels

Channel	Center Frequencies (Hz)	Lower Deviation Limit* (Hz)	Upper Deviation Limit* (Hz)	Nominal Frequency Response (Hz)	Nominal Rise Time (ms)	Maximum Frequency Response† (Hz)*	Minimum Rise Time† (ms)
			±7.5% Channels				
1	400	370	430	6	58.00	30	11.700
2	560	518	602	8	42.00	42	8.330
3	730	675	785	11	32.00	55	6.400
4	960	886	1,032	14	42.00	72	4.860
5	1,300	1,202	1,398	20	18.00	98	3.600
6	1,700	1,572	1,828	25	14.00	128	2.740
7	2,300	2,127	2,473	35	10.00	173	2.030
8	3,000	2,775	3,225	45	7.80	225	1.560
9	3,900	3,607	4,193	59	6.00	293	1.200
10	5,400	4,995	5,805	81	4.30	405	0.864
11	7,350	6,799	7,901	110	3.20	551	0.635
12	10,500	9.712	11,288	160	2.20	788	0.444
13	14,500	13,412	15,588	220	1.60	1,088	0.322
14	22,000	20,350	23,650	330	1.10	1,650	0.212
15	30,000	27,750	32,250	450	0.78	2,250	0.156
16	40,000	37,000	43,000	600	0.58	3,000	0.117
17	52,500	48,562	56,438	790	0.44	3,938	0.089
18	70,000	64,750	75,250	1050	0.33	5,250	0.067
19	93,000	86,025	99,975	1395	0.25	6,975	0.050
20	124,000	114,700	133,300	1860	0.19	9,300	0.038
21	165,000	152,624	177,375	2475	0.14	12,375	0.029
			±15% Channels‡				
A	22,000	18,700	25,300	660	0.53	3,330	0.106
B	30,000	25,500	34,500	900	0.39	4,500	0.078
C	40,000	34,000	46,000	1200	0.29	6,000	0.058
D	52,500	44,625	60,375	1575	0.22	7,875	0.044
E	70,000	59,500	80,500	2100	0.17	10,500	0.033
F	93,000	79,050	106,950	2790	0.13	13,950	0.025
G	124,000	105,400	142,600	3720	0.09	18,600	0.018
H	165,000	140,250	189,750	4950	0.07	24,750	0.014

*Rounded off to nearest Hz.

†The indicated data for maximum frequency response and minimum rise time are based on the maximum theoretical response that can be obtained in a bandwidth between the upper and lower frequency limits specified for the channels.

‡Channels A through H may be used by omitting adjacent lettered and numbered channels. Channels 13 and A may be used together with some increase in adjacent channel interference.

Twenty-one data channels with subcarrier center frequencies ranging from 400 Hz to 165 kHz (column 2 of Table 11.1) have been designated for telemetry use. *Subcarriers* are carrier frequencies in telemetry systems that are modulated by the measurement or intelligence signal. These center frequencies are designed with a ±7.5% frequency deviation. Thus, Table 11.1 is used for FM purposes.

The *nominal frequency response*, or data bandwidth, (column 5 of Table 11.1) designates the highest frequencies at which the data can vary for each channel. For example, channel 1's data cannot vary any faster than 6 Hz in order to stay in channel 1, assuming a modulation index of 5. The *modulation index* for FM is defined as the maximum frequency deviation of the signal divided by the modulating frequency. While modulation indices of 5 are recommended, indices as low as 1 or less may be used. For these low indices, low signal-to-noise ratios, increased harmonic distortion, and cross talk must be expected.

The relationship among the data bandwidth (data BW), the frequency deviation of the center frequency (400 ± 30 for channel 1), and the modulation index is expressed by the following equation:

$$\text{data } BW = \frac{\text{frequency deviation}}{\text{modulation index}}$$

For channel 1:

$$\text{data } BW = \frac{30}{5} = 6 \text{ Hz}$$

In Table 11.1 the center frequency deviation and the maximum frequency response are identical, since the maximum frequency response column is based on a modulation index of 1. The *rise times* t_r (nominal and minimum) given in Table 11.1 are related to the data BW by the following equation:

$$t_r = \frac{0.35}{BW}$$

Again, nominal rise time is based on a modulation index of 5, and minimum rise time is for an index of 1. The rise time equation can now relate pulse rise time to data bandwidth and therefore expands the use of Table 11.1 beyond FM applications to pulse applications.

The 21 bands were chosen to make the best use of present equipment and the frequency spectrum. There is a ratio of approximately 1.3:1 between center frequencies of adjacent bands, except between 14.5 and 22 kHz, where a larger gap was left to provide for a compensation tone for magnetic tape recording. The deviation has been kept at ±7.5% for all bands, with the option of ±15% deviation on the five higher bands to provide for transmission of higher-frequency data. When this option is used on any of these five bands, certain adjacent bands cannot be used, as noted in a footnote to Table 11.1.

The basic operation of an FDM system is illustrated in Figure 11.3. The measurement signals from the transducers are used to modulate subcarrier oscillators tuned to different frequencies. The outputs of the subcarrier oscillators are then linearly summed (linearly mixed), and the resulting composite signal is used to modulate the main transmitter. At the receiving site the composite signal is obtained from the receiver demodulator and input to a number of bandpass filters, which are tuned to the center frequencies of the subcarrier oscillators.

An example might be helpful at this point. Let us assume that the system given in Figure 11.3 has four input transducers and therefore four outputs at the receiver. We also will assume that transducer A varies no faster than 6 Hz, transducer B no faster than 8 Hz, transducer C no faster than 11 Hz, and transducer D no faster than 14 Hz. This arrangement allows

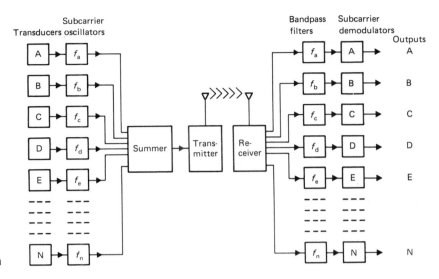

Figure 11.3   FDM System

us to use channels 1, 2, 3, and 4 from Table 11.1. The subcarrier oscillator frequencies will be 400, 560, 730, and 960 Hz, respectively. Figure 11.4 is a diagram of this arrangement.

The system bandwidth now varies from 370 to 1032 Hz, with guard bands in between each channel. This data bandwidth of 662 Hz is now sent to the transmitter to be transmitted. If we maintain a modulation index of 5, the main transmitter frequency deviation about its center frequency should be 3.31 kHz (622 Hz × 5, from the data *BW* equation). The receiver will demodulate these frequencies back into their original data signals. Both the subcarriers and the main carrier frequencies were frequency-modulated in this example.

All types of modulation can be used for both the subcarrier and the main carrier. The types of modulation are FM (frequency modulation), PM (phase modulation), AM (amplitude modulation), SC (suppressed-carrier amplitude modulation), and SS or SSB (single-sideband amplitude modulation). FDM systems normally are designated by listing the type of modulation used by the subcarriers, followed by the type of modulation used by the main carrier. Thus, FM/AM indicates an FDM system in which the subcarriers are frequency-modulated by the measurement, and the main carrier is amplitude-modulated by the composite subcarrier signals. Almost all combinations of subcarrier and main-carrier modulation techniques have been used in the past. However, FM/FM is by far the most common technique in use today.

The principal sources of errors in an FDM transmission system are drift, bandlimiting,

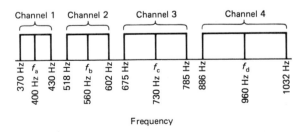

Figure 11.4   FDM System Using Channels 1, 2, 3, and 4 of Table 11.1

cross talk, distortion, and radio frequency (RF) link noise.

The errors due to drifts (slowly changing circuit parameters caused by heat, age, and so on) are those associated with modulation and demodulation of the subcarriers. If the drifts are slow, their effect can be greatly diminished by use of calibration signals. Their reduction depends on the circuit design.

Errors due to bandlimiting (limited frequency spectrum) occur throughout the system whenever the data are dynamic (changing). This error source cannot be eliminated regardless of the circuit design.

Cross talk errors are of two kinds. The first is caused by nonlinearities in the summing amplifier before the main transmitter and in the modulation or demodulation process of the main carrier. This type of cross talk error can be eliminated. The second type of cross talk error is due to overlap of frequencies from the bandpass filters of adjacent channels. Since no filter is perfect, some overlap will always be present in other channels. This error cannot be eliminated.

Distortion error is due to nonlinearities in the subcarrier modulator and demodulator (not to be confused with cross talk errors due to nonlinearities in the main-carrier modulator or demodulator). Calibration signals can help reduce or eliminate this error.

RF link noise (the RF link is the transmitter, receiver, and medium in between) causes errors that are generally random and come from three sources: receiving station front-end noise, noise from space, and technological interference. These types of noise will always be present but can be reduced by proper circuit design.

TIME-DIVISION MULTIPLEXING (TDM)

The major alternative to FDM is time-division multiplexing (TDM). Here the time available is divided up into small slots, and each of these slots is occupied by a piece of one of the signals to be sent. The multiplexing apparatus scans the input signals sequentially. Only one signal occupies the channel at one instant. TDM is thus quite different from FDM, in which all the signals are sent at the same time but each occupies a different frequency band.

The operation of a TDM system from a functional standpoint is illustrated in Figure 11.5. The signals from the transducers are input to a *commutator* (a switching device), which samples the channels sequentially. Thus the output of the commutator (also referred to as

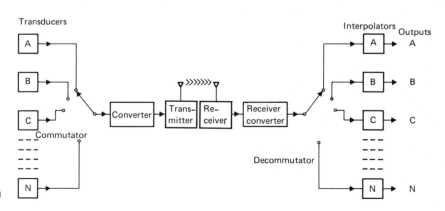

Figure 11.5   TDM System

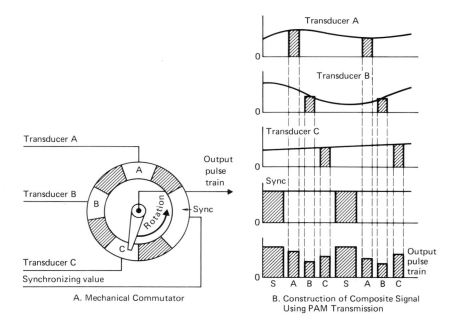

Figure 11.6 TDM Commutation

A. Mechanical Commutator

B. Construction of Composite Signal Using PAM Transmission

the multiplexer) is a series of pulses, the amplitudes of which correspond to the sampled values of the input channels from the transducers. This train of pulses is then passed through a device that converts it to a form suitable for modulating at the transmitter.

At the receiving station the process is reversed. The demodulated output from the receiver is passed through a converter, which reproduces the pulse train that existed at the commutator at the transmission site. This pulse train can then be decommutated to produce pulses with values corresponding to samples of the original measurement signals.

Let us look at the commutator section a little more closely. Figure 11.6 is a diagram of a mechanical commutator connected to three transducers. As the commutator arm or rotor rotates, it makes contact with conductors A, B, C, and sync. Each time a conductor is contacted, a voltage is output. Between each conductor is an insulator. When the commutator rotor is on an insulator, there is no output. The

insulator contact corresponds to the space between signals in the output pulse train diagram of Figure 11.6B.

Figure 11.6B shows a synchrogram of the operation occurring in Figure 11.6A. If the values of the voltages from the instruments are not varying too rapidly compared with the rotation time of the rotor, the individual inputs can be reconstructed from the composite signal.

For separation of the signals when they are received, a commutator similar to that illustrated in Figure 11.6A might be used, but with the input and output reversed. The receiving commutator must be exactly synchronized with the transmitting commutator. That is exactly what the sync pulse is for, to synchronize the receiver commutator with the transmitter commutator. Modern commutators operate at very high speeds and are electronic instead of mechanical.

As in FDM, TDM modulation of the transmitter may take any form, that is, AM, FM, PM, and so on. However, the principal distinc-

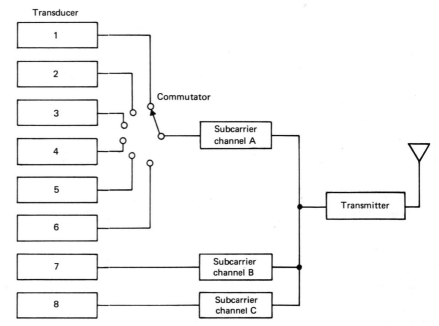

Figure 11.7 Combination of TDM and FDM in One System

tion among TDM systems lies in the form of the processor used (see Figure 11.1). For the processor any of the pulse modulation methods discussed in Chapter 10, such as PAM, PWM, PPM, PCM, and delta modulation, will work. The system designation normally lists the type of processor (or converter) first, followed by the type of main-carrier modulation. For example, a PWM/PM system has a pulse-width modulation processor, the output of which is used to phase-modulate the main carrier.

Submultiplexing

In FDM transmission systems it has been common to submultiplex some of the wider-band subcarrier channels. *Submultiplexing* usually involves the use of a TDM waveform as the modulation of one of the subcarrier channels, as shown in Figure 11.7. The use of submultiplexed channels of this nature allows the total number of measurements handled by the trans-

mission system to be increased considerably. The frequency content of the submultiplexed measurements must be low relative to the frequency content that normally could be transmitted over the subcarrier channel.

The most common types of submultiplexing are PAM/FM/FM and PWM/FM/FM, although almost all varieties have been used for some applications. Combined telemetry systems have not been widely used to date. However, there is some indication that they may receive increased attention in the future.

Pulse-Amplitude TDM

Let us return our attention now to pure TDM. The first modulation type we will consider is PAM. A PAM system is one in which the output of the commutator is used to modulate directly the main transmitter. The processor is simply a pair of wires, so PAM is the simplest of the TDM systems.

The PAM wave trains may take several

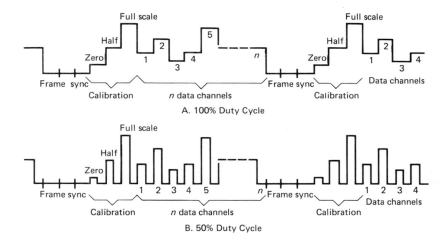

Figure 11.8 PAM Waveforms in a TDM System

forms, two of which are illustrated in Figure 11.8. The principal difference here lies in the duty cycle system (50% and 100%). The length of time necessary to sample all channels (including frame sync and calibration pulses) is normally referred to as *frame time*. For identification of the sample at the receiving station, it is necessary to insert frame synchronization. Several different methods can be used to designate a frame. The one illustrated in Figure 11.8 consists of forcing several consecutive channels to a level below the minimum allowable data value. Since drifts and nonlinearities in the system cause error directly, it is also common to transmit calibration pulses (zero, half, and full scale) as shown. The data are offset in such a fashion that channels with signal outputs that are capable of both positive and negative polarities are centered about the half-scale value. The frame sync pulses and calibration pulses are sometimes referred to as *housekeeping pulses,* since they are not part of the

data but are used to synchronize and calibrate the receiver.

In general, the 100% duty cycle system (Figure 11.8A) requires a smaller frequency spectrum than that of the 50% duty cycle system (Figure 11.8B). However, the 50% system was used to a greater extent in the past because of the relative ease of synchronizing at the receiving station. In addition, the dead time available in the 50% system allows the use of circuitry to get rid of transients (undesirable short-duration signals) and therefore reduce cross talk between successive channels.

Pulse-Width TDM

In PWM the width of pulses is varied in proportion to the modulation signal. The PWM pulse train, then, consists of a string of pulses with different widths, as illustrated in Figure 11.9.

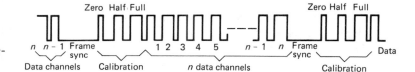

Figure 11.9 PWM Waveform in a TDM System

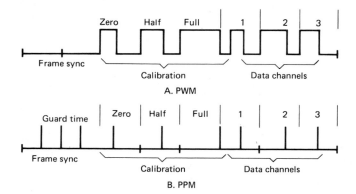

Figure 11.10 PPM Waveform Compared with PWM Waveform in TDM System

Guard time is allowed at both the beginning and the end of each pulse to reduce the difficulties associated with interchannel cross talk. Since the information is carried in terms of pulse width, drifts and nonlinearities in the system do not have as great an effect on data accuracy as they do in an equivalent PAM system.

Pulse-Position TDM

PPM is similar to PWM. Here, though, only the trailing edge of the pulses, rather than the entire duration of the pulse, is transmitted to identify the pulse width. See Figure 11.10. PPM is used principally in connection with an amplitude-modulated main carrier since its greatest advantage is in the small percentage of time that pulses are present. Thus, the transmitter operates at relatively high peak powers and low average powers. The wider system bandwidths required (as compared with PWM and PAM) makes its use in connection with a frequency-modulated main carrier undesirable, because the reduction in average power for the FM system could not be realized.

A typical PPM waveform is shown in Figure 11.10 along with the equivalent PWM waveform. The synchronization pulses shown may vary from system to system. Pulses corresponding to the leading edge may be transmitted in order to reduce synchronization problems at the receiving station, although Figure 11.10 does not illustrate this possibility.

PFM systems will not be discussed here since they are so similar to PPM systems.

Analog TDM Errors

The principal sources of error in analog TDM systems are closely related to those in FDM systems. These error sources are drift and nonlinearity, bandlimiting, cross talk, interpolation errors, and RF link noise.

With the exception of PAM, only drifts and nonlinearities associated with equipment prior to and including the remote converters (processors), and after and including the receiver converter, are of primary concern to analog TDM systems. As in FDM systems this drift and nonlinearity are entirely dependent on circuit design. Calibration signals can be of considerable help here.

Bandlimiting errors in TDM systems are produced only in the equipment preceding the multiplexer. Errors due to bandwidth restrictions in other parts of the system usually are classified under different names. Bandlimiting errors depend on the phase and the amplitude response of the filters involved, as well as on the character of the data.

Cross talk in a TDM system normally occurs from bandlimiting of the pulse signal, which causes transients from channel pulses to affect the pulses of the channels following. The relationship between bandlimiting and cross talk varies from system to system.

Interpolation error comes about when an attempt is made to reconstruct a continuous waveform from the sampled values of the original waveform available to the receiving station. The amount of interpolation error depends on the characteristic of the data that are sampled, the method of interpolation, and the sampling rate.

The RF link errors cannot be eliminated. They depend on the received signal power at the receiving station and many other system parameters.

Pulse-Code TDM

PCM comes in many varieties. The most common is binary modulation, the type shown in Figures 10.26 through 10.32 in Chapter 10. In *binary modulation* the transmission of a 1 or 0 value during a bit interval is done by the presence or absence of a pulse. In actual practice this waveform can have many different forms. Figure 11.11 shows a number of the different waveforms that can be used.

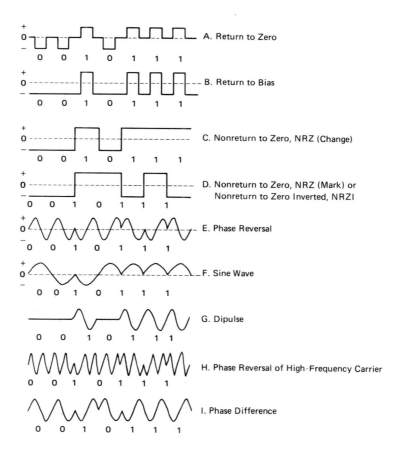

Figure 11.11 Different Ways to Represent 1 and 0 in Binary Modulation

The nonreturn-to-zero (NRZ) waveform is probably the most commonly used of those shown. Notice there are two types of NRZ signals. The NRZ (change) transitions only when the next bit of data changes value. NRZ (mark) transitions with each 1 bit, and does not change when a 0 bit is present. However, all waveforms shown in the figure either have been used or are to be used in modern systems. When the NRZ or return-to-zero (RZ) waveforms are used, it is common practice to employ a premodulation filter to round off the corners of the modulating waveform to limit the bandwidth requirement.

Even though the NRZ waveform is probably the most commonly used binary modulation, it is seriously lacking in at least one important area—that of clocking. As shown in Figure 11.11 a long string of 1s or 0s would result in a lengthy data level without transitions. It now becomes very important to know *when* to examine the NRZ signal to determine the logic level of the signal in order to reconstruct accurately the transmitted data. To accomplish this a separate clock signal is transmitted along with the NRZ data.

This additional clock signal has at least two drawbacks: two channels instead of one are needed to transmit the data and reconstruct it, and the phasing of the clock signal with the data can get out of synchronization. This phase shift is most likely to occur when the data are transmitted over long distances and the data and clock signals are on different channels.

To avoid transmitting a clock signal with the data, a self-clocking format may be used. A self-clocking code, such as the RZ type in Figure 11.11, is one in which one type of data bit is a frequency multiple of the other. A clock signal as well as data can be recovered from this type of modulation by using a PLL.

The arrangement of the code sequence in a binary transmission system is called the *format*. In general the format is defined by the arrangement of *bits* (binary digits) in each code group representing a sample value (sometimes called

a *word*) and the arrangement of these code groups with respect to one another.

The bits within a particular coded word may consist of information bits, parity bits, or synchronization (sync) bits. *Information bits* appear in a binary code that represents the sample value. A *parity bit* is either present or not present in order to make the total number of 1 bits per word either even or odd (depending on whether even or odd parity is to be used). Parity bits allow the detection of words with a single bit in error. Parity bits generally are not used in radio transmission systems since it is much more likely that two or more bits will be missing. However, in ground data-handling systems parity bits are nearly always used since single-bit errors are most likely. The *sync bits* are timing bits. They may appear after every word or after a certain small sequence of words, and they usually are called *word synchronization bits*. Either the parity bits, the sync bits, or both may be missing, depending on the requirements of the system. Today, virtually all operational binary PCM systems include some form of word synchronization, although some systems presently under development do not.

In the simplest cases of TDM of PCM, the information channels are sampled sequentially (as in Figure 11.5) with fixed word lengths. In this case each word represents the code for a separate information channel, appearing sequentially in time. The sequence is repeated each time the commutator completes a cycle. This entire cycle is called a *frame*.

In most recent PCM systems the programming of channels within a frame has been quite complex, with some channels being sampled more often than others. This system can be best visualized by considering a commutator with a very large number of terminals and with some of the terminals tied together so that some channels are sampled more often than others.

In most PCM systems it also is necessary to supply timing information to designate the start of the frame. Timing information usually is supplied by transmitting a unique or identifiable code pattern in one or more word positions.

Digital TDM Errors

Errors in a digital system can be categorized as analog errors (drifts, bandlimiting, nonlinearity, and so on, in analog portions of the system), digital dropouts, quantization errors, interpolation errors, and RF link errors.

Analog errors due to drifts, bandlimiting, and so on, are the same as those types discussed previously. Since most digital systems contain analog circuitry, they also have analog errors.

The errors associated with data after it has been digitized are called *dropouts*. In a binary-coded system the principal sources of dropouts are in the tape and disc recording processes and are due to tape or disc imperfections.

Recall from Chapter 10 that the largest quantization error occurring is equal to one-half the amplitude of the levels into which the signal is divided. Interpolation and RF link errors are the same as those previously discussed.

Cassette Recording Methods

Cassette magnetic tape recording is an inexpensive method for storing telemetry data and computer programs and is used extensively in industry. The cassette tape and cartridge are relatively rugged, have a large storage space, and are very economical. A disadvantage is that the data or programs are, of necessity, stored serially. This means that accessing a particular piece of data may take seconds, which may or may not be a problem, depending upon the application.

Many cassette recording techniques place data bytes sequentially along the length of the tape, leaving no blanks. Other methods leave blank spaces between blocks of data to facilitate searches, starts, and stops. Figure 11.12 illustrates nine popular recording methods. Note that these methods show basically two different waveshapes: sine and square waves. The square waves represent complete magnetic saturation, first in one direction and then in the other. The sine waves do not represent saturation but rather the normal recording of a frequency.

Double-frequency recording, DFR (Figure 11.12A), is used for recording data on fixed-head discs at high transfer rates. A 1 bit is represented by a reversal of magnetization and a 0 bit by the absence of a reversal. An additional reversal is inserted between each bit to provide a timing signal. This encoding requires a maximum of two reversals per bit. Unfortunately, trying to record on cassette tape with this method poses a major problem: The cassette tape bandwidth is too narrow. DFR requires a bandwidth obtainable on cassette tape only at low speeds. Thus, data rates greater than 800 bits per second (bps) produce unreliable recordings. Tape speed changes do not affect this method because of its self-clocking characteristic. In DFR, however, noise and signal strength variations can cause loss of information.

FSK, frequency shift keying (Figure 11.12B), is the digitizing technique used in most modems used for telecommunications. Initially FSK would appear to be an ideal method for recording data onto cassettes; however, there are a few drawbacks. Even though FSK is resistant to noise and amplitude variation, qualities important to data reliability, it is more costly to implement than DFR, and cannot tolerate more than a $\pm 6\%$ tape-speed variation. Since many cassette systems cannot maintain speed accuracies within this $\pm 6\%$ limit, tape recording using FSK generally has been limited to high-quality, expensive systems.

NRZ and NRZI (Figures 11.12C and 11.12D) find greatest use in 7-track tape drives that meet IBM compatibility standards. With

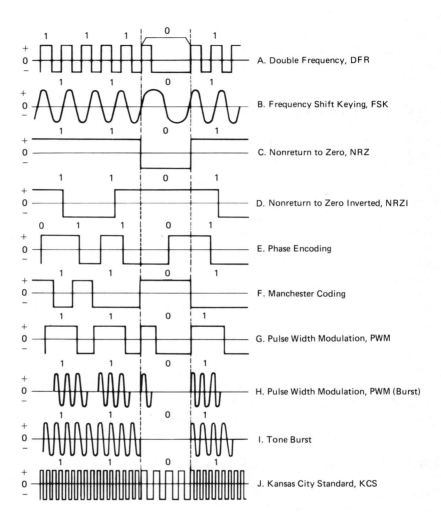

Figure 11.12 Cassette Tape Recording Methods

NRZ and NRZI methods, current through the write head of the tape recorder does not return to zero after a write, so each state of the magnetic recording corresponds to a particular binary bit, either 1 or 0. NRZ and NRZI recording methods are resistant to noise and amplitude variation because of magnetic saturation. Data rates of 1500 bps and greater can produce accurate recordings. NRZ and NRZI techniques are not self-clocking since the transitions from one flux direction to the other depend upon the data, and therefore may not occur at regular intervals necessary for gener-

ating the self-clocking signals. Because of this, most NRZ/NRZI recorders require a separate clock track to maintain a sync signal, and therefore need a separate read/write head, which usually adds expense and mechanical complexity.

Phase encoding, also known as biphase-L (Figure 11.12E), is one of two phase recording techniques that remain popular. This method is resistant to noise, amplitude variation, and frequency change, and is also self-clocking. If one clock cycle is defined as a bit cell, a 1 or 0 is determined by both a transition in the middle of the cell (like NRZI) *and* by the polarity of that transition (not like NRZI). Equipment necessary to generate this recording technique is easy to implement and use.

Manchester coding, also called biphase-M (Figure 11.12F), is similar to biphase-L. The difference is in the location of the transition in the bit cell, which is at the beginning of each cell. Manchester coding also is easy to generate and decode.

PWM (Figure 11.12G) for magnetic tape recording is a patented process, and is self-clocking and resistant to frequency, noise, and amplitude variation. Its major disadvantage is the strict bandwidth requirement, making it only slightly less desirable than the phase encoding methods.

PWM burst (Figure 11.12H) is similar to standard PWM with one exception. A tone burst replaces the magnetic saturation technique required to generate a 1 or 0 pulse.

Tone burst (Figure 11.12I), also called continuous-wave recording, is one of the simplest techniques to generate and decode, but probably the worst method for recording data on magnetic tape. It is nonclocking, and is not resistant to noise or speed variations. Reliable recording data rates over 110 bps are almost impossible.

Kansas City Standard, KCS (Figure 11.12J), is a variation on the Manchester code and is quite popular among hobbyists. The KCS method uses an 8-cycle burst of a 2400 Hz saturation frequency to represent a 1 and a 4-cycle burst of a 1200 Hz saturation frequency to represent a 0.

Table 11.2 gives a summary of cassette recording techniques and a very general evaluation of the more important specifications.

DATA COMMUNICATION

As discussed early in Chapter 10, data communication is the process of transmitting pulses, which are the output of some data source, from one point to another. Telemetry deals mainly with the transmission of measurements. With the rapid incursion of the microcomputer into the industrial workplace, however, data communications other than telemetry are becoming commonplace. Communication between computers in the same plant, between computers many miles apart, and storage of computer programs are just a few of the applications that will require tech-

Table 11.2 A Comparison of Cassette Tape Recording Methods

Recording Method	Performance under Speed Variations	Self-Clocking Code (Yes or No)	Performance under Noise Variations	Ease of Implementation	Maximum Transfer (bps)
DFR	Good	Yes	Fair	Good	800
FSK	Poor	No	Excellent	Good	450
NRZ	Excellent	No	Excellent	Excellent	1500+
NRZI	Excellent	No	Excellent	Excellent	1500+
Phase Encoding	Excellent	Yes	Excellent	Good	1500
Manchester Coding	Excellent	Yes	Excellent	Good	1500
PWM	Excellent	Yes	Good	Excellent	1500
PWM (Burst)	Excellent	Yes	Good	Excellent	1500
Tone Burst	Poor	No	Poor	Excellent	450
Kansas City Standard (KCS)	Excellent	Yes	Excellent	Excellent	1200

nical expertise in data communication systems in order to keep them operational. Because of the pervasiveness of data communication in the industrial workplace, some of the more common and important concepts will be presented here.

Transmission Classifications

There are basically two ways in which data can be moved from one place to another: serial and parallel transmission. Microcomputers (μC) generally deal with eight bits of data at a time. *Parallel transmission* is used for movement of data within the microcomputer and for short distances outside the computer. This method is efficient and quick, important considerations for the computer. *Serial transmission* is reserved for longer distances and slower data rates, such as storage and retrieval of data or programs, and data transmission over telephone lines. Serial transmission requires only two wires, whereas parallel transmission requires eight or more, which increases costs and the likelihood of problems caused by broken wires.

If serial transmission is used, the bits can be transmitted synchronously or asynchronously. In *synchronous transmission* all characters are sent at a constant rate, and timing is synchronized between the transmitter and receiver so the receiver can keep in step with the transmitter. (This concept was discussed in the pulse modulation chapter.) The synchronization can be done by sending a clock pulse train on a separate line, or recovering a clock signal from the data stream itself. One of the principal characteristics of synchronous data transfer is that data streams must be continuous, or one unit of data must immediately follow another, until the entire communication has been completed. Blank data generally terminate transmission. In *asynchronous transmission* charac-

ters can be sent at any time and the receiver is provided with a means for recognizing the start and end of each character. As a result each character is sent as a unit unto itself, and there can be any amount of time between characters. Both synchronous and asynchronous methods will be discussed in the following sections.

Synchronous Serial Transmission. As stated earlier the most important characteristic of synchronous data transmission is that data must be transmitted continuously. Both transmitter and receiver must agree upon a bits per second transfer rate. Thereafter, the data stream will be interpreted as one bit per recovered clock signal transition, as shown in Figure 11.13. Notice that this data stream uses the NRZ transmission method.

Some problems now arise. How can the receiving device determine the beginning and end of each data unit, or character? And we know that the binary digits 1 and 0 by themselves cannot represent all the letter and number possibilities. How many bits are needed to represent these letters and numbers?

To answer the second question first, there are two popular code formats used for communications. They are the ASCII, American Standards Committee on Information Interchange, code (pronounced ask-ee) and the EBCDIC, Extended Binary Coded Decimal Interchange Code (pronounced eb-see-dick). These codes are given in Appendixes A and B, respectively. The EBCDIC code is used by IBM

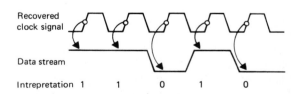

Figure 11.13   Recovery of Data from the Data Stream

and systems attached to IBM equipment, while ASCII is used in most other applications. We will use the ASCII code in this text.

Now back to the question posed earlier. How can the receiving device determine the beginning and end of each character? The answer is that synchronizing characters are first sent to alert the receiver that data are to follow. Thereafter, each specific number of bits in the string represent a character.

As you can see, many rules must be imposed upon the transmitter and receiver operation in order to ensure that information will be transferred correctly. The set of rules for each method of transmission is called its *communications protocol.*

The data unit in a synchronous serial data stream may contain 5, 6, 7, or 8 data bits that may be followed by a parity bit. Seven bits plus odd parity will be used here. Odd parity requires that the parity bit be a 1 or 0, whatever is required to make the data group have an odd number of 1s. Of course, even parity will have an even number of 1s, including the parity bit. An example of a transmission message might be:

Messageƀtoƀfollow.

where

ƀ represents a blank, or space character

The middle part of the message would appear as:

sync	char	char	sync	sync	char

-- 00101100 01000000 11101001 00101100 00101100 11011111 --

space   t   o

Since the data characters may not be available when the transmitter needs to send them, sync characters are inserted to keep the transmission flowing. Eight bits per character make up part of the communications protocol for this particular synchronous serial transmission system.

Other types of synchronous serial data transmission protocol that are in present use are Bisync, Serial Data Link Control (SDLC), and High-Level Data Link Control (HDLC). These protocols are IBM standards that are fast becoming world standards for synchronous serial data.

Compared to asynchronous serial transmission, synchronous transmission has somewhat better characteristics, even though it is more complex to implement. Synchronous transmission has the advantages of a good speed-to-distance ratio (a minimum of 1000 ft/ 333 thousand bps), reliability (error checking), expandability (able to communicate with over 200 stations), low cost (two conductor cables for connections), and efficiency (async uses up to 20% of each word transmission in start and stop bits).

Asynchronous Serial Transmission. In asynchronous transmission, characters can be sent at any time with any amount of time between characters. The protocol provides a means for enabling the receiver to recognize the start and end of each character. A standard format such as that shown in Figure 11.14 generally is used. This collection of pulses, including the start and stop bits, is called a *frame.*

The frame begins with a start bit and ends with one or more stop bits. These bits enable the receiver to recognize the beginning and end of the character. The transition from a high in the idle state to a low start bit informs the receiver that the next eight bits are data. The stop bit (or bits) marks the end of the character and gives the receiver time to reset its hardware and prepare for the next character. There may be one, one-and-one-half, or two stop bits. (Fractional bits are permissible since their period is just a resting time for the receiver.) Hardware that is completely electronic needs only one stop bit, but the early electromechanical machines needed two stop bits before the next

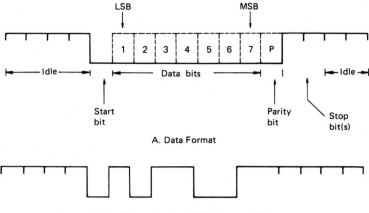

Figure 11.14 Data Frame Format

character. If noise pulses or other errors cause the receiver not to detect the stop pulse, a *framing error* is indicated. The transmitter then may retransmit the frame that had the error.

The bit time determines the maximum rate at which bits can be transmitted, and thus defines the *bit rate* at which transfer of data can occur. The bit rate is measured in bits per second (bps). The terms *baud* and *baud rate* are often used, somewhat inaccurately, in place of bps or data rate. Baud is a term used accurately in telegraphy and has been carried over into Teletype-like communications. Baud is not used accurately with the type of computer-based serial data communication for which it is generally applied. The correct term, bits per second, will be used in this text. Some common communication data rates are 110, 150, 300, 600, 1200, 2400, 4800, and 9600 bps.

Serial transmission also must consider the direction of signal travel on the data line. Figure 11.15 shows three techniques for dealing with data transmission direction. Simplex transmission (Figure 11.15A) is the simplest technique, but is limited; it allows communication in only one direction. Duplex transmis-

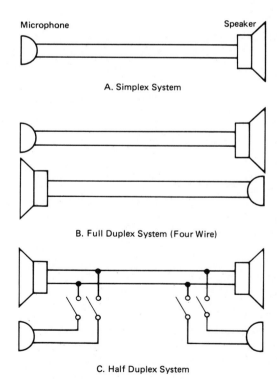

Figure 11.15 Simplex and Duplex Transmission

Table 11.3 Similar Signals and Functions of the Two Serial Standards: RS-232C and RS-449

RS-232C			RS-449 *		
Pin	Circuit	Description	Pin	Circuit	Description
7	AB	Signal ground (common return)	19	SG	Signal ground
			37	SC	Send common
			20	RC	Receive common
22	CE	Ring indicator	28	IS	Terminal in service
20	CD	Data terminal ready (DTR)	15	IC	Incoming call
			12	TR	Terminal ready
6	CC	Data set ready (DSR)	11	DM	Data mode
2	BA	Transmitted data (TxD)	4	SD	Send data
3	BB	Received data (RxD)	6	RD	Receive data
24	DA	Transmit signal element timing (DTE source)	7	TT	Terminal timing
			5	ST	Send timing
			8	RT	Receive timing
15	DB	Transmission signal element timing (DCE source)			
17	DD	Receiver signal element timing (DCE source)			
4	CA	Request to send (RTS)	7	RS	Request to send
5	CB	Clear to send (CTS)	9	CS	Clear to send
8	CF	Received line signal detector (DCD)	13	RR	Receiver ready
			33	SQ	Signal quality
21	CG	Signal quality detector	34	NS	New signal
23	CH	Data signal rate selector (DTE source)	16	SF	Select frequency
			16	SR	Signaling rate selector
23	CI	Data signal rate selector (DCE source)	2	SI	Signaling rate indicator
14	SBA	Secondary transmitted data	3†	SSD	Secondary send data
			4†	SRD	Secondary receive data
16	SBB	Secondary received data	7†	SRS	Secondary request to send
			8†	SCS	Secondary clear to send
19	SCA	Secondary request to send	2†	SRR	Secondary receiver ready
13	SCB	Secondary clear to send			
12	SCF	Secondary received line signal detector			
			10	LL	Local loopback
			14	RL	Remote loopback
			18	TM	Test mode
			32	SS	Select standby
			36	SB	Standby indicator

*37-pin connector except where noted.

†9-pin connector.

sion allows bidirectional communication, and comes in two varieties. Full duplex (Figure 11.15B) not only allows two-way communication, but both users can talk at the same time. The telephone is an example of such a process. The subtleties of human communication sometimes require the capabilities of full duplex transmission. Thus, it is important that the telephone have that capability. Full duplex, however, is somewhat inefficient because it generally requires four wires in order to be implemented. Figure 11.15C shows the half duplex method in which one user must listen while the other talks. Most radio voice communication uses this "press-to-talk" procedure.

After the rules, procedures, and protocol are presented, the hardware implementation cannot be far behind. That is the topic of the next section.

RS-232C Serial Interface

The RS-232C is an electrical interface standard for connecting system components such as modems, printers, terminals, and computers. This standard was established by the Electronic Industries Association (EIA), an industry trade organization, to be an interface between some form of data terminal equipment (DTE) and some form of data communication equipment (DCE). The *RS* in the RS-232C stands for *recommended standard;* the *C* indicates that this is the third change in the standard. The standard defines 20 distinct signals (Table 11.3) to be used to interface terminals and communication equipment.

What is a standard? A *standard* is a model or pattern laid down by some authority by which the quality or correctness of a thing may be determined. Authorities for interface standards may be some independent agency such as the Institute of Electrical and Electronic Engineers (IEEE), or a manufacturer such as IBM. Two things are important for the establishment

of a standard: An authority is involved, and the standard is published and widely available. Some so-called standards are not standards at all because one or both of these elements are missing. These are generally known as *de facto* standards because of their widespread use in industry.

Standards are an aid to the user, not the manufacturer. The larger the manufacturer, the less advantage is received from standardization. Many companies keep their standards and systems as private as possible. In addition, a standard does not have to be perfect, merely adequate. No matter what standard is produced, someone will disagree with the details; it is difficult to satisfy an entire industry.

The RS-232C standard also defines the voltages, the ranges for a logical 1 and a logical 0, used in all circuits. Figure 11.16A shows an RS-232C signal at the transmitter end of a cable just as it is being transmitted down the cable. A

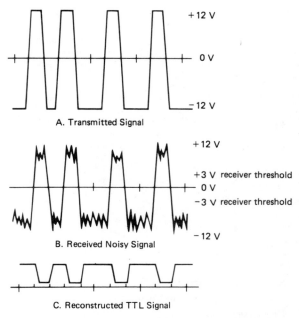

Figure 11.16 Reduced Effects of Noise and Transmission-Wire Length

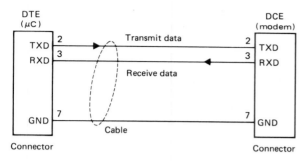

A. Minimum DTE to DCE Configuration

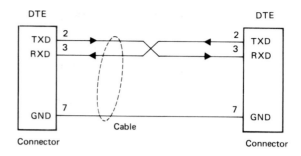

A. Simple Null Modem Connection

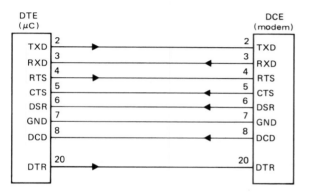

B. Complete DTE to DCE Configuration

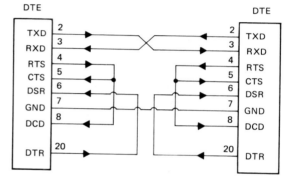

B. Full Null or General-Purpose Modem Connection

Figure 11.17 DTE to DCE Connection Configuration

Figure 11.18 DTE to DTE Connection

high, or 1, is represented by -12 V, and a low, or 0, by $+12$ V. Figure 11.16B shows the same signal after it has traveled the length of the cable and been received at the receiver end. Noise and transmission-line parameters cause the signal to be degraded. However, the signal need only exceed -3 V at the receiver for a 1, and $+3$ V for a 0. Figure 11.16C shows the reconstructed TTL signal. TTL signals vary from 0 to 0.8 V for a 0, and from 2.4 to 5 V for a 1.

Some devices do not use all of the RS-232C standard signals, so not all devices will work together without modification. The most important pins in the RS-232C standard are transmit data (TxD) and receive data (RxD), pins 2 and

3, respectively. These are the two wires over which serial data are simultaneously sent and received. The remaining pins, except for ground (pin 7) and pins 14 and 16 of the secondary group, are control circuits. The original RS-232C interface was designed to connect some type of data terminal equipment (DTE), such as a time-sharing terminal, to some type of data communication equipment (DCE), such as a modem. Figure 11.17A shows the minimum configuration for connecting a DTE and a DCE. The DTE in this case is a microcomputer (μC).

Notice in Figure 11.17A that the two transmit and two receive pins are connected. Which

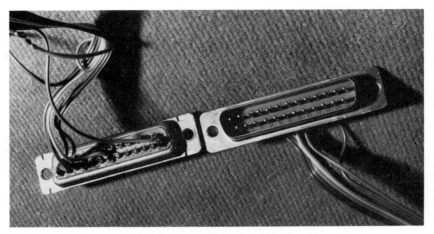

A. DB-25 Connectors, Male (Front and Back)

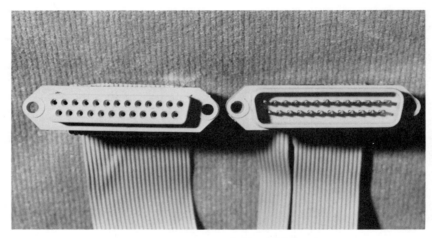

B. DB-25 Connectors, Female (Left) and Male (Right)

Figure 11.19   DB-25 Connectors

transmits and which receives? The answer is that just as directions in a computer system are from the computer's point of view, all the names of RS-232C signals are from the perspective of the DTE. Thus, the DTE transmits on the TxD line, and the DCE receives on it. Similarly, the DTE receives data on the RxD line, and the DCE transmits on it. Figure 11.17B shows the complete DTE to DCE connection. The following list is a summary of the

eight important pins on the RS-232C interface. Table 11.3 gives the pin description, while the following summary gives the general usage of the listed pins.

- Pin 2, Transmitted Data (TxD): This pin transmits data from the DTE to the DCE.
- Pin 3, Received Data (RxD): This pin transmits data from the DCE to the DTE.
- Pin 4, Request to Send (RTS): This pin

must be at logic zero to enable the DCE to transmit data.

- Pin 5, Clear to Send (CTS): A logic zero on this pin indicates that the DCE is ready to transmit.
- Pin 6, Data Set Ready (DSR): A logic zero on this pin signals the DTE that the DCE is powered and ready to operate.
- Pin 7, Signal Ground or Common Return: This line *must* be connected.
- Pin 8, Data Carrier Detect or Received Line Signal Detector (DCD): This input to the DTE may be used to disable data reception if, for example, the originating data channel is disconnected.
- Pin 20, Data Terminal Ready (DTR): A logic low to the DCE indicates that the DTE is powered and ready to operate.

The designations of DTE and DCE are sometimes referred to as the electronic sex of the equipment because the RS-232C standard requires that the female connector be associated with the DCE. Connecting a DTE and DCE would be termed as connecting devices of complementary sex.

What happens when the RS-232C interface is used to connect two DTEs, a use for which it was not intended? It turns out that by switching certain wires, the connection and operation can be completed successfully. Figure 11.18A is the simple null modem arrangement that is expanded into the general-purpose or full null modem arrangement of Figure 11.18B.

Even though the RS-232C standard specifies pin numbers, it does not specify a connector. The commonly used connector for this standard is the DB-25 connector. It is available in male and female "genders": the DB-25P (male) and the DB-25S (female). The connector's gender, however, has nothing to do with the device being a DTE or a DCE. Figure 11.19 shows these two connectors.

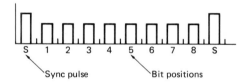

Figure 11.21 Diagram for Second Problem

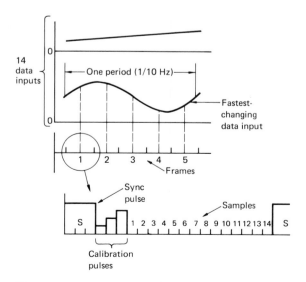

Figure 11.20 Diagram for First Problem

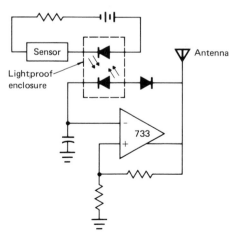

Figure 11.22 Proposed Transmitter for Heartbeat Monitor

One of the problem areas for the RS-232C interface standard is the electrical signal level used. This standard was developed before the TTL levels (0–5 V) became popular. TTL can use a single 5 V DC supply for its operation, but the RS-232C standard requires two equal voltages of opposite polarity in the 5 to 25 V range. To resolve this problem the EIA introduced two standards, the RS-422 and the RS-423, that specify TTL-compatible voltage levels. The new standards alter only the electrical characteristics of the interface signals; their functions remain the same.

The RS-232C standard has another prob-lem. Its speed is limited to about 20 thousand bps and about 50 ft (15 m) of length. For this reason the EIA published another standard, designated RS-449, that allows data rates as high as 2 million bps. In addition, a few more pins and signals have been added. Refer again to Table 11.3 for the pin designations. The RS-449 also uses two connectors, a 37-pin plug that carries the most frequently used signals and a 9-pin plug that contains the seldom-used secondary channel. The RS-449 standard is intended to gradually replace the RS-232C. In the meantime, the two standards are bound to cause confusion.

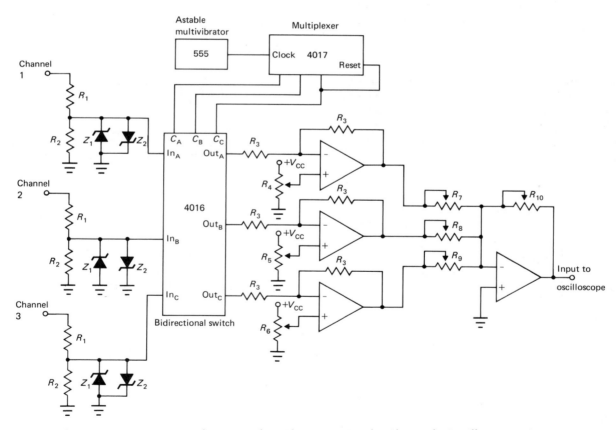

Figure 11.23 Basic Schematic for Changing Single-Channel Oscilloscope into Three-Channel Oscilloscope

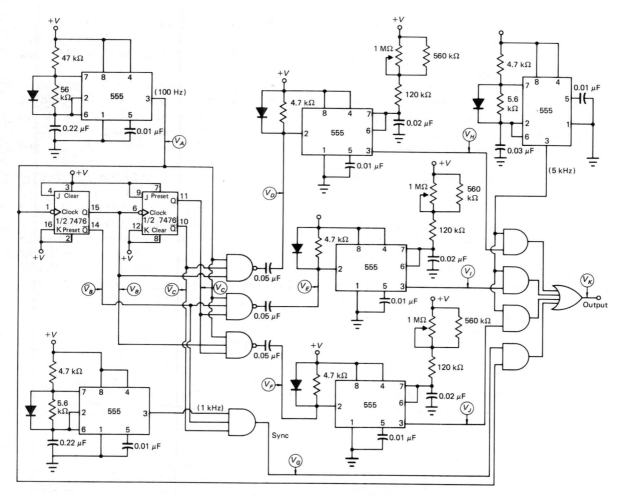

Figure 11.24   Three-Channel Multiplexer Using PWM

MULTIPLEXING SAMPLE PROBLEMS

We begin this section by presenting a few definitions that were given previously or were implied.

A *frame* is one set of data. The *frame time* is the time from the start of one sync pulse to the start of the next. Its unit of measure is the second. The *frame rate* is the reciprocal of frame time; its unit of measure is hertz(Hz). A

sample is the individual data input into the frame. The *sample time* is the time allotted for each sample. *Commutation* is the process of changing from one sample to the next. The *commutation rate* is the rate of changing from sample to sample; it is also the reciprocal of sample time. Its unit of measure is hertz(Hz).

Note that the circuits in this section are only proposed circuits. Some have been constructed, and some have not.

Now, let us work some problems. In a certain PAM TDM system there are 14 data inputs, three calibration pulses, and a sync pulse that takes three sample times. The sample time divisions are all equal and are at a 100% duty cycle (see Figure 11.8). All data inputs are sampled only once per frame. The fastest data input changes no faster than 10 Hz, and it is sampled 5 times during its period. Determine the following: frame time, frame rate, sample time, and commutation rate.

The first thing to do is draw a diagram to clarify the problem. The diagram is shown in Figure 11.20.

If the fastest data input changes no faster than 10 Hz, then this sample will determine the frame time. The period of the fastest-changing data input is then 0.1 s. Five frames occur during this time, so frame time is 0.1 s divided by 5, or 0.02 s. Therefore, the frame rate is 50 Hz. Since there are 14 samples, 3 calibration pulses, and a sync pulse that takes three sample times, then the sum of these yields 20 sample times occurring within the frame. Sample time is the number of samples divided into the frame time, or 0.001 s. Commutation rate is then 1 kHz (the reciprocal of sample time).

For the second problem we want to design a single-channel, voice PCM system. Voice sample rate generally is set at 8000 samples per second. We will use eight bits for each word, a 50% duty cycle, and a sync pulse whose amplitude is twice that of a data bit but has the same sample time. Determine the following: pulse width, commutator frequency, frame time, and frame rate.

The diagram for this problem is shown in Figure 11.21. Since there are 8000 samples per second, frame time, which is the reciprocal of 8000 samples per second, is 125 μs, and frame rate is 8 kHz. A 50% duty cycle will require the equivalent of 18 sample spaces per frame because there is one sample space between bits, and one-half sample space at each end of the frame. Thus, pulse width is frame time divided by 18, or 6.94 μs. Commutator frequency is 144.1 kHz.

If the circuit of Figure 10.25 (in Chapter 10) were to generate the output desired in this problem, the first 555 would be set to 8 kHz and the second 555 to 144.1 kHz.

For our third problem let us design a physiological monitor of some type. It should be lightweight, portable, and inexpensive. As an example we will make a heartbeat monitor. TDM is generally used for slowly changing data, so we will use PFM for this problem.

The PFM generator of Figure 10.20 (Chapter 10) would seem to be a prime candidate for this problem. And if we want an inexpensive system, we could use an FM radio as the receiver. The bandwidth for FM radio is from 88 to 108 MHz, but the frequency limit of the LM741 is 1 MHz, so this device won't do. However, other op amps may be used if they can operate at high frequencies.

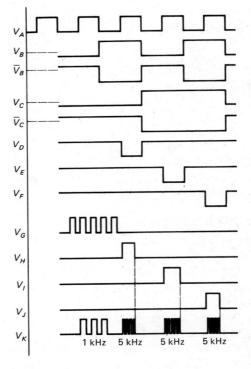

Figure 11.25 Waveforms for Schematic of Figure 11.24

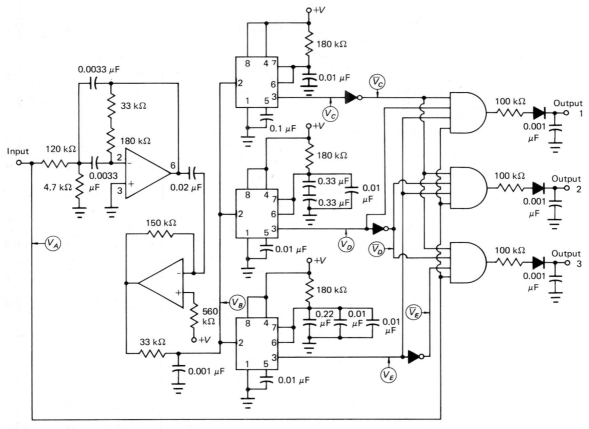

Figure 11.26 Three-Channel Demultiplexer Using PWM to Demultiplex Signal from Circuit of Figure 11.24

An example of an op amp that might meet the frequency requirements is the LM733 differential video amp. It has a bandwidth of 120 MHz and selectable gains of 10, 100, and 400, and no frequency compensation is required. Figure 11.22 shows how the transmitter might be connected.

An alternative might be to use a lower-frequency transmitter of the same design and a PLL IC as the receiver/demodulator. Since the IC is low-power, the transmission distances would be necessarily short.

For the fourth problem we will make a single-channel oscilloscope into a multichannel

oscilloscope. To do so we will use TDM principles but no transmitter or receiver. (The oscilloscope will be the receiver.)

The idea in this problem is to divide the inputs (called channels here) into PAM and give each input channel a different DC offset. These PAM TDM signals will be input to the single-channel oscilloscope, which will display all the inputs on the screen, multiplexed in time. If the multiplexing is done fast enough, the display will appear to be that of a multichannel oscilloscope.

The schematic of Figure 11.23 can be used to produce the desired results. The input resis-

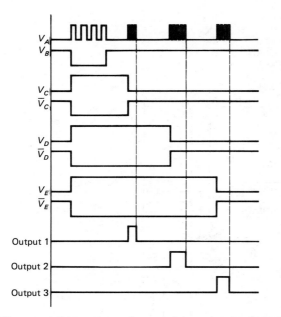

Figure 11.27 Waveforms for Schematic of Figure 11.26

tors R_1 and R_2 are used to attenuate the input signals to keep them within the voltage limits of the circuitry. Notice that the input resistors for the three channels have the same designation (R_1 and R_2). Thus, all R_1's are the same resistance and all R_2's are the same resistance. The input zener diodes, Z_1 and Z_2, protect the ICs if the maximum voltages are still exceeded.

The first line of op amps gives each channel a different DC level offset so that each input channel will be displayed on a different level on the screen. Resistors R_7, R_8, and R_9 adjust the individual channel gains, while R_{10} adjusts the overall gain. The final op amp sums all the signals and prevents each channel from interfering with the others.

The final problem is to make a three-channel multiplexer suitable for a radio-controlled airplane. The solution is represented in Figures 11.24, 11.25, 11.26, and 11.27.

Figures 11.24 and 11.26 are the schematics for the modulator and demodulator. Figures 11.25 and 11.27 are the synchrograms for the schematics. The receiver in the schematic of Figure 11.26 is producing pulses whose pulse widths correspond to the pulse widths being transmitted. The output from each channel can be attached to appropriate mechanical actuators in the airplane.

CONCLUSION

As shown in this chapter the principles of telemetry have many industrial applications. And, as in the case of most modern electronics applications, only the competence and imagination of the investigator will limit the results.

QUESTIONS

1. A telemetry system can be divided into three parts: The _____, which are at remote locations and transform the physical quantities to be monitored into electrical signals; the _____ system, which transforms the transducer signals, transmits and receives signals, and transforms receiver signals for display; and the _____ and _____ system, which displays received signals for interpretation.

2. Communication using different frequency channels at the same time is called _____.

3. Communication using the same frequency channel but at different times is called _____.

4. According to Table 11.1, in order to keep a low signal-to-noise ratio while transmitting telemetry signals, we need a modulation index of _____.

5. An FM/AM multiplexed telemetry system

designation indicates that the subcarriers are _____-modulated and the main carrier is _____-modulated.

6. _____/_____ is by far the most common modulated telemetry technique in use today.

7. Communication errors caused by equipment aging can be reduced by using _____ signals.

8. A 100% duty cycle telemetry system uses a _____ frequency spectrum than the 50% duty cycle system uses.

9. In Question 8 the 50% duty cycle system _____ cross talk between successive channels.

10. Pulse-position TDM uses the _____ edge of the pulse to carry the information.

11. Pulse-code TDM can transmit 1 and 0 information only by the _____ or _____ of a pulse.

12. Define multiplexing.

13. Name two types of multiplexing in common use.

14. Define subcarrier.

15. Define frame time.

16. What are housekeeping pulses?

17. Define binary transmission system format.

18. Define pulse-code TDM word.

19. Discuss when parity bits are and are not used.

20. If a data signal varies at a rate of 75 Hz, what would the lowest standard frequency band be that could be used to carry this information?

21. Band 14 would use a subcarrier of what frequency?

22. To demodulate an FM data signal on band 10, what discriminator center frequency would you use?

23. What is the lowest frequency the subcarrier can deviate to and still be in band 8?

24. Measurement of a signal indicates that its minimum rise time is 7 ms. What channels could be used to carry this information?

25. Sketch the two types of NRZ waveforms and explain how they differ.

26. What are two disadvantages of non-self-clocking binary modulation methods compared to self-clocking methods?

27. List nine magnetic medium recording methods.

28. Which magnetic recording format has the best characteristics other than the transfer rate?

29. Which two magnetic recording formats have the best characteristics that include the fastest transfer rates?

30. The two major classifications of data transmission are _____ and _____.

31. The two major classifications of serial data transmission are _____ and _____. Explain the differences between these classifications.

32. List two character code formats in common use today.

33. Define simplex, half duplex, and full duplex transmission.

34. What are the requirements for a proposed model to become a true standard?

35. What are the data speed and cable length limitations for the RS-232C serial interface standard?

36. What is the data speed for the RS-449 serial interface standard?

PROBLEMS

1. It is desired to transmit voice (20 kHz) on an FM system that has a modulation index of 5. What will be the maximum frequency deviation of this transmitter?

2. A telemetry system has seven different sensors to sample. Two additional sample spaces are set aside for the high-level and low-level references, and one sample is set aside for a sync pulse. In this telemetry system the highest-frequency input variable has a maximum frequency of 10 Hz. It is desired to sample this variable ten times in each waveform period. Each of the seven inputs is sampled only once during each frame. Determine (a) sample time, (b) frame time, (c) frame rate, and (d) commutation rate.

CHAPTER

12

SEQUENTIAL PROCESS CONTROL

OBJECTIVES

On completion of this chapter, you should be able to:

- Define sequential control;
- Distinguish a batch process from a continuous process;
- Identify batch process applications;
- Recognize and explain the parts of a ladder logic diagram;
- Generate a ladder logic diagram from a pictorial diagram and sequential description of a sequential process;
- Determine that a ladder logic diagram is accurate by "hand running the logic."

INTRODUCTION

Traditionally, industrial processes and control systems used relays, timers, and counters as well as sensors and actuators. The first three types of these devices constitute a class of control devices that are used to control processes known as sequential control processes (as opposed to continuous control processes). A *sequential process* is a process in which one event follows another until a job is completed. Sequential process control can be subdivided into two major classes: control of event-based sequential processes and control of time-based sequential processes. Of course, many sequential processes may contain elements of both of these classes.

An *event-based* sequential process is one in which the occurrence of an event causes a certain action to take place. For example, a limit switch closes to reverse a motor moving a part because the part has reached the furthest limits of safe operation. Or, the inputs to a logic device become true, such as the input connections to an AND gate going high, and cause the output of that device to activate some other device. In an event-based sequential process the action of the controller is being governed by events external to the controller. These processes are of the type that can be controlled by a programmable controller.

A *time-based* sequential process is one in which the events occur in a specific sequence, not the nearly random closure of switches external to the controller. The action of this controller basically is determined by the program in the controller and not by external occurrences. The controlling system for this type of process is called a programmable sequence controller or programmable process controller. An extensive treatment of time-based control processes is beyond the scope of this book. This chapter and the next will deal mainly with event-based control.

Another name for a sequential process is a *batch process*. A process is considered to be batch in nature if the process consists of a sequence of one or more *steps* or *phases* that must be performed in a defined order. The completion of this sequence of steps creates a finite quantity of finished product. If more of the product is to be created, the sequence must be repeated.

This chapter covers sequential control and how it can be implemented on automatic machines. The ladder diagram is the traditional method of choice for representing the logic necessary to help design and document these automatic processes. Sequential control and ladder diagrams form the base upon which the programming of the programmable controller, the subject of the next chapter, is founded.

CHARACTERISTICS

The following list outlines some key concepts that distinguish a batch process from a continuous process.

1. *Discrete* loads of raw materials normally are introduced into the line for processing, rather than the continuous flow of materials found in a continuous process.
2. Each load of material being processed normally can be identified all the way through the processing, because each is kept separate from all others being processed. In most continuous processes raw materials cannot be uniquely tracked.
3. Each load of raw material can be processed differently at the various equipment areas in the line. In continuous processes raw materials usually are processed in an identical fashion.
4. Movement of a load from one equipment area or step to the next can occur only when the step has been completed and the next equipment area has been vacated. In continuous processes materials flow steadily from one area to the next.
5. Batch processes generally have *recipes* associated with each load of raw product to be processed. These recipes direct the processing of the raw product. *Note:* Even though the word *recipe* is used here, it does not specifically refer to food preparation, although many foodstuffs are made by batch processing.
6. Batch processes usually require more *sequential logic* than continuous processes.

7. Batch processes often include processing steps that can fail, and include other special processing steps to be taken in that event.

APPLICATIONS

Batch control systems generally are configured to control four different types of applications. They are used: for the control of solids, which normally use input parameters from weighing equipment; for the control of liquids, with input parameters from flowmetering equipment; for the control of the environment, such as temperature, dew point, and other variables; and for the control of "universal" systems, which accept inputs from many different types of sensors.

When weighing solids two parameters of the system need to be considered. The first, called *dribble,* occurs when the loading chute discharge rate changes from high- to low-speed discharge so that the scale can have enough reaction time to measure the solid accurately and not overshoot the intended weight. The other parameter, sometimes called *preactuate,* considers the amount of solid still in the loading chute but not yet fallen to the scale after the discharge has been cut off. These two parameters will vary with each system but must be accounted for if accurate weighing is desired.

The accurate measurement of fluid will depend upon the type of flowmeter used and the density of the fluid. Because most fluids change density in a consistent manner with temperature, density can often be inferred from a temperature measurement.

Environmental control applications include controlling heat for sintering, carburizing, hardening, tempering, brazing, annealing, and gaseous diffusion processes such as that used in IC manufacturing. Other applications are freeze-drying vegetables and baking out mois-ture from large rolls of paper to reduce their shipping weight.

Universal control system applications require special functions such as floating point math, exponential functions, and various other functions. This class of processes covers such a wide range of applications that special sensors and devices may be required.

DEVICE SYMBOLS

Documentation of industrial processes is continuing to grow in importance, particularly because systems are becoming more complex and interdependent. Logical and easy-to-read documentation is important not only for the design and installation of a system, but for continued system maintenance and troubleshooting as well. For these reasons, standardized symbols and formats need to be understood and used. The symbols that follow generally are accepted by industry personnel when documenting sequential control systems using relay logic. The standardized format for documenting sequential control systems is the ladder logic diagram, the topic of the next section.

Some mention should be made at this point about the general differences between electrical and electronic symbol designations. These two industries matured somewhat independently of each other, with different ruling bodies, and therefore it is understandable that some differences should exist in their respective symbols for the same component. For example, the electronic symbol for a resistor is a zigzag line with an alphanumeric designation of R_1. The same symbol in the electrical or industrial world is a rectangle with lines out the ends and with an alphanumeric designation of 1R. These differences can sometimes be confusing. In this chapter we will use industrial symbols because designing and documenting the sequential process

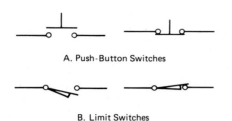

A. Push-Button Switches

B. Limit Switches

Figure 12.1 Common Switch Symbols

is predominantly an industrial process for industrial machines.

Switch Symbols

One of the most common symbols used in sequential control documentation is for switches. Two switches are particularly common: push-button and limit switches (Figure 12.1).

The two categories of each switch shown in Figure 12.1 are normally open (NO), shown on the left, and normally closed (NC), shown on the right. There are variations to these symbols, such as those in Figure 12.2, but the symbols shown are the ones in general use. For more information concerning the push-button and limit switches, refer to Chapter 6 on control devices. Push-button switches generally are actuated by the operator, whereas limit switches are actuated by the machine or system being operated.

A newer type of switch that is becoming more popular is the proximity switch. A proximity switch senses the presence or absence of a target without physical contact. The six basic types are: magnetic, capacitive, ultrasonic, inductive, air jet stream, and photoelectric. The symbols for proximity switches are the same as those for limit switches.

Temperature switches, shown in Figure 12.3A, open or close when a designated temperature is reached. It is now obvious from the symbols which switches are NO and which are NC; therefore they will not be labeled as such except for clarity. Level switches (Figure 12.3B) sense the height of a liquid and open or close their contacts when the predetermined height is reached. Pressure switches (Figure 12.3C) detect pressure and actuate (open or close) their contacts when the specified pressure is reached.

The most common type of flowmeter used in sequential control is one that produces pulses. These pulses can be counted by a counter, which in turn can cause an action when a specified count is reached. Counters will be discussed later, but the flowmeter symbol and its output are shown in Figure 12.4. Notice that the flowmeter output is a series of pulses with amplitudes of 24 V, since this particular flowmeter operates on 24 V. Other voltages can be used depending upon the requirements of the flowmeter.

The proximity, temperature, level, and pressure switches and the flowmeter are more likely to be called sensors, and as such are discussed in Chapter 8 on transducers.

Relay Symbols

Another very commonly used device in sequential control is the electromagnetic relay. The symbol for the relay (Figure 12.5) is a circle and parallel lines. The circle represents the coil of the relay and the parallel lines represent the relay contacts. A relay will have only one coil,

Figure 12.2 A Variation of Push-Button Switch Symbols

A. Normally Open

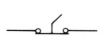

B. Normally Closed

C. A Variation

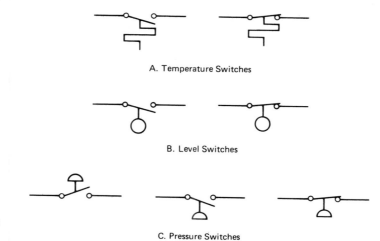

A. Temperature Switches

B. Level Switches

Figure 12.3 Temperature, Level, and Pressure Switch Symbols

C. Pressure Switches

but may have any number of contacts, either NO, NC, or a mix of both. The relay coil generally has an alphanumeric designation, such as 1CR, inside the coil circle, and the same designation next to the contacts that are controlled by that coil. The letters *CR* represent *control relay,* and the number distinguishes that relay from all others in the schematic. Electromagnetic relays are discussed in Chapter 6 on control devices.

Other devices whose symbols are similar to the control relay are the timer relay and the counter. The timer relay symbol (Figure 12.6) is also a circle, but there are two variations. The three-connection timer relay coil has the third, or bottom, connection permanently connected to the power in order to operate the timer motor. This connection may or may not be shown. Current flow through the other two con-

nections initiates the timing sequence. The timer relay switch contacts will actuate upon completion of the timing period.

The contacts in Figure 12.6B are known as *on-delay* timer contacts. The energizing of coil 1TR causes the NO contacts to time closed after a period of time, and the NC contacts to time open. De-energizing timer coil 1TR instantaneously opens the NO contacts and closes the NC contacts.

The contacts in Figure 12.6C are known as *off-delay* timer contacts. Energizing the coil of the off-delay timer instantaneously closes the NO contacts and opens the NC contacts. De-energizing the off-delay timer coil causes the NO contacts to open and the NC contacts to close *after a time delay.*

There are various types of timers, but generally the timer is reset and ready to start tim-

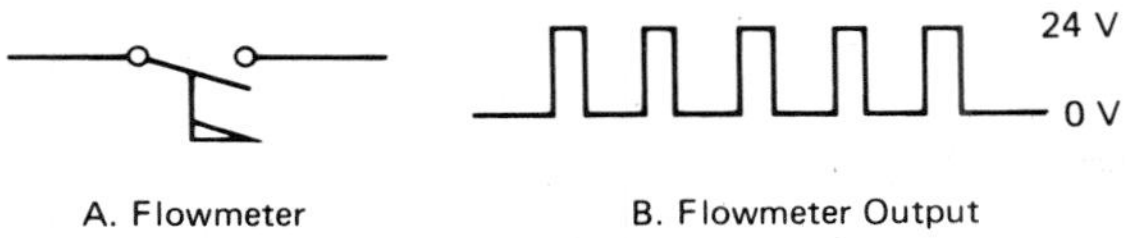

A. Flowmeter   B. Flowmeter Output

Figure 12.4 Flowmeter Symbol and Output Pulses

A. Control Relay Coil   B. NO Contacts   C. NC Contacts

Figure 12.5 Control Relay and Contacts Symbols

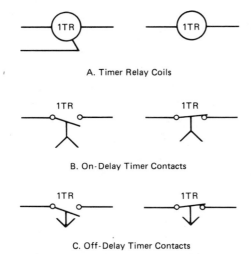

A. Timer Relay Coils

B. On-Delay Timer Contacts

C. Off-Delay Timer Contacts

Figure 12.6   Timer Relay and Contacts Symbols

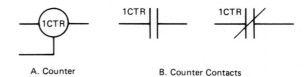

A. Counter

B. Counter Contacts

Figure 12.7   Counter and Contacts Symbols

Figure 12.8   Solenoid Symbol

ing again whenever the current flow to the timer is interrupted.

The counter symbol (Figure 12.7A) generally has three connections. The third, or bottom, connection is used to enable (allow it to operate) and then to reset the counter when desired. The remaining two connections carry the signal or pulses to be counted. Notice the counter contacts are no different than the control relay contacts, except for their alphanumeric designation. These contacts actuate when the predetermined, or preset, count is reached, and reset when the counter is reset.

Solenoid Symbol

Another useful sequential control device is the solenoid (Figure 12.8). A solenoid provides electrical control over materials such as liquids and gases. An electrical signal to the solenoid causes a valve to open and allow the liquid or gas to flow. The valve stays open until the electrical signal ceases. A solenoid is somewhat dif-

ferent from a relay. A solenoid generally has a coil of wire with a moving metal core in the coil, and the relay has a fixed metal core in the coil that, when activated, pulls an armature that makes the contact. Solenoids, as well as relays, can be used to switch heavy current circuits.

Miscellaneous Symbols

Some other device symbols are shown in Figure 12.9. These are self-explanatory.

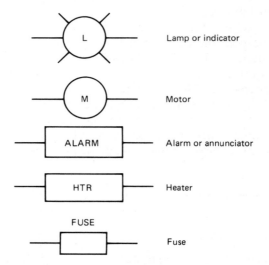

Lamp or indicator

Motor

Alarm or annunciator

Heater

Fuse

Figure 12.9   Miscellaneous Symbols

Dotted lines connecting symbols indicate that those devices are physically connected and are actuated together. Figure 12.10 illustrates push buttons and limit switches that are activated together.

LADDER LOGIC DIAGRAM

The control logic for a sequential process traditionally has been depicted in the language of a ladder logic diagram. A *ladder logic diagram* is a diagram with a vertical line (the power line) on each side. All the components are placed between these two lines, called *rails* or *legs,* connecting the two power lines with what look like rungs of a ladder—thus the name, ladder logic diagram.

Ladder logic diagrams are universally used and understood in industry, whether in the process industry, in manufacturing, on the assembly line, or inside electrical appliances and products. The programmable controller (PC) of the next chapter, which first appeared in 1969, increased its chances of success because it capitalized on well-established concepts and practices. Thus, the ladder logic diagram was the language of choice for programming the PC.

Power connections for ladder logic diagrams can be either AC or DC, as shown in Figure 12.11. The transformer (Figure 12.11A) and battery (Figure 12.11B) are the means by which power is introduced into their respective circuits. These power connections usually are not shown in ladder logic diagrams, but are under-

stood to be there. The alphanumeric designations on the transformer symbol in Figure 12.11A are also located on the transformer wires, or on the transformer case next to the wires on larger transformers. The H's designate the primary connections; the X's are the secondary connections. The standard features of a ladder logic diagram that are shown include the following:

- Left-hand rails (rails 1 and 3 in Figure 12.11),
- Right-hand rail (rails 2 and 4 in Figure 12.11),
- Ladder rungs that show the device symbol wiring,
- Wired-in device symbols.

Notice that if a circuit is completed on any rung, current will flow between the rails on that rung. If there is no resistance or impedance on that rung, the power supply will be shorted. For example, in Figure 12.11 if the heating device were left out of rung 2, and the 1CR contact closed, the 24 V DC supply would be shorted. This condition should never occur, except maybe with inexperienced personnel.

There are a number of additional important points about ladder logic diagrams illustrated in Figure 12.11.

1. By convention, all of the devices that represent a resistive or inductive load to the circuit are shown on the right, and the devices that

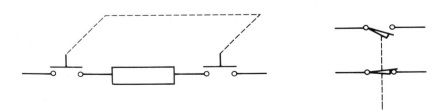

Figure 12.10 Symbols for Devices Acting Together

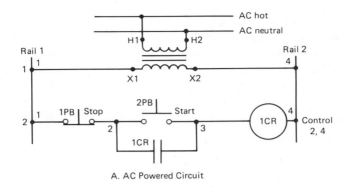

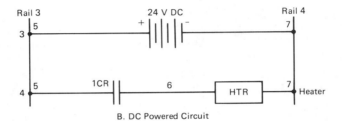

A. AC Powered Circuit

Figure 12.11   Basic Ladder Logic Diagrams

B. DC Powered Circuit

make or break electrical contact are shown on the left of the diagram.

2. The circuits connected in parallel, such as the 1CR contacts in parallel with the start push button 2PB, are sometimes referred to as *branches*. Some manufacturers, however, refer to each different horizontal line in the diagram as a rung. The rung concept will be used in this text because it makes it easier to identify and refer to each line in the diagram.

3. The numbers to the right of rails 1 and 3, to the left of rails 2 and 4, and at each point in the circuit where devices connect indicate *wiring labels*. These labels on the physical wires are numbered according to the diagram, and are used by the installation and maintenance personnel to help them trace wires and prevent wiring mistakes. Figure 12.12 shows an example of wiring labels.

4. The words to the right of the diagrams in Figure 12.11 indicate the major function or component in that rung. For example, rung 2 has a control relay in it and thus the word *con-*

Figure 12.12   Example of Wiring Labels

trol is to the right. The load in rung 4 is the *heater.*

5. Immediately under these words are another set of numbers that indicate the rung in which the contacts for that load device is located. Under the word *control* are the numbers "2, 4." These numbers show that the contacts for the control relay 1CR are located in rungs labeled 2 and 4. If these numbers were underlined, they would refer to NC contacts. Since these contacts are NO, the numbers are not underlined. The word *heater* at the end of rung 4 does not have numbers under it because the heater does not control other contacts. The practice of indicating the rung in which the contacts are located is very helpful in large system diagrams where the contacts may be spread throughout the diagram.

6. The 1CR contacts in parallel with the start switch in rung 2 illustrate a common industrial practice in ladder logic diagrams. The 1CR contacts are in the same rung as the control relay that activates these contacts. Thus, when the start button is pushed and the 1CR relay is energized (or pulled in or picked up), the 1CR contacts activate and keep the 1CR relay active even though the start button is released. This action *latches,* or *seals,* the relay on; therefore the circuit is called a latch, or seal, circuit.

The ladder logic diagrams of Figure 12.11 do nothing more than turn on a heater with no provision for controlling the heater automatically. This is an impractical system, but it does illustrate the elements in ladder logic diagrams. The following sections will show examples that are more practical.

Writing a Ladder Logic Diagram

There are as many ways to start writing ladder logic diagrams as there are manufacturers who sell systems that use these diagrams. In general

the following suggestions apply to writing ladder logic diagrams.

1. Look at the system to be analyzed. Draw a picture of the system, making sure all of the components of the system are present in the drawing. Accuracy is important. If anything is left out here, it will most likely be forgotten later also.

2. Determine the sequence of operations to be performed. Write these in sentences or put them in a table, whatever is best suited to the person making the diagram.

3. Write the ladder logic diagram from this sequence of desired operations.

4. Check the operation of the diagram in the following manner: Go through the logic process as if you were the machine, generally starting with the pressing of the start button. As each load is activated, write its designation on a piece of paper. When that device is deactivated, put a line through its designation. When the sequence of operations is completed and ready to start over, all the load designations should have been written down and crossed off. This procedure is sometimes called *hand running the logic.* If any loads were not used or were used but not crossed off, recheck the diagram and system to see if that operation was necessary or incomplete. If step 4 is successfully completed, an accurate ladder logic diagram should be generated.

Automatic Mixer Example.   These steps will be used to produce a ladder logic diagram for a simplified automatic mixing process. Step 1, a picture or diagram of the system, is shown in Figure 12.13A. For step 2, the sequence of operations, the following short description is given: The tank in Figure 12.13A is filled with a fluid, agitated for a specified length of time, and then emptied.

Another way of examining the order of operation of the process is illustrated in Figure 12.13B and is called a *state description* of the

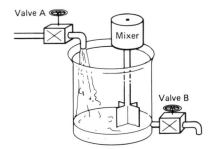

Valve A

Mixer

Valve B

A. Simplified Mixer Process

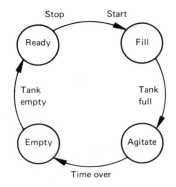

Stop   Start

Ready   Fill

Tank   Tank
empty   full

Empty   Agitate

Time over

B. State Description of Process

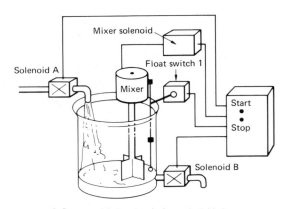

Mixer solenoid

Float switch 1

Solenoid A

Mixer

Start

Stop

Solenoid B

C. Process with Automatic Controls Added

Figure 12.13   Sequential Control Process

process. A state description is similar to a flow-chart in computer programming. It shows each action in a circle, and arrows indicate the order in which these actions occur.

Step 3, the ladder logic diagram, is shown in Figure 12.14. Now we will follow the series of events for the full control cycle of the automatic mixer process and check the logic (step 4) at the same time. Mark down each device when it is activated and cross it off when it is deactivated. Refer to Figures 12.13C and 12.14. The process is initiated by pressing the start push button 2PB. Activating the start button energizes the control relay 1CR located in the start/stop switch box of Figure 12.13C. (Write 1CR on a piece of paper or check sheet, since it was energized.)

When the control relay 1CR is energized, the 1CR relay contacts change state; in this case, since they are both NO, they close. When the 1CR contacts in rung 1 close, current is allowed to continue through the coil of the 1CR control relay, even though the start push button 2PB has been released. This circuit holds the 1CR relay in as long as the power line power is applied, the stop button 1PB is not pushed, and the timing relay 1TR has not timed out.

This latch circuit has another useful purpose. If the main power to the system is interrupted for any reason, the process cannot start up by itself. This safety feature is important to an operator, unaware that the main power is off, investigating the cause of a stoppage when the main power resumes. Without this safety feature, the operator could be injured.

When the 1CR contacts in rung 2 close, current can flow through solenoid A. Solenoid A is an electromechanical device that is electrically activated to open a valve mechanically, which allows fluid to flow into the tank. Fluid flows because the float switch 1FS in rung 2 is closed. This situation is indicated in Figure 12.13C by the empty position for the float switch. (Write solenoid A on the check sheet.)

When the tank has filled, the float switch 1FS changes to the filled position. This tank-full condition de-energizes solenoid A by opening 1FS in rung 2, starts the timer relay 1TR in rung 3 by closing 1FS in rung 3, and energizes the mixer relay MR. (On the check sheet, put a

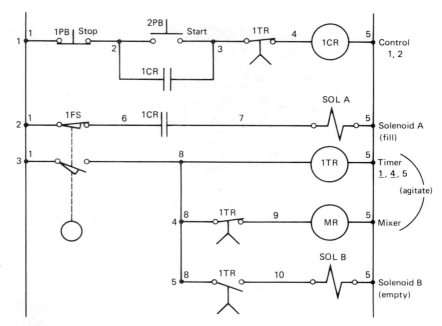

Figure 12.14 Ladder Logic Diagram for Control of Automatic Mixer of Figure 12.13C

line through solenoid A since it is de-energized, and add 1TR and MR to the sheet since they are energized.)

Notice here that each of the two limit switches does not operate the same for the same conditions. For example, 1FS in rung 2 opens when the tank is full, but does not close until the tank is empty, not just below the full level as the tank is emptying. Limit switch 1FS in rung 3 closes when the tank is full, but does not open until the tank is completely empty. In electronics this type of operation would be considered switching with hysteresis. If in practice this type of switch cannot be found, additional logic could be added to the ladder diagram to accomplish the desired result. However, a sump pump type of float switch will do the job.

After the timer has timed out, relay 1TR switches off the mixer in rung 4, energizes solenoid B in rung 5, which starts emptying the tank, and de-energizes the control relay 1CR in rung 1. (On the check sheet, put a line through MR and 1CR, and add solenoid B to the sheet.)

Because the control relay 1CR was de-energized by the timer, the 1CR contacts in rungs 1 and 2 also opened at this time. This action does not change anything on the check sheet, but does prevent solenoid A from turning on while the tank is being emptied.

When the tank is empty, float switch 1FS in rung 3 shuts off solenoid B, and resets the timer, placing the system in the ready position for the next manual start. (Now cross off solenoid B and 1TR on the check sheet. All devices have now been turned on and then turned off; thus step 4 has been successfully completed.)

The ladder logic diagram in Figure 12.14 is a basic configuration that will work properly even if the start button is pressed and held down. The 1TR contacts in rung 1 make sure that the control relay 1CR is not activated until the correct time in the cycle. Pressing the stop button will stop the fill process, and then later could be started again, but the stop button will not stop the process once the timer and mixer are started. The only way to emergency stop the

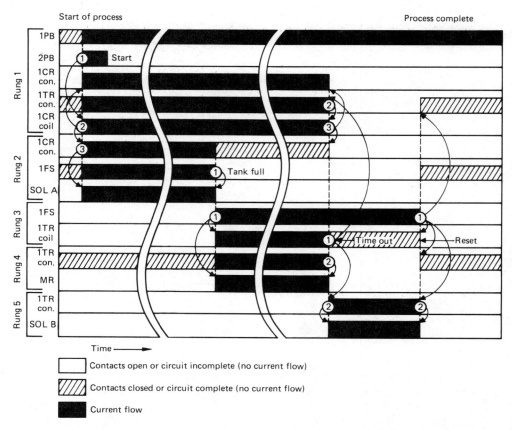

Figure 12.15   Timing Diagram for the Automatic Mixer Process

timer and mixer would be to throw the main power. The operating personnel must be made aware of this situation, or change the ladder logic diagram to accommodate this emergency condition.

Another helpful procedure that can be used to comprehend what is happening in the process is the timing diagram shown in Figure 12.15. This diagram shows along the left side all the ladder devices grouped by rung number, and shows along the bottom time as the independent variable. The start of the process is indicated at the top left, and completion of the process is shown at the top right of the diagram. The circled numbers in each row of the diagram indicate the order in which events happen. For

example, number 1 in the second row shows that pressing the start button 2PB is the initiating event that causes the 1CR coil to energize. This coil, in turn, causes the two 1CR contacts to change state, in this case close and conduct current. Closing the 1CR contacts in rung 2 causes solenoid A to energize, as shown in rows 6 and 8. The numbers in circles along with the arrows are called *cause-effect arrows*. In other words, the action produced where the circled number is located produces an action where the arrow points. It is not always easy to visualize these actions in the ladder diagram.

In general there are three states in which the ladder logic devices can exist. They are: contacts open or circuit not complete, contacts

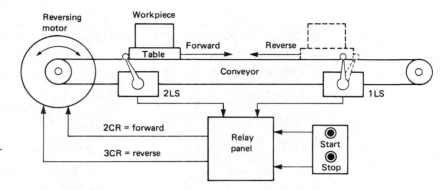

Figure 12.16 Reciprocating (Oscillatory) Motion Process

closed or circuit complete but no current flow, and current flow. These conditions are shown in the key for the timing diagram in Figure 12.15. Follow through the ladder logic operation again, but this time follow the events as they occur in the timing diagram. It is easy to see in the timing diagram and in the state description (Figure 12.13B) that there are only four places where device conditions may change state. This is not at all apparent in the ladder logic diagram.

Oscillatory Motion Example. Becoming proficient in writing ladder logic diagrams takes practice, so here is another example. Figure

12.16 shows an example of a reciprocating motion system. The workpiece starts on the left and moves to the right when the start button is pressed. When it reaches the rightmost limit, the drive motor reverses and brings the workpiece back to the leftmost position again, and the process repeats. This relatively uncomplicated process can be represented by the ladder logic diagram of Figure 12.17 and timing diagram of Figure 12.18.

The operation of the ladder logic is as follows (verification of the logic by means of the check sheet is left to the reader): While the process is stopped, limit switch 2LS is kept open by the table holding the workpiece. Limit

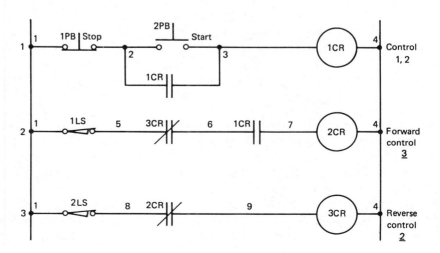

Figure 12.17 Ladder Logic Diagram for Reciprocating Motion Process

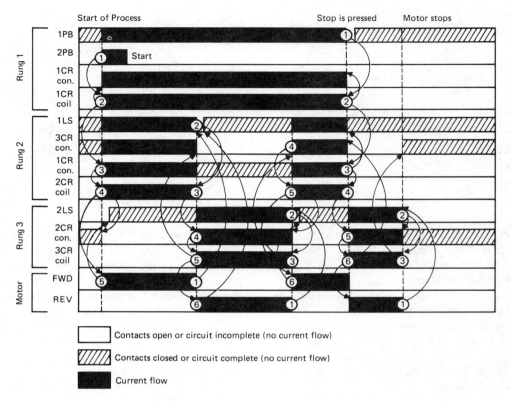

Figure 12.18   Timing Diagram for the Oscillatory Motion Process

switch 1LS is an NC switch and is in its normal position (closed). After the start button is pressed, the control relay 1CR is actuated, closing the two 1CR contacts in rungs 1 and 2. The 1CR contacts in rung 1 latch the 1CR relay on. Taking a look at rung 2, the 1LS switch is closed since the workpiece is not all the way to the right, the 3CR contacts are closed since the 3CR control relay in rung 3 is not energized, and the 1CR contacts in rung 2 are closed; thus the 2CR control relay in rung 2 is energized, opening the 2CR contacts in rung 3. In addition, the 2CR control relay activates the contacts to the belt motor (not shown) that causes the motor to turn in the forward direction. When the table leaves the 2LS switch, the switch closes in rung 3. As the belt moves the

workpiece along, the rightmost position is reached, causing switch 1LS to open. This breaks the power to the 2CR relay in rung 2, allowing the 2CR contacts in rung 3 to close. This now energizes the 3CR relay in rung 3, opens the 3CR contacts in rung 2, and activates the contacts to the belt motor (not shown) that causes the motor now to turn in the reverse direction. The workpiece now travels to the left, eventually opening switch 2LS, and the process repeats. It is left to the reader to determine that when the stop button is pressed, the workpiece will travel to the leftmost position and stop.

In Figure 12.16 all of the switch connections have arrowheads that are pointing to the relay panel, while the connections necessary to drive the motor are pointing out of the panel.

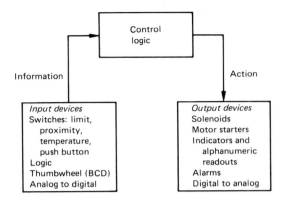

Figure 12.19   Input and Output Devices

The control relays are contained in the panel. This illustrates that in relay logic the switches are considered inputs, and the motor control signals are considered outputs. Figure 12.19 shows the categories that are considered input and output devices. Information coming from the input devices goes into the control logic that, in turn, drives the output devices.

Conveyor Belt Example.   One more example of writing a ladder logic diagram is given here. Try to write the ladder logic diagram from the information given; then check your answer. One possible solution to the ladder logic diagram is given at the end of this chapter in Figure 12.26.

This conveyor belt process example is illus-trated in Figure 12.20. In this process, 2LS is set at 0.9 in. and 3LS is set at 1.1 in. to accommo-date a part height between or equal to these val-ues. If the part trips 2LS but not 3LS, the part is greater than or equal to 0.9 in. and less than or equal to 1.1 in. tall. This is considered a "good" part. When this part trips 4LS, 5CR is actuated, dropping the part onto the good part conveyor.

If the part trips both 2LS and 3LS, or nei-ther, the part is too large or too small and is considered "bad." When either of these condi-tions occurs, the part travels down the con-veyor, not falling onto the good part conveyor when (or if) it trips 4LS, but falling into the bad part bin when it trips 5LS. Each time a part en-ters the bad part bin, a counter is incremented. When the bin is full (count complete), solenoid 3 (SOL 3) is energized, opening the bottom of the bin and emptying it. The counter will then reset automatically. If a part comes on the con-veyor, does not fall into either chute, and one minute elapses after 1LS is tripped, then an alarm sounds and the conveyor belt stops, in-dicating a malfunction.

Fail-Safe

The term *fail-safe* is used when considering the design of a system in which a failure (in a fail-safe system) would cause no or minimal dam-age and inconvenience. A perfect fail-safe de-

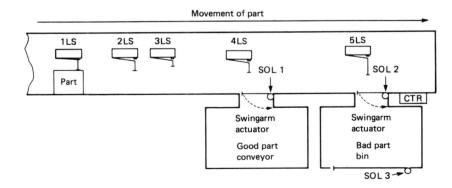

Figure 12.20   Conveyor Belt Process

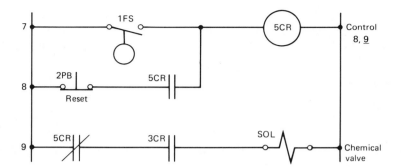

Figure 12.21 Partial Ladder Diagram for a Chemical Tank Filling System

vice would never fail for any reason. Of course this is not practical, but a low probability for failure is acceptable.

In general the best fail-safe circuit practice when using input devices such as start and stop buttons, limit switches, float switches, and pressure switches is to adhere to the following procedure: *Start a sequence by closing NO circuit contacts and stop a sequence by opening NC contacts.* The validity of this statement can be seen from examining the most common types of circuit failure:

• Dirty or oxidized contacts close but fail to conduct,
• A broken wire or loose connection opens the circuit,
• A different power source used in the input device is interrupted, causing the input device to go dead.

If one or more of these common circuit failures occur, either the circuitry wired with the fail-safe procedure would fail to start or, if started, the system would stop. These two failure consequences usually are safe, but each case must be analyzed to see what the consequences are. Use the type of contacts on the input device that minimize danger.

A less likely but possible mode of failure could occur if the contacts were to weld shut or short in some other way. If NO contacts on the input device were selected to protect from dirty contacts or an open circuit condition, then a failure could occur if the contacts welded closed. A way to overcome this difficulty would be to use two contacts in series in a redundant circuit, such that both contacts must work correctly to allow the sequence to operate. If the probability of one contact welding shut is one in two million, then the redundant circuit re-

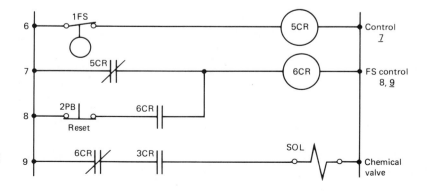

Figure 12.22 Increased Fail-Safe Capabilities for the Chemical Tank Filling System

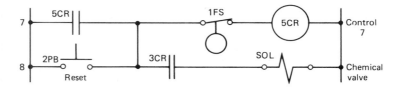

Figure 12.23 More Efficient Fail-Safe System

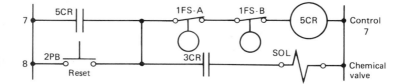

Figure 12.24 Redundant Fail-Safe System

sults in a probability of failure due to contacts welding of one in 4×10^{12}. The following examples will help clarify these concepts.

In a certain process a chemical starts to fill a tank when contacts 3CR close, as shown in Figure 12.21. The tank fills until float switch 1FS closes, which energizes 5CR, opening NC contacts 5CR, and closes the chemical valve. The system stays off until the reset button 2PB is pressed.

The problem with this circuit is that, although simple, it is not safe. If the contacts in the float switch failed or a wire broke, the system could not shut off and the chemical would overflow. To correct the fault, the float switch could use NC contacts instead of NO ones. An improved version of the circuit is shown in Figure 12.22.

An even more efficient circuit is illustrated in Figure 12.23. Here, the reset button is used to turn the control relay 5CR on, but 3CR is still used to turn on the solenoid. This circuit will protect against a loose or broken wire, bad contacts, and loss of power from the float switch, but would not be safe if the float switch contacts were to become welded shut. In that case a redundant circuit with two NC contacts on the float switch, as shown in Figure 12.24, would reduce the probability of failure of the system. *Note:* The line between 2PB and 1FS in Figures 12.23 and 12.24 connecting rungs 1 and 2 is called a *crossover.*

The problem in Figure 12.21 could have been solved using contacts of opposite type—NO and NC in a *parity* circuit, as shown in Figure 12.25. The redundancy solution in Figure

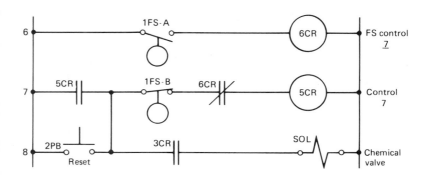

Figure 12.25 Fail-Safe System with Parity

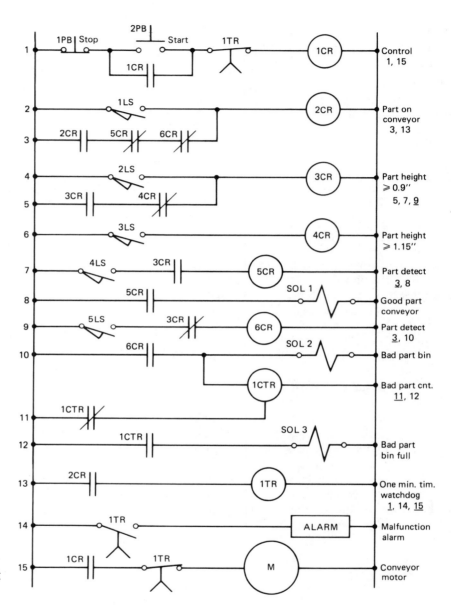

Figure 12.26 One Solution to the Conveyor Belt Problem

12.24 could have been defeated if both contacts had shorted simultaneously. The parity solution of Figure 12.25 could be defeated if the NC contacts were shorted and the NO contacts failed to close. The better solution depends upon which situation is less likely to occur.

CONCLUSION

The basis of the programmable controller's programming language has been, by popular demand, the ladder logic diagram with standard relay symbology. The reason for the ladder diagram's popularity is that plant personnel are very familiar with relay logic from their previous experience with sequential controls. Understanding how to generate a ladder logic diagram from a sequential process, the subject of this chapter, is the first step in being able to program most programmable controllers, the subject of the next chapter.

QUESTIONS

1. A programmable controller generally controls what type of process?
2. Define the following: sequential process, event-based sequential process, and time-based sequential process.
3. Batch control systems generally are used for what four types of applications?
4. List seven differences between batch processes and continuous processes.
5. Push-button switches generally are actuated by the _____, whereas limit switches are actuated by the _____.
6. List at least six general types of proximity switches.
7. In a ladder logic diagram the numbers to the left of the left rail give the _____ numbers, and the numbers to the right of the left rail indicate the _____ numbers.
8. Explain what the numbers to the right of the right rail in a ladder logic diagram indicate.
9. In a ladder logic diagram loads are shown on the _____ side, and devices that make or break electrical contact are shown on the _____ side.
10. What are the four steps in writing a ladder logic diagram for a process?
11. List five ladder logic input devices.
12. List five ladder logic output devices.
13. When designing a fail-safe circuit, a sequence is started by activating _____ contacts and stopped by activating _____ contacts.
14. Define the term fail-safe in your own words.

PROBLEMS

1. Draw a hierarchical flowchart using the following terms: (a) time-based sequential process, (b) sequential process, (c) event-based sequential process, (d) programmable sequence controller, (e) programmable controller, (f) batch process, and (g) programmable process controller. Start with sequential process at the top of the chart. Show how each process is broken down from that point, and which controller is used with which process.
2. A life science laboratory technician wishes to automate the feeding of laboratory animals somewhat by automatically measuring the feed going onto a scale. After the first batch of feed is weighed automatically, the technician will remove the feed from the scale and distribute it while the next batch of feed is being automatically weighed. Write an accurate ladder logic diagram for the process described, taking into consideration the following constraints:
(1) The process must have a start and a stop

button that can start or stop the process at any point and resume the process where it left off.

(2) When the process is started, food is augered onto the scale at regular speed. As the scale gets close to balance, the auger speed changes to half speed (or trickle). When the scale is almost balanced, the auger stops and turns on only for one second in every ten or so seconds until the scale is balanced. (This is called *jogging* the motor.)

(3) If no jogging occurs for one minute, a light comes on to indicate the scale is in balance.

(4) The technician locks the scale, removes the feed, and gently releases the scale to start the next batch.

CHAPTER
13

PROGRAMMABLE CONTROLLERS

OBJECTIVES

On completion of this chapter, you should be able to:

- State the historical definition for the programmable controller (PC);
- State the parts and functional operation of a PC;
- Determine the important factors to consider when selecting a PC for an application;
- Identify the approximate size of a PC;
- Identify the symbols for a PC ladder diagram;
- Write a PC ladder diagram.

INTRODUCTION

In the 1960s computers were considered by many in industry as the ultimate way to achieve increased efficiency, reliability, productivity, and automation of industrial processes. Computers possessed the ability to acquire and analyze data at extremely high speeds, make decisions, and then disseminate information back to the control process. However, there were disadvantages associated with computer control, such as high cost, complexity of programs, hesitancy on the part of industry personnel to rely on a machine, and lack of personnel trained in computer technology. Thus, computer applications throughout the 1960s were mostly in the area of data collection, on-line monitoring, and open-loop advisory control.

During the mid-1960s, however, a new concept in electronic controllers evolved: the *programmable controllers,* PCs. (Note that the abbreviation PC will be used in this text to denote the programmable controller and not the personal computer, which also is popularly represented by the abbreviation.) The concept of the PC developed from a mix of solid-state computer technology and traditional sequential controllers, such as the stepping drum (a mechanical rotating switching device) and the solid-state programmer with plug-in modules. The PC first came about as a result of problems faced by the auto industry, which had to scrap costly assembly line controls each time a new model went into production. The first PCs were installed in 1969 as electronic replacements of electromechanical relay controls. The PC presented the best compromise of existing relay ladder logic techniques and expanding solid-state technology. It increased the efficiency of the auto industry's manufacturing system by eliminating the costly job of rewiring relay controls used in the assembly line process. The PC reduced the changeover downtime, increased flexibility, and considerably reduced the space requirements formerly used by the relay controls.

Since the introduction of the PC into the *manufacturing* industry in 1969, the PC has become widely used in the *process* industry as well. A process industry generally performs the functions and operations necessary to change a material either physically or chemically. In this chapter we first will discuss the basic concepts of a PC. Then we will describe a specific PC, the Allen-Bradley Bulletin 1772 PLC-2. We will use this PC to learn the specifics of programming.

DEFINING A PC

In 1978 the National Electrical Manufacturers Association (NEMA) released a standard for PCs. This standard, NEMA Standard ICS3-1978, was the result of four years of work by a committee made up of representatives from PC manufacturers. Part ICS3-304 of the standard defines a PC as "a digitally operating electronic apparatus which uses a programmable memory for the internal storage of instructions for implementing specific functions such as logic, sequencing, timing, counting, and arithmetic to control, through digital or analog input/output modules, various types of machines or processes. A digital computer which is used to perform the functions of a programmable controller is considered to be within this scope. Excluded are drum and similar mechanical type sequencing controllers."

The official NEMA definition allows almost any computer-based controller to be considered a PC, including single-board computers, numerical controllers, and programmable process controllers or sequential controllers. Although PCs can be used to perform the jobs of numerical and sequential controllers, numerical and sequential controllers typically cannot perform the event-based control functions discussed in the previous chapter. Because of this official definition confusion, we will use the historical, or *de facto,* definition for the PC that is developed in the following text.

Regardless of size, cost, or complexity, all PCs share the same basic parts and functional characteristics, as shown in Figure 13.1. A PC will always consist of input and output interfaces, memory, a processor, a programming language and device, a power supply, and housings.

Functionally a PC examines the status of input interfaces and, in response, controls something through output interfaces. Several logic combinations or instructions usually are needed to carry out a control plan, or *program* as it is commonly called. This control plan is stored in memory using a programming device. All logic combinations stored in memory are

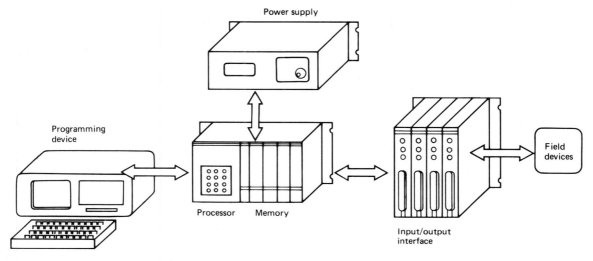

Figure 13.1   Basic Programmable Controller

periodically evaluated by the processor in a predetermined order. The period of time required to evaluate the status of input devices, output devices, and the control plan is called a *scan*. The input and output devices, such as switches, motors, lights, and so on, that attach to the input/output (I/O) interfaces are called *field devices*. During a scan all the inputs are examined, the control plan is evaluated, and outputs are updated.

PCs employ combinational logic, and frequently are programmed using (but not restricted to) relay ladder logic. Conventional PC inputs and outputs are made at industrial line voltages, for example, 120 V AC, 240 V AC, 24 V DC, and 5 V DC. PCs are designed to operate in industrial environments that are dirty, are electrically noisy, have a wide fluctuation in temperatures (0°–60°C), and have relative humidities of from 0% to 95% (noncondensing). Air conditioning, which generally is required for computers, is not required for PCs. PCs can be maintained by the plant electrician or technician with minimal training. Most of the

maintenance is done by replacing modules rather than components. Often the PC has a diagnostic program that assists the technician in locating bad modules. Computers, on the other hand, require highly trained electronics specialists to maintain and troubleshoot them.

In summary, the historical definition for the PC is basically a special-purpose computer whose special features set it apart from other controllers. A PC consists of I/O interfaces, memory, a processor, a programming device, and a power supply. The PC inputs and outputs operate at one of the industrial line voltages of 240 V AC, 120 V AC, 24 V DC, or 5 V DC. The PC also is designed to operate in the harsh industrial workplace and be maintained by plant personnel. Most PCs have additional capabilities, but these are the qualities that set it apart from other controllers.

Appendix C provides a listing of most of the PCs that are available today, together with some of their more important specifications. Of these, manufacturers caution that scan rate data is, at most, an estimate of typical performance. For any given program the scan rate will vary depending on the types of actions called for by the program.

We will discuss each basic part of a PC in the following sections.

INPUT/OUTPUT INTERFACES

Input interfaces, which are modular because they can be plugged into and unplugged from the system, accept signals from the machine or process devices (115 V AC) and convert them into signals (5 V DC) that can be used by the controller. *Output interfaces* or modules convert controller signals (5 V DC) into external signals (115 V AC) used to control the machine or process. Other voltage arrangements may be used, but these are the most common. There are usually 1, 2, 4, 8, or 16 circuits on an I/O interface.

Typically input and output interfaces are housed in a rigid card rack that can be mounted in an enclosure, either next to or thousands of feet away from the processor. In addition there is some means of connecting field wiring on the I/O interface or I/O housing. Connecting the field wiring to the I/O housing allows easier disconnection and reconnection of the wiring in order to change modules. Lights also are added to interface modules to indicate the ON or OFF status of each I/O circuit. Each circuit usually is isolated and fused. Some I/O interface modules have blown fuse indicators. The following are common I/O interface modules:

- AC voltage inputs and outputs,
- DC voltage inputs and outputs,
- Pulse inputs,
- BCD (binary-coded decimal) inputs and outputs,
- Low-level analog inputs (such as thermocouples),
- High-level analog inputs and outputs (in the 4–20 mA range),
- Special-purpose interfaces.

MEMORY

Memory is where the control plan or program is held or stored in the PC. The information stored in the memory relates to how the input and output data should be processed.

The complexity of the control plan determines the amount of memory required. Most PC memories are expandable in fixed increments. Memory elements store individual pieces of information called *bits* (for *bi*nary dig*its*). These memory modules are mounted on *printed circuit* boards (another use for the abbreviation PC). Memory is specified in thousand or "K" increments, where 1K is 1024 bytes of memory storage (a *byte* is 8 bits). PC memory capacity may vary from less than 100 bits to over 256K bytes, depending on the manufacturer.

Memory is categorized as either read-only memory (ROM) or random access memory (RAM). Both terms are somewhat misleading. ROM has to be written into at some time in order to enter the program, so it should probably be termed "read-mostly" memory. And both RAM and ROM can be accessed randomly, but RAM is the historical term that indicates memory that can be altered easily. Some ROM may be altered physically by ultraviolet light or electronically to change its contents. The most popular ROM used in PCs today is UV PROM (ultraviolet programmable read-only memory). These UV PROM ICs have a clear quartz window in the middle of the device for erasing the memory contents. They then may be reprogrammed. In the past the most common RAM used in PCs was magnetic core. The main advantage of core memory was that it could be easily altered, convenient for changing setpoints and so on that must be changed often, and yet was nonvolatile (did not lose its memory when power was lost). Today most PCs are using MOS IC memory that is more compact and less expensive. However, since this memory loses its stored data when power is removed, a backup battery supply is required to retain the memory contents.

PROCESSOR

The *processor,* sometimes called the *central processing unit* (CPU), is the heart of the PC and organizes all controller activity. The CPU causes the control plan logic stored in memory to be evaluated, along with the status of the inputs, and issues a specified command to the appropriate output.

A number or a combination of letters and numbers, such as I 150801, is used to code data locations referred to as *addresses* (locations in memory). The user selects the I/O address by assigning a specific input or output interface circuit to a specific field device. All programming of the field device is referenced to the number assigned.

In addition to straight logic processing, the processor may perform other functions such as timing, counting, latching, comparing, and retentive storage. It also can emulate stepping switches and shift registers. These additional processor functions may be either special hardware units that are part of the PC or software programs integrated into the memory.

In recent years many PC manufacturers have started using a microprocessor as the CPU, thus decreasing the PC's size and increasing its decision-making capabilities. Some controllers have the ability to perform complicated math well beyond the basic four functions of add, subtract, multiply, and divide.

PROGRAMMING LANGUAGE AND PROGRAMMING DEVICE

The *programming language* allows the user to communicate with the PC via a programming device. PC manufacturers use slightly different programming languages, but all languages are designed to convey to the PC, by means of instructions, how to carry out the control plan.

Figure 13.2 illustrates some of the more common PC programming languages available. The PC ladder diagram (Figure 13.2B), based on the relay ladder diagram, (Figure 13.2A), is by far the most common. Boolean statements (Figure 13.2C) relate logical inputs such as AND, OR, and INVERT to a single statement output, in this case solenoid A (SOL A). Another type of PC language is the code or mnemonic (pronounced nee-*mon*-ic) language, shown in Figure 13.2D. This language is very similar to computer assembly language.

The *programming device* is used to load the program into the PC memory. The most common programming device is the CRT (cathode ray tube) terminal, which provides a full alphanumeric and/or special function keyboard.

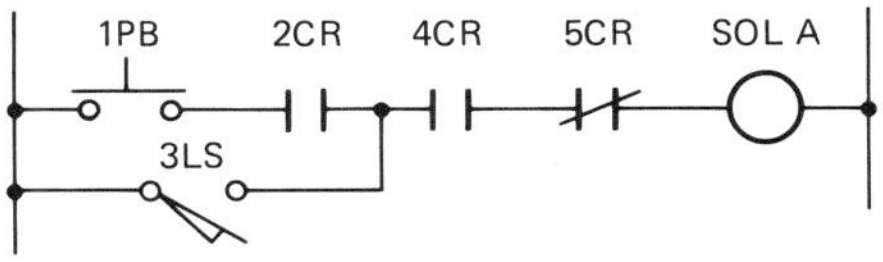

A. Relay Ladder Diagram

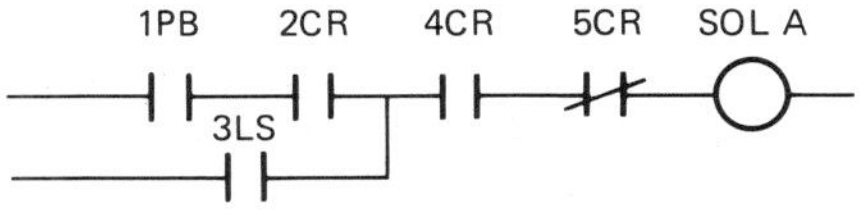

B. Free-Format-Equivalent PC Diagram

$$[(1PB \cdot 2CR) + 3LS] \cdot 4CR \cdot \overline{5CR} = SOL\ A$$

C. Boolean Statement

```
LOAD    1PB
AND     2CR
OR      3LS
AND     4CR
CAND    5CR
STORE   SOLA
```

D. Code or Mnemonic Language

Figure 13.2 Comparison of Programmable Languages Used with Various PCs

The CRT allows the user to see the ladder diagram or coded language as interpreted by the PC.

At the other end of the programming device spectrum is the small manual programmer. (All programming is manual, but this type of programming device has been referred to traditionally as a manual programmer.) In this case data and instructions are entered through thumbwheels and special function buttons or keypads. Usually only one logic statement can be entered or monitored at one time. Manual programmers are less expensive and more portable than CRT programming devices.

After program entry has been completed, the programming device continues to be used as a diagnostic tool. Even the manual programmers have the capability of interrogating the PC to determine I/O, memory, and CPU status. With the CRT terminal, "living pictures" of the PC's operation make troubleshooting easier and faster.

POWER SUPPLY

The PC power supply may be integrated with the CPU, memory, and I/Os into a single housing, or may be a separate unit connected to the main housing through a cable. As a system expands to include more I/O modules or special function modules, most PCs require an auxiliary power supply to meet the power demand. Power supplies are also the first line of defense against electrical noise generated over the power lines.

HOUSINGS

One of the most popular PC features is its modularity. Modularity makes repairs easier and reduces downtime. Most major PC components are mounted on printed circuit boards that can be inserted into a housing or card rack. One or more housings make up a PC system.

Housings may contain the CPU, memory, I/O modules, special interface modules, and a power supply, or in some cases just I/O modules. Housings may be rack-mounted in a control console or mounted to a subpanel in an enclosure. Most housings are designed to protect the PC control circuits from dirt, dust, electrical noise, and vibration.

PC SIZE

There is much variation in size identification of PCs, but they can be roughly divided into three sizes: small, medium, and large. The small size category covers units with up to 128 I/Os and memories up to 2K bytes. These PCs are capable of providing versatility and sophistication ranging from simple to advanced levels of machine control.

Medium-sized PCs generally have from 256 to 512 I/O modules and memories ranging from 4K to 7K bytes. Intelligent I/O cards make medium PCs adaptable to temperature, pressure, flow, weight, position, or any type of analog function commonly encountered in process control applications.

Large PCs, of course, are the most sophisticated units of the PC family. In general large units have 1024 to 4096 I/Os and memories from 8K to 192K bytes. This size of PC has virtually unlimited applications. Large PCs can control individual production processes or entire plants. Modular design permits systems to be expanded to control thousands of analog and digital points.

SELECTING A PC

The key factor in selecting a PC is establishing exactly what the unit is supposed to do. Current designs cover a broad range of sizes and capabilities. At the small end, PCs are primarily used as relay replacements to provide standard relay logic, timing, counting, and shift register functions. At the large end, analog I/O capability has made it possible for the units to become an integral part of process control systems.

Probably the most important step in selecting the proper PC system is determining what the I/O requirements are, including types, location, and quantity. If the application involves the replacement of relays, the user can determine I/O needs quickly. Establishing analog I/O needs is much more complicated and may require expert assistance. Other requirements to be evaluated include memory type and capacity, programming procedures, and peripheral equipment needs. Normally a 10%–20% spare or expansion capability should be allowed for each application.

Expense of the PC is another factor to consider. Determining the cost of a PC system is not easy, however. Many intangibles must be taken into consideration. System requirements dictate costs to a certain degree, but the true cost also depends on the value of increased production, improved quality, greater flexibility, and reduced downtime. Installation, operation,

and maintenance costs are important economic factors to consider. Servicing the PC can be expensive. However, if in-house repair capability is available, servicing costs may be reduced.

In general it is not advisable to buy a PC system that is larger than current needs dictate. All phases of the project must be considered, however, and future conditions must be accurately anticipated to ensure that the system is the proper size to fill the current, and possibly future, requirements of an application.

EXAMPLE OF A SMALL SYSTEM

Our main purpose in this chapter is to acquaint you with a PC and how it can be used. In order to do that, an example had to be chosen. The Allen-Bradley PLC-2 was chosen as a representative small PC system that has most of the capabilities desired and is one of the easier ones to learn and use.

The PC shown in Figure 13.3A is the Bulletin 1772 Mini-Processor PLC-2 Programma-

A. Bulletin 1772 Mini-Processor PLC-2 Programmable Controller

B. Bulletin 1770 Industrial Terminal System

Figure 13.3 Allen-Bradley's Programmable Controller and Terminal

ble Controller. It has a small space require-ment, powerful capabilities, compact design, and relatively low cost. The PC has a Zilog Z80 microprocessor unit (MPU) and an Intel 8251 interface IC. The left third of the device is the power supply. The power connections are made to the front of the panel. Notice the battery holder on the bottom right part of the power supply for memory backup. The center part of the device is the processor and memory. The connector on the front of the processor is how the terminal (Figure 13.3B) is attached in order to load the program. A key switch on the pro-cessor selects the program, test, or run modes. The right half of the unit is devoted to the I/O modules. This unit can have any combination of four modules attached to it. The screws on the bottom part of the modules are where the field wiring connections are made. The top part of each module contains the LEDs to show when an input or output device is activated. The controller allows interfacing with a maxi-mum of 128 I/O devices, although as few as 8 I/Os may be used. The chassis is self-contained and requires about a ninth of a cubic meter (3 ft^3) of installation space (assuming a 32 I/O configuration). If the application later demands a more powerful processor, hardware expan-sion is available.

The programming terminal shown in Fig-ure 13.3B is the Bulletin 1770 Industrial Ter-minal System (catalog number 1770-TA). The terminal is used to enter the programs, test them, run them, and then troubleshoot them if necessary. Figure 13.4 shows the 1770 terminal as a portable device. It weighs less than 16 kg (35 lb) and is rugged enough to adapt to most industrial environments.

Figure 13.5 shows the controls on the ter-minal, and the sealed touchpad keyboard. The keyboard overlays, one of which is being placed on the keyboard in Figure 13.5D, are illustrated in more detail in Figure 13.6.

Portions of the *PLC-2 Programming and*

A. Portability

B. Operation in Any Position

Figure 13.4   Allen-Bradley 1770 Industrial Terminal

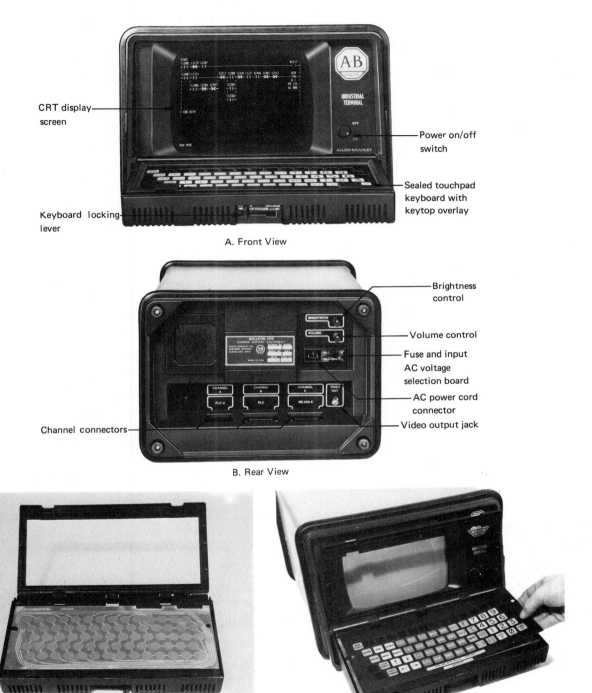

CRT display screen

Power on/off switch

Keyboard locking lever

Sealed touchpad keyboard with keytop overlay

A. Front View

Brightness control

Volume control

Fuse and input AC voltage selection board

AC power cord connector

Video output jack

Channel connectors

B. Rear View

C. Touchpad Keyboard

D. Installing Keytop Overlay

Figure 13.5 Close-Ups of Allen-Bradley 1770 Industrial Terminal

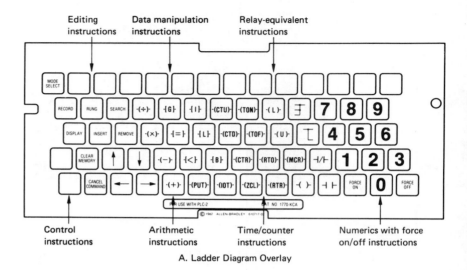

Editing instructions

Data manipulation instructions

Relay-equivalent instructions

Control instructions

Arithmetic instructions

Time/counter instructions

Numerics with force on/off instructions

A. Ladder Diagram Overlay

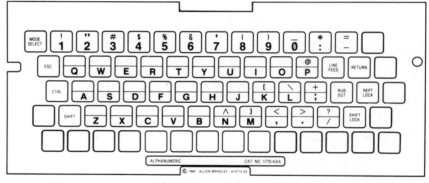

Figure 13.6 Keytop Overlays for Allen-Bradley 1770 Terminal

B. Alphanumeric Overlay

Operations Manual are provided in Appendix D. You will need to refer to it often in order to understand better the following discussion.

PC Ladder Instructions

The instructions for the PLC-2 are representative of and similar to the PC logic used in other PCs. Notice that there are some possibly unexpected differences between PC logic and relay ladder logic. A small group of instructions and an explanation of what they mean will be given first. Then, examples and tips will be provided to show you how to use those instructions.

Basic Instructions. The three basic instructions for the PLC-2 are shown in Figure 13.7. Even though these are symbols, they are also programming *instructions*. As you type these symbols on the CRT, you are programming the PC. Procedures for operating the CRT are given in Appendix D.

Program Panel Key	Display CRT Symbol	Description
┤├	┤ ├	*Examine on:* Examine an input for a closed condition, or examine an output for an energized state.
┤╱├	┤ ╱ ├	*Examine off:* Examine an input for an open condition, or examine an output for a de-energized state.
─○─	─( )─	*Energize:* Energize the output if its input conditions are TRUE.

Figure 13.7 Basic Instructions

Figure 13.8 shows a single relay ladder rung and its equivalent PC rung. The limit switch LS1 is replaced with the examine-on symbol, and the relay coil MS1 is replaced with the energize symbol. As you program each rung, you should ask yourself the following three questions:

1. What is the output? What you decide to be the output will go on the right side of the rung.
2. What inputs will control this output? These will go on the left side of the rung.
3. What state must these inputs be in order to control the output? That will determine whether you use the examine on, examine off, or some combination of these to control the output.

The relay logic in Figure 13.9 shows NO and NC limit switches in series, and both are represented by the examine-on symbol. Do not be confused by the normal state of an input (NO or NC); the normal state does not matter to the controller. Be concerned with the controlling state of an input. Must the contacts be open or closed to energize the output? If contacts need to close to energize the output, then use the examine-on instruction. Since both LS1 and LS2 in Figure 13.9 must be closed to energize CR1, use the examine-on instruction for each. If the contacts need to open to energize the output, then use the examine-off instruction. It is important to remember that the PC logic rung displayed on the CRT is not an electrical circuit; it is a *logic* circuit. What is seen on the display is a *line of computer instruction.*

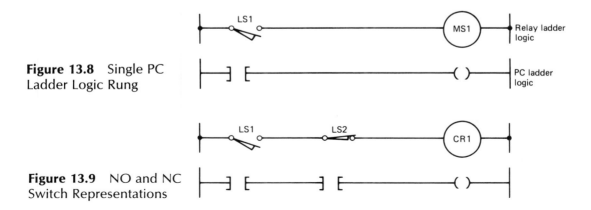

Figure 13.8 Single PC Ladder Logic Rung

Figure 13.9 NO and NC Switch Representations

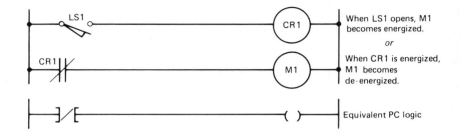

Figure 13.10 Examine-Off Representation

Then when is an examine-off instruction used if not for NC contacts? Figure 13.10 shows an example. Here LS1 must be opened in order to turn off M1. The examine-off instruction is used to represent this situation. Notice that this is a different concept than what the NC contacts of the relay logic represent. Also, a single rung in PC logic represents two rungs in relay logic in this example.

There are some situations in relay logic, however, that require more lines in PC logic. Figure 13.11 shows relay logic with two outputs on one rung. This representation uses three rungs in PC logic.

Memory Structure

As we have seen, each symbol in the PC diagram represents an instruction. In addition each instruction has reserved for it a location in memory that contains the status bit of the instruction. The status bit is shown by a binary 1 or 0, the 1 representing a TRUE condition and the 0 representing a FALSE condition. When the input instructions on a rung are logically TRUE (or the bits in memory for those input instructions are all 1s because its input devices have electrical continuity), logic continuity is established. This causes the output instruction to be TRUE and the output device to be turned on.

Each instruction status bit location in memory has a bit address represented by a five-digit octal number. The *octal* number system uses the digits 0 through 7 to represent all number combinations. (The decimal system uses the digits 0 through 9.) The numbers 0 through 7 are the same in octal as they are in decimal.

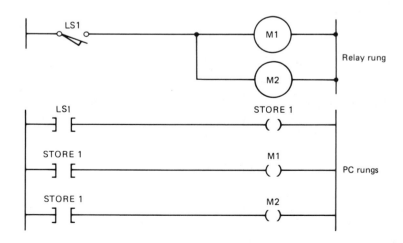

Figure 13.11 Only One PC Rung per Output

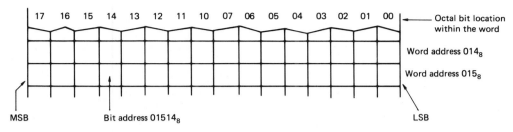

Figure 13.12 Memory Word and Bit Address Structure

However, the decimal number 8 is 10 in octal, the decimal 9 is octal 11, and so on. (A subscript of 8 beside an octal number distinguishes it from a decimal number.) In Figure 13.12 instruction status bit locations are shown in octal within each 16-bit *word* in memory. A word in this case is determined by the number of lines or connections on the data bus in the microcomputer system. This microcomputer has a 16-bit data bus. The word size for various PCs is shown in Appendix C by the column labeled "Number of Bits." There are six different word sizes for PCs represented in this table: 1, 4, 8, 12, 16, and 24. An 8- or 16-bit word size is the most common. A specific bit location can be identified by combining the three-digit word address and the two-digit bit number to form the five-digit bit address. Figure 13.12 shows the bit address of 01514_8 identified. Each input and output device is associated with a bit address that is displayed next to the device in the PC logic diagram, as shown in Figure 13.13.

Memory Organization

The organization of memory or memory map can now be shown. A *memory map,* as shown in Figure 13.14, is a picture of the memory space that shows the memory size and for what, if anything, each part of the memory is reserved. Every PC has a memory map, but it may not be like the one illustrated.

Figure 13.14 shows the memory space divided into two major areas: the data table and the user program area. The size of each area can be varied within limits to suit user needs, but the total cannot exceed the processor memory size. The data table stores the information needed in the execution of the user program, such as the status of input and output devices, timer/counter preset and accumulated values, data storage, and so on. Any instruction in the user program can address any word or bit in the data table except in the processor work area. The user program is the logic that controls the

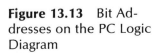

Figure 13.13 Bit Addresses on the PC Logic Diagram

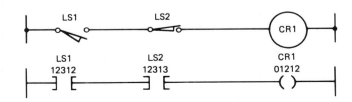

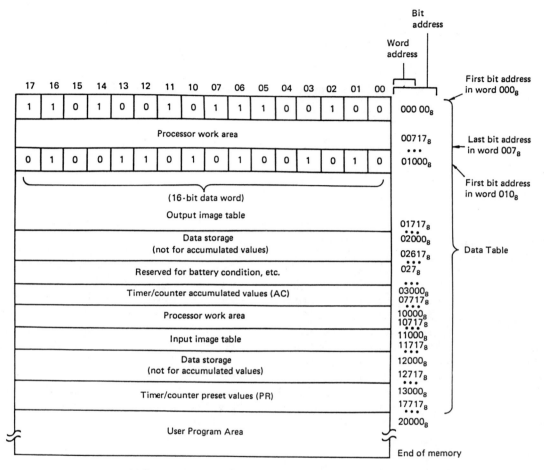

Figure 13.14   Memory Map of PC Memory

machine operation. The logic consists of instructions that are programmed in the PC ladder logic format. Each instruction requires one word of memory.

Hardware-to-Memory Interface

The processor monitors input conditions and controls output devices according to a user-entered program. The interface between the hardware and the program occurs in the I/O image table (Figure 13.14). The purpose of the input image table is to duplicate the status of the input devices wired to input module terminals. If an input device is electrically closed, its corresponding input image table bit is a 1. If an input device is open, its corresponding input image table bit is a 0. The input image table bits are *monitored* in conjunction with the user pro-

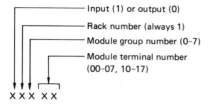

A. Generic Structure

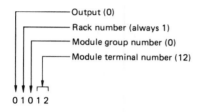

B. Specific Hardware Location

Figure 13.15 Hardware Address Structure

gram. The output image table contains the status of the output devices wired to the output module terminals. The output image table bits are *controlled* by the user program.

The instruction addresses in the I/O image table have a dual role: The five-digit bit address references both an I/O image table address and a hardware location, as shown in Figure 13.15. The most significant digit indicates an input or output device; the example in Figure 13.15B is for an output device. Referring to Figure 13.14, you can see why this is true. The output image table starts at address 01000_8 and ends at 01717_8. The most significant digit in these address locations is a 0. The most significant digit in the input image table memory area is a 1. Each five-digit bit address in the I/O image table directly relates to an I/O module terminal, as shown in Figure 13.16.

Figure 13.17 illustrates the hardware-to-memory interface. When an input device connected to terminal 11212_8 is closed, the input

module circuitry senses a voltage. The TRUE condition is entered into the input image table bit 11212_8. During the program execution or scan the processor examines bit 11212_8 for a TRUE condition. If the bit is TRUE (in this case it is), the examine-on instruction is logically TRUE. The rung is TRUE because bit 11212_8 is TRUE. The processor then sets output image table bit 01306_8 to TRUE. The processor turns on terminal 01306_8 during the next I/O scan, and the output device wired to this terminal becomes energized. This process is repeated as long as the processor is in the run mode. If the input device were to open, a 0 would be placed in the input image table, causing the output image table to go to 0 and, in turn, causing the output device to turn off.

Advanced Instructions

Some of the instructions entered are used to represent the *external* input and output devices connected to the I/O modules of the controller as previously discussed. Other instructions are *internal* and are used to establish the exact conditions under which the processor will energize or de-energize output devices in response to the status of input devices. An example of this is the energize instruction. Not only can this instruction be used to represent a physical output device, but it also can be used to turn on (set to 1) a storage bit for later use in the program. Figure 13.11 illustrates this point. In the figure the energize symbol is labeled STORE 1 and does not represent a physical output device. Instead, it is used to control two nonphysical input devices also labeled STORE 1. When the STORE 1 energize symbol becomes TRUE, the STORE 1 examine-on symbols also become TRUE.

There are two output instructions that are termed retentive: output latch and output unlatch. The term *retentive* means that if the rung condition becomes TRUE and then becomes

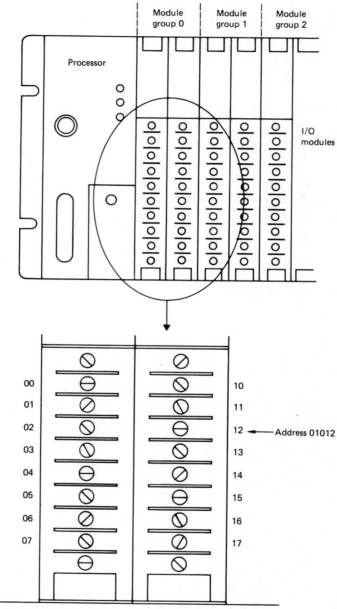

Figure 13.16 Address Bit Relationship to Hardware Location

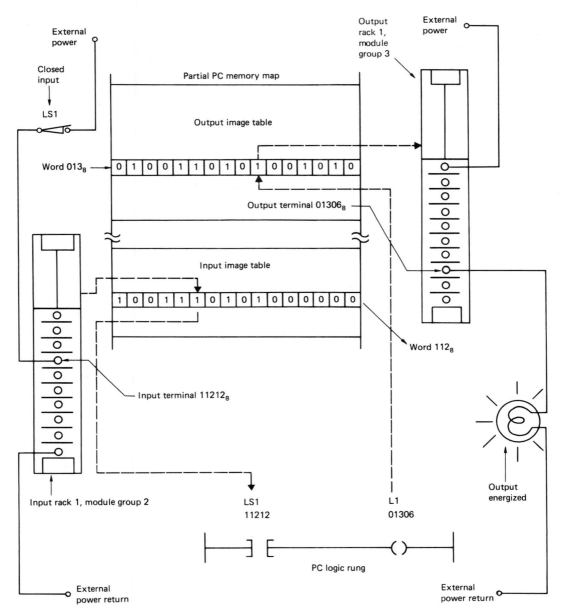

Figure 13.17   Hardware-to-Memory Interface

Keyboard Symbol	Description	
—(L)—	*Output latch:*	When the rung is TRUE, the addressed memory bit is latched on and remains on until it is unlatched.
—(U)—	*Output unlatch:*	When the rung is TRUE, the addressed bit is unlatched.

Figure 13.18   Output Latch and Unlatch Instructions

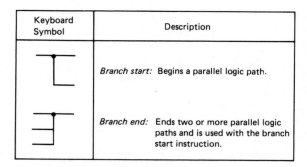

Keyboard Symbol	Description	
	Branch start:	Begins a parallel logic path.
	Branch end:	Ends two or more parallel logic paths and is used with the branch start instruction.

Figure 13.19   Branch Instructions

Keyboard Symbol	Display	Description	
—(TON)—	XXX —(TON)— TB PR YYY AC ZZZ	*Timer on delay:*	When the rung is TRUE, the timer begins to increment the accumulated value (AC) at a rate specified by the time base (TB). When the rung is FALSE, the timer resets the AC value to 000.
—(TOF)—	XXX —(TOF)— TB PR YYY AC ZZZ	*Timer off delay:*	When the rung is FALSE, the timer begins to increment the AC value. When the rung is TRUE, the timer resets the AC value to 000.
—(RTO)—	XXX —(RTO)— TB PR YYY AC ZZZ	*Retentive timer:*	When the rung is TRUE, the timer begins to increment the AC value. When the rung is FALSE, the AC value is retained. RTO is reset only by the RTR instruction.
—(RTR)—	XXX —(RTR)— TB PR YYY AC ZZZ	*Retentive timer reset:*	When the rung is TRUE, the AC value and status bit are reset to zero. XXX is the word address of the retentive timer that RTR is resetting.

Figure 13.20   Timer Instructions

Note: The timer word address, XXX, is assigned to the timer AC area of the data table. The time base, TB, is user-selectable and can be 1.0, 0.1, or 0.01 seconds. Preset (PR) values YYY and AC values ZZZ can vary from 000 to 999 (in decimal).

FALSE, the output will remain TRUE (or retain its condition) until reset. These instructions are given in Figure 13.18.

The output-latch instruction tells the processor to set an addressed memory bit on when the rung condition is TRUE. Once the rung condition goes FALSE, the latched bit remains on until reset by an output unlatch instruction. For this reason, these two instructions usually are used as a pair for any bit address they control.

The branch instructions, shown in Figure 13.19, allow more than one combination of input conditions to energize an output device. The branch-start instruction begins each parallel logic branch of a rung, whereas the branch-end instruction completes a set of parallel branches.

Timer and counter instructions are output instructions internal to the processor. They provide many of the capabilities available with timing relays and solid-state timing/counting devices. These instructions usually are conditioned by examine instructions and are used to keep track of timed intervals or counted events according to the logic continuity of the rung.

Each timer or counter instruction has two three-digit values associated with it, and thus requires two words of data table memory. These three-digit values are as follows:

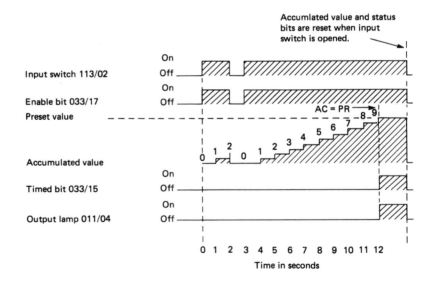

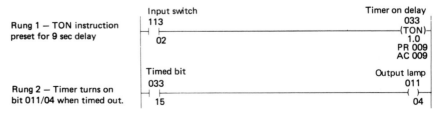

Figure 13.21 Timer On Delay

Rung 1 — TON instruction preset for 9 sec delay

Rung 2 — Timer turns on bit 011/04 when timed out.

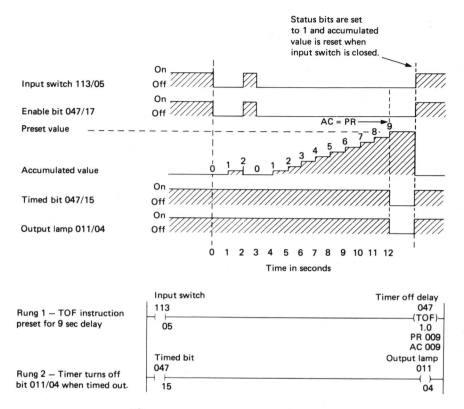

Figure 13.22   Timer Off Delay

1. The accumulated (AC) value is stored in the AC value area of the data table. For timers this is the number of timed intervals that have elapsed. For counters this is the number of events that have been counted.

2. The preset (PR) value is stored in a PR value area of the data table, always 100_8 words greater than its corresponding AC value. This value is entered into memory. The PR value is the number of timed intervals or events to be counted. When the AC value equals the PR value, a status bit is set to 1 and can be examined to turn on or off an output device.

The AC and PR values are stored in the data table in three-digit BCD (binary-coded

decimal) format and can range from 000 to 999. These numbers will be displayed in decimal while the bit addresses are displayed in octal.

A timer counts elapsed time-base intervals (1.0, 0.1, or 0.01 second) and stores this count in its AC value word. Four timer instructions are shown in Figure 13.20.

The timer-on-delay instruction (TON) can be used to turn a device on or off once an interval is timed out. Refer to Figure 13.21. When the rung condition becomes TRUE, the timer begins to count time-base intervals. Whenever the rung condition becomes FALSE, the AC value is reset to 000. As long as the rung condition remains TRUE, the timer increments its AC value for each interval. When the AC value equals the programmed PR value, the

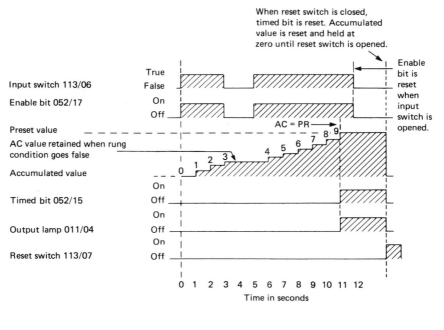

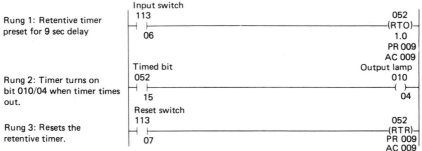

Figure 13.23 Retentive Timer with Reset

timer stops incrementing its AC value and sets the timed bit. This bit is then used to turn on or off an output device.

The timer-off-delay instruction (TOF) can be used to turn a device off or on after a timed interval. The TOF instruction works like the TON instruction except that the rung condition must be FALSE in order for the timer to operate, as shown in Figure 13.22.

The retentive-timer instruction (RTO), like the TON instruction, can be used to turn on or off a device once a programmed PR value is reached. However, unlike the TON instruction, the RTO instruction (Figure 13.23) retains its AC value when any of the following conditions occur:

1. The rung condition goes FALSE.
2. The mode select switch is changed to the program position.
3. A power outage occurs provided memory backup power is maintained.

Keyboard Symbol	Display	Description
—(CTU)—	XXX —(CTU)— PR YYY AC ZZZ	*Up counter:* Each time the rung goes TRUE, the AC value is incremented one count. The counter will continue counting after the PR value is reached. The AC value can be reset by the CTR instruction.
—(CTR)—	XXX —(CTR)— PR YYY AC ZZZ	*Counter reset:* When the rung is TRUE, the CTU AC value and status bits are reset to 000. XXX is the word address of the counter that CTR is resetting.
—(CTD)—	XXX —(CTD)— PR YYY AC ZZZ	*Down counter:* Each time the rung goes TRUE, the AC value is decreased one count.

Figure 13.24 Counter Instructions

Note: The counter word address, XXX, is assigned to the counter AC areas of the data table. PR values YYY and AC values ZZZ can vary from 000 to 999 (in decimal).

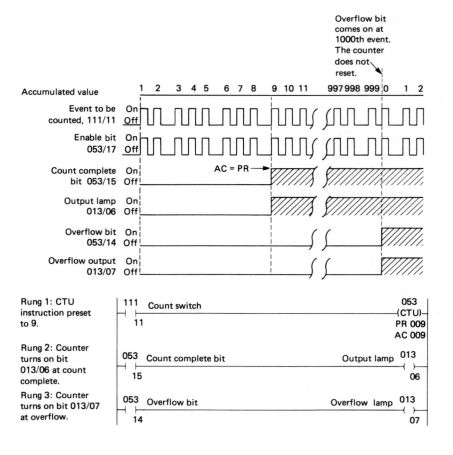

Figure 13.25 Up Counter

The retentive-timer-reset instruction (RTR) is used to reset the AC value and timed bit of the retentive timer to zero. This instruction is given the same word address as its corresponding RTO instruction, as shown in Figure 13.23. When the rung condition goes TRUE, the RTR instruction resets the AC value and status bits of the RTO instruction to zero.

A counter counts the number of events that occur and stores this count in its AC value word. Three types of counter instructions are shown in Figure 13.24. Counter instructions differ from timer instructions in that they have no time base. They count one event each FALSE-to-TRUE transition of the rung condition.

The up-counter instruction (CTU) increments its AC value for each FALSE-to-TRUE transition of the rung condition (Figure 13.25). The CTU instruction retains its AC value when:

1. The mode select switch is changed to the program position.
2. The rung condition goes FALSE.
3. A power outage occurs provided memory backup power is maintained.

When the AC value reaches the PR value, the count-complete bit is set to one. Unlike the TON instruction, the CTU instruction continues to increment its AC value after the PR value has been reached. If the AC value goes above 999, the overflow bit is set and the CTU continues up-counting from 000. Because this counter retains its AC value, it must be reset by a separate instruction, the counter reset instruction (CTR). The results can be seen in Figure 13.26.

The down-counter instruction (CTD) is like the CTU instruction except that one count is subtracted from its AC value for each FALSE-to-TRUE transition of its rung condi-

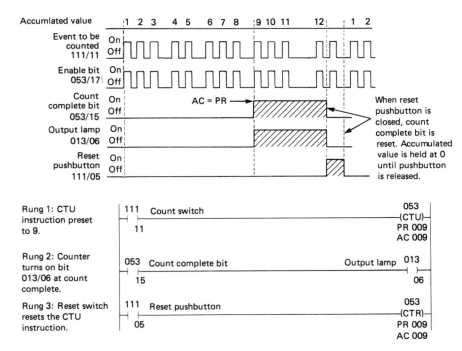

Figure 13.26 Up Counter with Reset

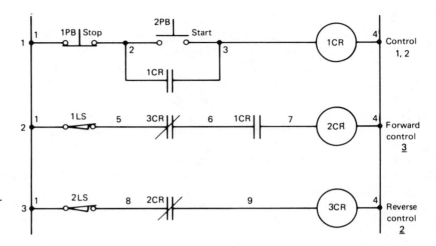

Figure 13.27 Reciprocating Motion Ladder Logic Diagram

tion. The CTD instruction may be paired with the CTU instruction to form an up/down counter, using the same word address, AC value, and PR value as the up counter.

CONCLUSION

These major PC logic instructions correspond to the symbols you learned in Chapter 12 on sequential control. As you have seen, some of the PC instructions are a little different than the relay ladder logic symbols. However, the con-

cepts used in programming the relay ladder logic systems still hold for program designing on the PC. With a little practice, programming PC logic diagrams will become second nature to you. With that in mind, try the following worked example.

The Reciprocating Motion Problem. We will take another look at the reciprocating motion problem that was presented in Chapter 12. The ladder logic for that problem is shown in Figure 13.27. Try to draw its PC logic equivalent. Check your answer with the solution given in Figure 13.28.

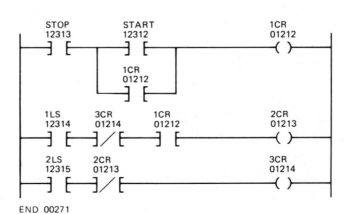

Figure 13.28 PC Ladder Representation of Reciprocating Motion Problem

END 00271

QUESTIONS

1. The first PCs were installed in the year _____.
2. The three advantages that the PC has over relay controls in the automobile manufacturing industry are _____, _____, and _____.
3. A process industry generally performs the operations necessary to change a material either _____ or _____.
4. List the six basic physical parts of a PC.
5. How is the PC control plan placed in memory?
6. Input and output devices are also referred to as _____ devices.
7. The period of time required for the PC to evaluate the I/O status and the control plan is referred to as _____ time.
8. From Appendix C, what are the six size categories for number of bits used in the various PCs?
9. From Appendix C, what are the maximum and minimum clock speeds used by the various PCs?
10. From Appendix C, what are the maximum and minimum scan rates used by the various PCs?
11. Are scan rates an accurate measure of PC performance?
12. From Appendix C, what are the maximum and minimum total system I/O connections available for the various PCs?
13. Field wiring connects to the _____ housing or modules.
14. Where is the control plan stored in the PC?
15. What do the letters RAM and ROM represent?
16. What part of the PC organizes all controller activity?
17. List three types of PC programming languages.
18. List two major factors to consider when selecting a PC.
19. What three questions should you ask yourself when programming each rung in the PC?

PROBLEMS

1. Draw the PC logic diagram for the relay logic diagram of Figure 12.14.
2. Draw the PC logic diagram for the relay logic diagram of Figure 12.26.
3. Draw the PC logic diagram for the relay logic diagram of Problem 2 at the end of Chapter 12.

BIBLIOGRAPHY

Chapter 1: Operational Amplifiers for Industrial Applications

*Berlin, H. W. *Design of Op-Amp Circuits with Experiments*. Indianapolis: H. W. Sams, 1977.
————. *Design of Active Filters with Experiments*. Indianapolis: H. W. Sams, 1977.
*Boyce, J. *Operational Amplifiers for Technicians*. North Scituate, MA: Breton Publishers. © 1983 by Wadsworth, Inc., Belmont, CA.
Burr-Brown Research Corporation. *Handbook of Operational Amplifier Applications*. Tucson, 1963.
*Cirovic, M. M. *Integrated Circuits: A User's Handbook*. Reston, VA: Reston, 1977.
*Coughlin, R. F., and Driscoll, F. F. *Operational Amplifiers and Linear Integrated Circuits*. 2nd ed. Englewood Cliffs, NJ: Prentice-Hall, 1982.
*Deboo, G. J., and Burrous, C. N. *Integrated Circuits and Semiconductor Devices*. New York: McGraw-Hill, 1977.
*Dungan, F. R. *Linear Integrated Circuits for Technicians*. North Scituate, MA: Breton Publishers. © 1984 by Wadsworth, Inc., Belmont, CA.
Faulkenberry, L. M. *An Introduction to Operational Amplifiers*. 2nd ed. New York: Wiley, 1984.
Garrett, P. H. *Analog I/O Design*. Reston, VA: Reston, 1981.
Graeme, J. G. *Designing with Operational Amplifiers*. New York: McGraw-Hill, 1977.
Hnatek, E. R. *Applications of Linear Integrated Circuits*. New York: Wiley, 1975.
Jacob, J. M. *Applications and Designs with Analog Integrated Circuits*. Reston, VA: Reston, 1982.

*Jung, W. C. *IC Op Amp Cookbook*. Indianapolis: H. W. Sams, 1974.
*Lancaster, D. *Active Filter Cookbook*. Indianapolis: H. W. Sams, 1974.
*Malcolm, D. R. *Fundamentals of Electronics*. 2nd ed. North Scituate, MA: Breton Publishers. © 1983 by Wadsworth, Inc., Belmont, CA.
National Semiconductor Corporation. *Databook*. Santa Clara, CA, 1980.
————. *Application Note AN-79*. Santa Clara, CA, 1980.
*Pasahow, E. *Principles of Integrated Circuits*. North Scituate, MA: Breton Publishers. © 1982 by Wadsworth, Inc., Belmont, CA.
*Rutkowski, G. B. *Handbook of Integrated Circuit Operational Amplifiers*. Englewood Cliffs, NJ: Prentice-Hall, 1982.
Wait, J. V. *Introduction to Operational Amplifier Theory and Applications*. New York: McGraw-Hill, 1975.

Chapter 2: Linear Integrated Circuits for Industrial Applications

*Cirovic, M. M. *Integrated Circuits: A User's Handbook*. Reston, VA: Reston, 1977.
*Coughlin, R. F., and Driscoll, F. F. *Operational Amplifiers and Linear Integrated Circuits*. 2nd ed. Englewood Cliffs, NJ: Prentice-Hall, 1982.
*Dungan, F. R. *Linear Integrated Circuits for Technicians*. North Scituate, MA: Breton Publishers. © 1984 by Wadsworth, Inc., Belmont, CA.
Garrett, P. H. *Analog I/O Design*. Reston, VA: Reston, 1981.
Hnatek, E. R. *Applications of Linear Integrated Circuits*. New York: Wiley, 1975.

*Books that have a technical rather than an engineering emphasis.

Jacob, J. M. *Applications and Designs with Analog Integrated Circuits.* Reston, VA: Reston, 1982.

*Malcolm, D. R. *Fundamentals of Electronics.* 2nd ed. North Scituate, MA: Breton Publishers. © 1983 by Wadsworth, Inc., Belmont, CA.

National Semiconductor Corporation. *Databook.* Santa Clara, CA, 1980.

*Pasahow, E. *Principles of Integrated Circuits.* North Scituate, MA: Breton Publishers. © 1982 by Wadsworth, Inc., Belmont, CA.

Signetics Corporation. *Signetics Linear LSI Data and Applications Manual Vol. 1.* Sunnyvale, CA, 1985.

————. *Signetics Linear LSI Data and Applications Manual Vol. 2.* Sunnyvale, CA, 1985.

————. *Analog Data Manual.* Sunnyvale, CA, 1985.

Chapter 3: Wound-Field DC Motors and Generators

Anderson, L. R. *Electric Machines and Transformers.* Reston, VA: Reston, 1981.

Emmanuel, P. *Motors, Generators, Transformers and Energy.* Englewood Cliffs, NJ: Prentice-Hall, 1985.

Fisher, F. "Convenient Comparison Charts Aid in Motor Selection." *EDN* 23 (August 5, 1978): 97–99.

Gingrich, H. W. *Electrical Machinery, Transformers, and Controls.* Englewood Cliffs, NJ: Prentice-Hall, 1979.

Headquarters, Department of the Army. *Electric Motor and Generator Repair.* Washington, D.C.: U.S. Government Printing Office, 1972.

Kosow, I. L. *Electric Machinery and Transformers.* Englewood Cliffs, NJ: Prentice-Hall, 1972.

*Lister, E. C. *Electric Circuits and Machines.* 6th ed. New York: McGraw-Hill, 1984.

Lloyd, T. C. *Electric Motors and Their Applications.* New York: Wiley, 1969.

*Naval Education and Training Support Command. *Electrician's Mate 3 & 2.* Washington, D.C.: U.S. Government Printing Office, 1972.

Richardson, D. V. *Rotating Electric Machinery and Transformer Technology.* Reston, VA: Reston, 1978.

Rosenblatt, J., and Friedman, M. H. *Direct and Alternating Current Machinery.* 2nd ed. Columbus, OH: Charles E. Merrill, 1984.

Wildi, T. *Electric Power Technology.* New York: Wiley, 1981.

Chapter 4: Brushless and Stepper DC Motors

Airpax Corporation. *Stepper Motor Handbook.* Cheshire, CT, 1985.

Electro-craft Corporation. *DC Motors, Speed Controls and Servo Systems.* 6th ed. Hopkins, MN, 1982.

International Rectifier Corporation, *HEXFET Power MOSFET Databook.* El Segundo, CA, 1985.

Kuo, B. C., and Tal, J. *DC Motors and Control Systems.* Champaign, IL: SRL Publishing Co., 1978.

Kuo, B. C. *Theory and Applications of Stepper Motors.* St. Paul, MN: West Publishing Co., 1974.

Oriental Motor U.S.A. Corporation. *Technical Information on Stepping Motors.* Torrance, CA, 1983.

Siliconix Corporation. *MOSPOWER Design Catalog.* Santa Clara, CA, 1983.

Chapter 5: AC Motors

Anderson, L. R. *Electric Machines and Transformers.* Reston, VA: Reston, 1981.

Emmanuel, P. *Motors, Generators, Transformers and Energy.* Englewood Cliffs, NJ: Prentice-Hall, 1985.

Gingrich, H. W. *Electrical Machinery, Transformers, and Controls.* Englewood Cliffs, NJ: Prentice-Hall, 1979.

Kosow, I. L. *Electric Machinery and Transformers.* Englewood Cliffs, NJ: Prentice-Hall, 1972.

*Lister, E. C. *Electric Circuits and Machines.* 6th ed. New York: McGraw-Hill, 1984.

Lloyd, T. C. *Electric Motors and Their Applications.* New York: Wiley, 1969.

*Naval Education and Training Support Command. *Electrician's Mate 3 & 2.* Washington, D.C.: U.S. Government Printing Office, 1972.

Richardson, D. V. *Rotating Electric Machinery and Transformer Technology.* Reston, VA: Reston, 1978.

Rosenblatt, J., and Friedman, M. H. *Direct and Alternating Current Machinery.* 2nd ed. Columbus, OH: Charles E. Merrill, 1984.

*Wildi, T. *Electric Power Technology.* New York: Wiley, 1981.

Chapter 6: Industrial Control Devices

*Bell, D. A. *Electronic Devices and Circuits.* 2nd ed. Reston, VA: Reston, 1978.

*Coughlin, R. F., and Driscoll, F. F. *Operational Amplifiers and Linear Integrated Circuits.* 2nd ed. Englewood Cliffs, NJ: Prentice-Hall, 1982.

*Deboo, G. J., and Burrous, C. N. *Integrated Circuits and Semiconductor Devices.* New York: McGraw-Hill, 1977.

General Electric Company. *Transistor Manual.* 2nd ed. Auburn, NY, 1969.

————. *SCR Manual.* 2nd ed. Auburn, NY, 1979.

Gottlieb, I. M. *Solid-State Power Electronics.* Indianapolis: H. W. Sams, 1977.

International Rectifier Corporation. *HEXFET Power MOSFET Databook.* El Segundo, CA, 1985.

————. *SCR Applications Handbook.* El Segundo, CA, 1985.

Maloney, T. J. *Industrial Solid-State Electronics.* 2nd ed. Englewood Cliffs, NJ: Prentice-Hall, 1986.

Malvino, A. P. *Transistor Circuit Approximations.* 3rd ed. New York: McGraw-Hill, 1980.

RCA Corporation. *Thyristor and Rectifier Manual.* Somerville, NJ, 1975.

Signetics Corporation. *Signetics Linear LSI Data and Applications Manual Vol. 1.* Sunnyvale, CA, 1985.

————. *Signetics Linear LSI Data and Applications Manual Vol. 2.* Sunnyvale, CA, 1985.

————. *Analog Data Manual.* Sunnyvale, CA, 1985.

Chapter 7: Power Control Circuits

Airpax Corporation. *Stepper Motor Handbook.* Cheshire, CT, 1985.

*Coughlin, R. F., and Driscoll, F. F. *Operational Amplifiers and Linear Integrated Circuits.* 2nd ed. Englewood Cliffs, NJ: Prentice-Hall, 1982.

Datta, S. *Power Electronics & Controls.* Reston, VA: Reston, 1978.

*Deboo, G. J., and Burrous, C. N. *Integrated Circuits and Semiconductor Devices.* New York: McGraw-Hill, 1977.

Electro-craft Corporation. *DC Motors, Speed Controls and Servo Systems.* 6th ed. Hopkins, MN, 1982.

General Electric Company. *Transistor Manual.* 2nd ed. Auburn, NY, 1969.

————. *SCR Manual.* 2nd ed. Auburn, NY, 1979.

*Gottlieb, I. M. *Solid-State Power Electronics.* Indianapolis: H. W. Sams, 1977.

International Rectifier Corporation. *HEXFET Power MOSFET Databook.* El Segundo, CA, 1985.

————. *SCR Applications Handbook.* El Segundo, CA, 1985.

Kuo, B. C., and Tal, J. *DC Motors and Control Systems.* Champaign, IL: SRL Publishing Co., 1978.

Kuo, B. C. *Theory and Applications of Stepper Motors.* St. Paul, MN: West Publishing Co., 1974.

Kusko, A. *Solid-State DC Motor Drives.* Cambridge, MA: MIT Press, 1969.

*Maloney, T. J. *Industrial Solid-State Electronics.* 2nd ed. Englewood Cliffs, NJ: Prentice-Hall, 1986.

Mazda, F. F. *Thyristor Control.* London: Butterworth & Co., 1973.

Murphy, J. M. D. *Thyristor Control of AC Motors.* Oxford: Pergamon Press, 1973.

Oriental Motor U.S.A. Corporation. *Technical Information on Stepping Motors.* Torrance, CA, 1983.

RCA Corporation. *Thyristor and Rectifier Manual.* Somerville, NJ, 1975.

Signetics Corporation. *Signetics Linear LSI Data and Applications Manual Vol. 1.* Sunnyvale, CA, 1985.

————. *Signetics Linear LSI Data and Applications Manual Vol. 2.* Sunnyvale, CA, 1985.

————. *Analog Data Manual.* Sunnyvale, CA, 1985.

Siliconix Corporation. *MOSPOWER Design Catalog.* Santa Clara, CA, 1983.

O'Higgins, P. J. *Basic Instrumentation.* New York: McGraw-Hill, 1966.

Sheingold, D. H. *Transducer Interfacing Handbook.* Norwood, MA: Analog Devices, 1980.

Chapter 8: Transducers

Anderson, N. A. *Instrumentation for Process Measurement and Control.* 2nd ed. Radnor, PA: Chilton, 1972.

Andrew, W. G., and Williams, H. B. *Applied Instrumentation in the Process Industries.* 2nd ed. Houston: Gulf, 1979.

Deboo, G. J., and Burrous, C. N. *Integrated Circuits and Semiconductor Devices.* New York: McGraw-Hill, 1977.

*Driscoll, E. F. *Industrial Electronics: Devices, Circuits and Applications.* Chicago: American Technical Society, 1976.

*Fribance, A. E. *Industrial Instrumentation Fundamentals.* New York: McGraw-Hill, 1962.

Herceg, E. E. *Schaevitz Handbook of Measurement and Control.* Camden, NJ: Shaevitz Engineering, 1976.

*Johnson, C. D. *Process Control Instrumentation Technology.* 2nd ed. New York: Wiley, 1982.

*Lenk, J. D. *Handbook of Controls and Instrumentation.* Englewood Cliffs, NJ: Prentice-Hall, 1980.

Mansfield, P. H. *Electrical Transducers for Industrial Measurement.* London: Buttterworth, 1973.

Micro Switch. *Solid State Sensors.* Freeport, IL: Honeywell.

Minnar, E. J., ed. *ISA Transducer Compendium.* Pittsburgh: Instrument Society of America, 1963.

Moore, R. L., ed. *Basic Instrumentation Lecture Notes and Study Guide.* 2nd ed. Pittsburgh: Instrument Society of America, 1976.

National Semiconductor Corporation. *Pressure Transducer Handbook.* Santa Clara, CA, 1980.

*Norton, H. M. *Handbook of Transducers for Electronic Measuring Systems.* Englewood Cliffs, NJ: Prentice-Hall, 1969.

Chapter 9: Industrial Process Control

Brewer, J. W. *Control Systems: Analysis, Design, and Simulation.* Englewood Cliffs, NJ: Prentice-Hall, 1974.

Hougen, J. O. *Measurements and Control Applications.* Research Park Triangle, NC: Instrument Society of America, 1979.

*Hunter, R. P. *Automatic Process Control Systems: Concepts and Hardware.* Englewood Cliffs, NJ: Prentice-Hall, 1978.

*Johnson, C. D. *Process Control Instrumentation Technology.* 2nd ed. New York: Wiley, 1982.

Kuo, B. C. *Automatic Control Systems.* Englewood Cliffs, NJ: Prentice-Hall, 1978.

Needler, M. A., and Baker, D. E. *Digital and Analog Controls.* Reston, VA: Reston, 1985.

Pearman, R. A. *Solid-State Industrial Electronics.* Reston, VA: Reston, 1984.

Pericles, E., and Leff, E. *Introduction to Feedback Control Systems.* New York: McGraw-Hill, 1979.

Weyrick, R. C. *Fundamentals of Automatic Control.* New York: McGraw-Hill, 1975. Reston, VA: Reston, 1978.

Chapter 10: Pulse Modulation

*Hnatek, E. R. *Applications of Linear Integrated Circuits.* New York: Wiley, 1975.

Journal Ministere Russie Defense 47, section 7, 25 (1845).

*Kennedy, G. *Electronic Communication Systems.* 2nd ed. New York: McGraw-Hill, 1977.

Miller, G. M. *Modern Electronic Communication.* Englewood Cliffs, NJ: Prentice-Hall, 1978.

*National Semiconductor Corporation. *Special Functions Databook.* Santa Clara, CA, 1979.

Shannon, C. E., and Weaver, W. *The Mathematical Theory of Communications.* Urbana: University of Illinois Press, 1949.

Taub, H., and Schilling, D. L. *Principles of Communication Systems.* New York: McGraw-Hill, 1971.

Chapter 11: Industrial Telemetry and Data Communication

*Campbell, J. *The RS-232 Solution.* Berkeley, CA: Sybex, 1984.

Fisher, H. F. *Telemetry Transducer Handbook.* Technical Report no. WADD-TR-61-67, vol. I, rev. I. Wright-Patterson Air Force Base, OH: Air Force Systems Command, 1963.

*Gruenberg, E. L., ed. *Handbook of Telemetry and Remote Control.* New York: McGraw-Hill, 1967.

Inter-Range Instrumentation Group. *Telemetry Standards.* Document no. 106-73, rev. White Sands Missile Range, NM: Secretariat, Range Commanders Council, 1973.

Kennedy, G. *Electronic Communication Systems.* 2nd ed. New York: McGraw-Hill, 1977.

Martin, J. *Telecommunications and the Computer.* Englewood Cliffs, NJ: Prentice-Hall, 1969.

Miller, G. M. *Modern Electronic Communication.* Englewood Cliffs, NJ: Prentice-Hall, 1978.

*Zanger, H. *Electronic Systems Theory and Applications.* Englewood Cliffs, NJ: Prentice-Hall, 1977.

Chapter 12: Sequential Process Control

Brewer, J. W. *Control Systems: Analysis, Design, and Simulation.* Englewood Cliffs, NJ: Prentice-Hall, 1974.

Hougen, J. O. *Measurements and Control Applications.* Research Park Triangle, NC: Instrument Society of America, 1979.

*Hunter, R. P. *Automatic Process Control Systems: Concepts and Hardware.* Englewood Cliffs, NJ: Prentice-Hall, 1978.

*Johnson, C. D. *Process Control Instrumentation Technology.* 2nd ed. New York: Wiley, 1982.

Kuo, B. C. *Automatic Control Systems.* Englewood Cliffs, NJ: Prentice-Hall, 1978.

*Maloney, T. J. *Industrial Solid-State Electronics.* 2nd ed. Englewood Cliffs, NJ: Prentice-Hall, 1986.

Needler, M. A., and Baker, D. E. *Digital and Analog Controls.* Reston, VA: Reston, 1985.

Pearman, R. A. *Solid-State Industrial Electronics.* Reston, VA: Reston, 1984.

Pericles, E., and Leff, E. *Introduction to Feedback Control Systems.* New York: McGraw-Hill, 1979.

Weyrick, R. C. *Fundamentals of Automatic Control.* New York: McGraw-Hill, 1975. Reston, VA: Reston, 1978.

Chapter 13: Programmable Controllers

*Allen-Bradley Company, *Bulletin 1770 Industrial Terminal Systems User's Manual.* Cleveland.

*————. *Bulletin 1772 Mini-PLC-2 Programmable Controller.* Cleveland.

Andrew, W. G., and Williams, H. B. *Applied Instrumentation in the Process Industries.* Houston: Gulf, 1979.

Deltano, D. "Programming Your PC." *Instruments & Control Systems* 53 (July 1980): 37–40.

Hickey, J. "Programmable Controller Roundup." *Instruments & Control Systems* 54 (July 1981): 57–64.

Jannotta, K. "What is a PC?" *Instruments & Control Systems* 53 (February 1980): 21–25.

DATA SHEETS

- 741 Operational Amplifier
- 3900 Current Differencing Amplifier
- SCR
- Triac
- 335 Temperature Sensor
- 555 Timer
- 565 Phase-Locked Loop
- ADC 0801 Analog-to-Digital Converter
- DAC 0808 Digital-to-Analog Converter

**National Semiconductor**

Operational Amplifiers/Buffers

LM741/LM741A/LM741C/LM741E Operational Amplifier

General Description

The LM741 series are general purpose operational amplifiers which feature improved performance over industry standards like the LM709. They are direct, plug-in replacements for the 709C, LM201, MC1439 and 748 in most applications.

The amplifiers offer many features which make their application nearly foolproof: overload pro-

tection on the input and output, no latch-up when the common mode range is exceeded, as well as freedom from oscillations.

The LM741C/LM741E are identical to the LM741/LM741A except that the LM741C/LM741E have their performance guaranteed over a 0°C to +70°C temperature range, instead of −55°C to +125°C.

Schematic and Connection Diagrams (Top Views)

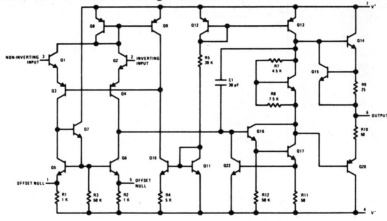

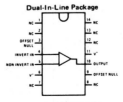

Metal Can Package

Order Number LM741H, LM741AH,
LM741CH or LM741EH
See NS Package H08C

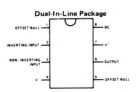

Dual-In-Line Package

Order Number LM741CN or LM741EN
See NS Package N08B
Order Number LM741CJ
See NS Package J08A

Dual-In-Line Package

Order Number LM741CN-14
See NS Package N14A
Order Number LM741J-14, LM741AJ-14
or LM741CJ-14
See NS Package J14A

Absolute Maximum Ratings

	LM741A	LM741E	LM741	LM741C
Supply Voltage	±22V	±22V	±22V	±18V
Power Dissipation (Note 1)	500 mW	500 mW	500 mW	500 mW
Differential Input Voltage	±30V	±30V	±30V	±30V
Input Voltage (Note 2)	±15V	±15V	±15V	±15V
Output Short Circuit Duration	Indefinite	Indefinite	Indefinite	Indefinite
Operating Temperature Range	−55°C to +125°C	0°C to +70°C	−55°C to +125°C	0°C to +70°C
Storage Temperature Range	−65°C to +150°C	−65°C to +150°C	−65°C to +150°C	−65°C to +150°C
Lead Temperature (Soldering, 10 seconds)	300°C	300°C	300°C	300°C

Electrical Characteristics (Note 3)

PARAMETER	CONDITIONS	LM741A/LM741E			LM741			LM741C			UNITS
		MIN	TYP	MAX	MIN	TYP	MAX	MIN	TYP	MAX	
Input Offset Voltage	$T_A = 25°C$										
	$R_S \leq 10\ k\Omega$					1.0	5.0		2.0	6.0	mV
	$R_S \leq 50\Omega$		0.8	3.0							mV
	$T_{AMIN} \leq T_A \leq T_{AMAX}$										
	$R_S \leq 50\Omega$			4.0							mV
	$R_S \leq 10\ k\Omega$						6.0			7.5	mV
Average Input Offset Voltage Drift				15							µV/°C
Input Offset Voltage Adjustment Range	$T_A = 25°C$, $V_S = ±20V$	±10				±15			±15		mV
Input Offset Current	$T_A = 25°C$		3.0	30		20	200		20	200	nA
	$T_{AMIN} \leq T_A \leq T_{AMAX}$			70		85	500			300	nA
Average Input Offset Current Drift				0.5							nA/°C
Input Bias Current	$T_A = 25°C$		30	80		80	500		80	500	nA
	$T_{AMIN} \leq T_A \leq T_{AMAX}$			0.210			1.5			0.8	µA
Input Resistance	$T_A = 25°C$, $V_S = ±20V$	1.0	6.0		0.3	2.0		0.3	2.0		MΩ
	$T_{AMIN} \leq T_A \leq T_{AMAX}$, $V_S = ±20V$	0.5									MΩ
Input Voltage Range	$T_A = 25°C$							±12	±13		V
	$T_{AMIN} \leq T_A \leq T_{AMAX}$				±12	±13					V
Large Signal Voltage Gain	$T_A = 25°C$, $R_L \geq 2\ k\Omega$										
	$V_S = ±20V$, $V_O = ±15V$	50									V/mV
	$V_S = ±15V$, $V_O = ±10V$				50	200		20	200		V/mV
	$T_{AMIN} \leq T_A \leq T_{AMAX}$, $R_L \geq 2\ k\Omega$,										
	$V_S = ±20V$, $V_O = ±15V$	32									V/mV
	$V_S = ±15V$, $V_O = ±10V$				25			15			V/mV
	$V_S = ±5V$, $V_O = ±2V$	10									V/mV
Output Voltage Swing	$V_S = ±20V$										
	$R_L \geq 10\ k\Omega$	±16									V
	$R_L \geq 2\ k\Omega$	±15									V
	$V_S = ±15V$										
	$R_L \geq 10\ k\Omega$				±12	±14		±12	±14		V
	$R_L \geq 2\ k\Omega$				±10	±13		±10	±13		V
Output Short Circuit Current	$T_A = 25°C$	10	25	35		25			25		mA
	$T_{AMIN} \leq T_A \leq T_{AMAX}$	10		40							mA
Common-Mode Rejection Ratio	$T_{AMIN} \leq T_A \leq T_{AMAX}$										
	$R_S \leq 10\ k\Omega$, $V_{CM} = ±12V$				70	90		70	90		dB
	$R_S \leq 50\ k\Omega$, $V_{CM} = ±12V$	80	95								dB

Electrical Characteristics (Continued)

PARAMETER	CONDITIONS	LM741A/LM741E			LM741			LM741C			UNITS
		MIN	TYP	MAX	MIN	TYP	MAX	MIN	TYP	MAX	
Supply Voltage Rejection Ratio	$T_{AMIN} \leq T_A \leq T_{AMAX}$, $V_S = \pm 20V$ to $V_S = \pm 5V$										
	$R_S \leq 50\Omega$	86	96								dB
	$R_S \leq 10\ k\Omega$				77	96		77	96		dB
Transient Response	$T_A = 25°C$, Unity Gain										
Rise Time			0.25	0.8		0.3			0.3		μs
Overshoot			6.0	20		5			5		%
Bandwidth (Note 4)	$T_A = 25°C$	0.437	1.5								MHz
Slew Rate	$T_A = 25°C$, Unity Gain	0.3	0.7			0.5			0.5		V/μs
Supply Current	$T_A = 25°C$					1.7	2.8		1.7	2.8	mA
Power Consumption	$T_A = 25°C$										
	$V_S = \pm 20V$		80	150							mW
	$V_S = \pm 15V$					50	85		50	85	mW
LM741A	$V_S = \pm 20V$										
	$T_A = T_{AMIN}$			165							mW
	$T_A = T_{AMAX}$			135							mW
LM741E	$V_S = \pm 20V$			150							mW
	$T_A = T_{AMIN}$			150							mW
	$T_A = T_{AMAX}$			150							mW
LM741	$V_S = \pm 15V$										
	$T_A = T_{AMIN}$					60	100				mW
	$T_A = T_{AMAX}$					45	75				mW

Note 1: The maximum junction temperature of the LM741/LM741A is 150°C, while that of the LM741C/LM741E is 100°C. For operation at elevated temperatures, devices in the TO-5 package must be derated based on a thermal resistance of 150°C/W junction to ambient, or 45°C/W junction to case. The thermal resistance of the dual-in-line package is 100°C/W junction to ambient.

Note 2: For supply voltages less than ±15V, the absolute maximum input voltage is equal to the supply voltage.

Note 3: Unless otherwise specified, these specifications apply for $V_S = \pm 15V$, $-55°C \leq T_A \leq +125°C$ (LM741/LM741A). For the LM741C/LM741E, these specifications are limited to $0°C \leq T_A \leq +70°C$.

Note 4: Calculated value from: BW (MHz) = 0.35/Rise Time(μs).

National Semiconductor

Operational Amplifiers/Buffers

LM2900/LM3900, LM3301, LM3401 Quad Amplifiers

General Description

The LM2900 series consists of four independent, dual input, internally compensated amplifiers which were designed specifically to operate off of a single power supply voltage and to provide a large output voltage swing. These amplifiers make use of a current mirror to achieve the non-inverting input function. Application areas include: ac amplifiers, RC active filters, low frequency triangle, squarewave and pulse waveform generation circuits, tachometers and low speed, high voltage digital logic gates.

Features

- Wide single supply voltage range or dual supplies   4 V$_{DC}$ to 36 V$_{DC}$ ±2 V$_{DC}$ to ±18 V$_{DC}$
- Supply current drain independent of supply voltage
- Low input biasing current   30 nA
- High open-loop gain   70 dB
- Wide bandwidth   2.5 MHz (Unity Gain)
- Large output voltage swing   (V$^+$ −1) Vp-p
- Internally frequency compensated for unity gain
- Output short-circuit protection

Schematic and Connection Diagrams

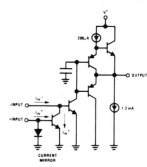

Order Number LM2900J
See NS Package J14A
Order Number LM2900N,
LM3900N, LM3301N
or LM3401N
See NS Package N14A

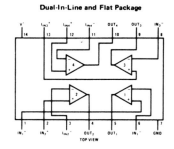

Dual-In-Line and Flat Package

TOP VIEW

Typical Applications   (V$^+$ = 15 V$_{DC}$)

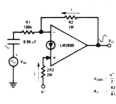

Inverting Amplifier

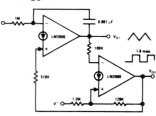

Triangle/Square Generator

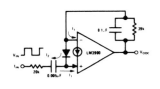

Frequency-Doubling Tachometer

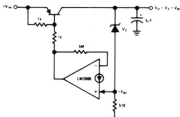

Low V$_{IN}$ − V$_{OUT}$ Voltage Regulator

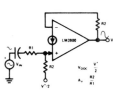

Non-Inverting Amplifier

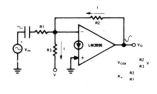

Negative Supply Biasing

Absolute Maximum Ratings

	LM2900/LM3900	LM3301	LM3401
Supply Voltage	32 VDC / ±16 VDC	28 VDC / ±14 VDC	18 VDC / ±9 VDC
Power Dissipation (T_A = 25°C) (Note 1)			
Cavity DIP	900 mW		
Flat Pack	800 mW		
Molded DIP	570 mW	570 mW	570 mW
Input Currents, I_{IN}^{+} or I_{IN}^{-}	20 mADC	20 mADC	20 mADC
Output Short-Circuit Duration – One Amplifier			
T_A = 25°C (See Application Hints)	Continuous	Continuous	Continuous
Operating Temperature Range			
LM2900	−40°C to +85°C		
LM3900	0°C to +70°C	−40°C to +85°C	0°C to +75°C
Storage Temperature Range	−65°C to +150°C	−65°C to +150°C	−65°C to +150°C
Lead Temperature (Soldering, 10 seconds)	300°C	300°C	300°C

Electrical Characteristics (Note 6)

PARAMETER	CONDITIONS	LM2900			LM3900			LM3301			LM3401			UNITS
		MIN	TYP	MAX	MIN	TYP	MAX	MIN	TYP	MAX	MIN	TYP	MAX	
Open Loop														
Voltage Gain	T_A = 25°C, f = 100 Hz										800			V/mV
Voltage Gain	T_A = 25°C, Inverting Input	1.2	2.8		1.2	2.8		1.2	2.8		1.2	2.8		V/mV
Input Resistance	T_A = 25°C, Inverting Input		1			1			1		0.1	1		MΩ
Output Resistance			8			8			8			8		kΩ
Unity Gain Bandwidth	T_A = 25°C, Inverting Input		2.5			2.5			2.5			2.5		MHz
Input Bias Current	T_A = 25°C, Inverting Input		30	200		30	200		30	300		30	300	nA
	Inverting Input												500	nA
Slew Rate	T_A = 25°C, Positive Output Swing		0.5			0.5			0.5			0.5		V/μs
	T_A = 25°C, Negative Output Swing		20			20			20			20		V/μs
Supply Current	T_A = 25°C, R_L = ∞ On All Amplifiers		6.2	10		6.2	10		6.2	10		6.2	10	mADC
Output Voltage Swing	T_A = 25°C, R_L = 2k, V_{CC} = 15.0 VDC													
VOUT High	I_{IN}^{-} = 0, I_{IN}^{+} = 0	13.5			13.5			13.5			13.5			VDC
VOUT Low	I_{IN}^{-} = 10μA, I_{IN}^{+} = 0		0.09	0.2		0.09	0.2		0.09	0.2		0.09	0.2	VDC
VOUT High	I_{IN}^{-} = 0, I_{IN}^{+} = 0 R_L = ∞, V_{CC} = Absolute Maximum Ratings		29.5			29.5			25.5			15.5		VDC
Output Current Capability	T_A = 25°C													
Source		6	18		6	10		5	18		5	10		mADC
Sink	(Note 2)	0.5	1.3		0.5	1.3		0.5	1.3		0.5	1.3		mADC
ISINK	V_{OL} = 1V, I_{IN} = 5μA		5			5			5			5		mADC

Electrical Characteristics (Continued) (Note 6)

PARAMETER	CONDITIONS	LM2900			LM3900			LM3301			LM3401			UNITS
		MIN	TYP	MAX	MIN	TYP	MAX	MIN	TYP	MAX	MIN	TYP	MAX	
Power Supply Rejection	T_A = 25°C, f = 100 Hz		70			70			70			70		dB
Mirror Gain	@ 20μA (Note 3)	0.90	1.0	1.1	0.90	1.0	1.1	0.90	1	1.10	0.90	1	1.10	μA/μA
	@ 200μA (Note 3)	0.90	1.0	1.1	0.90	1.0	1.1	0.90	1	1.10	0.90	1	1.10	μA/μA
ΔMirror Gain	@ 20μA To 200μA (Note 3)		2	5		2	5		2	5		2	5	%
Mirror Current	(Note 4)		10	500		10	500		10	500		10	500	μA$_{DC}$
Negative Input Current	T_A = 25°C (Note 5)		1.0			1.0			1.0			1.0		mA$_{DC}$
Input Bias Current	Inverting Input		300			300								nA

Note 1: For operating at high temperatures, the device must be derated based on a 125°C maximum junction temperature and a thermal resistance of 175°C/W which applies for the device soldered in a printed circuit board, operating in a still air ambient.

Note 2: The output current sink capability can be increased for large signal conditions by overdriving the inverting input. This is shown in the section on Typical Characteristics.

Note 3: This spec indicates the current gain of the current mirror which is used as the non-inverting input.

Note 4: Input V_{BE} match between the non-inverting and the inverting inputs occurs for a mirror current (non-inverting input current) of approximately 10μA. This is therefore a typical design center for many of the application circuits.

Note 5: Clamp transistors are included on the IC to prevent the input voltages from swinging below ground more than approximately −0.3 V_{DS}. The negative input currents which may result from large signal overdrive with capacitance input coupling need to be externally limited to values of approximately 1 mA. Negative input currents in excess of 4 mA will cause the output voltage to drop to a low voltage. This maximum current applies to any one of the input terminals. If more than one of the input terminals are simultaneously driven negative smaller maximum currents are allowed. Common-mode current biasing can be used to prevent negative input voltages; see for example, the "Differentiator Circuit" in the applications section.

Note 6: These specs apply for −55° ≤ T_A ≤ +125°C, unless otherwise stated.

Typical Performance Characteristics

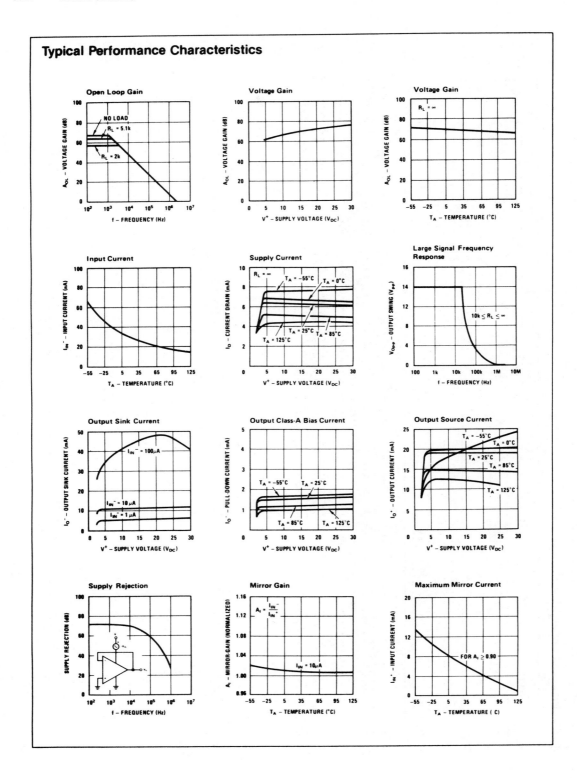

Application Hints

When driving either input from a low-impedance source, a limiting resistor should be placed in series with the input lead to limit the peak input current. Currents as large as 20 mA will not damage the device, but the current mirror on the non-inverting input will saturate and cause a loss of mirror gain at mA current levels—especially at high operating temperatures.

Precautions should be taken to insure that the power supply for the integrated circuit never becomes reversed in polarity or that the unit is not inadvertently installed backwards in a test socket as an unlimited current surge through the resulting forward diode within the IC could cause fuzing of the internal conductors and result in a destroyed unit.

Output short circuits either to ground or to the positive power supply should be of short time duration. Units can be destroyed, not as a result of the short circuit current causing metal fuzing, but rather due to the large increase in IC chip dissipation which will cause eventual failure due to excessive junction temperatures. For example, when operating from a well-regulated +5 V_{DC} power supply at T_A = 25°C with a 100 kΩ shunt-feedback resistor (from the output to the inverting input) a short directly to the power supply will not cause catastrophic failure but the current magnitude will be approximately 50 mA and the junction temperature will be above T_J max. Larger feedback resistors will reduce the current, 11 MΩ provides approximately 30 mA, an open circuit provides 1.3 mA, and a direct connection from the output to the non-inverting input will result in catastrophic failure when the output is shorted to V^+ as this then places the base-emitter junction of the input transistor directly across the power supply. Short-circuits to ground will have magnitudes of approximately 30 mA and will not cause catastrophic failure at T_A = 25°C.

Unintentional signal coupling from the output to the non-inverting input can cause oscillations. This is likely only in breadboard hook-ups with long component leads and can be prevented by a more careful lead dress or by locating the non-inverting input biasing resistor close to the IC. A quick check of this condition is to bypass the non-inverting input to ground with a capacitor. High impedance biasing resistors used in the non-inverting input circuit make this input lead highly susceptible to unintentional ac signal pickup.

Operation of this amplifier can be best understood by noticing that input currents are differenced at the inverting-input terminal and this difference current then flows through the external feedback resistor to produce the output voltage. Common-mode current biasing is generally useful to allow operating with signal levels near ground or even negative as this maintains the inputs biased at $+V_{BE}$. Internal clamp transistors (see note 5) catch negative input voltages at approximately $-0.3\ V_{DC}$ but the magnitude of current flow has to be limited by the external input network. For operation at high temperature, this limit should be approximately 100μA.

This new "Norton" current-differencing amplifier can be used in most of the applications of a standard IC op amp. Performance as a dc amplifier using only a single supply is not as precise as a standard IC op amp operating with split supplies but is adequate in many less critical applications. New functions are made possible with this amplifier which are useful in single power supply systems. For example, biasing can be designed separately from the ac gain as was shown in the "inverting amplifier," the "difference integrator" allows controlling the charging and the discharging of the integrating capacitor both with positive voltages, and the "frequency doubling tachometer" provides a simple circuit which reduces the ripple voltage on a tachometer output dc voltage.

Typical Applications (Continued)

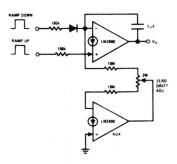

Low-Drift Ramp and Hold Circuit

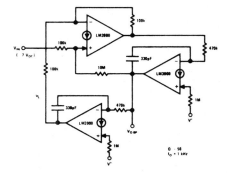

Bi-Quad Active Filter
(2nd Degree State-Variable Network)

Typical Applications (Continued)

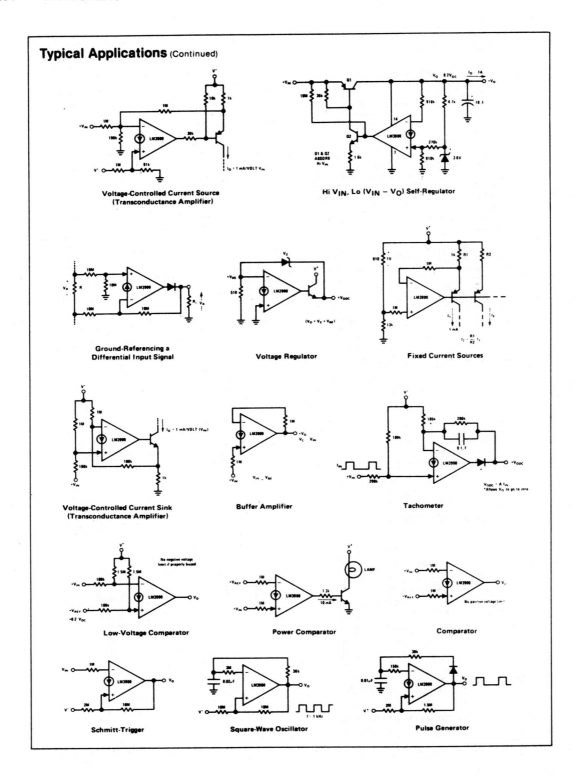

Voltage-Controlled Current Source
(Transconductance Amplifier)

Hi V_{IN}, Lo ($V_{IN} - V_O$) Self-Regulator

Ground-Referencing a
Differential Input Signal

Voltage Regulator

Fixed Current Sources

Voltage-Controlled Current Sink
(Transconductance Amplifier)

Buffer Amplifier

Tachometer

Low-Voltage Comparator

Power Comparator

Comparator

Schmitt-Trigger

Square-Wave Oscillator

Pulse Generator

Typical Applications (Continued)

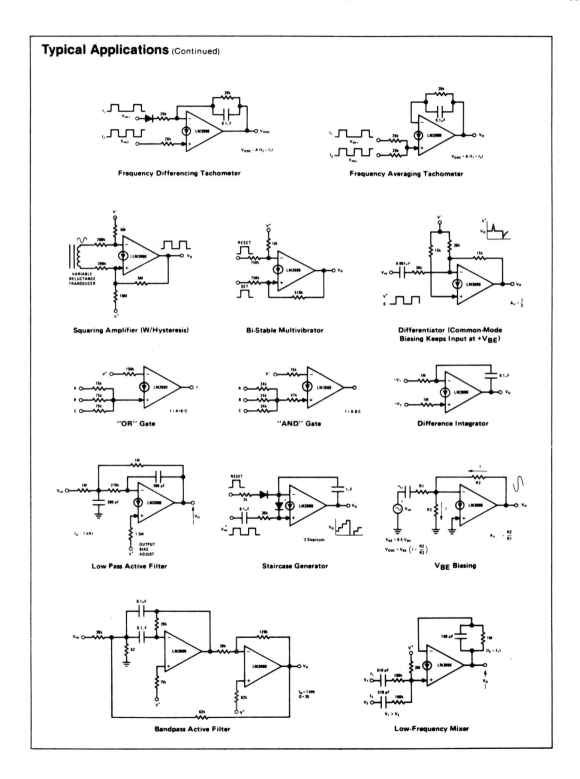

Frequency Differencing Tachometer

Frequency Averaging Tachometer

Squaring Amplifier (W/Hysteresis)

Bi-Stable Multivibrator

Differentiator (Common-Mode Biasing Keeps Input at $+V_{BE}$)

"OR" Gate

"AND" Gate

Difference Integrator

Low Pass Active Filter

Staircase Generator

V_{BE} Biasing

Bandpass Active Filter

Low-Frequency Mixer

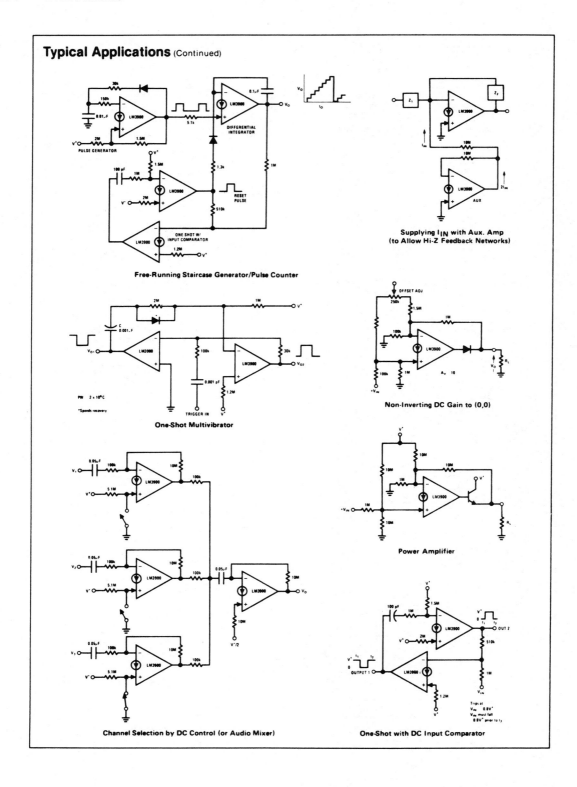

Typical Applications (Continued)

Free-Running Staircase Generator/Pulse Counter

Supplying I_{IN} with Aux. Amp
(to Allow Hi-Z Feedback Networks)

One-Shot Multivibrator

Non-Inverting DC Gain to (0,0)

Channel Selection by DC Control (or Audio Mixer)

Power Amplifier

One-Shot with DC Input Comparator

Typical Applications (Continued)

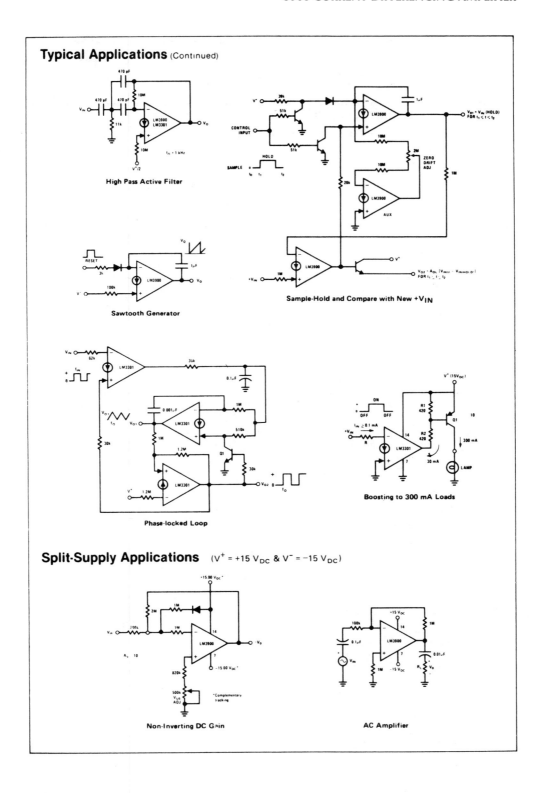

High Pass Active Filter

Sawtooth Generator

Sample-Hold and Compare with New +V$_{IN}$

Phase-locked Loop

Boosting to 300 mA Loads

Split-Supply Applications (V$^+$ = +15 V$_{DC}$ & V$^-$ = −15 V$_{DC}$)

Non-Inverting DC Gain

AC Amplifier

TYPES TIC35, TIC36
P-N-P-N PLANAR EPITAXIAL SILICON REVERSE-BLOCKING TRIODE THYRISTORS

RADIATION-TOLERANT THYRISTORS
400 mA DC • 15 and 30 VOLTS

- Max I_{GT} of 5 mA after 1×10^{14} Fast Neutrons/cm^2
- Max V_{TM} of 1.6 V at I_{TM} of 1 A after 1×10^{14} Fast Neutrons/cm^2

description

The TIC35, TIC36 thyristors offer a significant advance in radiation-tolerant-device technology. Unique construction techniques produce thyristors which maintain useful characteristics after fast-neutron radiation fluences through 10^{15} n/cm^2.

mechanical data

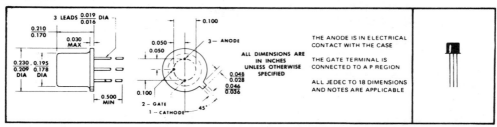

absolute maximum ratings over operating free-air temperature range (unless otherwise noted)

	TIC35	TIC36	UNIT
Continuous Off-State Voltage, V_D (See Note 1)	15	30	V
Repetitive Peak Off-State Voltage, V_{DRM} (See Note 1)	15	30	V
Continuous Reverse Voltage, V_R	5		V
Repetitive Peak Reverse Voltage, V_{RRM}	5		V
Nonrepetitive Peak Reverse Voltage, V_{RSM} (See Note 2)	5		V
Continuous On-State Current at (or below) 55°C Case Temperature (See Note 3)	400		mA
Continuous On-State Current at (or below) 25°C Free-Air Temperature (See Note 4)	225		mA
Average On-State Current (180° Conduction Angle) at (or below) 55°C Case Temperature (See Note 5)	320		mA
Surge On-State Current (See Note 6)	3		A
Peak Negative Gate Voltage	−4		V
Peak Positive Gate Current (Pulse Width ⩽300 μs)	250		mA
Peak Gate Power Dissipation (Pulse Width ⩽300 μs)	500		mW
Average Gate Power Dissipation	10		mW
Operating Free-Air or Case Temperature Range	−55 to 125		°C
Storage Temperature Range	−65 to 200		°C
Lead Temperature 1/16 Inch from Case for 10 Seconds	260		°C

NOTES: 1. These values apply when the gate-cathode resistance R_{GK} = 1 kΩ.
2. This value applies for a 5-ms rectangular pulse when the device is operating at (or below) rated values of peak reverse voltage and on-state current. Surge may be repeated after the device has returned to original thermal equilibrium.
3. These values apply for continuous d-c operation with resistive load. Above 55°C derate according to Figure 2.
4. These values apply for continuous d-c operation with resistive load. Above 25°C derate according to Figure 3.
5. This value may be applied continuously under single-phase, 60-Hz, half-sine-wave operation with resistive load. Above 55°C derate according to Figure 2.
6. This value applies for one 60-Hz half sine wave when the device is operating at (or below) rated values of peak reverse voltage and on-state current. Surge may be repeated after the device has returned to original thermal equilibrium.

TYPES TIC35, TIC36
P-N-P-N PLANAR EPITAXIAL SILICON REVERSE-BLOCKING TRIODE THYRISTORS

electrical characteristics at 25°C free-air temperature (unless otherwise noted)

	PARAMETER	TEST CONDITIONS		MIN	TYP	MAX	UNIT
I_D	Static Off-State Current	V_D = Rated V_D,	R_{GK} = 1 kΩ, T_A = 125°C			20	μA
I_R	Static Reverse Current	V_R = 5 V,	R_{GK} = 1 kΩ, T_A = 125°C			100	μA
I_{GR}	Gate Reverse Current	V_{KG} = 4 V,	I_A = 0			5	μA
I_{GT}	Gate Trigger Current	V_{AA} = 6 V, $R_{G(source)} \geqslant$ 10 kΩ, $t_{p(g)} \geqslant$ 20 μs, T_A = −55°C	R_L = 100 Ω, V_{GG} = 6 V,			100	μA
		V_{AA} = 6 V, $R_{G(source)} \geqslant$ 10 kΩ, $t_{p(g)} \geqslant$ 20 μs	R_L = 100 Ω, V_{GG} = 6 V,			20	
V_{GT}	Gate Trigger Voltage	V_{AA} = 6 V, $t_{p(g)} \geqslant$ 20 μs,	R_L = 100 Ω, R_{GK} = 1 kΩ, T_A = −55°C			0.9	V
		V_{AA} = Rated V_D, $t_{p(g)} \geqslant$ 20 μs,	R_L = 100 Ω, R_{GK} = 1 kΩ, T_A = 125°C	0.2			
		V_{AA} = 6 V, $t_{p(g)} \geqslant$ 20 μs	R_L = 100 Ω, R_{GK} = 1 kΩ,			0.75	
I_H	Holding Current	V_{AA} = 6 V,	R_{GK} = 1 kΩ, Initiating I_T = 10 mA, T_A = −55°C			4	mA
		V_{AA} = 6 V,	R_{GK} = 1 kΩ, Initiating I_T = 10 mA			2	
V_{TM}	Peak On-State Voltage	I_{TM} = 1 A,	See Note 7			1.6	V
dv/dt	Critical Rate of Rise of Off-State Voltage	V_D = Rated V_D,	R_{GK} = 1 kΩ		12		V/μs

post-irradiation electrical characteristics at 25°C free-air temperature

	PARAMETER	TEST CONDITIONS	RADIATION FLUENCE[†]	MIN	TYP	MAX	UNIT
I_{GT}	Gate Trigger Current	V_{AA} = 6 V, R_L = 100 Ω	1 x 10^{14} n/cm^2			5	mA
V_{TM}	Peak On-State Voltage	I_{TM} = 1 A, See Note 7	1 x 10^{14} n/cm^2			1.6	V

† Radiation is fast neutrons (n) at E $\geqslant$ 10 keV (reactor spectrum).

thermal characteristics

	PARAMETER	MIN	TYP	MAX	UNIT
$\theta_{J\text{-}C}$	Junction-to-Case Thermal Resistance			124	°C/W
$\theta_{J\text{-}A}$	Junction-to-Free-Air Thermal Resistance			345	

NOTE: 7. These parameters must be measured using pulse techniques. t_w = 300 μs, duty cycle $\leqslant$ 2%. Voltage-sensing contacts, separate from the current-carrying contacts, are used.

TYPES TIC226B, TIC226D
SILICON BIDIRECTIONAL TRIODE THYRISTORS

8 A RMS • 200 V and 400 V
TRIACS
for
HIGH-TEMPERATURE, HIGH-CURRENT, and HIGH-VOLTAGE APPLICATIONS
• Typ dv/dt of 500 V/μs at 25°C

description

These devices are bidirectional triode thyristors (triacs) which may be triggered from the off-state to the on-state by either polarity of gate signal with Main Terminal 2 at either polarity.

mechanical data

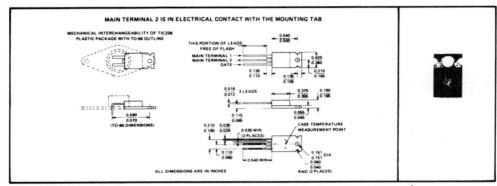

MAIN TERMINAL 2 IS IN ELECTRICAL CONTACT WITH THE MOUNTING TAB

ALL DIMENSIONS ARE IN INCHES

absolute maximum ratings over operating case temperature range (unless otherwise noted) [†]

			UNIT
Repetitive Peak Off-State Voltage, V$_{DRM}$ (See Note 1)	TIC226B	200	V
	TIC226D	400	
Full-Cycle RMS On-State Current at (or below) 85°C Case Temperature, I$_{T(RMS)}$ (See Note 2)		8	A
Peak On-State Surge Current, Full-Sine-Wave, I$_{TSM}$ (See Note 3)		70	A
Peak On-State Surge Current, Half-Sine-Wave, I$_{TSM}$ (See Note 4)		80	A
Peak Gate Current, I$_{GM}$		1	A
Peak Gate Power Dissipation, P$_{GM}$, at (or below) 85°C Case Temperature (Pulse Width ⩽ 200 μs)		2.2	W
Average Gate Power Dissipation, P$_{G(av)}$, at (or below) 85°C Case Temperature (See Note 5)		0.9	W
Operating Case Temperature Range		−40 to 110	°C
Storage Temperature Range		−40 to 125	°C
Lead Temperature 1/16 Inch from Case for 10 Seconds		230	°C

NOTES: 1. These values apply bidirectionally for any value of resistance between the gate and Main Terminal 1.
2. This value applies for 50-Hz to 60-Hz full-sine-wave operation with resistive load. Above 85°C derate according to Figure 2.
3. This value applies for one 60-Hz full sine wave when the device is operating at (or below) the rated value of on-state current. Surge may be repeated after the device has returned to original thermal equilibrium. During the surge, gate control may be lost.
4. This value applies for one 60-Hz half sine wave when the device is operating at (or below) the rated value of on-state current. Surge may be repeated after the device has returned to original thermal equilibrium. During the surge, gate control may be lost.
5. This value applies for a maximum averaging time of 16.6 ms.

[†]All voltage values are with respect to Main Terminal 1.

TYPES TIC226B, TIC226D
SILICON BIDIRECTIONAL TRIODE THYRISTORS

electrical characteristics at 25°C case temperature (unless otherwise noted)†

	PARAMETER	TEST CONDITIONS			MIN	TYP	MAX	UNIT
I_{DRM}	Repetitive Peak Off-State Current	V_{DRM} = Rated V_{DRM}, I_G = 0	T_C = 110°C				±2	mA
I_{GTM}	Peak Gate Trigger Current	V_{supply} = +12 V†,	R_L = 10 Ω,	$t_{p(g)}$ ⩾ 20 μs		15	50	mA
		V_{supply} = +12 V†,	R_L = 10 Ω,	$t_{p(g)}$ ⩾ 20 μs		−25	−50	
		V_{supply} = −12 V†,	R_L = 10 Ω,	$t_{p(g)}$ ⩾ 20 μs		−30	−50	
		V_{supply} = −12 V†,	R_L = 10 Ω,	$t_{p(g)}$ ⩾ 20 μs		75		
V_{GTM}	Peak Gate Trigger Voltage	V_{supply} = +12 V†,	R_L = 10 Ω,	$t_{p(g)}$ ⩾ 20 μs		0.9	2.5	V
		V_{supply} = +12 V†,	R_L = 10 Ω,	$t_{p(g)}$ ⩾ 20 μs		−1.2	−2.5	
		V_{supply} = −12 V†,	R_L = 10 Ω,	$t_{p(g)}$ ⩾ 20 μs		−1.2	−2.5	
		V_{supply} = −12 V†,	R_L = 10 Ω,	$t_{p(g)}$ ⩾ 20 μs		1.2		
V_{TM}	Peak On-State Voltage	I_{TM} = ±12 A,	I_G = 100 mA, See Note 6				±2.1	V
I_H	Holding Current	V_{supply} = +12 V†,	I_G = 0,	Initiating I_{TM} = 500 mA		20	60	mA
		V_{supply} = −12 V†,	I_G = 0,	Initiating I_{TM} = −500 mA		−30	−60	
I_L	Latching Current	V_{supply} = +12 V†,	See Note 7			30	70	mA
		V_{supply} = −12 V†,	See Note 7			−40	−70	
dv/dt	Critical Rate of Rise of Off-State Voltage	V_{DRM} = Rated V_{DRM},	I_G = 0,	T_C = 110°C		500		V/μs
dv/dt	Critical Rate of Rise of Commutation Voltage	V_{DRM} = Rated V_{DRM}, I_{TRM} = ±12 A, T_C = 85°C, See Figure 3			5			V/μs

†All voltage values are with respect to Main Terminal 1.

NOTES: 6. This parameter must be measured using pulse techniques. t_w ⩽ 1 ms, duty cycle ⩽ 2%. Voltage-sensing contacts, separate from the current-carrying contacts, are located within 0.125 inch from the device body.

7. The triacs are triggered by a 15-V (open-circuit amplitude) pulse supplied by a generator with the following characteristics: R_G = 100 Ω, t_w = 20 μs, t_r ⩽ 15 ns, t_f ⩽ 15 ns, f = 1 kHz.

thermal characteristics

	PARAMETER	MAX	UNIT
$R_{\theta JC}$	Junction-to-Case Thermal Resistance	1.8	°C/W
$R_{\theta JA}$	Junction-to-Free-Air Thermal Resistance	62.5	

Industrial Blocks

LM135/LM235/LM335, LM135A/LM235A/LM335A
Precision Temperature Sensors

General Description

The LM135 series are precision, easily-calibrated, integrated circuit temperature sensors. Operating as a 2-terminal zener, the LM135 has a breakdown voltage directly proportional to absolute temperature at +10 mV/°K. With less than 1Ω dynamic impedance the device operates over a current range of 400 μA to 5 mA with virtually no change in performance. When calibrated at 25°C the LM135 has typically less than 1°C error over a 100°C temperature range. Unlike other sensors the LM135 has a linear output.

Applications for the LM135 include almost any type of temperature sensing over a −55°C to +150°C temperature range. The low impedance and linear output make interfacing to readout or control circuitry especially easy.

The LM135 operates over a −55°C to +150°C temperature range while the LM235 operates over a −40°C

to +125°C temperature range. The LM335 operates from −40°C to +100°C. The LM135/LM235/LM335 are available packaged in hermetic TO-46 transistor packages while the LM335 is also available in plastic TO-92 packages.

Features

- Directly calibrated in °Kelvin
- 1°C initial accuracy available
- Operates from 400 μA to 5 mA
- Less than 1Ω dynamic impedance
- Easily calibrated
- Wide operating temperature range
- 200°C overrange
- Low cost

Schematic Diagram

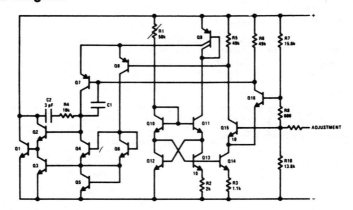

Typical Applications

Basic Temperature Sensor

Calibrated Sensor

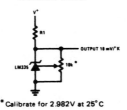

*Calibrate for 2.982V at 25°C

Wide Operating Supply

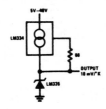

Absolute Maximum Ratings

Reverse Current	15 mA
Forward Current	10 mA
Storage Temperature	
TO-46 Package	-60°C to $+180^\circ$C
TO-92 Package	-60°C to $+150^\circ$C

Specified Operating Temperature Range

	Continuous	Intermittent (Note 2)
LM135, LM135A	-55°C to $+150^\circ$C	150°C to 200°C
LM235, LM235A	-40°C to $+125^\circ$C	125°C to 150°C
LM335, LM335A	-40°C to $+100^\circ$C	100°C to 125°C
Lead Temperature (Soldering, 10 seconds)		300°C

Temperature Accuracy LM135/LM235, LM135A/LM235A (Note 1)

PARAMETER	CONDITIONS	LM135A/LM235A			LM135/LM235			UNITS
		MIN	TYP	MAX	MIN	TYP	MAX	
Operating Output Voltage	$T_C = 25^\circ$C, $I_R = 1$ mA	2.97	2.98	2.99	2.95	2.98	3.01	V
Uncalibrated Temperature Error	$T_C = 25^\circ$C, $I_R = 1$ mA		0.5	1		1	3	$^\circ$C
Uncalibrated Temperature Error	$T_{MIN} < T_C < T_{MAX}$, $I_R = 1$ mA		1.3	2.7		2	5	$^\circ$C
Temperature Error with 25°C Calibration	$T_{MIN} < T_C < T_{MAX}$, $I_R = 1$ mA		0.3	1		0.5	1.5	$^\circ$C
Calibrated Error at Extended Temperatures	$T_C = T_{MAX}$ (Intermittent)		2			2		$^\circ$C
Non-Linearity	$I_R = 1$ mA		0.3	0.5		0.3	1	$^\circ$C

Temperature Accuracy LM335, LM335A (Note 1)

PARAMETER	CONDITIONS	LM335A			LM335			UNITS
		MIN	TYP	MAX	MIN	TYP	MAX	
Operating Output Voltage	$T_C = 25^\circ$C, $I_R = 1$ mA	2.95	2.98	3.01	2.92	2.98	3.04	V
Uncalibrated Temperature Error	$T_C = 25^\circ$C, $I_R = 1$ mA		1	3		2	6	$^\circ$C
Uncalibrated Temperature Error	$T_{MIN} < T_C < T_{MAX}$, $I_R = 1$ mA		2	5		4	9	$^\circ$C
Temperature Error with 25°C Calibration	$T_{MIN} < T_C < T_{MAX}$, $I_R = 1$ mA		0.5	1		1	2	$^\circ$C
Calibrated Error at Extended Temperatures	$T_C = T_{MAX}$ (Intermittent)		2			2		$^\circ$C
Non-Linearity	$I_R = 1$ mA		0.3	1.5		0.3	1.5	$^\circ$C

Electrical Characteristics (Note 1)

PARAMETER	CONDITIONS	LM135/LM235 LM135A/LM235A			LM335 LM335A			UNITS
		MIN	TYP	MAX	MIN	TYP	MAX	
Operating Output Voltage Change with Current	$400\ \mu$A $< I_R <$ 5 mA At Constant Temperature		2.5	10		3	14	mV
Dynamic Impedance	$I_R = 1$ mA		0.5			0.6		Ω
Output Voltage Temperature Drift			+10			+10		mV/$^\circ$C
Time Constant	Still Air		80			80		sec
	100 ft/Min Air		10			10		sec
	Stirred Oil		1			1		sec
Time Stability	$T_C = 125^\circ$C		0.2			0.2		$^\circ$C/khr

Note 1: Accuracy measurements are made in a well-stirred oil bath. For other conditions, self heating must be considered.

Note 2: Continuous operation at these temperatures for 10,000 hours for H package and 5,000 hours for Z package may decrease life expectancy of the device.

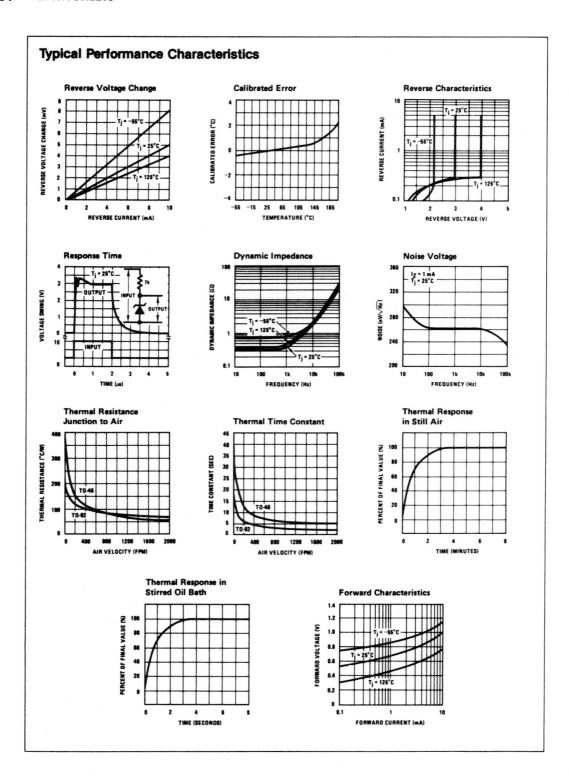

Typical Performance Characteristics

 **National Semiconductor**

LM555/LM555C Timer

General Description

The LM555 is a highly stable device for generating accurate time delays or oscillation. Additional terminals are provided for triggering or resetting if desired. In the time delay mode of operation, the time is precisely controlled by one external resistor and capacitor. For astable operation as an oscillator, the free running frequency and duty cycle are accurately controlled with two external resistors and one capacitor. The circuit may be triggered and reset on falling waveforms, and the output circuit can source or sink up to 200 mA or drive TTL circuits.

Features

- Direct replacement for SE555/NE555
- Timing from microseconds through hours
- Operates in both astable and monostable modes

- Adjustable duty cycle
- Output can source or sink 200 mA
- Output and supply TTL compatible
- Temperature stability better than 0.005% per $^\circ$C
- Normally on and normally off output

Applications

- Precision timing
- Pulse generation
- Sequential timing
- Time delay generation
- Pulse width modulation
- Pulse position modulation
- Linear ramp generator

Schematic Diagram

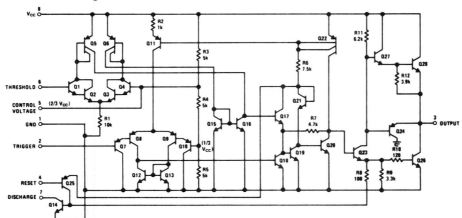

Connection Diagrams

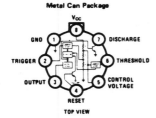

Metal Can Package

TOP VIEW

Order Number LM555H, LM555CH
See NS Package H08C

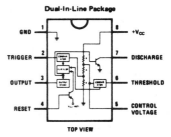

Dual-In-Line Package

TOP VIEW

Order Number LM555CN
See NS Package N08B
Order Number LM555J or LM555CJ
See NS Package J08A

LM555/LM555C

Absolute Maximum Ratings

Supply Voltage	+18V
Power Dissipation (Note 1)	600 mW
Operating Temperature Ranges	
LM555C	0°C to +70°C
LM555	−55°C to +125°C
Storage Temperature Range	−65°C to +150°C
Lead Temperature (Soldering, 10 seconds)	300°C

Electrical Characteristics (T_A = 25°C, V_{CC} = +5V to +15V, unless otherwise specified)

PARAMETER	CONDITIONS	LIMITS						UNITS
		LM555			LM555C			
		MIN	TYP	MAX	MIN	TYP	MAX	
Supply Voltage		4.5		18	4.5		16	V
Supply Current	V_{CC} = 5V, R_L = ∞		3	5		3	6	mA
	V_{CC} = 15V, R_L = ∞		10	12		10	15	mA
	(Low State) (Note 2)							
Timing Error, Monostable								
Initial Accuracy			0.5			1		%
Drift with Temperature	R_A, R_B = 1k to 100 k,		30			50		ppm/°C
	C = 0.1μF, (Note 3)							
Accuracy over Temperature			1.5			1.5		%
Drift with Supply			0.05			0.1		%/V
Timing Error, Astable								
Initial Accuracy			1.5			2.25		%
Drift with Temperature			90			150		ppm/°C
Accuracy over Temperature			2.5			3.0		%
Drift with Supply			0.15			0.30		%/V
Threshold Voltage			0.667			0.667		x V_{CC}
Trigger Voltage	V_{CC} = 15V	4.8	5	5.2		5		V
	V_{CC} = 5V	1.45	1.67	1.9		1.67		V
Trigger Current			0.01	0.5		0.5	0.9	μA
Reset Voltage		0.4	0.5	1	0.4	0.5	1	V
Reset Current			0.1	0.4		0.1	0.4	mA
Threshold Current	(Note 4)		0.1	0.25		0.1	0.25	μA
Control Voltage Level	V_{CC} = 15V	9.6	10	10.4	9	10	11	V
	V_{CC} = 5V	2.9	3.33	3.8	2.6	3.33	4	V
Pin 7 Leakage Output High			1	100		1	100	nA
Pin 7 Sat (Note 5)								
Output Low	V_{CC} = 15V, I_7 = 15 mA		150			180		mV
Output Low	V_{CC} = 4.5V, I_7 = 4.5 mA		70	100		80	200	mV
Output Voltage Drop (Low)	V_{CC} = 15V							
	I_{SINK} = 10 mA		0.1	0.15		0.1	0.25	V
	I_{SINK} = 50 mA		0.4	0.5		0.4	0.75	V
	I_{SINK} = 100 mA		2	2.2		2	2.5	V
	I_{SINK} = 200 mA		2.5			2.5		V
	V_{CC} = 5V							
	I_{SINK} = 8 mA		0.1	0.25				V
	I_{SINK} = 5 mA					0.25	0.35	V
Output Voltage Drop (High)	I_{SOURCE} = 200 mA, V_{CC} = 15V		12.5			12.5		V
	I_{SOURCE} = 100 mA, V_{CC} = 15V	13	13.3		12.75	13.3		V
	V_{CC} = 5V	3	3.3		2.75	3.3		V
Rise Time of Output			100			100		ns
Fall Time of Output			100			100		ns

Note 1: For operating at elevated temperatures the device must be derated based on a +150°C maximum junction temperature and a thermal resistance of +45°C/W junction to case for TO-5 and +150°C/W junction to ambient for both packages.

Note 2: Supply current when output high typically 1 mA less at V_{CC} = 5V.

Note 3: Tested at V_{CC} = 5V and V_{CC} = 15V.

Note 4: This will determine the maximum value of R_A + R_B for 15V operation. The maximum total (R_A + R_B) is 20 MΩ.

Note 5: No protection against excessive pin 7 current is necessary providing the package dissipation rating will not be exceeded.

Typical Performance Characteristics

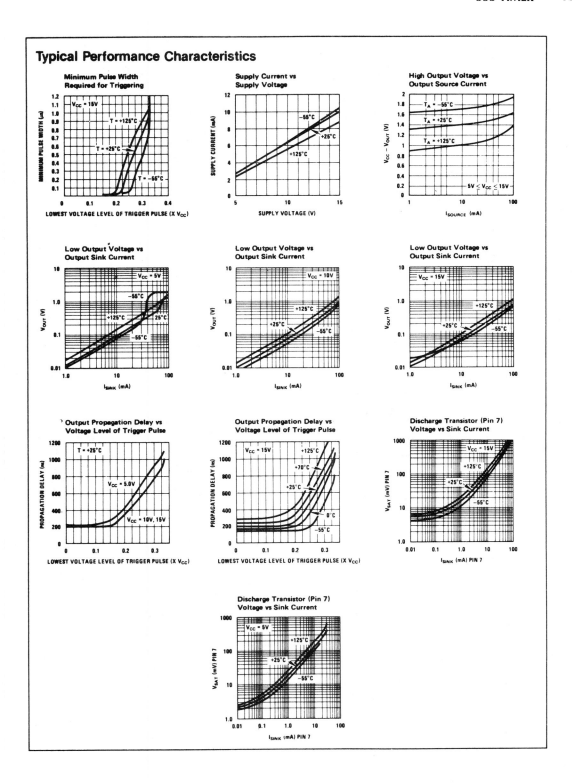

Applications Information

MONOSTABLE OPERATION

In this mode of operation, the timer functions as a one-shot (*Figure 1*). The external capacitor is initially held discharged by a transistor inside the timer. Upon application of a negative trigger pulse of less than 1/3 V_{CC} to pin 2, the flip-flop is set which both releases the short circuit across the capacitor and drives the output high.

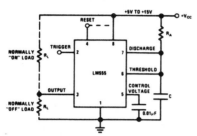

FIGURE 1. Monostable

The voltage across the capacitor then increases exponentially for a period of t = 1.1 R_AC, at the end of which time the voltage equals 2/3 V_{CC}. The comparator then resets the flip-flop which in turn discharges the capacitor and drives the output to its low state. *Figure 2* shows the waveforms generated in this mode of operation. Since the charge and the threshold level of the comparator are both directly proportional to supply voltage, the timing internal is independent of supply.

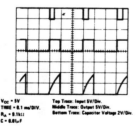

V_{CC} = 5V
TIME = 0.1 ms/DIV.
R_A = 9.1kΩ
C = 0.01μF

Top Trace: Input 5V/Div.
Middle Trace: Output 5V/Div.
Bottom Trace: Capacitor Voltage 2V/Div.

FIGURE 2. Monostable Waveforms

During the timing cycle when the output is high, the further application of a trigger pulse will not effect the circuit. However the circuit can be reset during this time by the application of a negative pulse to the reset terminal (pin 4). The output will then remain in the low state until a trigger pulse is again applied.

When the reset function is not in use, it is recommended that it be connected to V_{CC} to avoid any possibility of false triggering.

Figure 3 is a nomograph for easy determination of R, C values for various time delays.

NOTE: In monostable operation, the trigger should be driven high before the end of timing cycle.

ASTABLE OPERATION

If the circuit is connected as shown in *Figure 4* (pins 2 and 6 connected) it will trigger itself and free run as a

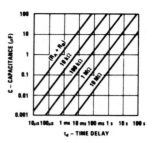

FIGURE 3. Time Delay

multivibrator. The external capacitor charges through R_A + R_B and discharges through R_B. Thus the duty cycle may be precisely set by the ratio of these two resistors.

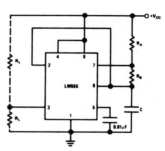

FIGURE 4. Astable

In this mode of operation, the capacitor charges and discharges between 1/3 V_{CC} and 2/3 V_{CC}. As in the triggered mode, the charge and discharge times, and therefore the frequency are independent of the supply voltage.

Figure 5 shows the waveforms generated in this mode of operation.

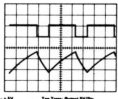

V_{CC} = 5V
TIME = 20μs/DIV.
R_A = 3.9 kΩ
R_B = 3 kΩ
C = 0.01μF

Top Trace: Output 5V/Div.
Bottom Trace: Capacitor Voltage 1V/Div.

FIGURE 5. Astable Waveforms

The charge time (output high) is given by:
$$t_1 = 0.693 (R_A + R_B) C$$

And the discharge time (output low) by:
$$t_2 = 0.693 (R_B) C$$

Thus the total period is:
$$T = t_1 + t_2 = 0.693 (R_A + 2R_B) C$$

Applications Information (Continued)

The frequency of oscillation is:

$$f = \frac{1}{T} = \frac{1.44}{(R_A + 2R_B)\,C}$$

Figure 6 may be used for quick determination of these RC values.

The duty cycle is: $$D = \frac{R_B}{R_A + 2R_B}$$

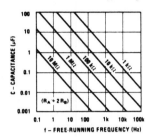

FIGURE 6. Free Running Frequency

FREQUENCY DIVIDER

The monostable circuit of *Figure 1* can be used as a frequency divider by adjusting the length of the timing cycle. *Figure 7* shows the waveforms generated in a divide by three circuit.

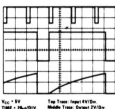

V_{CC} = 5V
TIME = 20μs/DIV.
R_A = 9.1 kΩ
C = 0.01μF

Top Trace: Input 4V/Div.
Middle Trace: Output 2V/Div.
Bottom Trace: Capacitor 2V/Div.

FIGURE 7. Frequency Divider

PULSE WIDTH MODULATOR

When the timer is connected in the monostable mode and triggered with a continuous pulse train, the output pulse width can be modulated by a signal applied to pin 5. *Figure 8* shows the circuit, and in *Figure 9* are some waveform examples.

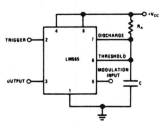

FIGURE 8. Pulse Width Modulator

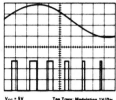

V_{CC} = 5V
TIME = 0.2 ms/DIV.
R_A = 9.1 kΩ
C = 0.01μF

Top Trace: Modulation 1V/Div.
Bottom Trace: Output 2V/Div.

FIGURE 9. Pulse Width Modulator

PULSE POSITION MODULATOR

This application uses the timer connected for astable operation, as in *Figure 10*, with a modulating signal again applied to the control voltage terminal. The pulse position varies with the modulating signal, since the threshold voltage and hence the time delay is varied. *Figure 11* shows the waveforms generated for a triangle wave modulation signal.

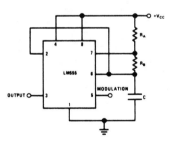

FIGURE 10. Pulse Position Modulator

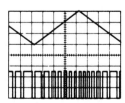

V_{CC} = 5V
TIME = 0.1 ms/DIV.
R_A = 3.9 kΩ
R_B = 3 kΩ
C = 0.01μF

Top Trace: Modulation Input 1V/Div.
Bottom Trace: Output 2V/Div.

FIGURE 11. Pulse Position Modulator

LINEAR RAMP

When the pullup resistor, R_A, in the monostable circuit is replaced by a constant current source, a linear ramp is

Applications Information (Continued)

generated. *Figure 12* shows a circuit configuration that will perform this function.

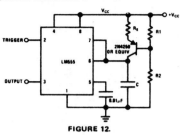

FIGURE 12.

Figure 13 shows waveforms generated by the linear ramp.

The time interval is given by:

$$T = \frac{2/3\ V_{CC}\ R_E\ (R_1 + R_2)\ C}{R_1\ V_{CC} - V_{BE}\ (R_1 + R_2)}$$

$$V_{BE} \simeq 0.6V$$

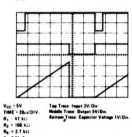

$V_{CC} = 5V$
TIME = 20μs/DIV.
$R_1 = 47$ kΩ
$R_2 = 100$ kΩ
$R_E = 2.7$ kΩ
$C = 0.01μF$

Top Trace: Input 3V/Div.
Middle Trace: Output 5V/Div.
Bottom Trace: Capacitor Voltage 1V/Div.

FIGURE 13. Linear Ramp

50% DUTY CYCLE OSCILLATOR

For a 50% duty cycle, the resistors R_A and R_B may be connected as in *Figure 14*. The time period for the out-put high is the same as previous, $t_1 = 0.693\ R_A\ C$. For the output low it is $t_2 =$

$$[(R_A\ R_B)/(R_A + R_B)]\ C Ln \left[\frac{R_B - 2R_A}{2R_B - R_A} \right]$$

Thus the frequency of oscillation is $f = \dfrac{1}{t_1 + t_2}$

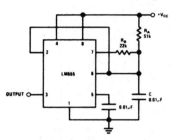

FIGURE 14. 50% Duty Cycle Oscillator

Note that this circuit will not oscillate if R_B is greater than $1/2\ R_A$ because the junction of R_A and R_B cannot bring pin 2 down to $1/3\ V_{CC}$ and trigger the lower comparator.

ADDITIONAL INFORMATION

Adequate power supply bypassing is necessary to protect associated circuitry. Minimum recommended is $0.1\mu F$ in parallel with $1\mu F$ electrolytic.

Lower comparator storage time can be as long as $10\mu s$ when pin 2 is driven fully to ground for triggering. This limits the monostable pulse width to $10\mu s$ minimum.

Delay time reset to output is $0.47\mu s$ typical. Minimum reset pulse width must be $0.3\mu s$, typical.

Pin 7 current switches within 30 ns of the output (pin 3) voltage.

National Semiconductor

Industrial Blocks

LM565/LM565C Phase Locked Loop

General Description

The LM565 and LM565C are general purpose phase locked loops containing a stable, highly linear voltage controlled oscillator for low distortion FM demodulation, and a double balanced phase detector with good carrier suppression. The VCO frequency is set with an external resistor and capacitor, and a tuning range of 10:1 can be obtained with the same capacitor. The characteristics of the closed loop system—bandwidth, response speed, capture and pull in range—may be adjusted over a wide range with an external resistor and capacitor. The loop may be broken between the VCO and the phase detector for insertion of a digital frequency divider to obtain frequency multiplication.

The LM565H is specified for operation over the −55°C to +125°C military temperature range. The LM565CH and LM565CN are specified for operation over the 0°C to +70°C temperature range.

Features

- 200 ppm/°C frequency stability of the VCO
- Power supply range of ±5 to ±12 volts with 100 ppm/% typical
- 0.2% linearity of demodulated output
- Linear triangle wave with in phase zero crossings available
- TTL and DTL compatible phase detector input and square wave output
- Adjustable hold in range from ±1% to > ±60%.

Applications

- Data and tape synchronization
- Modems
- FSK demodulation
- FM demodulation
- Frequency synthesizer
- Tone decoding
- Frequency multiplication and division
- SCA demodulators
- Telemetry receivers
- Signal regeneration
- Coherent demodulators.

Schematic and Connection Diagrams

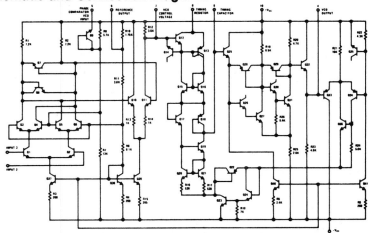

Metal Can Package

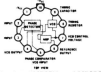

Order Number LM565H or LM565CH
See NS Package H10C

Dual-In-Line Package

Order Number LM565CN
See NS Package N14A

Absolute Maximum Ratings

Supply Voltage	±12V
Power Dissipation (Note 1)	300 mW
Differential Input Voltage	±1V
Operating Temperature Range LM565H	−55°C to +125°C
LM565CH, LM565CN	0°C to 70°C
Storage Temperature Range	−65°C to +150°C
Lead Temperature (Soldering, 10 sec)	300°C

Electrical Characteristics (AC Test Circuit, T_A = 25°C, V_C = ±6V)

PARAMETER	CONDITIONS	LM565			LM565C			UNITS
		MIN	TYP	MAX	MIN	TYP	MAX	
Power Supply Current			8.0	12.5		8.0	12.5	mA
Input Impedance (Pins 2, 3)	−4V < V_2, V_3 < 0V	7	10			5		kΩ
VCO Maximum Operating Frequency	C_o = 2.7 pF	300	500		250	500		kHz
Operating Frequency Temperature Coefficient			−100	300		−200	500	ppm/°C
Frequency Drift with Supply Voltage			0.01	0.1		0.05	0.2	%/V
Triangle Wave Output Voltage		2	2.4	3	2	2.4	3	V_{p-p}
Triangle Wave Output Linearity			0.2	0.75		0.5	1	%
Square Wave Output Level		4.7	5.4		4.7	5.4		V_{p-p}
Output Impedance (Pin 4)			5			5		kΩ
Square Wave Duty Cycle		45	50	55	40	50	60	%
Square Wave Rise Time			20	100		20		ns
Square Wave Fall Time			50	200		50		ns
Output Current Sink (Pin 4)		0.6	1		0.6	1		mA
VCO Sensitivity	f_o = 10 kHz	6400	6600	6800	6000	6600	7200	Hz/V
Demodulated Output Voltage (Pin 7)	±10% Frequency Deviation	250	300	350	200	300	400	mV_{pp}
Total Harmonic Distortion	±10% Frequency Deviation		0.2	0.75		0.2	1.5	%
Output Impedance (Pin 7)			3.5			3.5		kΩ
DC Level (Pin 7)		4.25	4.5	4.75	4.0	4.5	5.0	V
Output Offset Voltage $|V_7 - V_6|$			30	100		50	200	mV
Temperature Drift of $|V_7 - V_6|$			500			500		μV/°C
AM Rejection		30	40			40		dB
Phase Detector Sensitivity K_D		0.6	.68	0.9	0.55	.68	0.95	V/radian

Note 1: The maximum junction temperature of the LM565 is 150°C, while that of the LM565C and LM565CN is 100°C. For operation at elevated temperatures, devices in the TO-5 package must be derated based on a thermal resistance of 150°C/W junction to ambient or 45°C/W junction to case. Thermal resistance of the dual-in-line package is 100°C/W.

Typical Performance Characteristics

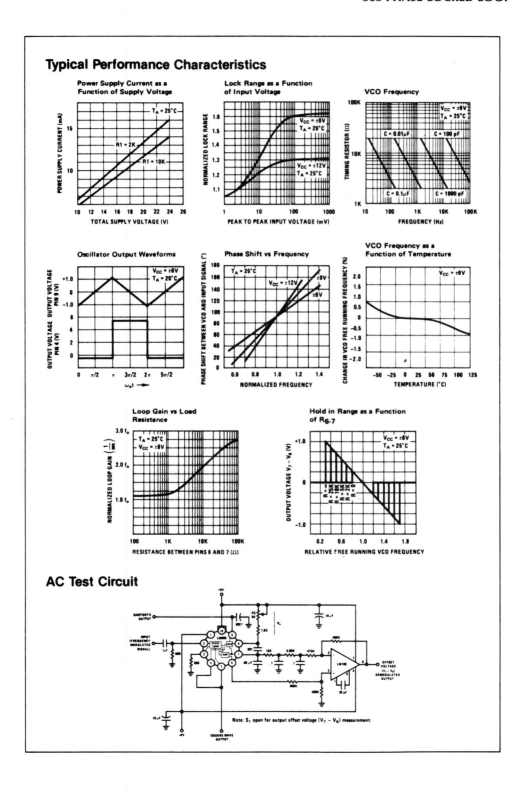

AC Test Circuit

Note: S₁ open for output offset voltage (V₇ − V₆) measurement.

Typical Applications

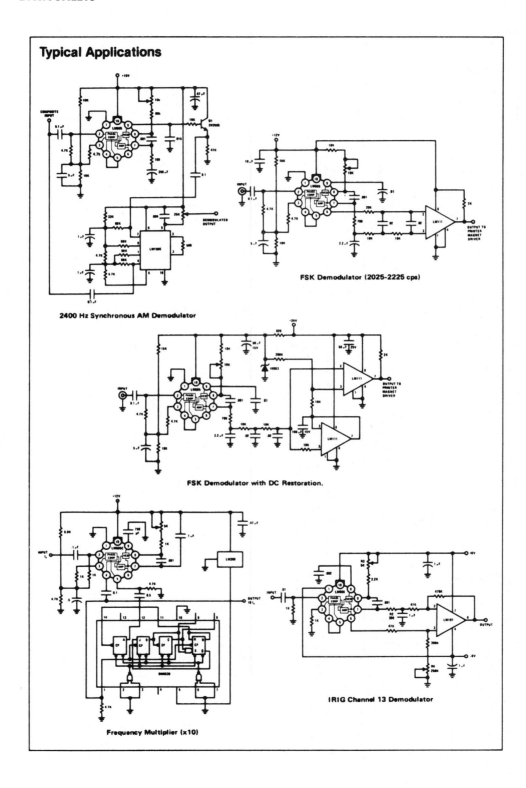

2400 Hz Synchronous AM Demodulator

FSK Demodulator (2025-2225 cps)

FSK Demodulator with DC Restoration.

Frequency Multiplier (x10)

IRIG Channel 13 Demodulator

Applications Information

In designing with phase locked loops such as the LM565, the important parameters of interest are:

FREE RUNNING FREQUENCY

$$f_o \cong \frac{1}{3.7\,R_0 C_0}$$

LOOP GAIN: relates the amount of phase change between the input signal and the VCO signal for a shift in input signal frequency (assuming the loop remains in lock). In servo theory, this is called the "velocity error coefficient".

$$\text{Loop gain} = K_o K_D \left(\frac{1}{sec}\right)$$

$$K_o = \text{oscillator sensitivity} \left(\frac{radians/sec}{volt}\right)$$

$$K_D = \text{phase detector sensitivity} \left(\frac{volts}{radian}\right)$$

The loop gain of the LM565 is dependent on supply voltage, and may be found from:

$$K_o K_D = \frac{33.6\,f_o}{V_c}$$

$f_o = $ VCO frequency in Hz

$V_c = $ total supply voltage to circuit.

Loop gain may be reduced by connecting a resistor between pins 6 and 7; this reduces the load impedance on the output amplifier and hence the loop gain.

HOLD IN RANGE: the range of frequencies that the loop will remain in lock after initially being locked.

$$f_H = \pm \frac{8\,f_o}{V_c}$$

$f_o = $ free running frequency of VCO

$V_c = $ total supply voltage to the circuit.

THE LOOP FILTER

In almost all applications, it will be desirable to filter the signal at the output of the phase detector (pin 7) this filter may take one of two forms:

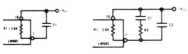

Simple Lag Filter   **Lag-Lead Filter**

A simple lag filter may be used for wide closed loop bandwidth applications such as modulation following where the frequency deviation of the carrier is fairly high (greater than 10%), or where wideband modulating signals must be followed.

The natural bandwidth of the closed loop response may be found from:

$$f_n = \frac{1}{2\pi} \sqrt{\frac{K_o K_D}{R_1 C_1}}$$

Associated with this is a damping factor:

$$\delta = \frac{1}{2} \sqrt{\frac{1}{R_1 C_1 K_o K_D}}$$

For narrow band applications where a narrow noise bandwidth is desired, such as applications involving tracking a slowly varying carrier, a lead lag filter should be used. In general, if $1/R_1 C_1 < K_o K_d$, the damping factor for the loop becomes quite small resulting in large overshoot and possible instability in the transient response of the loop. In this case, the natural frequency of the loop may be found from

$$f_n = \frac{1}{2\pi} \sqrt{\frac{K_o K_D}{\tau_1 + \tau_2}}$$

$$\tau_1 + \tau_2 = (R_1 + R_2)\,C_1$$

R_2 is selected to produce a desired damping factor δ, usually between 0.5 and 1.0. The damping factor is found from the approximation:

$$\delta \simeq \pi\,\tau_2 f_n$$

These two equations are plotted for convenience.

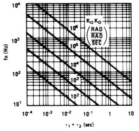

Filter Time Constant vs Natural Frequency

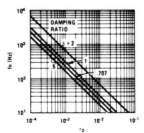

Damping Time Constant vs Natural Frequency

Capacitor C_2 should be much smaller than C_1 since its function is to provide filtering of carrier. In general $C_2 \leq 0.1\,C_1$.

 **National Semiconductor**

ADC0801, ADC0802, ADC0803, ADC0804, ADC0805 8-Bit μP Compatible A/D Converters

General Description

The ADC0801, ADC0802, ADC0803, ADC0804 and ADC0805 are CMOS 8-bit successive approximation A/D converters which use a differential potentiometric ladder—similar to the 256R products. These converters are designed to allow operation with the NSC800 and INS8080A derivative control bus, and TRI-STATE® output latches directly drive the data bus. These A/Ds appear like memory locations or I/O ports to the microprocessor and no interfacing logic is needed.

A new differential analog voltage input allows increasing the common-mode rejection and offsetting the analog zero input voltage value. In addition, the voltage reference input can be adjusted to allow encoding any smaller analog voltage span to the full 8 bits of resolution.

- Differential analog voltage inputs
- Logic inputs and outputs meet both MOS and T^2L voltage level specifications
- Works with 2.5V (LM336) voltage reference
- On-chip clock generator
- 0V to 5V analog input voltage range with single 5V supply
- No zero adjust required
- 0.3" standard width 20-pin DIP package
- Operates ratiometrically or with 5 V_{DC}, 2.5 V_{DC}, or analog span adjusted voltage reference

Key Specifications

- Resolution   8 bits
- Total error   ±1/4 LSB, ±1/2 LSB and ±1 LSB
- Conversion time   100 μs

Features

- Compatible with 8080 μP derivatives—no interfacing logic needed — access time — 135 ns
- Easy interface to all microprocessors, or operates "stand alone"

Typical Applications

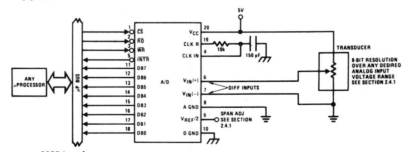

8080 Interface

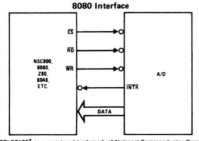

ERROR SPECIFICATION (INCLUDES FULL-SCALE, ZERO ERROR, AND NON-LINEARITY)			
PART NUMBER	FULL-SCALE ADJUSTED	$V_{REF}/2$ = 2.500 V_{DC} (NO ADJUSTMENTS)	$V_{REF}/2$ = NO CONNECTION (NO ADJUSTMENTS)
ADC0801	±1/4 LSB		
ADC0802		±1/2 LSB	
ADC0803	±1/2 LSB		
ADC0804		±1 LSB	
ADC0805			±1 LSB

TRI-STATE® is a registered trademark of National Semiconductor Corp.

Absolute Maximum Ratings (Notes 1 and 2)

Supply Voltage (V_{CC}) (Note 3)	6.5V
Voltage	
Logic Control Inputs	−0.3V to +18V
At Other Input and Outputs	−0.3V to (V_{CC} + 0.3V)
Storage Temperature Range	−65°C to +150°C
Package Dissipation at T_A = 25°C	875 mW
Lead Temperature (Soldering, 10 seconds)	300°C

Operating Ratings (Notes 1 and 2)

Temperature Range	$T_{MIN} \leq T_A \leq T_{MAX}$
ADC0801/02LD	−55°C $\leq T_A \leq$ +125°C
ADC0801/02/03/04LCD	−40°C $\leq T_A \leq$ +85°C
ADC0801/02/03/05LCN	−40°C $\leq T_A \leq$ +85°C
ADC0804LCN	0°C $\leq T_A \leq$ +70°C
Range of V_{CC}	4.5 V_{DC} to 6.3 V_{DC}

Electrical Characteristics

The following specifications apply for V_{CC} = 5 V_{DC}, $T_{MIN} \leq T_A \leq T_{MAX}$ and f_{CLK} = 640 kHz unless otherwise specified.

PARAMETER	CONDITIONS	MIN	TYP	MAX	UNITS
ADC0801:					
Total Adjusted Error	With Full-Scale Adj.			±1/4	LSB
(Note 8)	(See Section 2.5.2)				
ADC0802:					
Total Unadjusted Error	$V_{REF}/2$ = 2.500 V_{DC}			±1/2	LSB
(Note 8)					
ADC0803:					
Total Adjusted Error	With Full-Scale Adj.			±1/2	LSB
(Note 8)	(See Section 2.5.2)				
ADC0804:					
Total Unadjusted Error	$V_{REF}/2$ = 2.500 V_{DC}			±1	LSB
(Note 8)					
ADC0805:					
Total Unadjusted Error	$V_{REF}/2$ − No Connection			±1	LSB
(Note 8)					
$V_{REF}/2$ Input Resistance (Pin 9)	ADC0801/02/03/05	2.5	8.0		kΩ
	ADC0804 (Note 9)	1.0	1.3		kΩ
Analog Input Voltage Range	(Note 4) V(+) or V(−)	Gnd−0.05		V_{CC}+0.05	V_{DC}
DC Common-Mode Error	Over Analog Input Voltage Range		±1/16	±1/8	LSB
Power Supply Sensitivity	V_{CC} = 5 V_{DC} ±10% Over Allowed V_{IN}(+) and V_{IN}(−) Voltage Range (Note 4)		±1/16	±1/8	LSB

AC Electrical Characteristics

The following specifications apply for V_{CC} = 5 V_{DC} and T_A = 25°C unless otherwise specified.

	PARAMETER	CONDITIONS	MIN	TYP	MAX	UNITS
T_c	Conversion Time	f_{CLK} = 640 kHz (Note 6)	103		114	µs
T_c	Conversion Time	(Note 5, 6)	66		73	$1/f_{CLK}$
f_{CLK}	Clock Frequency	V_{CC} = 5V, (Note 5)	100	640	1460	kHz
	Clock Duty Cycle	(Note 5)	40		60	%
CR	Conversion Rate In Free-Running Mode	$\overline{INTR}$ tied to $\overline{WR}$ with $\overline{CS}$ = 0 V_{DC}, f_{CLK} = 640 kHz			8770	conv/s
$t_{W(\overline{WR})L}$	Width of $\overline{WR}$ Input (Start Pulse Width)	$\overline{CS}$ = 0 V_{DC} (Note 7)	100			ns
t_{ACC}	Access Time (Delay from Falling Edge of $\overline{RD}$ to Output Data Valid)	C_L = 100 pF		135	200	ns
t_{1H}, t_{0H}	TRI-STATE Control (Delay from Rising Edge of $\overline{RD}$ to Hi-Z State)	C_L = 10 pF, R_L = 10k (See TRI-STATE Test Circuits)		125	200	ns
t_{WI}, t_{RI}	Delay from Falling Edge of $\overline{WR}$ or $\overline{RD}$ to Reset of $\overline{INTR}$			300	450	ns
C_{IN}	Input Capacitance of Logic Control Inputs			5	7.5	pF
C_{OUT}	TRI-STATE Output Capacitance (Data Buffers)			5	7.5	pF

Electrical Characteristics

The following specifications apply for V_{CC} = 5 V_{DC} and $T_{MIN} \leq T_A \leq T_{MAX}$, unless otherwise specified.

PARAMETER		CONDITIONS	MIN	TYP	MAX	UNITS
CONTROL INPUTS [Note: CLK IN (Pin 4) is the input of a Schmitt trigger circuit and is therefore specified separately]						
V_{IN} (1)	Logical "1" Input Voltage (Except Pin 4 CLK IN)	V_{CC} = 5.25 V_{DC}	2.0		15	V_{DC}
V_{IN} (0)	Logical "0" Input Voltage (Except Pin 4 CLK IN)	V_{CC} = 4.75 V_{DC}			0.8	V_{DC}
I_{IN} (1)	Logical "1" Input Current (All Inputs)	V_{IN} = 5 V_{DC}		0.005	1	μA_{DC}
I_{IN} (0)	Logical "0" Input Current (All Inputs)	V_{IN} = 0 V_{DC}	−1	−0.005		μA_{DC}
CLOCK IN AND CLOCK R						
V_{T+}	CLK IN (Pin 4) Positive Going Threshold Voltage		2.7	3.1	3.5	V_{DC}
V_{T-}	CLK IN (Pin 4) Negative Going Threshold Voltage		1.5	1.8	2.1	V_{DC}
V_H	CLK IN (Pin 4) Hysteresis $(V_{T+}) - (V_{T-})$		0.6	1.3	2.0	V_{DC}
V_{OUT} (0)	Logical "0" CLK R Output Voltage	I_O = 360 μA, V_{CC} = 4.75 V_{DC}			0.4	V_{DC}
V_{OUT} (1)	Logical "1" CLK R Output Voltage	I_O = −360 μA, V_{CC} = 4.75 V_{DC}	2.4			V_{DC}
DATA OUTPUTS AND $\overline{INTR}$						
V_{OUT}(0)	Logical "0" Output Voltage					
	Data Outputs	I_{OUT} = 1.6 mA, V_{CC} = 4.75 V_{DC}			0.4	V_{DC}
	$\overline{INTR}$ Output	I_{OUT} = 1.0 mA, V_{CC} = 4.75 V_{DC}			0.4	V_{DC}
V_{OUT} (1)	Logical "1" Output Voltage	I_O = −360 μA, V_{CC} = 4.75 V_{DC}	2.4			V_{DC}
V_{OUT} (1)	Logical "1" Output Voltage	I_O = −10 μA, V_{CC} = 4.75 V_{DC}	4.5			V_{DC}
I_{OUT}	TRI-STATE Disabled Output Leakage (All Data Buffers)	V_{OUT} = 0 V_{DC}	−3			μA_{DC}
		V_{OUT} = 5 V_{DC}			3	μA_{DC}
I_{SOURCE}		V_{OUT} Short to Gnd, T_A = 25°C	4.5	6		mA_{DC}
I_{SINK}		V_{OUT} Short to V_{CC}, T_A = 25°C	9.0	16		mA_{DC}
POWER SUPPLY						
I_{CC}	Supply Current (Includes Ladder Current)	f_{CLK} = 640 kHz, $V_{REF}/2$ = NC, T_A = 25°C and $\overline{CS}$ = "1"				
		ADC0801/02/03/05		1.1	1.8	mA
		ADC0804 (Note 9)		1.9	2.5	mA

Note 1: Absolute maximum ratings are those values beyond which the life of the device may be impaired.

Note 2: All voltages are measured with respect to Gnd, unless otherwise specified. The separate A Gnd point should always be wired to the D Gnd.

Note 3: A zener diode exists, internally, from V_{CC} to Gnd and has a typical breakdown voltage of 7 V_{DC}.

Note 4: For $V_{IN}(-) \geq V_{IN}(+)$ the digital output code will be 0000 0000. Two on-chip diodes are tied to each analog input (see block diagram) which will forward conduct for analog input voltages one diode drop below ground or one diode drop greater than the V_{CC} supply. Be careful, during testing at low V_{CC} levels (4.5V), as high level analog inputs (5V) can cause this input diode to conduct—especially at elevated temperatures, and cause errors for analog inputs near full-scale. The spec allows 50 mV forward bias of either diode. This means that as long as the analog V_{IN} does not exceed the supply voltage by more than 50 mV, the output code will be correct. To achieve an absolute 0 V_{DC} to 5 V_{DC} input voltage range will therefore require a minimum supply voltage of 4.950 V_{DC} over temperature variations, initial tolerance and loading.

Note 5: Accuracy is guaranteed at f_{CLK} = 640 kHz. At higher clock frequencies accuracy can degrade. For lower clock frequencies, the duty cycle limits can be extended so long as the minimum clock high time interval or minimum clock low time interval is no less than 275 ns.

Note 6: With an asynchronous start pulse, up to 8 clock periods may be required before the internal clock phases are proper to start the conversion process. The start request is internally latched, see *Figure 2* and section 2.0.

Note 7: The $\overline{CS}$ input is assumed to bracket the $\overline{WR}$ strobe input and therefore timing is dependent on the $\overline{WR}$ pulse width. An arbitrarily wide pulse width will hold the converter in a reset mode and the start of conversion is initiated by the low to high transition of the $\overline{WR}$ pulse (see timing diagrams).

Note 8: None of these A/Ds requires a zero adjust (see section 2.5.1). To obtain zero code at other analog input voltages see section 2.5 and *Figure 5*.

Note 9: For ADC0804LCD typical value of $V_{REF}/2$ input resistance is 8 kΩ and of I_{CC} is 1.1 mA.

1.0 UNDERSTANDING A/D ERROR SPECS

A perfect A/D transfer characteristic (staircase waveform) is shown in *Figure 1a*. The horizontal scale is analog input voltage and the particular points labeled are in steps of 1 LSB (19.53 mV with 2.5V tied to the $V_{REF}/2$ pin). The digital output codes which correspond to these inputs are shown as D−1, D, and D+1. For the perfect A/D, not only will center-value (A−1, A, A+1, . . .) analog inputs produce the correct output digital codes, but also each riser (the transitions between adjacent output codes) will be located ±1/2 LSB away from each center-value. As shown, the risers are ideal and have no width. Correct digital output codes will be provided for a range of analog input voltages which extend ±1/2 LSB from the ideal center-values. Each tread (the range of analog input voltage which provides the same digital output code) is therefore 1 LSB wide.

Figure 1b shows a worst case error plot for the ADC0801. All center-valued inputs are guaranteed to produce the correct output codes and the adjacent risers are guaranteed to be no closer to the center-value points than

±1/4 LSB. In other words, if we apply an analog input equal to the center-value ±1/4 LSB, *we guarantee* that the A/D will produce the correct digital code. The maximum range of the position of the code transition is indicated by the horizontal arrow and it is guaranteed to be no more than 1/2 LSB.

The error curve of *Figure 1c* shows a worst case error plot for the ADC0802. Here we guarantee that if we apply an analog input equal to the LSB analog voltage center-value the A/D will produce the correct digital code.

Next to each transfer function is shown the corresponding error plot. Many people may be more familiar with error plots than transfer functions. The analog input voltage to the A/D is provided by either a linear ramp or by the discrete output steps of a high resolution DAC. Notice that the error is continuously displayed and includes the quantization uncertainty of the A/D. For example the error at point 1 of *Figure 1a* is +1/2 LSB because the digital code appeared 1/2 LSB in advance of the center-value of the tread. The error plots always have a constant negative slope and the abrupt upside steps are always 1 LSB in magnitude.

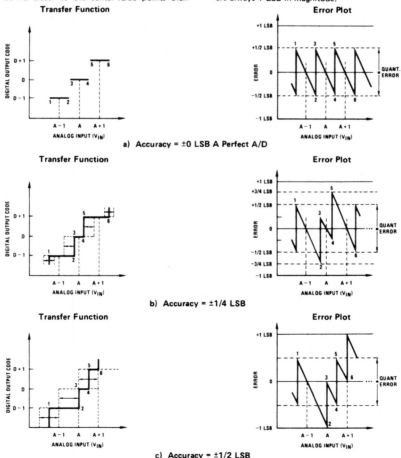

a) Accuracy = ±0 LSB A Perfect A/D

b) Accuracy = ±1/4 LSB

c) Accuracy = ±1/2 LSB

FIGURE 1. Clarifying the Error Specs of an A/D Converter

2.0   FUNCTIONAL DESCRIPTION

The ADC0801 series contains a circuit equivalent of the 256R network. Analog switches are sequenced by successive approximation logic to match the analog difference input voltage $[V_{IN}(+) - V_{IN}(-)]$ to a corresponding tap on the R network. The most significant bit is tested first and after 8 comparisons (64 clock cycles) a digital 8-bit binary code (1111 1111 = full-scale) is transferred to an output latch and then an interrupt is asserted ($\overline{INTR}$ makes a high-to-low transition). A conversion in process can be interrupted by issuing a second start command. The device may be operated in the free-running mode by connecting $\overline{INTR}$ to the $\overline{WR}$ input with $\overline{CS} = 0$. To insure start-up under all possible conditions, an external $\overline{WR}$ pulse is required during the first power-up cycle.

On the high-to-low transition of the $\overline{WR}$ input the internal SAR latches and the shift register stages are reset. As long as the $\overline{CS}$ input and $\overline{WR}$ input remain low, the A/D will remain in a reset state. *Conversion will start from 1 to 8 clock periods after at least one of these inputs makes a low-to-high transition.*

A functional diagram of the A/D converter is shown in *Figure 2.* All of the package pinouts are shown and the major logic control paths are drawn in heavier weight lines.

The converter is started by having $\overline{CS}$ and $\overline{WR}$ simultaneously low. This sets the start flip-flop (F/F) and the resulting "1" level resets the 8-bit shift register, resets the Interrupt (INTR) F/F and inputs a "1" to the D flop, F/F1, which is at the input end of the 8-bit shift register. Internal clock signals then transfer this "1" to the Q output of F/F1. The AND gate, G1, combines this "1" output with a clock signal to provide a reset signal to the start F/F. If the set signal is no longer present (either $\overline{WR}$ or $\overline{CS}$ is a "1") the start F/F is reset and the 8-bit shift register then can have the "1" clocked in, which starts the conversion process. If the set signal were to still be present, this reset pulse would have no effect (both outputs of the start F/F would momentarily be at a "1" level) and the 8-bit shift register would continue to be held in the reset mode. This logic therefore allows for wide $\overline{CS}$ and $\overline{WR}$ signals and the converter will start after at least one of these signals returns high and the internal clocks again provide a reset signal for the start F/F.

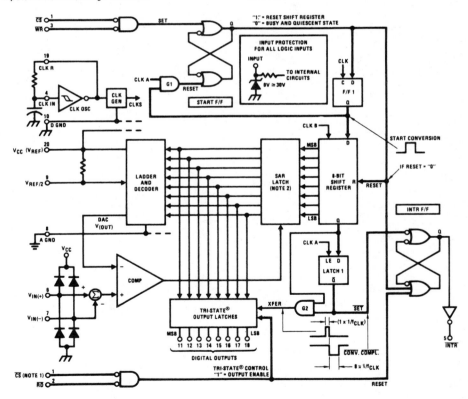

Note 1: $\overline{CS}$ shown twice for clarity.
Note 2: SAR = Successive Approximation Register.

FIGURE 2. Block Diagram

After the "1" is clocked through the 8-bit shift register (which completes the SAR search) it appears as the input to the D-type latch, LATCH 1. As soon as this "1" is output from the shift register, the AND gate, G2, causes the new digital word to transfer to the TRI-STATE output latches. When LATCH 1 is subsequently enabled, the Q output makes a high-to-low transition which causes the INTR F/F to set. An inverting buffer then supplies the $\overline{INTR}$ output signal.

Note that this $\overline{SET}$ control of the INTR F/F remains low for 8 of the external clock periods (as the internal clocks run at 1/8 of the frequency of the external clock). If the data output is continuously enabled ($\overline{CS}$ and $\overline{RD}$ both held low), the $\overline{INTR}$ output will still signal the end of conversion (by a high-to-low transition), because the $\overline{SET}$ input can control the Q output of the INTR F/F even though the RESET input is constantly at a "1" level in this operating mode. This $\overline{INTR}$ output will therefore stay low for the duration of the $\overline{SET}$ signal, which is 8 periods of the external clock frequency (assuming the A/D is not started during this interval).

When operating in the free-running or continuous conversion mode ($\overline{INTR}$ pin tied to $\overline{WR}$ and $\overline{CS}$ wired low—see also section 2.8), the START F/F is SET by the high-to-low transition of the $\overline{INTR}$ signal. This resets the SHIFT REGISTER which causes the input to the D-type latch, LATCH 1, to go low. As the latch enable input is still present, the $\overline{Q}$ output will go high, which then allows the INTR F/F to be RESET. This reduces the width of the resulting $\overline{INTR}$ output pulse to only a few propagation delays (approximately 300 ns).

When data is to be read, the combination of both $\overline{CS}$ and $\overline{RD}$ being low will cause the INTR F/F to be reset and the TRI-STATE output latches will be enabled to provide the 8-bit digital outputs.

2.1 Digital Control Inputs

The digital control inputs ($\overline{CS}$, $\overline{RD}$, and $\overline{WR}$) meet standard T^2L logic voltage levels. These signals have been renamed when compared to the standard A/D Start and Output Enable labels. In addition, these inputs are active low to allow an easy interface to microprocessor control busses. For non-microprocessor based applications, the $\overline{CS}$ input (pin 1) can be grounded and the standard A/D Start function is obtained by an active low pulse applied at the $\overline{WR}$ input (pin 3) and the Output Enable function is caused by an active low pulse at the $\overline{RD}$ input (pin 2).

2.2 Analog Differential Voltage Inputs and Common-Mode Rejection

This A/D has additional applications flexibility due to the analog differential voltage input. The $V_{IN}(-)$ input (pin 7) can be used to automatically subtract a fixed voltage value from the input reading (tare correction). This is also useful in 4 mA–20 mA current loop conversion. In addition, common-mode noise can be reduced by use of the differential input.

The time interval between sampling $V_{IN}(+)$ and $V_{IN}(-)$ is 4-1/2 clock periods. The maximum error voltage due

to this slight time difference between the input voltage samples is given by:

$$\Delta V_e(MAX) = (V_P)\,(2\pi f_{cm})\left(\frac{4.5}{f_{CLK}}\right)$$

where:

ΔV_e is the error voltage due to sampling delay

V_P is the peak value of the common-mode voltage

f_{cm} is the common-mode frequency

As an example, to keep this error to 1/4 LSB (~5 mV) when operating with a 60 Hz common-mode frequency, f_{cm}, and using a 640 kHz A/D clock, f_{CLK}, would allow a peak value of the common-mode voltage, V_P, which is given by:

$$V_P = \frac{[\Delta V_e(MAX)\,(f_{CLK})]}{(2\pi f_{cm})\,(4.5)}$$

or

$$V_P = \frac{(5 \times 10^{-3})\,(640 \times 10^3)}{(6.28)\,(60)\,(4.5)}$$

which gives

$$V_P \cong 1.9V.$$

The allowed range of analog input voltages usually places more severe restrictions on input common-mode noise levels.

An analog input voltage with a reduced span and a relatively large zero offset can be easily handled by making use of the differential input (see section 2.4 Reference Voltage).

2.3 Analog Inputs

2.3.1 Input Current

Normal Mode

Due to the internal switching action, displacement currents will flow at the analog inputs. This is due to on-chip stray capacitance to ground as shown in *Figure 3*.

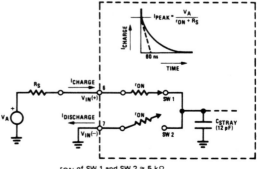

r_{ON} of SW 1 and SW 2 $\cong$ 5 kΩ

$\tau = r_{ON}\,C_{STRAY} \cong$ 5 kΩ x 12 pF = 60 ns

FIGURE 3. Analog Input Impedance

The voltage on this capacitance is switched and will result in currents entering the $V_{IN}(+)$ input pin and leaving the $V_{IN}(-)$ input which will depend on the analog differential input voltage levels. These current transients occur at the leading edge of the internal clocks. They rapidly decay and *do not cause errors* as the on-chip comparator is strobed at the end of the clock period.

Fault Mode

If the voltage source which is applied to the $V_{IN}(+)$ pin exceeds the allowed operating range of V_{CC} + 50 mV, large input currents can flow through a parasitic diode to the V_{CC} pin. If these currents could exceed the 1 mA max allowed spec, an external diode (1N914) should be added to bypass this current to the V_{CC} pin (with the current bypassed with this diode, the voltage at the $V_{IN}(+)$ pin can exceed the V_{CC} voltage by the forward voltage of this diode).

2.3.2 Input Bypass Capacitors

Bypass capacitors at the inputs will average these charges and cause a DC current to flow through the output resistances of the analog signal sources. This charge pumping action is worse for continuous conversions with the $V_{IN}(+)$ input voltage at full-scale. For continuous conversions with a 640 kHz clock frequency with the $V_{IN}(+)$ input at 5V, this DC current is at a maximum of approximately 5 μA. Therefore, *bypass capacitors should not be used at the analog inputs or the $V_{REF}/2$ pin* for high resistance sources (> 1 kΩ). If input bypass capacitors are necessary for noise filtering and high source resistance is desirable to minimize capacitor size, the detrimental effects of the voltage drop across this input resistance, which is due to the average value of the input current, can be eliminated with a full-scale adjustment while the given source resistor and input bypass capacitor are both in place. This is possible because the average value of the input current is a precise linear function of the differential input voltage.

2.3.3 Input Source Resistance

Large values of source resistance where an input bypass capacitor is not used, *will not cause errors* as the input currents settle out prior to the comparison time. If a low pass filter is required in the system, use a low valued series resistor (≤ 1 kΩ) for a passive RC section or add an op amp RC active low pass filter. For low source resistance applications, (≤ 1 kΩ), a 0.1 μF bypass capacitor at the inputs will prevent pickup due to series lead inductance of a long wire. A 100Ω series resistor can be used to isolate this capacitor—both the R and C are placed outside the feedback loop—from the output of an op amp, if used.

2.3.4 Noise

The leads to the analog inputs (pins 6 and 7) should be kept as short as possible to minimize input noise coupling. Both noise and undesired digital clock coupling to these inputs can cause system errors. The source resistance for these inputs should, in general, be kept below 5 kΩ. Larger values of source resistance can cause undesired system noise pickup. Input bypass capacitors, placed from the analog inputs to ground, will eliminate

system noise pickup but can create analog scale errors as these capacitors will average the transient input switching currents of the A/D (see section 2.3.1). This scale error depends on both a large source resistance and the use of an input bypass capacitor. This error can be eliminated by doing a full-scale adjustment of the A/D (adjust $V_{REF}/2$ for a proper full-scale reading—see section 2.5.2 on Full-Scale Adjustment) with the source resistance and input bypass capacitor in place.

2.4   Reference Voltage

2.4.1 Span Adjust

For maximum applications flexibility, these A/Ds have been designed to accommodate a 5 V_{DC}, 2.5 V_{DC} or an adjusted voltage reference. This has been achieved in the design of the IC as shown in *Figure 4*.

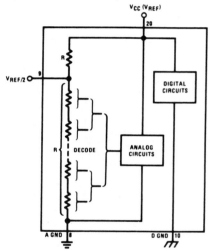

FIGURE 4. The $V_{REFERENCE}$ Design on the IC

Notice that the reference voltage for the IC is either 1/2 of the voltage which is applied to the V_{CC} supply pin, or is equal to the voltage which is externally forced at the $V_{REF}/2$ pin. This allows for a ratiometric voltage reference using the V_{CC} supply, a 5 V_{DC} reference voltage can be used for the V_{CC} supply or a voltage less than 2.5 V_{DC} can be applied to the $V_{REF}/2$ input for increased application flexibility. The internal gain to the $V_{REF}/2$ input is 2 making the full-scale differential input voltage twice the voltage at pin 9.

An example of the use of an adjusted reference voltage is to accommodate a reduced span—or dynamic voltage range of the analog input voltage. If the analog input voltage were to range from 0.5 V_{DC} to 3.5 V_{DC}, instead of 0V to 5 V_{DC}, the span would be 3V as shown in *Figure 5*. With 0.5 V_{DC} applied to the $V_{IN}(-)$ pin to absorb the offset, the reference voltage can be made equal to 1/2 of the 3V span or 1.5 V_{DC}. The A/D now will encode the $V_{IN}(+)$ signal from 0.5V to 3.5V with the 0.5V input corresponding to zero and the 3.5 V_{DC} input corresponding to full-scale. The full 8 bits of resolution are therefore applied over this reduced analog input voltage range.

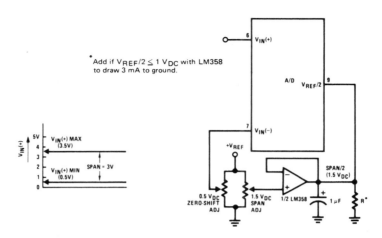

a) Analog Input Signal Example

b) Accommodating an Analog Input from
   0.5V (Digital Out = 00$_{HEX}$) to 3.5V
   (Digital Out = FF$_{HEX}$)

FIGURE 5. Adapting the A/D Analog Input Voltages to Match an Arbitrary Input Signal Range

2.4.2 Reference Accuracy Requirements

The converter can be operated in a ratiometric mode or an absolute mode. In ratiometric converter applications, the magnitude of the reference voltage is a factor in both the output of the source transducer and the output of the A/D converter and therefore cancels out in the final digital output code. The ADC0805 is specified particularly for use in ratiometric applications with no adjustments required. In absolute conversion applications, both the initial value and the temperature stability of the reference voltage are important accuracy factors in the operation of the A/D converter. For $V_{REF}/2$ voltages of 2.5 V_{DC} nominal value, initial errors of ±10 mV_{DC} will cause conversion errors of ±1 LSB due to the gain of 2 of the $V_{REF}/2$ input. In reduced span applications, the initial value and the stability of the $V_{REF}/2$ input voltage become even more important. For example, if the span is reduced to 2.5V, the analog input LSB voltage value is correspondingly reduced from 20 mV (5V span) to 10 mV and 1 LSB at the $V_{REF}/2$ input becomes 5 mV. As can be seen, this reduces the allowed initial tolerance of the reference voltage and requires correspondingly less absolute change with temperature variations. Note that spans smaller than 2.5V place even tighter requirements on the initial accuracy and stability of the reference source.

In general, the magnitude of the reference voltage will require an initial adjustment. Errors due to an improper value of reference voltage appear as full-scale errors in the A/D transfer function. IC voltage regulators may be used for references if the ambient temperature changes are not excessive. The LM336B 2.5V IC reference diode

(from National Semiconductor) is available which has a temperature stability of 1.8 mV typ (6 mV max) over $0°C \leq T_A \leq +70°C$. Other temperature range parts are also available.

2.5 Errors and Reference Voltage Adjustments

2.5.1 Zero Error

The zero of the A/D does not require adjustment. If the minimum analog input voltage value, $V_{IN(MIN)}$, is not ground, a zero offset can be done. The converter can be made to output 0000 0000 digital code for this minimum input voltage by biasing the A/D V_{IN} (−) input at this $V_{IN(MIN)}$ value (see Applications section). This utilizes the differential mode operation of the A/D.

The zero error of the A/D converter relates to the location of the first riser of the transfer function and can be measured by grounding the V (−) input and applying a small magnitude positive voltage to the V (+) input. Zero error is the difference between the actual DC input voltage which is necessary to just cause an output digital code transition from 0000 0000 to 0000 0001 and the ideal 1/2 LSB value (1/2 LSB = 9.8 mV for $V_{REF}/2$ = 2.500 V_{DC}).

2.5.2 Full-Scale

The full-scale adjustment can be made by applying a differential input voltage which is 1-1/2 LSB down from the desired analog full-scale voltage range and then adjusting the magnitude of the $V_{REF}/2$ input (pin 9 or the V_{CC} supply if pin 9 is not used) for a digital output code which is just changing from 1111 1110 to 1111 1111.

2.5.3 Adjusting for an Arbitrary Analog Input Voltage Range

If the analog zero voltage of the A/D is shifted away from ground (for example, to accommodate an analog input signal which does not go to ground) this new zero reference should be properly adjusted first: A $V_{IN}(+)$ voltage which equals this desired zero reference plus 1/2 LSB (where the LSB is calculated for the desired analog span, 1 LSB = analog span/256) is applied to pin 6 and the zero reference voltage at pin 7 should then be adjusted to just obtain the 00_{HEX} to 01_{HEX} code transition.

The full-scale adjustment should then be made (with the proper $V_{IN}(-)$ voltage applied) by forcing a voltage to the $V_{IN}(+)$ input which is given by:

$$V_{IN}(+) \text{ fs adj} = V_{MAX} - 1.5 \left[\frac{(V_{MAX} - V_{MIN})}{256} \right]$$

where:

V_{MAX} = The high end of the analog input range

and

V_{MIN} = the low end (the offset zero) of the analog range. (Both are ground referenced.)

The $V_{REF}/2$ (or V_{CC}) voltage is then adjusted to provide a code change from FE_{HEX} to FF_{HEX}. This completes the adjustment procedure.

2.6 Clocking Option

The clock for the A/D can be derived from the CPU clock or an external RC can be added to provide self-clocking. The CLK IN (pin 4) makes use of a Schmitt trigger as shown in *Figure 6*.

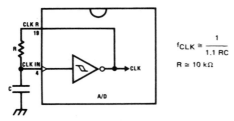

$$f_{CLK} \cong \frac{1}{1.1 \, RC}$$

$$R \cong 10 \, k\Omega$$

FIGURE 6. Self-Clocking the A/D

Heavy capacitive or DC loading of the clock R pin should be avoided as this will disturb normal converter operation. Loads less than 50 pF, such as driving up to 7 A/D converter clock inputs from a single clock R pin of 1 converter, are allowed. For larger clock line loading, a CMOS or low power T^2L buffer or PNP input logic should be used to minimize the loading on the clock R pin (do not use a standard T^2L buffer).

2.7 Restart During a Conversion

If the A/D is restarted ($\overline{CS}$ and $\overline{WR}$ go low and return high) during a conversion, the converter is reset and a new conversion is started. The output data latch is not updated if the conversion in process is not allowed to be completed, therefore the data of the previous conversion remains in this latch. The $\overline{INTR}$ output also simply remains at the "1" level.

2.8 Continuous Conversions

For operation in the free-running mode an initializing pulse should be used, following power-up, to insure circuit operation. In this application, the $\overline{CS}$ input is grounded and the $\overline{WR}$ input is tied to the $\overline{INTR}$ output. This $\overline{WR}$ and $\overline{INTR}$ node should be momentarily forced to logic low following a power-up cycle to guarantee operation.

2.9 Driving the Data Bus

This MOS A/D, like MOS microprocessors and memories, will require a bus driver when the total capacitance of the data bus gets large. Other circuitry, which is tied to the data bus, will add to the total capacitive loading, even in TRI-STATE (high impedance mode). Backplane bussing also greatly adds to the stray capacitance of the data bus.

There are some alternatives available to the designer to handle this problem. Basically, the capacitive loading of the data bus slows down the response time, even though DC specifications are still met. For systems operating with a relatively slow CPU clock frequency, more time is available in which to establish proper logic levels on the bus and therefore higher capacitive loads can be driven (see typical characteristics curves).

At higher CPU clock frequencies time can be extended for I/O reads (and/or writes) by inserting wait states (8080) or using clock extending circuits (6800).

Finally, if time is short and capacitive loading is high, external bus drivers must be used. These can be TRI-STATE buffers (low power Schottky is recommended such as the DM74LS240 series) or special higher drive current products which are designed as bus drivers. High current bipolar bus drivers with PNP inputs are recommended.

2.10 Power Supplies

Noise spikes on the V_{CC} supply line can cause conversion errors as the comparator will respond to this noise. A low inductance tantalum filter capacitor should be used close to the converter V_{CC} pin and values of 1 μF or greater are recommended. If an unregulated voltage is available in the system, a separate LM340LAZ-5.0, TO-92, 5V voltage regulator for the converter (and other analog circuitry) will greatly reduce digital noise on the V_{CC} supply.

2.11 Wiring and Hook-Up Precautions

Standard digital wire wrap sockets are not satisfactory for breadboarding this A/D converter. Sockets on PC boards can be used and all logic signal wires and leads should be grouped and kept as far away as possible from the analog signal leads. Exposed leads to the analog inputs can cause undesired digital noise and hum pickup, therefore shielded leads may be necessary in many applications.

A single point analog ground should be used which is separate from the logic ground points. The power supply bypass capacitor and the self-clocking capacitor (if used) should both be returned to digital ground. Any $V_{REF}/2$ bypass capacitors, analog input filter capacitors, or input signal shielding should be returned to the analog ground point. A test for proper grounding is to measure the zero error of the A/D converter. Zero errors in excess of 1/4 LSB can usually be traced to improper board layout and wiring (see section 2.5.1 for measuring the zero error).

3.0 TESTING THE A/D CONVERTER

There are many degrees of complexity associated with testing an A/D converter. One of the simplest tests is to apply a known analog input voltage to the converter and use LEDs to display the resulting digital output code as shown in *Figure 7*.

For ease of testing, the $V_{REF}/2$ (pin 9) should be supplied with 2.560 V_{DC} and a V_{CC} supply voltage of 5.12 V_{DC} should be used. This provides an LSB value of 20 mV.

If a full-scale adjustment is to be made, an analog input voltage of 5.090 V_{DC} (5.120 − 1 1/2 LSB) should be applied to the $V_{IN}(+)$ pin with the $V_{IN}(-)$ pin grounded. The value of the $V_{REF}/2$ input voltage should then be adjusted until the digital output code is just changing from 1111 1110 to 1111 1111. This value of $V_{REF}/2$ should then be used for all the tests.

The digital output LED display can be decoded by dividing the 8 bits into 2 hex characters, the 4 most significant (MS) and the 4 least significant (LS). Table I shows the fractional binary equivalent of these two 4-bit groups. By adding the decoded voltages which are obtained from the column: Input voltage value for a 2.560 $V_{REF}/2$ of both the MS and the LS groups, the value of

the digital display can be determined. For example, for an output LED display of 1011 0110 or B6 (in hex), the voltage values from the table are 3.520 + 0.120 or 3.640 V_{DC}. These voltage values represent the center-values of a perfect A/D converter. The effects of quantization error have to be accounted for in the interpretation of the test results.

For a higher speed test system, or to obtain plotted data, a digital-to-analog converter is needed for the test set-up. An accurate 10-bit DAC can serve as the precision voltage source for the A/D. Errors of the A/D under test can be provided as either analog voltages or differences in 2 digital words.

A basic A/D tester which uses a DAC and provides the error as an analog output voltage is shown in *Figure 8*. The 2 op amps can be eliminated if a lab DVM with a numerical subtraction feature is available to directly readout the difference voltage, "A−C". The analog input voltage can be supplied by a low frequency ramp generator and an X-Y plotter can be used to provide analog error (Y axis) versus analog input (X axis). The construction details of a tester of this type are provided in the NSC application note AN-179, "Analog-to-Digital Converter Testing".

For operation with a microprocessor or a computer-based test system, it is more convenient to present the errors digitally. This can be done with the circuit of *Figure 9*, where the output code transitions can be detected as the 10-bit DAC is incremented. This provides 1/4 LSB steps for the 8-bit A/D under test. If the results of this test are automatically plotted with the analog input on the X axis and the error (in LSB's) as the Y axis, a useful transfer function of the A/D under test results. For acceptance testing, the plot is not necessary and the testing speed can be increased by establishing internal limits on the allowed error for each code.

4.0 MICROPROCESSOR INTERFACING

To discuss the interface with 8080A and 6800 microprocessors, a common sample subroutine structure is used. The microprocessor starts the A/D, reads and stores the results of 16 successive conversions, then returns to the user's program. The 16 data bytes are stored in 16 successive memory locations. All Data and Addresses will be given in hexadecimal form. Software and hardware details are provided separately for each type of microprocessor.

4.1 Interfacing 8080 Microprocessor Derivatives (8048, 8085)

This converter has been designed to directly interface with derivatives of the 8080 microprocessor. The A/D can be mapped into memory space (using standard memory address decoding for $\overline{CS}$ and the $\overline{MEMR}$ and $\overline{MEMW}$ strobes) or it can be controlled as an I/O device by using the $\overline{I/O\ R}$ and $\overline{I/O\ W}$ strobes and decoding the address bits A0 → A7 (or address bits A8 → A15 as they will contain the same 8-bit address information) to obtain the $\overline{CS}$ input. Using the I/O space provides 256 additional addresses and may allow a simpler 8-bit address decoder but the data can only be input to the accumulator. To make use of the additional memory reference instructions, the A/D should be mapped into memory space. An example of an A/D in I/O space is shown in *Figure 10*.

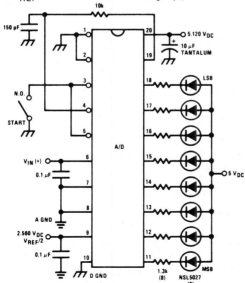

FIGURE 7. Basic A/D Tester

 **National Semiconductor**

A to D, D to A

DAC0808, DAC0807, DAC0806 8-Bit D/A Converters

General Description

The DAC0808 series is an 8-bit monolithic digital-to-analog converter (DAC) featuring a full scale output current settling time of 150 ns while dissipating only 33 mW with ±5V supplies. No reference current (I_{REF}) trimming is required for most applications since the full scale output current is typically ±1 LSB of 255 I_{REF}/256. Relative accuracies of better than ±0.19% assure 8-bit monotonicity and linearity while zero level output current of less than 4 μA provides 8-bit zero accuracy for $I_{REF} \geq 2$ mA. The power supply currents of the DAC0808 series are independent of bit codes, and exhibits essentially constant device characteristics over the entire supply voltage range.

The DAC0808 will interface directly with popular TTL, DTL or CMOS logic levels, and is a direct replacement for the MC1508/MC1408. For higher speed applications, see DAC0800 data sheet.

Features

- Relative accuracy: ±0.19% error maximum (DAC0808)
- Full scale current match: ±1 LSB typ
- 7 and 6-bit accuracy available (DAC0807, DAC0806)
- Fast settling time: 150 ns typ
- Noninverting digital inputs are TTL and CMOS compatible
- High speed multiplying input slew rate: 8 mA/μs
- Power supply voltage range: ±4.5V to ±18V
- Low power consumption: 33 mW @ ±5V

Block and Connection Diagrams

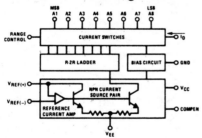

Dual-In-Line Package

TOP VIEW

Typical Application

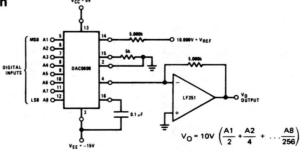

$$V_O = 10V \left(\frac{A1}{2} + \frac{A2}{4} + \ldots \frac{A8}{256} \right)$$

FIGURE 1. +10V Output Digital to Analog Converter

Ordering Information

ACCURACY	OPERATING TEMPERATURE RANGE	ORDER NUMBERS*					
		D PACKAGE (D16C)		J PACKAGE (J16A)		N PACKAGE (N16A)	
8-bit	−55°C ≤ T_A ≤ +125°C	DAC0808LD	MC1508L8				
8-bit	0°C ≤ T_A ≤ +75°C			DAC0808LCJ	MC1408L8	DAC0808LCN	MC1408P8
7-bit	0°C ≤ T_A ≤ +75°C			DAC0807LCJ	MC1408L7	DAC0807LCN	MC1408P7
6-bit	0°C ≤ T_A ≤ +75°C			DAC0806LCJ	MC1408L6	DAC0806LCN	MC1408P6

*Note. Devices may be ordered by using either order number.

Absolute Maximum Ratings

Power Supply Voltage		Power Dissipation (Package Limitation)	1000 mW
V_{CC}	$+18\ V_{DC}$	Derate above $T_A = 25°C$	$6.7\ mW/°C$
V_{EE}	$-18\ V_{DC}$	Operating Temperature Range	
Digital Input Voltage, V5–V12	$-10\ V_{DC}$ to $+18\ V_{DC}$	DAC0808L	$-55°C \leq T_A \leq +125°C$
Applied Output Voltage, V_O	$-11\ V_{DC}$ to $+18\ V_{DC}$	DAC0808LC Series	$0 \leq T_A \leq +75°C$
Reference Current, I_{14}	5 mA	Storage Temperature Range	$-65°C$ to $+150°C$
Reference Amplifier Inputs, V14, V15	V_{CC}, V_{EE}		

Electrical Characteristics

($V_{CC} = 5V$, $V_{EE} = -15\ V_{DC}$, $V_{REF}/R14 = 2$ mA, DAC0808: $T_A = -55°C$ to $+125°C$, DAC0808C, DAC0807C, DAC0806C, $T_A = 0°C$ to $+75°C$, and all digital inputs at high logic level unless otherwise noted.)

	PARAMETER	CONDITIONS	MIN	TYP	MAX	UNITS
E_r	Relative Accuracy (Error Relative	(Figure 4)				%
	to Full Scale I_O)					
	DAC0808L (LM1508-8),				±0.19	%
	DAC0808LC (LM1408-8)					
	DAC0807LC (LM1408-7), (Note 1)				±0.39	%
	DAC0806LC (LM1408-6), (Note 1)				±0.78	%
	Settling Time to Within 1/2 LSB	$T_A = 25°C$ (Note 2),		150		ns
	(Includes t_{PLH})	(Figure 5)				
t_{PLH},	Propagation Delay Time	$T_A = 25°C$, (Figure 5)		30	100	ns
t_{PHL}						
TCI_O	Output Full Scale Current Drift			±20		ppm/°C
MSB	Digital Input Logic Levels	(Figure 3)				
V_{IH}	High Level, Logic "1"		2			V_{DC}
V_{IL}	Low Level, Logic "0"				0.8	V_{DC}
MSB	Digital Input Current	(Figure 3)				
	High Level	$V_{IH} = 5V$		0	0.040	mA
	Low Level	$V_{IL} = 0.8V$		-0.003	-0.8	mA
I_{15}	Reference Input Bias Current	(Figure 3)		-1	-3	μA
	Output Current Range	(Figure 3)				
		$V_{EE} = -5V$	0	2.0	2.1	mA
		$V_{EE} = -15V$, $T_A = 25°C$	0	2.0	4.2	mA
I_O	Output Current	$V_{REF} = 2.000V$, $R14 = 1000\Omega$,				
		(Figure 3)	1.9	1.99	2.1	mA
	Output Current, All Bits Low	(Figure 3)		0	4	μA
	Output Voltage Compliance	$E_r \leq 0.19\%$, $T_A = 25°C$				
	Pin 1 Grounded,				-0.55, +0.4	V_{DC}
	V_{EE} Below $-10V$				-5.0, +0.4	V_{DC}
SRI_{REF}	Reference Current Slew Rate	(Figure 6)	4	8		mA/μs
	Output Current Power Supply Sensitivity	$-5V \leq V_{EE} \leq -16.5V$		0.05	2.7	μA/V
	Power Supply Current (All Bits Low)	(Figure 3)				
I_{CC}				2.3	22	mA
I_{EE}				-4.3	-13	mA
	Power Supply Voltage Range	$T_A = 25°C$, (Figure 3)				
V_{CC}			4.5	5.0	5.5	V_{DC}
V_{EE}			-4.5	-15	-16.5	V_{DC}
	Power Dissipation					
	All Bits Low	$V_{CC} = 5V$, $V_{EE} = -5V$		33	170	mW
		$V_{CC} = 5V$, $V_{EE} = -15V$		106	305	mW
	All Bits High	$V_{CC} = 15V$, $V_{EE} = -5V$		90		mW
		$V_{CC} = 15V$, $V_{EE} = -15V$		160		mW

Note 1: All current switches are tested to guarantee at least 50% of rated current.

Note 2: All bits switched.

Note 3: Range control is not required.

Test Circuits

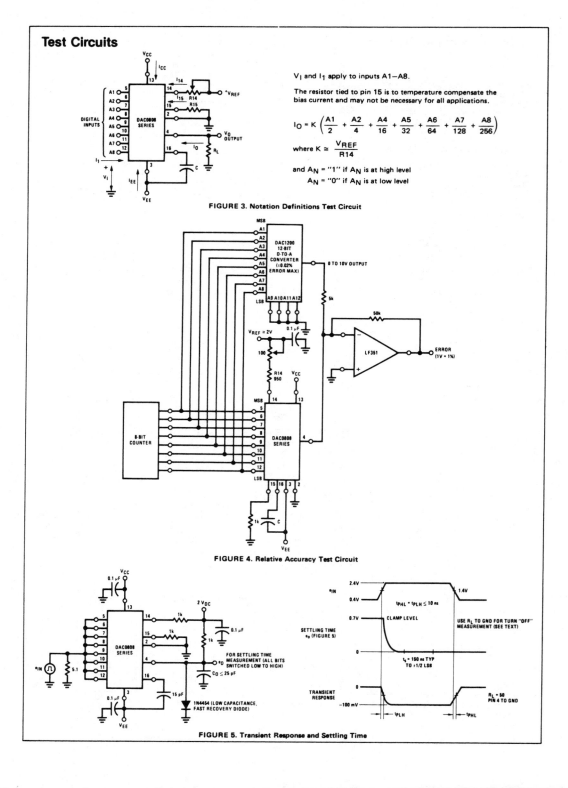

V_I and I_1 apply to inputs A1—A8.

The resistor tied to pin 15 is to temperature compensate the bias current and may not be necessary for all applications.

$$I_O = K \left(\frac{A1}{2} + \frac{A2}{4} + \frac{A4}{16} + \frac{A5}{32} + \frac{A6}{64} + \frac{A7}{128} + \frac{A8}{256} \right)$$

where $K \cong \dfrac{V_{REF}}{R14}$

and $A_N = $ "1" if A_N is at high level
$A_N = $ "0" if A_N is at low level

FIGURE 3. Notation Definitions Test Circuit

FIGURE 4. Relative Accuracy Test Circuit

FIGURE 5. Transient Response and Settling Time

Test Circuits (Continued)

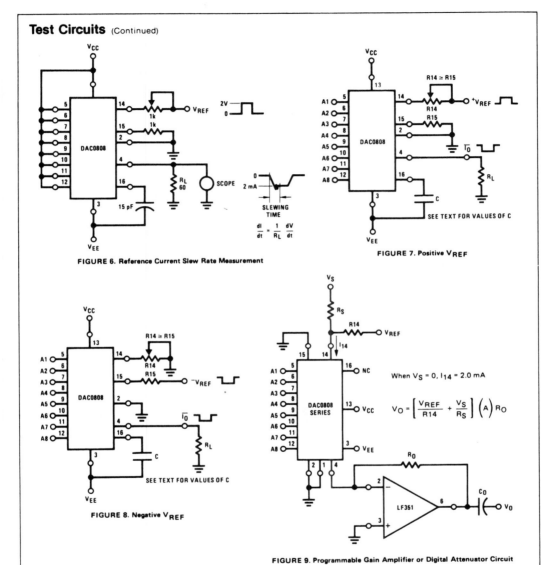

FIGURE 6. Reference Current Slew Rate Measurement

FIGURE 7. Positive V_{REF}

FIGURE 8. Negative V_{REF}

When $V_S = 0$, $I_{14} = 2.0$ mA

$$V_O = \left[\frac{V_{REF}}{R14} + \frac{V_S}{R_S} \right] (A) R_O$$

FIGURE 9. Programmable Gain Amplifier or Digital Attenuator Circuit

Application Hints

REFERENCE AMPLIFIER DRIVE AND COMPENSATION

The reference amplifier provides a voltage at pin 14 for converting the reference voltage to a current, and a turn-around circuit or current mirror for feeding the ladder. The reference amplifier input current, I_{14}, must always flow into pin 14, regardless of the set-up method or reference voltage polarity.

Connections for a positive voltage are shown in *Figure 7*. The reference voltage source supplies the full current I_{14}. For bipolar reference signals, as in the multiplying mode, R15 can be tied to a negative voltage corresponding to the minimum input level. It is possible to eliminate R15 with only a small sacrifice in accuracy and temperature drift.

The compensation capacitor value must be increased with increases in R14 to maintain proper phase margin; for R14 values of 1, 2.5 and 5 kΩ, minimum capacitor values are 15, 37 and 75 pF. The capacitor may be tied to either V_{EE} or ground, but using V_{EE} increases negative supply rejection.

Application Hints (Continued)

A negative reference voltage may be used if R14 is grounded and the reference voltage is applied to R15 as shown in *Figure 8*. A high input impedance is the main advantage of this method. Compensation involves a capacitor to V_{EE} on pin 16, using the values of the previous paragraph. The negative reference voltage must be at least 4V above the V_{EE} supply. Bipolar input signals may be handled by connecting R14 to a positive reference voltage equal to the peak positive input level at pin 15.

When a DC reference voltage is used, capacitive bypass to ground is recommended. The 5V logic supply is not recommended as a reference voltage. If a well regulated 5V supply which drives logic is to be used as the reference, R14 should be decoupled by connecting it to 5V through another resistor and bypassing the junction of the 2 resistors with 0.1 μF to ground. For reference voltages greater than 5V, a clamp diode is recommended between pin 14 and ground.

If pin 14 is driven by a high impedance such as a transistor current source, none of the above compensation methods apply and the amplifier must be heavily compensated, decreasing the overall bandwidth.

OUTPUT VOLTAGE RANGE

The voltage on pin 4 is restricted to a range of −0.6 to 0.5V when V_{EE} = −5V due to the current switching methods employed in the DAC0808.

The negative output voltage compliance of the DAC0808 is extended to −5V where the negative supply voltage is more negative than −10V. Using a full-scale current of 1.992 mA and load resistor of 2.5 kΩ between pin 4 and ground will yield a voltage output of 256 levels between 0 and −4.980V. Floating pin 1 does not affect the converter speed or power dissipation. However, the value of the load resistor determines the switching time due to increased voltage swing. Values of R_L up to 500Ω do not significantly affect performance, but a 2.5 kΩ load increases worst-case settling time to 1.2 μs (when all bits are switched ON). Refer to the subsequent text section on Settling Time for more details on output loading.

OUTPUT CURRENT RANGE

The output current maximum rating of 4.2 mA may be used only for negative supply voltages more negative than −7V, due to the increased voltage drop across the resistors in the reference current amplifier.

ACCURACY

Absolute accuracy is the measure of each output current level with respect to its intended value, and is dependent upon relative accuracy and full-scale current drift. Relative accuracy is the measure of each output current level as a fraction of the full-scale current. The relative accuracy of the DAC0808 is essentially constant with temperature due to the excellent temperature tracking of the monolithic resistor ladder. The reference current may drift with temperature, causing a change in the absolute accuracy of output current. However, the DAC0808 has a very low full-scale current drift with temperature.

The DAC0808 series is guaranteed accurate to within ±1/2 LSB at a full-scale output current of 1.992 mA. This corresponds to a reference amplifier output current drive to the ladder network of 2 mA, with the loss of 1 LSB (8 μA) which is the ladder remainder shunted to ground. The input current to pin 14 has a guaranteed value of between 1.9 and 2.1 mA, allowing some mismatch in the NPN current source pair. The accuracy test circuit is shown in *Figure 4*. The 12-bit converter is calibrated for a full-scale output current of 1.992 mA. This is an optional step since the DAC0808 accuracy is essentially the same between 1.5 and 2.5 mA. Then the DAC0808 circuits' full-scale current is trimmed to the same value with R14 so that a zero value appears at the error amplifier output. The counter is activated and the error band may be displayed on an oscilloscope, detected by comparators, or stored in a peak detector.

Two 8-bit D-to-A converters may not be used to construct a 16-bit accuracy D-to-A converter. 16-bit accuracy implies a total error of ±1/2 of one part in 65,536, or ±0.00076%, which is much more accurate than the ±0.019% specification provided by the DAC0808.

MULTIPLYING ACCURACY

The DAC0808 may be used in the multiplying mode with 8-bit accuracy when the reference current is varied over a range of 256:1. If the reference current in the multiplying mode ranges from 16 μA to 4 mA, the additional error contributions are less than 1.6 μA. This is well within 8-bit accuracy when referred to full-scale.

A monotonic converter is one which supplies an increase in current for each increment in the binary word. Typically, the DAC0808 is monotonic for all values of reference current above 0.5 mA. The recommended range for operation with a DC reference current is 0.5 to 4 mA.

SETTLING TIME

The worst-case switching condition occurs when all bits are switched ON, which corresponds to a low-to-high transition for all bits. This time is typically 150 ns for settling to within ±1/2 LSB, for 8-bit accuracy, and 100 ns to 1/2 LSB for 7 and 6-bit accuracy. The turn OFF is typically under 100 ns. These times apply when $R_L \leq 500\Omega$ and $C_O \leq 25$ pF.

Extra care must be taken in board layout since this is usually the dominant factor in satisfactoy test results when measuring settling time. Short leads, 100 μF supply bypassing for low frequencies, and minimum scope lead length are all mandatory.

APPENDIXES

- Appendix A: ASCII Character Codes
- Appendix B: EBCDIC Code
- Appendix C: PC Manufacturers' Products
- Appendix D: Programming and Operations Manual for Mini-PLC-2 Programmable Controller

APPENDIX A: ASCII CHARACTER CODES

b4	b3	b2	b1	HEX LSD \\ HEX MSD	0	1	2	3	4	5	6	7
				b7 b6 b5 →	0 0 0	0 0 1	0 1 0	0 1 1	1 0 0	1 0 1	1 1 0	1 1 1
0	0	0	0	0	NUL	DLE	SP	0	@	P	`	p
0	0	0	1	1	SOH	DC1	!	1	A	Q	a	q
0	0	1	0	2	STX	DC2	"	2	B	R	b	r
0	0	1	1	3	ETX	DC3	#	3	C	S	c	s
0	1	0	0	4	EOT	DC4	$	4	D	T	d	t
0	1	0	1	5	ENQ	NAK	%	5	E	U	e	u
0	1	1	0	6	ACK	SYN	&	6	F	V	f	v
0	1	1	1	7	BEL	ETB	'	7	G	W	g	w
1	0	0	0	8	BS	CAN	(	8	H	X	h	x
1	0	0	1	9	HT	EM	)	9	I	Y	i	y
1	0	1	0	A(10)	LF	SUB	*	:	J	Z	j	z
1	0	1	1	B(11)	VT	ESC	+	;	K	[	k	{
1	1	0	0	C(12)	FF	FS	,	<	L	\	l	|
1	1	0	1	D(13)	CR	GS	-	=	M	]	m	}
1	1	1	0	E(14)	SO	RS	.	>	N	^	n	~
1	1	1	1	F(15)	SI	US	/	?	O	_	o	DEL

Bit Numbers (arrows to b5, b6, b7)

NUL Null
SOH Start of heading
STN Start of text
ETX End of text
EOT End of transmission
ENQ Enquiry
ACK Acknowledge
BEL Bell, or alarm
BS   Backspace

HT   Horizontal tabulation
LF   Line feed
VT   Vertical tabulation
FF   Form feed
CR   Carriage return
SO   Shift out
SI   Shift in
DLE Data link escape
DC1 Device control 1

DC2 Device control 2
DC3 Device control 3
DC4 Device control 4
NAK Negative acknowledge
SYN Synchronous idle
ETB End of transmission block
CAN Cancel
EM   End of medium
SUB Substitute

ESC Escape
FS   File separator
GS   Group separator
RS   Record separator
US   Unit separator
SP   Space
DEL Delete

APPENDIX B: EBCDIC CODE

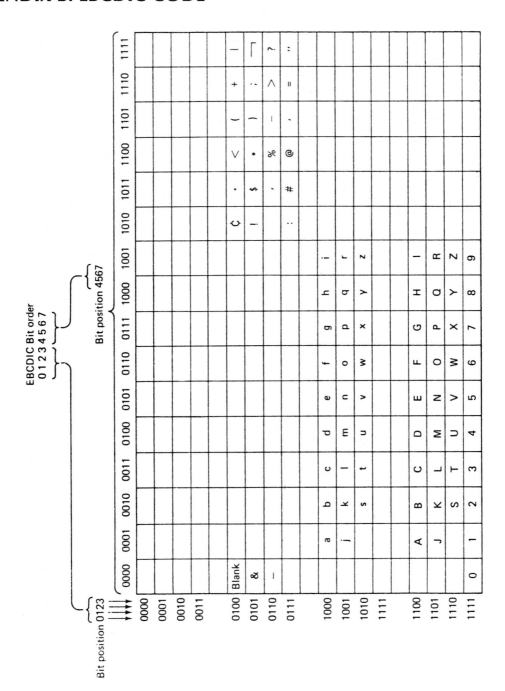

APPENDIX C: PC MANUFACTURERS' PRODUCTS

The following table is a listing of the known products of PC manufacturers and some of the more important PC specifications, as of July 1983. These devices meet the historical definition of a conventional PC. There are many factors that cannot be grouped into a table, and some of these factors may alter your perception about which system will best suit your needs; therefore, it is recommended that you check with the manufacturers for specific data.

The manufacturers caution that scan rate data is, at most, an *estimate* of typical performance. For any given program, the scan rate will vary depending on the types of actions called for by the program.

Most systems can accept any combination of local or remote I/Os. However, some do have dedicated quantities of inputs versus outputs, such as 20 inputs and 12 outputs. Only the totals are shown.

Where high-level language usage is indicated, both the high-level languages, such as TINY BASIC, and additional commands beyond normal ladder logic diagramming are available.

The category "Documentation" is not to be confused with manufacturer-generated documentation such as owners' manuals and service manuals. The term "Documentation" in the table means the ability of the system to provide a hard copy of its configuration by means of a built-in or external printer.

MANUFACTURER (City, State) MODEL	Type CPU	Number of bits	Clock speed (MHz)	Scan rate (msec)	System total I/O	Local total I/O	Total remote I/O	High level language	Relay ladder logic	No. of instructions	PID capabilities	Motion control	Documentation	PC data highway
ADATEK, INC. (Sandpoint, ID)														
CP73A	INS8073	16	4	20	240	48	216	YES		N/A	YES	YES	YES	
CP73P	INS8073	16	4	20	240	48	216	YES		8	YES	YES	YES	
ADATEK, INC. (Sandpoint, ID)														
System 1	INS8073	8	4	25	48	24	24	Yes		8			Yes	
System 2	INS8073	8	4	N/A	240	24	216	Yes		8			Yes	
System 3	INS8073	8	4	25	240	24	216	Yes		8		Yes	Yes	Yes
ALLEN-BRADLEY CO. IND'L. CONTROLS DIV. (Milwaukee, WI)														
1742 Modular Automation Cont.	8051	8	12	25	32	32			Yes	N/A				
SYSTEMS DIVISION PC BUSINESS UNIT (Highland Heights, OH)														
PLC-4 Microtrol	8031	8	12	25	256	32	224	Yes	Yes	17	Yes	Yes	Yes	Yes
MINI-PLC-2/15	Z-80A	8	2.5	12	128	128		Yes	Yes	41	Yes	Yes	Yes	Yes
PLC-2/30	AMD2900	8	5	5	896	896	896	Yes	Yes	60	Yes	Yes	Yes	Yes
PLC-3	AMD2900	16	20	2	8192	8192	8192	Yes	Yes	74	Yes	Yes	Yes	Yes
APPLIED SYS. CORP. (St. Clair Shores, MI)														
ASC 80	8051	8	5	20	256	256	256	YES	YES	24	OPT.	OPT.	OPT.	YES
ASEA IND'L. SYST., INC. (Milwaukee, WI)														
Masterpiece 010	8084	8	6	N/A	24/16	24/16				14				
Masterpiece 100	6803	8	1.2	10	10000	128	9872	Yes		60	Yes	Yes	Yes	Yes
Masterpiece 200	68000	16	8	10	10000	1300	8700	Yes		70	Yes	Yes	Yes	Yes
AUGUST SYS. (Salem, OR)														
PCS-300	86/30	16	8	80	4096	4096	4096		YES	20			YES	
AUTOMATION SYS., INC. (Eldridge, IA)														
IMP PLC-2	MFR.OWN	4/8	0.2	5m	128	128		YES	YES	16	YES	YES	YES	YES
IMP PFC-2	MFR.OWN	4/8	0.2	5m	128	128		YES	YES	16	YES	YES	YES	YES
PFC-4	MFR.OWN	16	0.2	5m	2048	2048		YES	YES	256	YES	YES	YES	YES
BARBER-COLMAN CO. IND'L. INST. DIV. (Rockford, IL)														
MACO III B	8085A	8	4	25	160	160		YES	YES	9				
MACO III B AMD	8085A	8	4	25	160	160		YES	YES	9				
CINCINNATI-MILACRON, INC. ELECTR. SYS. DIV. (Lebanon, OH)														
ACRAMATIC APC-1	8088	16	5/8	ADJ.	512	512			YES	45	YES	YES		
ACRAMATIC APC-2	8088	16	5/8	ADJ.	2048	2048		YES		N/A	YES	YES	YES	

Programmable controllers														
MANUFACTURER (City, State) — MODEL	Type CPU	Number of bits	Clock speed (MHz)	Scan rate (msec)	System total I/O	Local total I/O	Total remote I/O	High level language	Relay ladder logic	No. of instructions	PID capabilities	Motion control	Documentation	PC data highway
CONRAC CORP. CRAMER DIV. (Old Saybrook, CT)														
TROLL 200	6803	8	2	10	8	8			YES	24				
TROLL 210	6803	8	2	20	168	168	128		YES	512			YES	
TROLL 310	6803	8	2	25	528	528	512		YES	2k	YES		YES	YES
CONRAC CORP. CRAMER DIV. (Old Saybrook, CT)														
TROLL 300	6803	8	2	N/A	192	64	128	Yes	Yes	47	Yes		Yes	
DIVELBISS CORP. (Fredericktown, OH)														
BEAR BONES	MC14500	1	0.2	5m	248	248			YES	15			YES	
ICM320A	MC14500	1	0.1	10m	250	250			YES	15			YES	
DYNAGE, INC. (Bloomfield, CT)														
MPC-8/01	3000	8	20	2	2176	2176	2176	YES		48		OPT.	OPT.	
MPC-8/02	3000	8	20	2	2176	2176	2176	YES		63	YES	OPT.	OPT.	
CMS-2001	6502	8	1	250	2176	2176	2176	YES		120	YES	OPT.	YES	
GULF + WESTERN MFG. CO. EAGLE SIGNAL DIV. (Davenport, IA)														
EPTAK 200	8048	8	N/A	20	128	128			YES	30			YES	
EPTAK 210	8049	8	N/A	10	32	32			YES	36			YES	
EPTAK 220	8049	8	N/A	20	128	128			YES	41			YES	
EPTAK 240	8051	8	N/A	20	128	128			YES	50	YES	YES	YES	
EPTAK 700	8080A	8	2	10	2048	2048	2048	YES	YES	235	YES	YES	YES	YES
EATON CORP. CONTROL COMP. DIV. (Milwaukee, WI) CUTLER-HAMMER														
MPC-1	8031	8	10	12	128	128	96		YES	500			YES	
COUNT CTLS./SYS. DIV. (Watertown, WI)														
DURANT 6500	1802	8	N/A	N/A	96	96	64	YES		222		YES	YES	YES
DURANT 6550	1802	8	N/A	N/A	96	96	64	YES		222		YES	YES	YES
EMICC (Macon, GA)														
GEM 80/100- 2K	8086	8	4	30	3072	512	3584		Yes	24			Yes	Yes
GEM 80/100- 6K	8086	8	4	30	3072	512	3584		Yes	24			Yes	Yes
GEM 80/100-12K	8086	8	4	30	3072	512	3584		Yes	32			Yes	Yes
GEM 80/130- 6K	8086	8	4	30	3072	512	3584		Yes	42	Yes		Yes	Yes
GEM 80/130-12K	8086	8	4	30	3072	512	3584		Yes	43	Yes		Yes	Yes
GEM 80/225-32K	8086	8	4	30	3072	512	3584	Yes	Yes	103	Yes	Yes	Yes	Yes
GEM 80/235-32K	8086	8	4	30	3072	512	3584	Yes	Yes	103	Yes	Yes	Yes	Yes
GEM 80/245-32K	8086	8	4	30	3072	512	3584	Yes	Yes	103	Yes	Yes	Yes	Yes
ENCODER PROD. CO. (Sandpoint, ID)														
EPC-7	INS8073	8	4	N/A	1120	96	1024	YES		75		YES	YES	

Programmable controllers

MANUFACTURER (City, State) / MODEL	Type CPU	Number of bits	Clock speed (MHz)	Scan rate (msec)	System total I/O	Local total I/O	Total remote I/O	High level language	Relay ladder logic	No. of instructions	PID capabilities	Motion control	Documentation	PC data highway
ENERTRON IND., INC. (Gasport, NY)														
SK-1600	6802	8	N/A	80	64	64	64	YES	YES	N/A		YES		
FESTO CORP. (Hauppauge, NY)														
FPC	IM6100A	12	4	5	512	512		YES	YES	90			YES	
GTE ELECTR. CTL. OPER. GTE PROD. CORP. (Waltham, MA)														
SYLVANIA I^2R 1110	8039	8	6	N/A	12	12		YES		51				
SYLVANIA I^2R 1111	8039	8	6	N/A	24	24		YES		51				
SYLVANIA MECA MC-2	8039	8	6	N/A	96	96		YES		51			YES	YES
SYLVANIA MECA MC-4	8039	8	6	N/A	96	96		YES		51			YES	YES
SYLVANIA MECA MC-8	8031	8	11	N/A	96	96		YES		51	YES	YES	YES	YES
GENERAL ELEC. CO. IND'L. CONTROL DEPT. (Charlottesville, VA)														
SERIES SIX 60	AMD2903	16	4	1	512	512	512	YES	YES	76	OPT.	YES	YES	YES
SERIES SIX 600	AMD2903	16	4	1	2000	2000	2000	YES	YES	76	OPT.	YES	YES	YES
SERIES SIX 6000	AMD2903	16	4	1	4000	4000	4000	YES	YES	76	OPT.	YES	YES	YES
GIDDINGS & LEWIS, INC. ELECTR. DIV. (Fond du Lac, WI)														
PC400	6800	8	1	5m	3056	2032	1024	YES	YES	104		YES	YES	YES
GOULD, INC. ELECTR. SYS. GROUP (Andover, MA)														
MODICON MICRO 84	Z8	4		10	64	64			YES	9	YES		YES	YES
MODICON 184	MFR.OWN	16		75	1024	1024	1024		YES	34	YES		YES	YES
MODICON 384	MFR.OWN	16		30	1024	1024	1024		YES	38	YES		YES	YES
MODICON 484	8X300	8	6.7	25	512	512			YES	18		YES	YES	YES
MODICON 584	AMD2901	16	16	1.5	8192	8192	8192		YES	32	YES	YES	YES	YES
MODICON 584L	2901	16	10	60	8092	1024	8092	Yes		30	Yes	Yes	Yes	Yes
MODICON 584M	2901	16	10	100	2048	1024	2048		Yes	30	Yes	Yes	Yes	Yes
MODICON 884	8086	16	8	10	256	256	N/A		Yes	35			Yes	Yes
HOLMOR, INC. (Flanders, NJ)														
AMICON 1000	8085	8	6.1	50m	32	32		OPT.	YES	256			OPT.	
AMICON 900	8085	8	6.1	100m	104	104		OPT.	YES	256	YES	YES	OPT.	YES
AMICON 800	8085	8	4	150m	976	720	256	OPT.	YES	1024	YES	YES	OPT.	YES
IDEC IZUMI IDEC SYS. & CTLS. CORP. (Santa Clara, CA)														
PLE-30R	8085	8	4.9	30	30	30			YES	10				
INCON ELECTR., INC. (Mississauga, ON)														
C-800	6802	8	1	20	16	16			YES	16				

Programmable controllers

MANUFACTURER (City, State) / MODEL	Type CPU	Number of bits	Clock speed (MHz)	Scan rate (msec)	System total I/O	Local total I/O	Total remote I/O	High level language	Relay ladder logic	No. of instructions	PID capabilities	Motion control	Documentation	PC data highway
INCONIX CORP. (Natick, MA)														
CONTROL LOGIC CINCH	Z80A	8	3	N/A	1876	1876	1876	YES		125	YES	YES	YES	YES
CONTROL LOGIC CINCHPAC	8031	8	11	N/A	8000	32	7968	YES		40	YES	YES	YES	YES
IND'L. SOLID STATE CTLS. (York, PA)														
620-10	6809	8	6.5	10	256	256			YES	19			YES	YES
620-15	6809	8	6.5	10	256	256		YES	YES	32		YES	YES	YES
620-20	6809	8	6.5	10	512	512	512	YES	YES	32		YES	YES	YES
620-30	6809	8	6.5	10	2048	2048	1536	YES	YES	49	YES	YES	YES	YES
KLOCKNER-MOELLER CORP. (Natick, MA)														
SUCOS PS21	MFR.OWN	16	3	1.8	56	56		YES	YES	24		YES	YES	
SUCOS PS22	MFR.OWN	16	3	1.8	1024	1024	1024	YES	YES	24		YES	YES	
SUCOS PS24	MFR.OWN	16	3	1	2048	2048	2048	YES	YES	24		YES	YES	
MTS SYSTEMS CORP. (Minneapolis, MN)														
INCOL 470	8085	8	5	2	255	255			YES	128		YES	YES	YES
McGILL MFG. CO. ELEC. DIV. (Valparaiso, IN)														
1701	38P70	8	4	2	512	512			YES	11		YES	YES	
MITSUBISHI ELECTRIC (Compton, CA)														
F20M	8049	8	6	0.1	40				YES	14			YES	
F40	8039	8	9.6	0.045	80				YES	14			YES	
K0E	8085A2	8	5	30m	480	480	480		YES	26	YES		YES	
K0J	8085A2	8	5	30m	480	480	480		YES	26			YES	
K2E	8085A2	8	5	5.6m	480	480	480		YES	26	YES	YES	YES	YES
OMRON ELECTR., INC. CONTROL COMP. DIV. (Schaumburg, IL)														
SYSMAC M0/CZ	6800	8	8	30	32	32			YES	14				
SYSMAC M1R	6800	8	8	15	320	320			YES	17			YES	
RELIANCE ELEC. CO. CONTROL SYS. DIV. (Stone Mountain, GA)														
AutoMate 15	8031	8	12	7	64	64			YES	35	YES		YES	YES
AutoMate 35	1802	8	4	4	2048	128	1920	YES	YES	83	YES		YES	YES
SIEMENS-ALLIS (Wichita Falls, TX)														
SIMPTIC S5-110A	MFR.OWN	16	1	20	256	256			YES	40				
SIMPTIC S5-110S	8X300	16	6	8	512	512			YES	200			YES	YES
SIMPTIC S5-130W	8X300	16	6	3	1024	1024			YES	200			YES	YES
SIMPTIC S5-150A	AMD2900	16	10	5	2048	2048	2048	YES	YES	400	YES	YES	YES	YES
SIMPTIC S5-150K	AMD2900	16	10	5	2048	2048	2048	YES	YES	400	YES	YES	YES	YES
SIMPTIC S5-150S	AMD2900	16	10	3	2048	2048	2048	YES	YES	600	YES	YES	YES	YES
SOLID CONTROLS, INC. (Minneapolis, MN)														
EPIC	6802/9	8	1	2.5	192	192	64	YES		12	OPT.	OPT.	YES	
SYSTEM 10	6802/9	8	1	2.5	512	512			YES	12	OPT.	OPT.	YES	

Programmable controllers

MANUFACTURER (City, State) MODEL	Type CPU	Number of bits	Clock speed (MHz)	Scan rate (msec)	System total I/O	Local total I/O	Total remote I/O	High level language	Relay ladder logic	No. of instructions	PID capabilities	Motion control	Documentation	PC data highway
SQUARE D CO. (Milwaukee, WI)														
SY/MAX 20	AMD2901	16	1	10	512	512			YES	69	YES		YES	
SY/MAX 100	8031	16	4	20	40	40			YES	136			YES	
SY/MAX 300	Z80	16	4	32	128	128	128	YES	YES	220			YES	OPT.
CLASS 8881	6800	16	I	5	2048	2048	2048	YES	YES	58	YES		YES	
STRUTHERS-DUNN, INC. SYSTEMS DIV. (Bettendorf, IA)														
DIRECTOR 4001	6502	8	1	40	384	384	384	YES	YES	32	YES	YES	YES	YES
DIRECTOR 4002	6502	8	1	25	64	64			YES	32	YES	YES	OPT.	YES
TEXAS INSTRUMENTS, INC. IND'L. CONTROLS DEPT. (Johnson City, TN)														
5TI	MFR.OWN	16	8	8.3	512	512	512	YES	YES	13			YES	YES
TI510	6512	8	2	16.7	20	20	20	YES	YES	11			YES	YES
TI520	TMS9995	16	12	8	128	128	128	YES	YES	37			YES	YES
TI530	TMS9995	16	12	8	1024	1024	1024	YES	YES	37			YES	YES
PM550	TMS9900	16	7.5	20	512	512	512	YES	YES	33	YES		YES	YES
TRIUS CORP. (Santa Clara, CA)														
2020	6502	8	1	50	32	8	24	YES	YES	16	"P"	YES	YES	YES
WESTINGHOUSE ELECTRIC INDUSTRY ELECTR. DIV. (Madison Hts. MI)														
NUMA-LOGIC PC700	8X300	16	16	8	512	512	512	YES	YES	40	YES	YES	YES	YES
NUMA-LOGIC PC900B	8085	16	10	20	255	255	255	YES	YES	39	YES	YES	YES	YES
NUMA-LOGIC PC1100	Z80A	16	4	20	128	128			YES	49	YES		YES	YES
THE FOXBORO CO. (Foxboro, MA)														
FOX 3PC1	8088	16	5/8	ADJ.	512	512			Yes	45	Yes	Yes		Yes
FOX 3PC2	8088	16	5/8	ADJ.	2048	2048		Yes		N/A	Yes	Yes	Yes	Yes
GTE PROD. CORP. GTE ELECTR. CTL. OPER. (Waltham, MA)														
MC-16	8088	16	8	5	56,896	224	56,896	Yes	Yes	47	Yes		Yes	Yes
GENERAL ELEC. CO. PROG. CONTROL DEPT. (Charlottesville, VA)														
Series One	Z-80A	16	3.2	30	112	112			Yes	18				future
Series Three	Z-80A	16	3.9	7.5	400	400			Yes	46				future
GIDDINGS & LEWIS, INC. ELECTR. DIV. (Fond du Lac, WI)														
PC409	6809	8	8	N/A	3056	2032	1024	Yes	Yes	4	Yes	Yes	Yes	Yes
MITSUBISHI ELECTRIC (Compton, CA)														
MELSEC F-40M	8039	8	9.6	N/A	80				Yes	N/A			Yes	

MANUFACTURER (City, State) MODEL	Programmable controllers													
	Type CPU	Number of bits	Clock speed (MHz)	Scan rate (msec)	System total I/O	Local total I/O	Total remote I/O	High level language	Relay ladder logic	No. of instructions	PID capabilities	Motion control	Documentation	PC data highway
OMRON ELECTR., INC. CONTROL COMP. DIV. (Schaumburg, IL)														
SYSMAC-M5R	68B00	16	12	15	256	256			Yes	26			Yes	
SYSMAC-S6	8051	24	12	10	64	64			Yes	17		Yes		
SQUARE D CO. (Milwaukee, WI)														
SY/MAX 500	8088	16	12	2	2000	2000	2000	Yes	Yes	245	Yes	Yes	Yes	Yes
TEXAS INSTRUMENTS, INC. IND'L. CONTROLS DEPT. (Johnson City, TN)														
PM550-116	TMS9900	16	7.5	10	512	512	512	Yes	Yes	36	Yes		Yes	Yes
KLOCKNER-MOELLER CORP. (Natick, MA)														
SUCOS PS21	MFR.OWN	16	3.3	2	32	32			Yes	24		Yes	Yes	

APPENDIX D: PROGRAMMING AND OPERATIONS MANUAL FOR MINI-PLC-2 PROGRAMMABLE CONTROLLER

Section 1
INTRODUCTION

1.0 OVERVIEW

The Bulletin 1772 Mini-PLC-2 Programmable Controller is a rugged, solid state programmable controller that consists of the Mini-PLC-2 Processor (Cat. No. 1772-LN1, -LN2 or -LN3) and the 1771 I/O Family of racks and modules. Refer to Figure 1-1.

FIGURE 1-1 — Mini-PLC-2 Programmable Controller

With a user-written program and appropriate I/O modules, the Mini-PLC-2 Controller can be used to control many types of industrial applications such as:

- Process control
- Material handling
- Palletizing
- Measurement and gaging
- Pollution control and monitoring

The Mini-PLC-2 Processor has a Read/Write CMOS memory that stores User Program instructions, numeric values and I/O device status.

The User Program is a set of instructions in a particular order that describes the operations to be performed and the operating conditions. It is entered into memory, instruction by instruction, in a ladder diagram format display from the keyboard of the Industrial Terminal (Cat. No. 1770-T1, -T2 or -T3). Some ladder diagram symbols closely resemble the relay symbols used in hardwired relay control systems.

During program operation, the Mini-PLC-2 Processor continuously monitors the status of input devices and, based on User Program instructions, either energizes or de-energizes output devices. Because the memory is programmable, the User Program can be readily changed if required by the application.

In addition to ON/OFF control, the Mini-PLC-2 Controller can perform additional functions such as:

- Timing/Counting operations
- Arithmetic (+, -) operations
- Arithmetic (x, ÷) operations (1772-LN3 Processor)
- Data comparisons
- Block Transfer (1772-LN3 Processor Module)
- I/O Forcing
- Data Highway and RS-232-C interfacing

When the Industrial Terminal (Cat. No. 1770-T1, -T2 or -T3) and the Mini-PLC-2 Programmable Controller are used together or with additional peripheral devices, application data can be recorded or displayed using a variety of peripheral functions:

- Report generation
- Contact histogram (the ON/OFF history of a bit in memory)
- Hard-copy printout of the User Program or the complete memory

• Recording/loading/verifying the User Program using magnetic tape

1.1 LADDER DIAGRAM LOGIC

PC ladder diagram logic closely resembles hardwired relay logic. Hardwired relay control systems require electrical continuity to turn output devices ON and OFF. For example, the relay diagram in Figure 1-2 shows that limit switch LS1 and relay contact CR2 must be closed to energize relay coil CR4.

Similarly, in each rung of ladder diagram program, logic continuity is needed to energize or de-energize the output instruction, and ultimately the output device. For example, the ladder diagram rung in Figure 1-3 shows the input devices and the output device with their respective Data Table bit addresses. The bit addresses correspond to the location of the I/O devices wired to the I/O modules. When the two input instructions are logically TRUE, or the bits in memory are ON, logic continuity is established. This causes the output instruction to be TRUE and the output device to be turned ON.

1.2 MEMORY STRUCTURE

The Data Table of the Mini-PLC-2 Processor memory is made up of an arrangement of storage points called bits (BInary digiTs). A bit is the smallest unit of memory and can store information as a "1" or a "0" (Figure 1-4). When a "1" occupies a bit, that bit is ON; when a "0" occupies a bit, that bit is OFF.

A group of 8 bits forms a single byte. Two bytes, or 16 bits, make up one word. All Data Table words are identified by their word address which is a 3-digit octal number. The octal numbering system is explained in Section 13.

Similarly, each bit in a word is identified by a two-digit number using the octal numbering system. The memory bits are numbered 00 through 07 and 10 through 17, with the most significant bit 17 (MSB) at the left and the least significant bit 00 (LSB) at the right.

A specific bit in Data Table can be identified by combining the 3-digit word address and 2-digit bit number to form the bit address, such as 03012 or 030/12. The difference depends on the Industrial Terminal used. The 1770-T1 and -T2 display the 5-digit bit address centered above the instruction symbol. The 1770-T3 displays the bit address by placing the 3-digit word address above and the 2-digit bit number below the instruction symbol. Programming examples will be illustrated using the 1770-T3 display. However, both displays will be illustrated in the Instruction Summary Table at the end of each programming section in this manual.

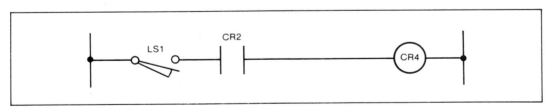

FIGURE 1-2 — Relay Diagram

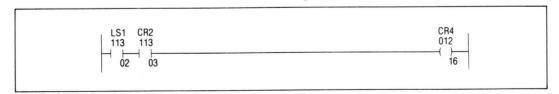

FIGURE 1-3 — Ladder Diagram Rung

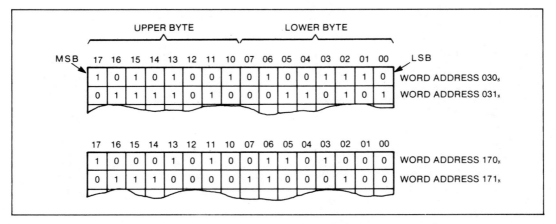

FIGURE 1-4 — Memory Word Structure

1.3 MEMORY ORGANIZATION

The Mini-PLC-2 Processors have either a 512 word memory size (Cat. No. 1770-LN1) or a 1024 word memory size (Cat. No. 1772-LN2 or -LN3). These memory words are organized by their word addresses and are divided into the following major areas. The size of each area can be varied within limits to suit user needs, but the total cannot exceed the Processor memory size.

- Data Table
- User Program
- Message Storage (if used)

The Data Table stores the information needed in the execution of the User Program such as the status of input and output devices, timer/counter Preset and Accumulated values, bit/word storage, etc. Any instruction in the User Program can address any word or bit in the Data Table except in the Processor Work Areas.

The Data Table is factory configured to 128 words. The words reserved for timers and counters can be decreased to any even-number value down to 48 words so that storage capacity for User Program and/or messages can be increased. Figures 1-5 shows the organization of a factory configured Data Table.

The User Program follows the Data Table in memory. The User Program is the logic that controls the machine operation. The logic consists of instructions that are programmed in ladder diagram format. Each instruction requires 1 word of memory.

Message Storage area begins after the END statement of the User Program. This area stores the alphanumeric characters of the messages. Two characters can be stored in one word.

For a detailed description of memory, refer to Publication 1772-700, the Organization and Structure of the Mini-PLC-2 Memory.

1.4 HARDWARE/PROGRAM INTERFACE

The Processor monitors input conditions and controls output devices according to a user-entered program. The interface between hardware and program occurs in the Input/Output Image Table.

1.4.1 Image Table

The primary purpose of the Input Image Table is to duplicate the status (ON or OFF) of the input devices wired to input module terminals. If an input device is ON (closed), its corresponding Input Image Table bit is ON ("1"). If an input is OFF (open), its corresponding Input Image Table bit is OFF ("0"). Input Image Table bits are MONITORED by User Program instructions.

The primary purpose of the Output Image Table is to control the status (ON or OFF) of the output devices wired to output module terminals. If an Output Image Table bit is ON ("1"), its corresponding output device is ON (energized). If a bit is OFF ("0"), its corresponding output device is OFF (de-energized). Output Image Table bits are CONTROLLED by User Program instructions.

1.4.2 Instruction Address

Instruction addresses in the Input/Output Image Table have a dual role. The 5-digit bit address references both an I/O Image Table

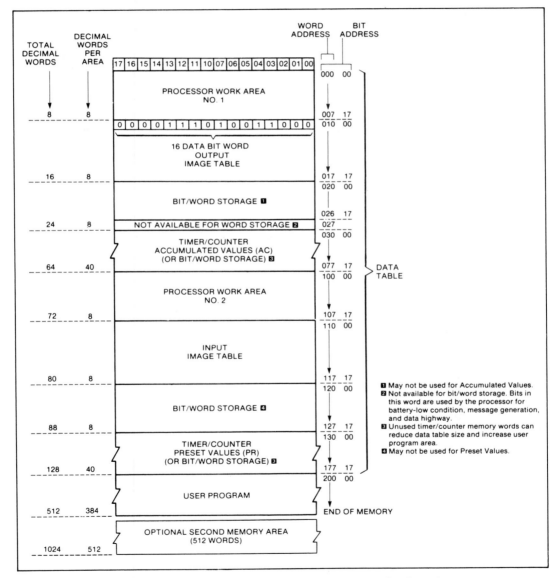

FIGURE 1-5 — Data Table Organization, Factory Configured

address and a hardware location. An I/O Image Table word corresponds to two I/O modules located in a Module Group in the I/O Rack. Both are represented by the upper 3 digits of the 5-digit bit address. The lower byte of the I/O Image Table word corresponds to an I/O module in the left slot of the Module Group. The upper byte corresponds to an I/O module in the right slot of the Module Group. See Figure 1-6. The remaining two digits represent the bit number in the I/O Image Table word and the terminal number in the Module Group. Each 5-digit bit address in the I/O Image Table directly relates to an I/O module terminal as shown in Figure 1-7.

1.4.3 Fundamental Operation

The hardware-program interface is illustrated in Figure 1-8 by showing the operational relationship between the input and output devices, the Input/Output Image Table and the User Program.

When an input device connected to terminal 112/12 is closed, the input module circuitry senses a voltage. The ON condition is reflected in the Input Image Table bit 112/12. During the Program scan, the Processor examines bit 112/12 for an ON (1) condition. If the bit is ON (1), the EXAMINE ON instruction is logically TRUE. A TRUE condition is displayed as an intensified instruction. A path of logic continuity is established and causes the rung to be TRUE. The Processor then sets Output Image Table bit 013/06 to ON (1). The Processor turns ON terminal 013/06 during the next I/O scan and the output device wired to this terminal becomes energized. When the rung condition is TRUE, the output instruction is intensified.

When the input device wired to terminal 112/12 opens, the input module senses no voltage. The OFF condition is reflected in the Input Image Table bit 112/12. During the program scan, the Processor examines bit 112/12 for an ON (1) condition. Since the bit is OFF (0), logic continuity is not established and the rung is FALSE. The Processor then sets Output Image Table bit 013/06 to OFF (0). In the next I/O scan, the Processor turns OFF terminal 013/06 and the output device wired to this terminal is turned OFF.

1.5 COMPATIBILITY

The Mini-PLC-2 Processors are compatible with the Industrial Terminal, the Data Highway and RS-232-C interfacing. The 1771-LN3 Processor is compatible with Bulletin 1771 Block Transfer I/O modules.

1.5.1 Industrial Terminal Compatibility

The Mini-PLC-2 Controller can be programmed using the Industrial Terminal (Cat. No. 1770-T1, T2, or T3). It can also be programmed with the PLC-2 Program Panel (Cat. No.1772-T1) or with the combination of the PLC Program Panel with PLC-2 Program Panel Adapter (Cat. No. 1774-TA and 1772-T4).

This manual will illustrate programming examples using the Industrial Terminal. The first edition of this manual, dated January 1980, illustrates programming examples using the (obsolete) PLC-2 Program Panel.

1.5.2 Data Highway Compatibility

The Mini-PLC-2 Controller can be connected to the Allen-Bradley Data Highway using the Communication Adapter Module (Cat. No. 1771-KA). Data Highway messages and the Data Highway Communication zone of User Program must reference only the addresses within the user-configured Data Table.

1.5.3 Block Transfer Compatibility

The Mini-PLC-2 Processor module (Cat. No. 1772-LN3) can be programmed to communicate with intelligent Bulletin 1771 I/O modules having Block Transfer capability. These include the Thermocouple, Analog Input, Analog Output, Encoder/Counter, PD, etc. modules. Section 11.2 of this manual covers Block Transfer Programming.

1.5.4 User Program Compatibility

User Programs written for the Mini-PLC-2/15, PLC-2/20 or PLC-2/30 should not be loaded into the Mini-PLC-2 Controller unless the User Program is compatible in the following areas:

- The Data Table is 128 words or less.

- The words of memory used do not exceed 512 for the 1772-LN1 Processor; or 1024 for the 1772-LN2 or -LN3 Processor.

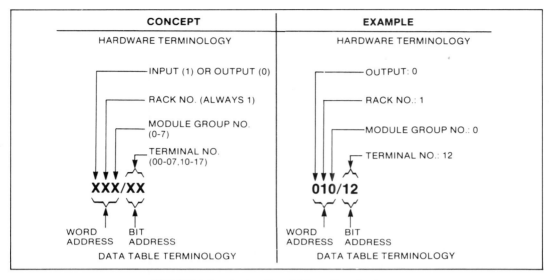

FIGURE 1-6 — Instruction Address Terminology

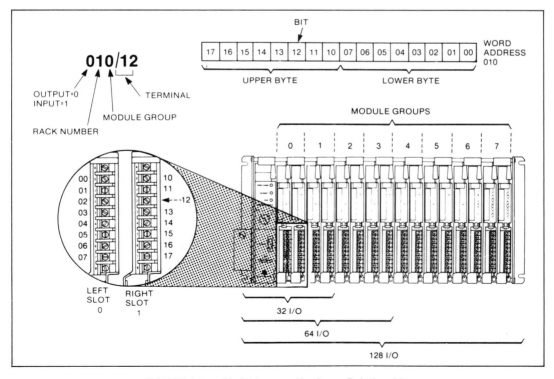

FIGURE 1-7 — Bit Address to Hardware Relationship

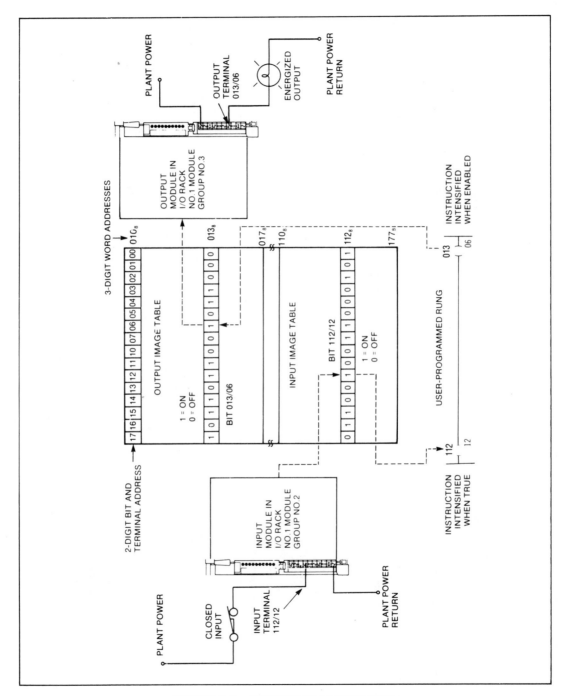

FIGURE 1-8 — Hardware-Program Interface

- The instruction set contains compatible instructions: block instructions and subroutine programming are absent; MULTIPLY or DIVIDE instructions can only be entered in the 1771-LN3 Processor; divide by 0 is compatible as stated in Section 6.4.

The Industrial Terminal will prevent the loading of memory from tape or data cartridge if the Data Table size or the number of memory words exceed the capacity of the Processor.

If a User Program has a compatible Data Table and memory capacity but contains incompatible instructions, it could be loaded into a Mini-PLC-2 Controller. When using a 1772-LN3 Processor, the incompatible instructions would be displayed on a 1770-T3 Industrial Terminal when the Processor was in PROGRAM mode. Upon switching to TEST or RUN mode, the Processor would fault before outputs could be energized. The message "PROCESSOR FAULT" would appear at the top of a blank screen. Upon switching back to PROGRAM mode, the Industrial Terminal would be initialized to the Mode Selection display. Options then available include:

a) Remove the incompatible User Program using the Total Memory Clear function [CLEAR MEMORY][9][9]

b) Salvage the User Program by replacing the incompatible instructions with equivalent programming that the Mini-PLC-2 can handle.

If a User Program with an incompatible instruction set were to be loaded into a 1772-LN3 Processor having a 1770-T1 or -T2 Industrial Terminal, ERR messages would appear randomly throughout the program when the Processor was in PROGRAM mode. They would be located adjacent to any instruction that the Processor was not capable of handling. Upon switching to TEST or RUN mode, the Processor would fault before outputs could be energized. The options available would be the same as those stated above. The 1770-T1 Industrial Terminal would display the message "COMMUNICATION FAULT-CHECK CABLES FIRST." Two messages would be alternately displayed by the 1770-T2 Industrial Terminal: "PROCESSOR FAULT" and INVALID INSTRUCTION ENCOUNTERED.

Incompatible instructions would be ignored by a 1772-LN1 or -LN2 Processor. The display would depend on the Industrial Terminal. A 1770-T1 or -T2 would display "ERR" messages in place of incompatible instructions. The 1770-T3 would display incompatible instructions although they could not executed.

1.6 SUPPORT DOCUMENTATION

The following support documents contain additional information regarding Mini-PLC-2 Controller components.

- Mini-PLC-2 Programmable Controller Assembly and Installation Manual (Publication 1772-820): contains necessary information on installation, assembly, maintenance and troubleshooting

- PLC-2 Family Support Documentation Manual (Publication 1772-803-1): contains useful information on memory organization, Data Table expansion, system features, wiring, module keying and various features of Mini-PLC-2 Controller components and 1771 I/O.

Section 2
HARDWARE CONSIDERATIONS

2.0 GENERAL

This section will only describe the hardware features of the Mini-PLC-2 Programmable Controller that are used when inputting or debugging the User Program. For information on installation, start-up, troubleshooting, etc, refer to Publication 1772-820, the Mini-PLC-2 Programmable Controller Assembly and Installation Manual.

2.1 MODE SELECT SWITCH

The Mini-PLC-2 Processor has a three-position keylock Mode Select Switch (Figure 2-1) that places the Processor in one of three operating modes:

- PROGRAM
- TEST
- RUN

FIGURE 2-1 — Mode Select Keyswitch

2.1.1 PROG Position

This switch position places the Processor in the PROGRAM mode. User Program instructions are entered in this mode. They can be entered from the Industrial Terminal, or entered from the Digital Cassette Recorder (Cat. No. 1770-SA) or the Data Cartridge Recorder (Cat. No. 1770-SB) when connected to the Industrial Terminal. All outputs are de-energized in this switch position and the machine controlled by the Mini-PLC-2 will not operate.

2.1.2 TEST Position

This switch position places the Processor in the TEST mode. The User Program is tested under simulated operating conditions. Inputs are active and recognized by the Processor, but user output devices are not energized. All outputs are disabled in this switch position. Changes to the User Program are NOT permitted, but Data Table values can be changed using the On-Line Data Change function or the Bit Manipulation function.

2.1.3 RUN Position

This switch position places the Processor in the RUN mode. The User Program will be executed and outputs are controlled by the program. Changes to the User Program are not permitted, but Data Table values can be changed using the On-Line Data Change function or the Bit Manipulation function.

This is the only switch position that allows removal of the Mode Select Switch key.

2.2 PROCESSOR DIAGNOSTIC INDICATORS

Indicators on the front panel of the Processor Module aid in analyzing controller status. (Refer to Figure 2-2.) During operation in any mode, the Processor continuously monitors its own status through checks on timing and data parity. In addition, the Processor receives a signal from the power supply if user AC power goes low for longer than one-half cycle.

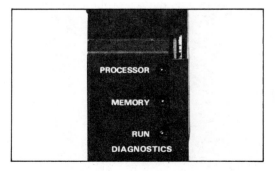

FIGURE 2-2 — Diagnostic Indicators

2.2.1 PROCESSOR Indicator

This red indicator illuminates if the Processor is unable to scan the User Program and Data Table. It is normally OFF. When ON, the Processor has stopped communication with I/O modules. If this occurs, the Last State Switch determines the status of energized controller outputs. (Refer to paragraph 2.4.)

Reset may be attempted for this type of fault by changing the Mode Select Switch to the PROGRAM mode, then back to RUN. Reset may also be accomplished by cycling line power to the system power supply or by reloading the User Program.

2.2.2 MEMORY Indicator

This red indicator illuminates if the Processor detects loss of User Program, a discrepancy in memory data, or a parity error. It is normally OFF.

The Processor stops communication with I/O modules if this type of fault is detected. The Last State Switch determines the status of controller outputs if this fault occurs. (Refer to paragraph 2.4.)

This error may be reset by turning the Mode Select Switch to the PROG position then back to RUN, by cycling line power to the system power supply, or by reloading the User Program.

2.2.3 RUN Indicator

This green indicator illuminates when the Processor is operating with the Mode Select Switch in the RUN mode. When this indicator is ON, controller outputs are enabled. This also implies that no Processor-related fault has been detected.

This indicator turns OFF in the RUN mode if the system power supply detects that voltage on the user AC line has dropped below 98V or 196V for 120V or 220/240V operation, respectively. In this event, the Processor disables all output devices and stops receiving input module data. This prevents the Processor from storing input data which might be inaccurate due to a low voltage level. The indicator also goes OFF when a fault occurs.

In the event of user AC power failure, the restart of the Processor is automatic with recovery of the line to the normal voltage range.

2.3 POWER SUPPLY DIAGNOSTIC INDICATORS

The system Power Supply (Figure 2-3) has two diagnostic indicators on the front panel: The BATTERY LOW and DC ON indicators.

FIGURE 2-3 — System Power Supply (Without Battery Pack)

2.3.1 BATTERY LOW Indicator

When the batteries for memory backup are low, this red indicator flashes ON and OFF. The Battery Low Bit, bit 027/00, will cycle ON and OFF when a battery-low condition is detected and the Mode Select Switch is in the RUN or TEST position. Programming Techniques can be used to examine this bit and to control some type of alerting device when a battery-low condition exists. The battery will continue to provide memory backup for about one week after the indicator begins to flash.

2.3.2 DC ON Indicator

The DC ON is a red indicator that monitors the 5.1V DC line to the logic circuitry in the

Processor, Processor memory and I/O modules. It is ON when 5.1V DC is present. If the line drops below 5.1V DC, the indicator turns OFF and the controller shuts down.

2.4 SWITCH GROUP ASSEMBLY

When the Processor detects an internal fault, communication with the I/O rack is terminated. The last state of output terminals in this situation is user-selectable by the settings on the Switch Group Assembly. It is located on the left side of the I/O chassis backplane. (Refer to Figure 2-4.) Switch #1 must be set to determine output response to a Processor fault. Switch numbers 2-8 are not used. There are two switch settings:

- ON — Outputs remain in their last state, energized or de-energized, when a fault is detected.

- OFF — Outputs are de-energized when a fault is detected.

> **WARNING:** Switch #1 should be set to OFF for most applications. This allows the Processor to turn controlled devices OFF when a fault is detected. If this switch is set ON, machine operation can continue after fault detection and damage to equipment and/or injury to personnel could result.

2.5 INDUSTRIAL TERMINAL

The Industrial Terminal (Cat. No. 1770-T1,-T2 or -T3) can be used to program the Mini-PLC-2.

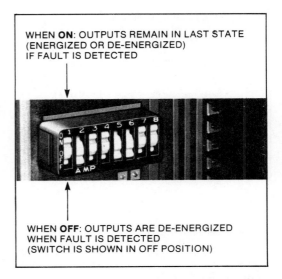

WHEN **ON**: OUTPUTS REMAIN IN LAST STATE (ENERGIZED OR DE-ENERGIZED) IF FAULT IS DETECTED

WHEN **OFF**: OUTPUTS ARE DE-ENERGIZED WHEN FAULT IS DETECTED (SWITCH IS SHOWN IN OFF POSITION)

FIGURE 2-4 — Switch Group Assembly

2.5.1 Hook-Up

Perform the following steps to connect the Industrial Terminal to the Mini-PLC-2 Processor: See Figure 2-5.

1. Plug the AC power cord of the Industrial Terminal into a grounded AC outlet.

2. Connect one end of the PLC-2 Program Panel Interconnect Cable (Cat. No. 1772-TC) to Channel A on the rear of the Industrial Terminal.

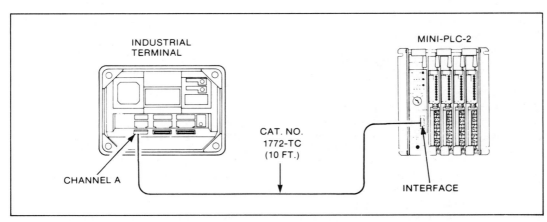

INDUSTRIAL TERMINAL

MINI-PLC-2

CAT. NO. 1772-TC (10 FT.)

CHANNEL A

INTERFACE

FIGURE 2-5 — Mini-PLC-2 Connection Diagram

3. Connect the other end of the cable to the socket labeled INTERFACE on the front of the Mini-PLC-2 Processor.

4. Insert the PLC-2 Keytop Overlay (Cat. No. 1770-KCA), Figure 2-6, on the Keyboard Module (Cat. No. 1770-FDC).

5. Turn the power switch on the front of the Industrial Terminal to the ON position. Mode Select display will appear.

For additional information on the Industrial Terminal, refer to the Industrial Terminal

System User's Manual, Publication No. 1770-805

2.5.2 Mode Selection and Initialization of the Industrial Terminal

When the Industrial Terminal is turned ON or when communication between the Industrial Terminal and Processor is interrupted for any reason, the Mode Selection display will appear. See Figure 2-7 for the 1770-T1 or -T2 display and Figure 2-8 for the 1770-T3 display. Any of the following occurrences can cause

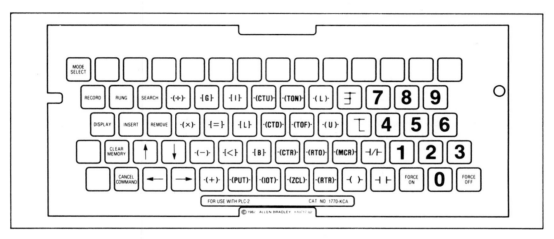

FIGURE 2-6 — PLC-2 Keytop Overlay (Cat. No. 1770-KCA)

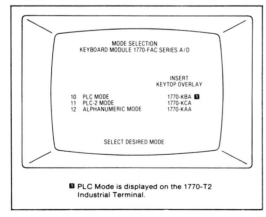

■ PLC Mode is displayed on the 1770-T2 Industrial Terminal.

FIGURE 2-7 — Mode Selection Display 1770-T1 or T2 Industrial Terminal

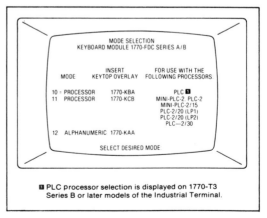

■ PLC processor selection is displayed on 1770-T3 Series B or later models of the Industrial Terminal.

FIGURE 2-8 — Mode Selection Display 1770-T3 Industrial Terminal

an interruption in communication and initialization of the Industrial Terminal

• Pressing the [MODE SELECT] Key

• Loss of power to the Processor or Industrial Terminal

• Disconnecting the Program Panel Interface Cable

To initialize the Industrial Terminal, one of the operating modes shown in Figure 2-7 or 2-8 must be chosen. If the Industrial Terminal is connected to a PLC-2 Family Processor, the Processor type (i.e. Mini-PLC-2, Mini-PLC-2/15, PLC-2/20, PLC-2/30) will be intensified.

To enter Ladder Diagram (PLC-2) mode, press [1][1] on the PLC-2 Keytop Overlay (Cat No. 1770-KCA).

When the Industrial Terminal is to be used as an alphanumeric data terminal, insert the Alphanumeric Keytop Overlay (Cat. No. 1770-KAA) and press [1][2]. Operation of the Industrial Terminal with the Alphanumeric Keytop Overlay is described in Section 10.3, Report Generation.

2.5.3 Keytop Overlay

The Mini-PLC-2 Processor should be programmed using the PLC-2 Keytop Overlay (1770-KCA) shown in Figure 2-6. All keys in this overlay are functional with the 1770-T1, -T2 or -T3 Industrial Terminal. The functions of the keys will be described in detail starting with Section 3 of this manual.

The PLC-2 Family Keytop Overlay (1770-KCB) can be used with any of the Processors in the PLC-2 Family. It should be used with the 1770-T3 Industrial Terminal when programming either the Mini-PLC-2/15 or the PLC-2/30. This overlay contains keys for some functions that are not possible with the Mini-PLC-2. When any one of these keys is pressed, the message "FUNCTION NOT AVAILABLE WITH THIS PROCESSOR" or "INVALID KEY" will appear on the screen.

Section 3
RELAY-TYPE INSTRUCTIONS

3.0 GENERAL Programmable controllers have many of the capabilities of hardwired relay control systems. Control functions similar to those available with relays are provided by the following relay-type instructions:

- Examine instructions
- Output instructions
- Branch instructions

3.1 EXAMINE INSTRUCTIONS

There are two Examine instructions:

- ON -| |-
- OFF -|/|-

The Examine instructions can examine the status of bits in any Data Table area except for Processor Work Areas. When an EXAMINE ON or EXAMINE OFF instruction is given an address in the I/O Image Table, the instruction can indirectly examine the status of a corresponding I/O device. The status of the I/O Image Table bit will be a 1 or 0 reflecting the ON or OFF condition, respectively, of the I/O device. The I/O device and the I/O Image Table bit have the same address. (See Hardware/Program Interface, Section 1.4).

The condition of the instruction can be either TRUE or FALSE depending on the status of the examined bit. If the Image Table bit is in the desired state, the instruction is TRUE. The TRUE-FALSE conditions of the Examine instructions are as follows:

- The EXAMINE ON instruction is TRUE when the addressed memory bit is a 1, meaning that the corresponding I/O device or bit is ON.

- The EXAMINE ON instruction is FALSE when the addressed memory bit is a 0, meaning that the corresponding I/O device or bit is OFF.

- The EXAMINE OFF instruction is TRUE when the addressed memory bit is a 0, meaning that the corresponding I/O device or bit is OFF.

- The EXAMINE OFF instruction is FALSE when the addressed memory bit is a 1, meaning that the corresponding I/O device or bit is ON.

The EXAMINE ON and EXAMINE OFF instructions are illustrated in an example rung in Figures 3-1 and 3-2, respectively.

FIGURE 3-1 — EXAMINE ON Instruction

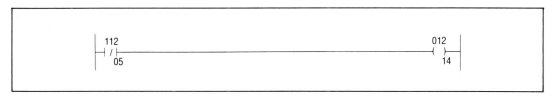

FIGURE 3-2 — EXAMINE OFF Instruction

3.2 OUTPUT INSTRUCTIONS

The output instructions set an addressed memory bit to "1" (ON) or reset it to "0" (OFF). An Output Image Table bit, as a "1" or "0", can cause an output device to be turned ON or OFF, respectively.

Output instructions are programmed at the end of the ladder-diagram rungs. Only one output instruction can be programmed on each rung. The output instruction will be performed only if the Condition (input) instructions preceding it provide a path of logical continuity (Figure 3-3).

These output instructions are:

• ENERGIZE -( )-

• LATCH -(L)-

• UNLATCH -(U)-

These instructions are used to set memory bits ON or OFF in any area of the Data Table, excluding the Processor Work Areas. Generally, they should NOT be assigned Input Image Table addresses because Input Image Table words are reset by the I/O scan.

3.2.1 OUTPUT ENERGIZE Instruction

The OUTPUT ENERGIZE instruction tells the Processor to turn an addressed memory bit ON when rung conditions are TRUE. This memory bit will determine the ON or OFF status of an output device when addressed to an output terminal. This instruction can also be used to turn ON a storage bit for later use in the program.

The OUTPUT ENERGIZE instruction tells the Processor to turn the addressed memory bit OFF when rung conditions go FALSE. Refer to Figure 3-4.

The OUTPUT ENERGIZE instruction can be programmed unconditionally for some types of specialized programming. Its use should be limited to storage bits for these special purposes. An unconditional OUTPUT ENERGIZE instruction (Figure 3-5) causes the output instruction to remain energized continuously. This is not advisable in output device programming for safety reasons, because the device cannot be turned OFF by program logic. Care should be taken not to inadvertently enter an unconditional output instruction.

CONDITIONS

OUTPUT INSTRUCTION

A CONTINUOUS PATH IS NEEDED FOR LOGICAL CONTINUITY

FIGURE 3-3 — Ladder-Diagram Rung

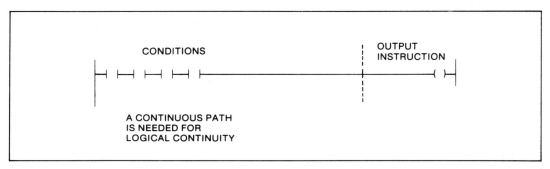

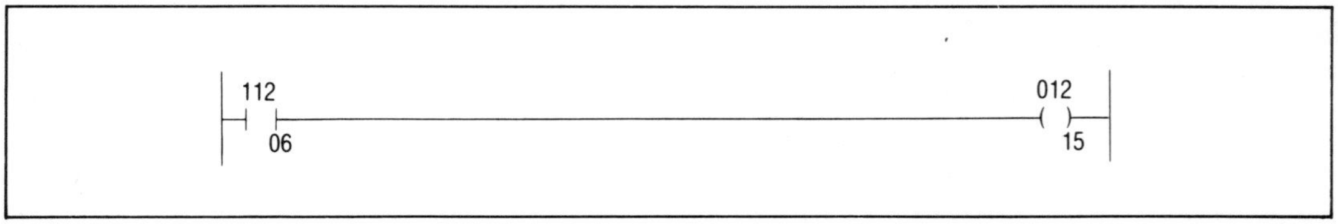

FIGURE 3-4 — OUTPUT ENERGIZE Instruction

FIGURE 3-5 — Unconditional OUTPUT ENERGIZE Instruction

3.2.2 OUTPUT LATCH and UNLATCH Instructions

There are two output instructions that are termed "retentive." These instructions are:

- OUTPUT LATCH

- OUTPUT UNLATCH

These instructions are usually used as a pair for any bit address they control.

The OUTPUT LATCH instruction (L) is somewhat similar to the OUTPUT ENERGIZE instruction. The OUTPUT LATCH instruction tells the Processor to set an addressed memory bit ON when the rung condition is TRUE. Unlike the OUTPUT ENERGIZE instruction, the OUTPUT LATCH instruction is "retentive." This means that once the rung condition goes FALSE, the latched bit remains ON until reset by an OUTPUT UNLATCH instruction. If power is lost but Processor back-up battery is maintained, all latched bits will remain ON. Outputs associated with the latched bits will be OFF with the power OFF. However, they will turn ON immediately when power is restored.

The OUTPUT UNLATCH instruction (U) is used to turn OFF a memory bit that has been latched ON. The OUTPUT UNLATCH instruction addresses the same memory bit that has been latched ON (Figure 3-6). When the rung condition for the OUTPUT UNLATCH instruction goes TRUE, the addressed memory bit is reset to zero (OFF). Refer to Figure 3-7. The OUTPUT UNLATCH is also "retentive." This means that once the rung condition goes FALSE, the unlatched bit remains OFF until reset by an OUTPUT LATCH instruction.

When the Mode Select Switch is changed from the RUN position, the last LATCH or UNLATCH instruction continues to control the addressed memory bit, but the output device is de-energized by the Processor. When the Mode Select Switch is turned back to RUN a latched output device will be energized.

The OUTPUT LATCH and UNLATCH instructions, when entered, automatically set the bit to OFF. The bit can be initially preset ON by entering the number [1] immediately after the address. The ON or OFF condition will be displayed below the instruction when the Processor is in the PROGRAM mode (Figure 3-8). When the Mode Select Switch is turned to the RUN position, the addressed memory bit and output device, if latched ON, will

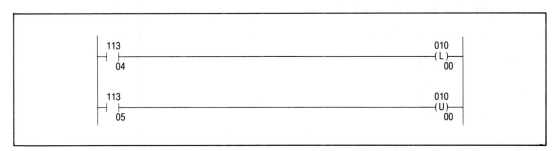

FIGURE 3-6 — LATCH/UNLATCH Instructions

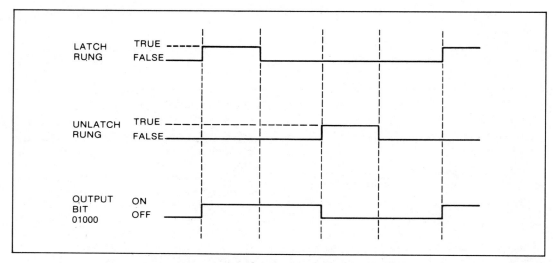

FIGURE 3-7 — LATCH/UNLATCH Timing Diagram

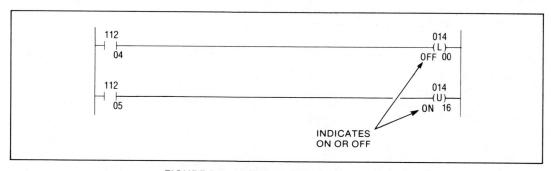

FIGURE 3-8 — LATCH/UNLATCH Indication

immediately be energized, regardless of the rung condition.

> **WARNING:** Do not preset a bit ON controlled by LATCH/UNLATCH instructions if it controls potentially hazardous machine motion. If the bit is preset ON by the LATCH/UNLATCH instructions, the output device controlled by that bit is energized immediately when the Mode Select Switch is turned to the RUN position. Injury to personnel near the machine could result.

Both LATCH and UNLATCH instructions can be programmed unconditionally. This programming technique is generally used with storage bits and should not be used to control output devices.

3.3 BRANCH INSTRUCTIONS

The branch instructions allow more than one combination of input conditions to energize an output device (Figure 3-9).

There are two branch instructions:

- BRANCH START
- BRANCH END

BRANCH START — This instruction begins each parallel logic branch of a rung. The BRANCH START is programmed immediately before the first instruction of each parallel logic path.

BRANCH END — This instruction completes a set of parallel branches. The BRANCH END is entered after the last instruction of the last branch to end a set of parallel branches.

Branch instructions must be entered in the correct order for proper logic function. The only limitation is that a "nested" branch (a branch within a branch) cannot be programmed directly (Figure 3-10). A total of seven (7) branches can be programmed in one rung and properly displayed on the Industrial Terminal.

3.4 PROGRAMMING RELAY-TYPE INSTRUCTIONS

All relay-type instructions are entered from the Industrial Terminal Keyboard with the Processor in the PROGRAM mode. When a relay type instruction is initially entered, it will appear intensified on the screen to indicate the cursor's present position. When a bit address is required, the instruction will blink to indicate information is needed to complete the instruction. The default bit address, 010/00, is displayed automatically with the instruction. (NOTE: The term "default" simply means that data is to be added.) A reverse-video character cursor is positioned at the first digit. This cursor indicates where information is needed and moves to the next digit as information is entered. When all information is entered, the instruction stops blinking and remains intensified until the next instruction is entered.

Refer to Table 3-1 for a complete summary of relay-type instructions.

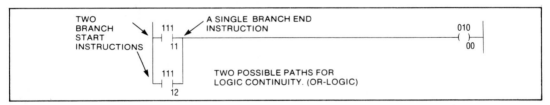

FIGURE 3-9 — Branching Instructions

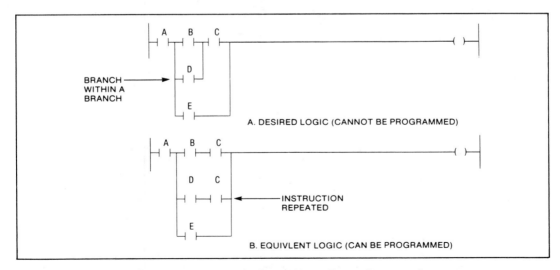

FIGURE 3-10 — Nested Branching vs. Proper Programming

TABLE 3-1 — Relay Type Instructions

Note: Examine and Output addresses, XXX/XX, can be assigned to any location in the Data Table, excluding the Processor Work Areas and as noted below.

KEYTOP SYMBOL	INSTRUCTION NAME	-T1, -T2 DISPLAY	1770-T3 DISPLAY	DESCRIPTION
-| |-	EXAMINE ON	XXXXX -| |-	XXX -| |- XX	When the addressed memory bit is ON, the instruction is TRUE.
-|/|-	EXAMINE OFF	XXXXX -|/|-	XXX -|/|- XX	When the addressed memory bit is OFF, the instruction it TRUE.
-( )-	ENERGIZE	XXXXX -( )-	XXX -( )- XX	❶When the rung is TRUE, the addressed memory bit is set ON. If the bit controls an output device that output device will be ON.
-(L)-	OUTPUT LATCH	XXXXX -(L)- ON or OFF	XXX -(L)- ON XX or OFF	❶When the rung is TRUE, the addressed memory bit is latched ON and remains ON until it is unlatched. The OUTPUT LATCH instruction is initially OFF when entered, as indicated below the instruction. It can be preset ON by pressing a [1] after entering the bit address. An ON will then be indicated below the instruction in PROGRAM mode.
-(U)-	OUTPUT UNLATCH	XXXXX -(U)- ON or OFF	XXX -(U)- ON XX or OFF	❶When the rung is TRUE, the addressed bit is unlatched. If the bit controls an output device that device is deenergized. ON or OFF will appear below the instruction indicating the status of the bit in PROGRAM mode only.
┌	BRANCH START			This instruction begins a parallel logic path and is entered at the beginning of each parallel path.
┘	BRANCH END			This instruction ends two or more parallel logic paths and is used with BRANCH START instructions.

❶ These instructions should not be assigned Input Image Table addresses because Input Image Table words are reset by the I/O scan.

Section 4
TIMER AND COUNTER INSTRUCTIONS

4.0 GENERAL

Timer and Counter instructions are output instructions internal to the Processor. They provide many of the capabilities available with timing relays and solid state timing/counting devices. Usually conditioned by Examine instructions, timers and counters keep track of timed intervals or counted events according to the logic continuity of the rung. Up to 40 internal timers and/or counters can be programmed.

Each Timer or Counter instruction has two 3-digit values associated with it, and thus requires two words of Data Table memory. These 3-digit values are:

- Accumulated (AC) Value — Stored in the Accumulated Value area of the Data Table starting at word address 030_8. For timers, this is the number of timed intervals that have elapsed. For counters, this is the number of events that have been counted.

- Preset (PR) Value — Stored in a Preset Value area of the Data Table, always 100_8 words greater than its corresponding AC Value. This value is entered into memory. The Preset value is the number of timed intervals or events to be counted. When the Accumulated value equals the Preset value, a status bit is set ON and can be examined to turn ON or OFF an output device.

The Accumulated and Preset values are stored in the Data Table in 3-digit BCD (Binary Coded Decimal) format. BCD numbers can range from 000 to 999 when stored in the lower 12 bits of a memory word (Figure 4-1). Each BCD digit is represented by a group of 4 bits. The arrangement of "1's" and "0's" in a group of 4 bits corresponds to a decimal number from 0 to 9. For more information on numbering systems, refer to Section 13.

The remaining 4 bits in a word (bits 14-17) are not used to form a BCD number. In the Accumulated Value word, they are used as status bits. In the Preset Value word, they are not used and are available for internal storage. With .01 second timers, these bits are used for internal timing functions and cannot be used for storage. For more information on bit storage, refer to Section 8.3.4.

4.1 TIMER INSTRUCTIONS

A timer counts elapsed time-base intervals (1.0, 0.1 or .01 seconds) and stores this count in its Accumulated Value word. When timing is complete (when AC = PR), bit 15 is either set ON or OFF depending on the type of timer instruction. For all timers, bit 17 is set ON when rung conditions are TRUE and is set OFF when they are FALSE. Both status bits are located in the Accumulated Value word as shown in Figure 4-2.

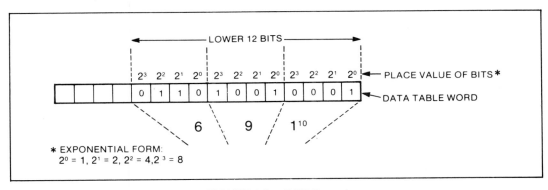

FIGURE 4-1 — BCD Format

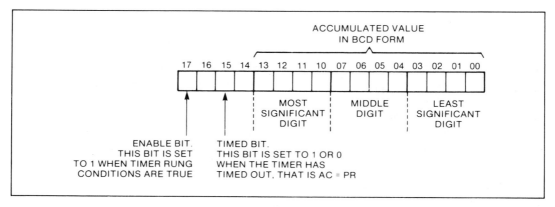

FIGURE 4-2 — Timer Accumulated Value Word

The four timer instructions available with the Mini-PLC-2 Controller are:

- TIMER ON-DELAY -(TON)-
- TIMER OFF-DELAY -(TOF)-
- RETENTIVE TIMER -(RTO)-
- RETENTIVE TIMER RESET -(RTR)-

The timers differ in the way they set and reset status bits, respond to rung logic continuity and reset the Accumulated Value. They are similar in time base selection. One of the following time bases must be selected when entering the instruction.

- 1.0 second
- 0.1 second
- 0.01 second (10 milliseconds)

Bit 16 of the timer Accumulated value word reflects the time base. It will go ON and OFF at the selected time base rate. Therefore, do not use bit 16 of a timer instruction in User Program as an output or storage bit.

4.1.1 TIMER ON-DELAY Instruction

The TIMER ON-DELAY instruction (TON) can be used to turn a device ON or OFF once an interval is timed out.

Refer to Figure 4-3. When the rung condition for a TIMER ON-DELAY instruction becomes TRUE, the timer begins to count time-base intervals. The "Enable" bit, bit 17, is set ON whenever the rung condition is TRUE and the

timer is enabled (rung 1). As long as the rung condition remains TRUE, the timer increments its Accumulated value for each interval. When the Accumulated value equals the programmed Preset value, the timer stops incrementing its Accumulated value and sets the "timed" bit, bit 15, of this word ON. Bit 15 is then used to turn an output device ON or OFF, as a condition for program logic (rung 2).

Whenever the rung condition for the TON instruction goes FALSE, the Accumulated value is reset to 000 and bits 15 and 17 of that word are reset to zero. The Accumulated value and status bits are also reset when the Mode Select Switch is turned to the PROG position or when there is a loss of power.

4.1.2 TIMER OFF-DELAY Instruction

The TIMER OFF-DELAY instruction (TOF) can be used to turn a device OFF or ON after a timed interval. Like the other timer instructions, the TOF instruction counts time-base intervals which are stored in its Accumulated Value word. The TOF instruction, however, varies from the other instructions in significant ways.

Refer to Figure 4-4. The TIMER OFF-DELAY instruction begins to time an interval as soon as its rung condition goes FALSE. The Enable bit, bit 17, goes FALSE when the timer begins (rung 1). As long as its rung condition remains FALSE, the TOF instruction continues to time, until the Accumulated Value equals the Preset

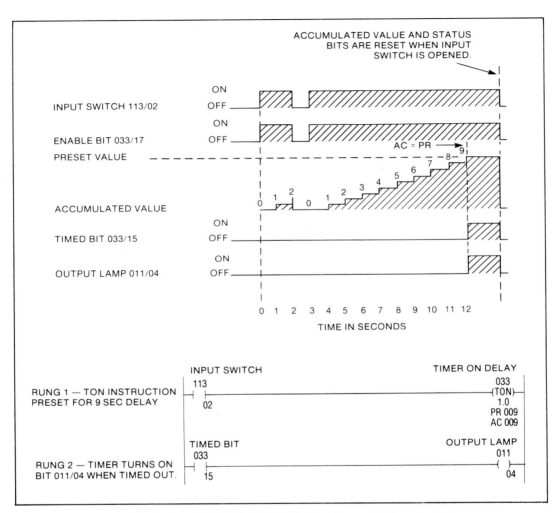

FIGURE 4-3 — TIMER ON-DELAY Timing Diagram & Programming

Value. When the TOF instruction times out, bit 15 is set to zero (OFF) which turns OFF the output (rung 2).

The Accumulated Value is reset to 000 and bit 15 is set ON when the rung condition again goes TRUE. The next timed interval begins when the rung condition goes FALSE.

Bit 17, the enabled bit, is controlled by the logic continuity of the rung. When the rung is TRUE, bit 17 is set to ONE (ON); when it is FALSE, bit 17 is set to zero (OFF).

4.1.3 RETENTIVE TIMER Instruction

The RETENTIVE TIMER instruction (RTO), like the TON instruction, can be used to turn a device ON or OFF once a programmed Preset value is reached.

Unlike the TIMER ON-DELAY instruction, the RETENTIVE TIMER instruction retains its Accumulated value when any of the following conditions occur:

• Rung condition goes FALSE.

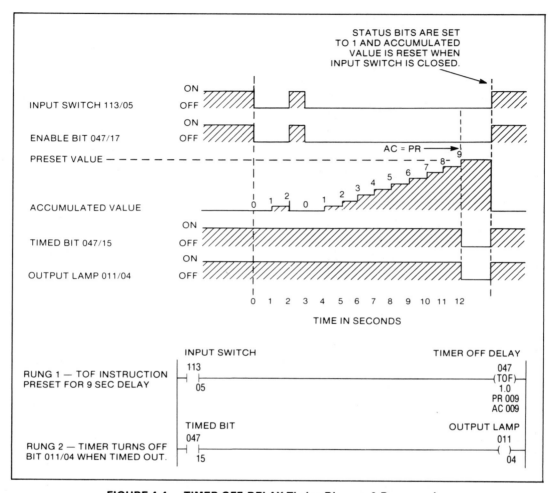

STATUS BITS ARE SET
TO 1 AND ACCUMULATED
VALUE IS RESET WHEN
INPUT SWITCH IS CLOSED.

INPUT SWITCH 113/05

ENABLE BIT 047/17

PRESET VALUE

AC = PR → 9

ACCUMULATED VALUE

TIMED BIT 047/15

OUTPUT LAMP 011/04

TIME IN SECONDS

RUNG 1 — TOF INSTRUCTION
PRESET FOR 9 SEC DELAY

INPUT SWITCH
113
05

TIMER OFF DELAY
047
—(TOF)—
1.0
PR 009
AC 009

RUNG 2 — TIMER TURNS OFF
BIT 011/04 WHEN TIMED OUT.

TIMED BIT
047
15

OUTPUT LAMP
011
—( )—
04

FIGURE 4-4 — TIMER OFF-DELAY Timing Diagram & Programming

- The Mode Select Switch is changed to the PROG position.

- A power outage occurs provided memory backup power is maintained.

Refer to Figure 4-5. When the rung condition goes TRUE, the enabled bit (bit 17) is set ON and the timer starts counting time-base intervals (rung 1). Any time the rung goes FALSE, bit 17 is set OFF but the Accumulated value is retained. When the timer times out, the timed bit (bit 15) is set ON which turns ON an output (rung 2).

By retaining its Accumulated value, the RTO instruction measures the cumulative period during which the rung condition is TRUE. Because this timer retains its Accumulated value, it must be reset by a separate instruction, the RETENTIVE TIMER RESET (RTR) instruction (rung 3).

4.1.4 RETENTIVE TIMER RESET Instruction

The RETENTIVE TIMER RESET instruction (RTR) is used to reset the Accumulated value and timed bit of the Retentive Timer to zero. This instruction is given the same word address as its corresponding RTO instruction as shown in figure 4-5. When the rung condition goes true, the RTR instruction resets the Accumulated value and status bits of the RTO instruction to zero.

4.1.5 Timer Accuracy for 10 msec Timers

The accuracy of the 10 msec timer is related to nominal scan time. When scan times are 9 msec or less, the 10 msec timer is accurate to

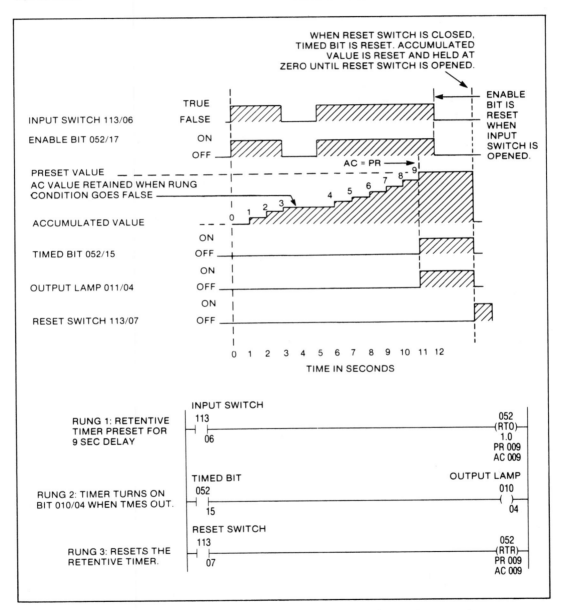

FIGURE 4-5 — RETENTIVE TIMER with RESET Timing Diagram & Programming

plus or minus one time base (±10 msec). When the scan time is greater than 9 msec, accuracy of ±10 msec can be achieved through special programming techniques described in Publication No. 1772-702, Programming 0.01 Sec. Timers With the Mini-PLC-2 Controller (included in Publication 1772-803-1 PLC-2 Family Support Documentation).

4.2 COUNTER INSTRUCTIONS

Three types of counter instructions are available with the Mini-PLC-2 Controller. They are:

• UP-COUNTER -(CTU)-

• COUNTER RESET -(CTR)-

• DOWN-COUNTER -CTD)-

A counter counts the number of events that occur and stores this count in its Accumulated Value word. The remaining four bits in the Accumulated Value word are used as status bits. See Figure 4-6.

• Bit 14 is the Overflow/Underflow bit. It is set to 1 when the AC value of the CTU instruction exceeds 999 or the AC value of the CTD instruction goes below 000.

• Bit 15 is set to 1 when a count has been reached or exceeded, that is, when the AC value is ≥ PR value.

• Bit 16 is the Enable bit for a CTD instruction. It is set ON when the rung condition is TRUE.

• Bit 17 is the Enable bit for a CTU instruction. It is set ON when the rung condition is TRUE.

Counter instructions differ from Timer instructions because they have no time-base. They count one event each FALSE-to-TRUE transition of the rung condition.

4.2.1 UP-COUNTER Instruction

The UP-COUNTER instruction (CTU) increments its Accumulated value for each FALSE-to-TRUE transition of the rung condition. Because only the FALSE-to-TRUE transition causes a count to be made, the rung condition must go from TRUE to FALSE and back to TRUE before the next count is registered. The CTU instruction retains its Accumulated value when:

• Mode Select Switch is changed to the PROG position.

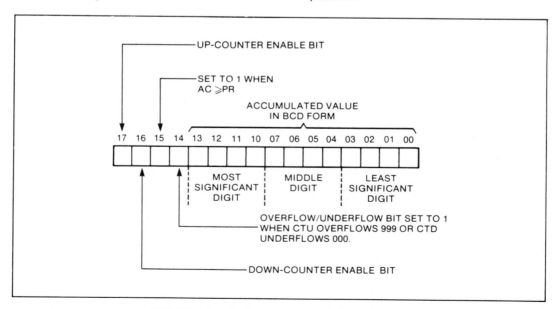

FIGURE 4-6 — Counter Accumulated Value Word

- The rung condition goes FALSE.

- A power outage occurs provided memory backup power is maintained.

Refer to Figure 4-7. Each time the CTU rung goes TRUE, bit 17, the Enable bit, is set ON (rung 1). When the Accumulated value reaches the Preset value, the Count Complete bit, bit 15, is set ON (rung 2). Unlike a timer, the CTU instruction continues to increment its Accumu-lated value after the Preset value has been reached. If the Accumulated value goes above 999, bit 14 is set ON to indicate an overflow condition and the CTU continues up-counting from 000 (rung 3). Bit 14 stays ON until the counter is reset. Bit 14 can be examined to cascade counters for counts greater than 999 (Section 4.3). Because this counter retains its Accumulated value, it must be reset by a separate instruction, the COUNTER RESET (CTR) instruction.

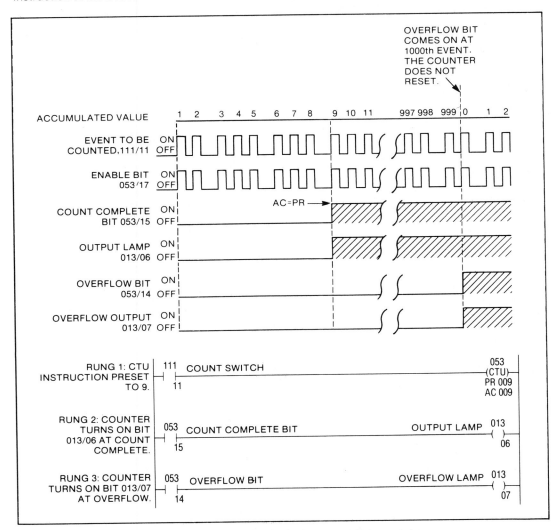

FIGURE 4-7 — UP-COUNTER Diagram & Programming

4.2.2 COUNTER RESET Instruction

The COUNTER RESET instruction (CTR) is an output instruction that resets the CTU Accumulated value and status bits to zero when the reset rung goes TRUE.

Refer to Figure 4-8. The counter operates in the same manner as described for the CTU instruction, with the addition of the Reset instruction (rung 3).

In this example, the reset pushbutton is pressed after count 12. When the pushbutton is released, the next event starts the sequence at count 1.

The CTR instruction is given the same word address as the CTU instruction. The Preset and Accumulated values are automatically displayed when the word address is entered.

4.2.3 DOWN-COUNTER Instruction

The DOWN-COUNTER instruction (CTD) subtracts one from its Accumulated value for each FALSE-to-TRUE transition of its rung condition. Because only the FALSE-to-TRUE transition causes a count to be made, the rung condition must go from TRUE to FALSE and back to TRUE before the next count is registered.

The CTD instruction Accumulated value is retained when:

• Mode Select Switch is changed to the PROG position.

• The rung condition goes FALSE.

• A power outage occurs provided memory backup power is maintained.

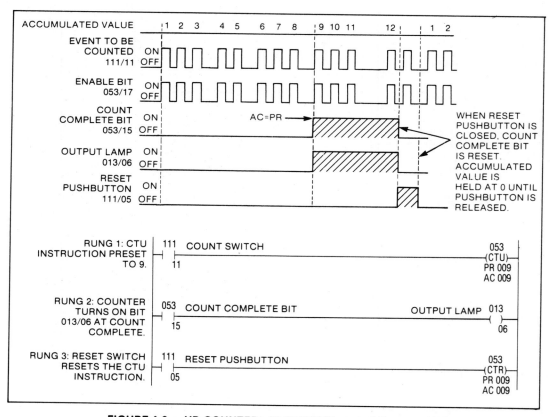

FIGURE 4-8 — UP-COUNTER with RESET Diagram & Programming

Each time the CTD instruction rung goes TRUE, bit 16, the Enable bit, is set ON. When the Accumulated value is greater than or equal to the Preset value, bit 15 is set ON. When the Accumulated value goes below 000, bit 14 is set ON to indicate an underflow condition and the CTD instruction continues down-counting from 999.

Normally, the DOWN-COUNTER instruction is paired with the UP-COUNTER instruction to form an up/down counter, using the same word address, AC value and PR value as shown in Figure 4-9.

NOTE: Bit 14 of the Accumulated Value word is set ON when the Accumulated value either "overflows" or "underflows." Because of this, bit 14 may require monitoring in some applications. When a DOWN-COUNTER Preset is set to 000, Underflow bit 14 will not be set ON when the count goes below zero.

When used alone, the CTD instruction's Accumulated value may need to be "reset" in the program to its original value (usually a value other than 000). For this reason, a GET/PUT transfer (described in Section 5.1) rather than a CTR instruction is usually used to load a value in the CTD instruction's Accumulated Value word.

4.3 CASCADING TIMERS OR COUNTERS

An individual timer or counter can time or count up to 999 intervals or events. By "cas-cading" two or more timers or counters, the timing or counting capability within the program can be increased beyond three digits.

To cascade timers or counters, each timer or counter is assigned a different word address (Figure 4-10). The status bit of the first timer (bit 15) changes status each time the Preset value is reached. The status bit of a counter (bit 14) is set ON each time a counter overflows. The status bit of the timer or counter is then used to increment the second timer or counter and reset the first to 000.

4.4 PROGRAMMING TIMER AND COUNTER INSTRUCTIONS

Timer and Counter instructions are programmed from the Industrial Terminal keyboard with the Processor in the PROGRAM mode. Allowable addresses are 030_8 through 077_8.

Timer instructions are programmed by entering the word address of the Accumulated value, a time base and a Preset value. With the RTO instruction, an Accumulated value can also be entered. The time base of 1.0, 0.1 or 0.01 second is entered as [1][0], [0][1], or [0][0] respectively.

Counter instructions are programmed by entering the word address of the Accumulated value, a Preset value, and if desired, an Accumulated value. Press [CANCEL COMMAND] if no Accumulated value is desired.

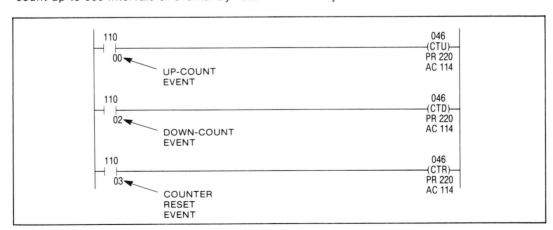

FIGURE 4-9 — UP/DOWN COUNTER Example

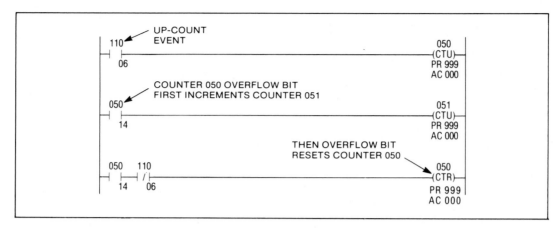

FIGURE 4-10 — Cascading Counters Example

When entered, these instructions will be displayed as intensified and blinking. The default word address above the instruction will have a reverse-video cursor positioned at the first digit. The instruction will continue to blink until all the data is entered.

Refer to Table 4-1 and 4-2 for a complete summary of Timer and Counter instructions.

TABLE 4-1 — Timer Instructions

Note: The Timer word address, XXX, is assigned to the timer Accumulated area of the Data Table.
The time base, TB, is user-selectable and can be 1.0, 0.1 or 0.01 second. Preset values YYY and Accumulated values ZZZ can vary from 000 to 999.

KEYTOP SYMBOL	INSTRUCTION NAME	DISPLAY	DESCRIPTION
-(TON)-	TIMER ON DELAY	XXX -(TON)- TB PR YYY AC ZZZ	When the rung is TRUE, the timer begins to increment the Accumulated Value at a rate specified by the time base. When the rung is FALSE, the timer resets the Accumulated Value to 000.
-(TOF)-	TIMER OFF DELAY	XXX -(TOF)- TB PR YYY AC ZZZ	When the rung is FALSE, the timer begins to increment the Accumulated Value. When the rung is TRUE, the timer resets the Accumulated Value to 000.
-(RTO)-	RETENTIVE TIMER	XXX -(RTO)- TB PR YYY AC ZZZ	When the rung is TRUE, the timer begins to increment the Accumulated Value. When FALSE, the Accumulated Value is retained. It is reset only by the RTR instruction.
-(RTR)-	RETENTIVE TIMER RESET	XXX -(RTR)- PR YYY AC ZZZ	XXX — Word address of the retentive timer it is resetting. YYY, ZZZ — Preset and Accumulated Values are automatically entered by the Industrial Terminal. When the rung is TRUE, the Accumulated Value and status bit are reset to zero.

TABLE 4-2 — Counter Instructions

Note: The Counter word address, XXX, is assigned to the counter Accumulated areas of the Data Table. Preset values YYY and Accumulated values ZZZ can vary from 000-999.

KEYTOP SYMBOL	INSTRUCTION NAME	DISPLAY	DESCRIPTION
-(CTU)-	UP COUNTER	XXX -(CTU)- PR YYY AC ZZZ	Each time the rung goes TRUE, the Accumulated Value is incremented one count. The counter will continue counting after the Preset Value is reached. The Accumulated Value can be reset by the CTR instruction. The Accumulated Value "Overflow" bit is bit 14.
-(CTR)-	COUNTER RESET	XXX -(CTR)- PR YYY AC ZZZ	XXX - Word address of the CTU it is resetting. Preset and Accumulated Values are automatically entered by the Industrial Terminal. When the rung is TRUE, the CTU Accumulated Value and status bits are reset to 000.
-(CTD)-	DOWN COUNTER	XXX -(CTD)- PR YYY AC ZZZ	Each time the rung goes TRUE, the Accumulated Value is decreased one count. The Accumulated Value "Underflow" bit is bit 14. The Enable bit is bit 16.

Section 5
DATA MANIPULATION INSTRUCTIONS

5.0 GENERAL

The Data Manipulation instructions are used to transfer or compare data that is stored in Data Table words and bytes. There are six Data Manipulation instructions:

- GET
- PUT
- LES
- EQU
- GET BYTE
- LIMIT TEST

The GET and PUT instructions are used together to transfer 16 bits of data from one word location in the Data Table to another word location. Data can be in the form of 3-digit Binary Coded Decimal numbers.

The LES and EQU instructions compare data such as 3-digit numberic values in BCD format using the first 12 bits of a Data Table word (Figure 5-1). This 3-digit value can be a decimal number ranging from 000 to 999.

The GET BYTE and LIMIT TEST instructions compare 3-digit values in octal format using eight bits (one byte) of a Data Table word (Figure 5-2). This 3-digit value is an octal number ranging from 000_8 to 377_8. Note that two 3-digit values can be stored in a word: one in the upper byte (bits 10-17) and one in the lower byte (bits 00-07).

A Data Manipulation instruction can address any word in the Data Table excluding the Processor Work Areas.

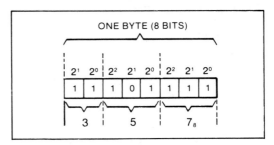

FIGURE 5-2 — Octal Representation

5.1 DATA TRANSFER INSTRUCTIONS

There are two Data Transfer instructions. They are:

- GET -[G]-
- PUT -(PUT)-

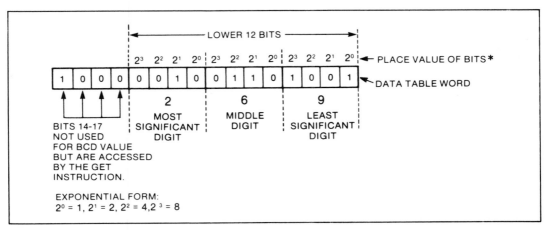

FIGURE 5-1 — BCD Word Format

5.1.1 GET Instruction

GET instructions -[G]- are programmed in the condition area of the ladder diagram rung. When the rung containing the GET/PUT instructions goes TRUE, the data (16 bits) in the word address of the GET instruction is duplicated and transferred to the word address of the PUT instruction. (Figure 5-3).

If the word addressed by a GET instruction already contains data, the lower 12 bits of the data are displayed automatically beneath the instruction after the word address is entered. Entry of new data such as a BCD value writes over the data previously stored in the addressed word.

Although each Data Table word can store data such as one BCD value, the word address can be assigned to more than one GET instruction in the same program. This allows the program to perform several different functions with the same data.

The GET instruction is NOT a "condition" that determines rung logic continuity. When the Processor is in the RUN or TEST mode, the GET instruction is always intensified regardless of rung logic continuity. This does not mean that data transfer will occur. Data transfer occurs only when the rung is TRUE.

The GET instruction can be programmed either at the beginning of a rung or with one or more Condition instructions preceding it. Condition instructions, however, should not be programmed after a GET instruction. When one or more Condition instructions precede the GET instruction, they determine whether the rung is TRUE or FALSE. Parallel branches of GET instructions cannot be programmed unless they are paired with a LES or EQU instruction.

5.1.2 PUT Instruction

The PUT instruction (PUT) is an output instruction. It receives 16 bits of data from its corresonding GET instruction and stores the data at its address as shown in Figure 5-3. A PUT instruction can have the same address as other instructions in the program. For example, a PUT instruction having the same address as a counter Preset will change the counter Preset value to that transferred from the GET instruction. See Figure 5-4.

The lower 12 bits of transferred data are displayed in BCD beneath the PUT instruction.

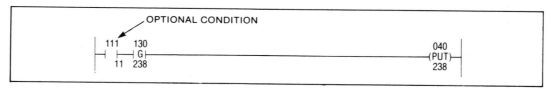

FIGURE 5-3 — GET and PUT Instructions

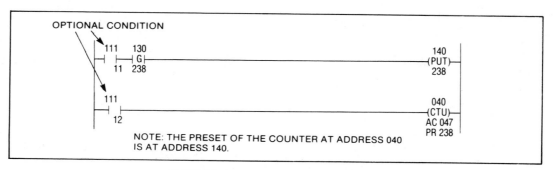

FIGURE 5-4 — Changing a Counter Preset

Bits 14-17 are not displayed but are transferred. While the rung is TRUE, any change in the data of the GET instruction also changes the data of the PUT instruction. However, the PUT instruction is retentive, which means that while the rung is FALSE, any change in the data of the GET instruction does not change the data of the PUT instruction. Also, during a power loss, the data is retained.

5.2 DATA COMPARISON INSTRUCTIONS

The Data Comparison instructions are:

- LESS THAN -[<]-
- EQUAL TO -[=]-
- GET BYTE -[B]-
- LIMIT TEST -[L]-

Data Comparison operations differ from Data Transfer operations because Data Table values are not transferred. Instead, the values at different word locations are compared.

Data Comparison instructions operate with either BCD values or octal values. With the LES and EQU instructions, only 12 bits of a word (the BCD values) are compared. Bits 14-17 are not compared. With the GET BYTE and LIMIT TEST instructions, 8 bits (one byte) of a word are compared.

5.2.1 LES and EQU Instructions

The LES (less than) and EQU (equal to) instructions, [<] and [=] are used with the GET instruction to perform data comparisons. They compare BCD values and are programmed in the condition area of the ladder diagram rung.

A GET/LES or GET/EQU pair of instructions forms a single condition for logic continuity. Alone or with other conditions, each pair can be used to energize an output device or other output instruction. In all cases, the GET instruction must be programmed before the LES or EQU instruction. If other conditions are also programmed, they should be entered before the GET instruction or after the LES or EQU instruction.

Data comparisons are made by comparing a changing BCD value to a reference BCD value. The reference value need not be fixed. The following types of data comparisons of BCD values can be made:

- Less than
- Greater than
- Equal to
- Less than or equal to
- Greater than or equal to

LESS THAN — A less-than comparison is made with the GET/LES pair of instructions. The BCD value of the GET instruction is the changing value. It is compared to the BCD value of the LES instruction which is the reference value (Figure 5-5). When the GET value is less than the LES value, the comparison is TRUE and logic continuity is established.

GREATER THAN — A greater-than comparison is also made with the GET/LES pair of instructions. This time the GET instruction BCD value is the reference and the LES instruction BCD value is the changing value.

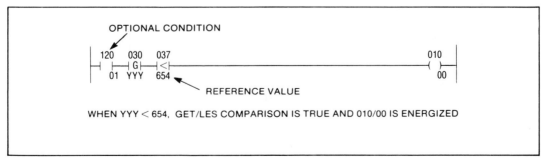

FIGURE 5-5 — LESS THAN Comparison

The LES value is compared to the GET value for a greater-than condition (Figure 5-6). When the LES value is greater than the GET value, the comparison is TRUE and logic continuity is established.

EQUAL TO — An equal-to comparison is made with the GET and EQU instructions (Figure 5-7). The GET value is the changing variable and is compared to the reference value of the EQU instruction for an equal to condition. When the GET value equals the EQU value, the comparison is TRUE and logic continuity is established.

LESS THAN OR EQUAL TO — This comparison is made using the GET, LES and EQU instructions. The GET value is the changing value. The LES and EQU instructions are assigned a reference value (Figure 5-8). When the GET value is either less than or equal to the value at the LES and EQU instructions, the comparison is TRUE and logic continuity is established.

NOTE: Only one GET instruction is required for a parallel comparison. The LES and EQU instructions are programmed on parallel branches.

GREATER THAN OR EQUAL TO — This comparison is made using the GET, LES and EQU instructions. The GET value is assigned a reference value. The LES and EQU values are changing values that are compared to the GET value (Figure 5-9). When the LES and EQU values are greater than or equal to the GET value, the comparison is TRUE and logic continuity is established.

NOTE: Only one GET instruction is required for this parallel comparison. The LES and EQU instructions are programmed on parallel branches.

5.2.2 GET BYTE and LIMIT TEST Instructions

The GET BYTE and LIMIT TEST instructions [B] and [L] are used together to compare an octal value to upper and lower limits that are

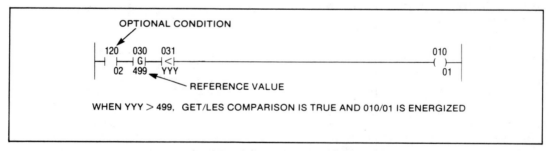

FIGURE 5-6 — GREATER THAN Comparison

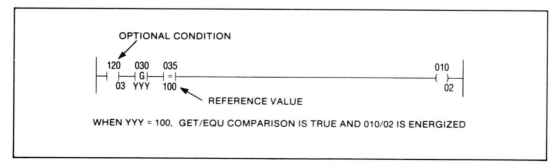

FIGURE 5-7 — EQUAL TO Comparison

also octal values. These values can range from 000_8 to 377_8.

The GET BYTE and LIMIT TEST instructions are programmed in the condition area of the ladder diagram rung. Together they form a single condition for logic continuity. Condition instructions can be programmed before the GET BYTE instruction or after the LIMIT TEST instruction but not between them.

The GET BYTE instruction addresses either the upper or lower byte of a Data Table word. A "1" is entered after the word address for an upper byte; a "0" is entered for the lower byte.

The LIMIT TEST instruction addresses one Data Table word that stores both the upper and lower limits. The upper limit is stored in the upper byte and the lower limit is stored in the lower byte. The upper byte of word 045_8 would be addressed as 0451. See Figure 5-10.

The Processor makes a duplicate of the upper or lower byte of the word addressed by the GET BYTE instruction. The octal value stored at that byte is then compared to the upper and lower octal values of the LIMIT TEST instruction. If the GET BYTE value is equal to or between the LIMIT TEST values, the comparison is TRUE and logic continuity is established.

5.3 PROGRAMMING DATA MANIPULATION INSTRUCTIONS

The Data Manipulation instructions are programmed from the Industrial Terminal keyboard with the Processor in the PROGRAM mode. When entered, they are displayed intensified and blinking. The default word address above the instruction will have a reverse-video cursor positioned at the first digit. The instruction will continue to blink until all information is entered.

Refer to Table 5-1 for a summarized description of these instructions.

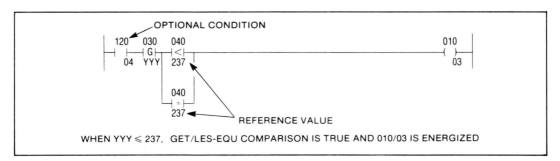

FIGURE 5-8 — LESS THAN or EQUAL TO Comparison

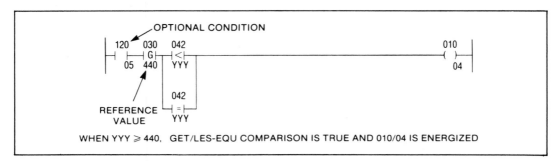

FIGURE 5-9 — GREATER THAN or EQUAL TO Comparison

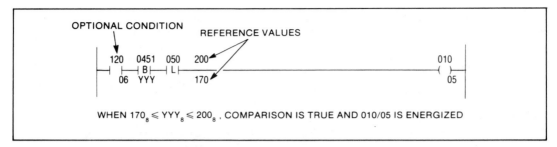

FIGURE 5-10 — GET BYTE/LIMIT TEST Comparison

TABLE 5-1 — Data Manipulation Instructions

Note: Data Manipulation instructions operate upon BCD values and/or 16 bit data in the Data Table. The word address XXX is displayed above the instruction; the BCD value or data operated upon YYY is displayed beneath it. The BCD value is stored in the lower 12 bits of the word address and can be any value from 000 to 999, except as noted.

KEYTOP SYMBOL	INSTRUCTION NAME	DISPLAY	DESCRIPTION
-[G]-	GET	XXX -[G]- YYY	The GET instruction is used with other Data Manipulation or Arithmetic instructions. When the rung is TRUE, all 16 bits of the GET instruction are duplicated and the operation of of the instruction following it is performed.
-(PUT)-	PUT	XXX -(PUT)- YYY	The PUT instruction should be preceded by the GET instruction. When the rung is TRUE, all 16 bits at the GET instruction address are transferred to the PUT instruction address.
-[<]-	LESS THAN	XXX -[<]- YYY	The LESS THAN instruction should be preceded by a GET instruction. 3-Digit BCD values at the GET and LESS THAN word addresses are compared. If the logic is TRUE, the rung is enabled.
-[=]-	EQUAL TO	XXX -[=]- YYY	The EQUAL TO instruction should be preceded by a GET instruction. 3-digit BCD values at the GET and EQUAL TO word addresses are compared. If equal, the rung is enabled.

TABLE 5-1 — Data Manipulation Instructions (cont.)

KEYTOP SYMBOL	INSTRUCTION NAME	DISPLAY	DESCRIPTION
-[B]-	GET BYTE	XXXD -[B]- YYY	D - Designates the upper or lower byte of the word. 1 = upper byte, 0 = lower byte. YYY - Octal value from 000_8 to 377_8 is stored in the upper or lower byte of the word address. The GET BYTE instruction should be followed by a LIMIT TEST instruction. A duplicate of the designated byte is made and compared with the upper and lower limits of the LIMIT TEST instruction.
-[L]-	LIMIT TEST	XXX AAA -[L]— BBB	AAA - Upper limit of LIMIT TEST, an octal value from 000_8 to 377_8 BBB - Lower limit of LIMIT TEST, an octal value from 000_8 to 377_8. The LIMIT TEST instruction should be preceded by a GET BYTE instruction. Compares the value at the GET BYTE instruction with the values at the LIMIT TEST instruction. If found to be between or equal to the limits, the rung is enabled.

Section 6
ARITHMETIC INSTRUCTIONS

6.0 GENERAL

The Mini-PLC-2 Processor can be programmed to perform arithmetic operations with two 3-digit BCD values using a set of Arithmetic instructions. These output instructions are:

- ADD -(+)-

- SUBTRACT -(-)-

- MULTIPLY -(x)-(x)- (1772-LN3 Processor)

- DIVIDE -(:)-(:)- (1772-LN3 Processor)

The two 3-digit BCD values to be computed are stored in two GET instruction words. The GET instructions, programmed in the condition area of the ladder diagram rung, should be followed by the Arithmetic instruction. Other condition instructions, if used, should be programmed before the GET instructions.

The Arithmetic instructions are programmed in the output position of the ladder diagram rung. They are assigned either one or two Data Table words to store the computed result, depending on the arithmetic operation performed. The ADD and SUBTRACT instructions use one Data Table word to store the result. The MULTIPLY and DIVIDE use two Data Table words to store the result.

The computed result is stored in BCD format in the lower 12 bits of the Arithmetic instruction word (Figure 6-1). Two of the remaining bits (bits 14 and 16) are used to indicate overflow and underflow conditions.

6.1 ADD INSTRUCTION

The ADD instruction (+) tells the Processor to add the two values stored in the GET words. The sum is then stored at the ADD instruction word address. When the sum exceeds 999, the overflow bit (bit 14) in the ADD instruction word is set ON (Figure 6-2). In the RUN or TEST mode, the overflow condition is displayed on the Industrial Terminal screen as a "1" preceding the sum.

NOTE: If an overflowed value (4 digits) is used for subsequent comparisons or other arithmetic operations, inaccurate results could occur. The Processor performs arithmetic and data manipulation operations with 3-digit BCD values, only.

6.2 SUBTRACT INSTRUCTION

The SUBTRACT instruction (–) tells the Processor to subtract the second GET word value from the first GET word value (Figure 6-3).

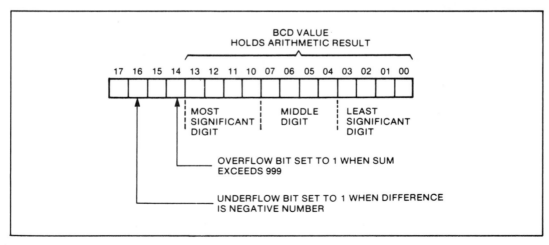

FIGURE 6-1 — Arithmetic Instruction Word

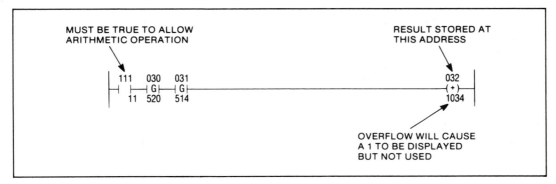

FIGURE 6-2 — ADD Instruction

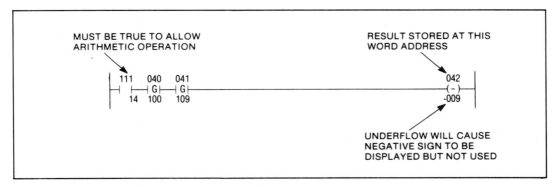

FIGURE 6-3 — SUBTRACT Instruction

The difference is then stored at the Data Table word addressed by the SUBTRACT instruction.

If the difference is a negative number, the underflow bit of the SUBTRACT word (bit 16) is set ON. In the RUN or TEST mode, the negative sign will appear on the Industrial Terminal screen preceding the difference.

NOTE: If a negative BCD value is used for subsequent operations, inaccurate results could occur. The Processor only compares, transfers and computes the absolute BCD value.

6.3 MULTIPLY INSTRUCTION (1772-LN3 Processor Module)

The MULTIPLY (X) instruction tells the Processor to multiply the two BCD values stored at the GET instruction words. The result is then stored in two Data Table words addressed by the MULTIPLY instruction (Figure 6-4).

For ease of programming, two consecutive Data Table words should be chosen to store the product. If the product is less than 6 digits, leading zeros will appear in the product.

6.4 DIVIDE INSTRUCTION (1772-LN3 Processor Module)

The DIVIDE instruction (:) tells the Processor to divide the first GET instruction value by the second GET instruction value. The result is stored in two Data Table words addressed by the DIVIDE instruction (Figure 6-5). Usually two consecutive Data Table words are chosen to store the quotient for ease of programming.

The quotient is rounded off and expressed as a decimal number. The decimal point is auto-

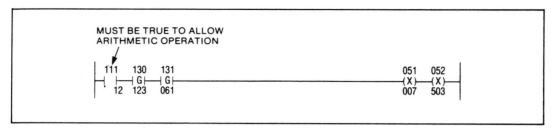

FIGURE 6-4 — MULTIPLY Instruction

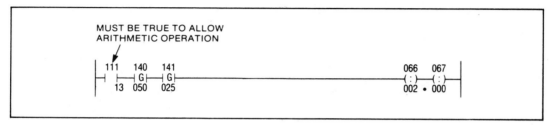

FIGURE 6-5 — DIVIDE Instruction

matically inserted between the two DIVIDE instruction values by the Industrial Terminal. Leading and trailing zeros in the quotient are also entered automatically by the Industrial Terminal.

Although division by 0 is undefined mathematically, the division of a number including zero by 0 will give the result of 999.999. (This differs from the PLC-2/20 and PLC-2/30 where $0 \div 0 = 1.000$.)

6.5 PROGRAMMING ARITHMETIC INSTRUCTIONS

Arithmetic instructions are programmed from the Industrial Terminal keyboard with the Mini-PLC-2 Processor in the PROGRAM mode. When entered, these instructions will be intensified and blinking. The default word address above the instuction will have a reverse-video cursor positioned at the first digit. The instruction will continue to blink until the word address is entered.

Refer to Table 6-1 for a summarized description of these instructions.

TABLE 6-1 — Arithmetic Instructions

Note: Arithmetic instructions operate on BCD values in the Data Table. The word address XXX is displayed above the instruction; the BCD value YYY which is the result of the arithmetic operation is displayed beneath it. The BCD value is stored in the lower 12 bits of the word address and can be any value from 000 to 999.

KEYTOP SYMBOL	INSTRUCTION NAME	DISPLAY	DESCRIPTION
-(+)-	ADD	XXX -(+)- YYY	The ADD instruction is an output instruction. It is always preceded by two GET instructions which store the BCD values to be added. When the sum exceeds 999, bit 14 is set to 1, and a 1 is displayed in front of the result YYY.
-(-)-	SUBTRACT	XXX -(-)- YYY	The SUBTRACT instruction is an output instruction. It is always preceded by two GET instructions. The value in the second GET address is subtracted from the value in the first. When the difference is negative, bit 16 is set to 1, and a minus sign is displayed in front of the result YYY.
-(X)-	MULTIPLY (1772-LN3 Processor)	XXX XXX -(X)—(X)- YYY YYY	The MULTIPLY instruction is an output instruction. It is always preceded by two GET instructions which store the values to be multiplied. Two word addresses are required to store the 6 digit product.
-(÷)-	DIVIDE (1772-LN3 Processor)	XXX XXX -(:)—(:)- YYY.YYY	The DIVIDE instruction is an output instruction. It is always preceded by two GET instructions. The value of the first is divided by the value of the second. Two word addresses are required to store the 6 digit quotient. Its decimal point is placed automatically by the Industrial Terminal.

Section 7
OUTPUT OVERRIDE AND I/O UPDATE INSTRUCTIONS

7.0 GENERAL

Programming instructions may be needed for certain applications requiring output overrides or I/O updates. They are:

- MASTER CONTROL RESEST Instruction
- ZONE CONTROL LAST STATE Instruction
- IMMEDIATE INPUT Instruction
- IMMEDIATE OUTPUT Instruction

7.1 OUTPUT OVERRIDE INSTRUCTIONS

The two output instructions that can be used to override a group of outputs are:

- MASTER CONTROL RESET -(MCR)-
- ZONE CONTROL LAST STATE -(ZCL)-

These instructions are similar to a hardwired Master Control Relay in that they can affect a group of outputs in the User Program. The MCR and ZCL instructions, however, are NOT a substitute for a hard-wired relay, which provides emergency stop capabilities for all I/O devices.

> **WARNING:** A PC system should not be operated without a hard-wired Master Control Relay and Emergency Stop switches to provide emergency I/O power shutdown. Emergency Stop switches can be monitored but should not be controlled by the User Program. These devices should be wired as described in the Mini-PLC-2 Assembly and Installation Manual (Publication 1772-820). The purpose of these devices is to guard against damage to equipment and/or injury to personnel.

The MCR and ZCL instructions control the zoned outputs differently:

- MCR — When FALSE, all nonretentive outputs within the MCR zone are de-energized or turned OFF.
- ZCL — When FALSE, the outputs within the ZCL zone are held in their last state: either ON or OFF.

To override a group of output devices, two MCR or ZCL instructions are required: one to begin the zone and one to end the zone (Figure 7-1). The Start Fence begins the zone and is always programmed with a set of input conditions. The End Fence ends the zone and must be programmed unconditionally.

When the MCR or ZCL Start Fence is TRUE, all outputs within the zone are controlled by their respective rung conditions. When the MCR or ZCL Start Fence is FALSE, the outputs within the zone are controlled by the MCR or ZCL Start Fence as stated above.

> **WARNING:** MCR or ZCL zones must not be overlapped or nested. Each zone must be separate and complete. Common outputs must not be shared between MCR zones. Overlapping MCR or ZCL zones could result in unpredictable machine operation with possible damage to equipment and/or injury to personnel.

Sharing common outputs in more than one ZCL zone is permitted, provided that only one ZCL zone is enabled at a time. Common outputs can be examined in more than one MCR or ZCL zone.

7.2 I/O UPDATE INSTRUCTIONS

Two instructions used to accelerate the update of I/O data during the execution of the User Program are:

- IMMEDIATE INPUT -[I]-
- IMMEDIATE OUTPUT -(IOT)-

These instructions are used to transfer critical I/O data ahead of the normal scan sequence. The status of inputs is made immediately available to User Program and output decisions are accelerated to the output devices.

The IMMEDIATE I/O instructions are usually used where I/O modules interface with I/O devices that operate in a shorter period than the Processor scan time. These may include TTL-logic or fast response input or output devices.

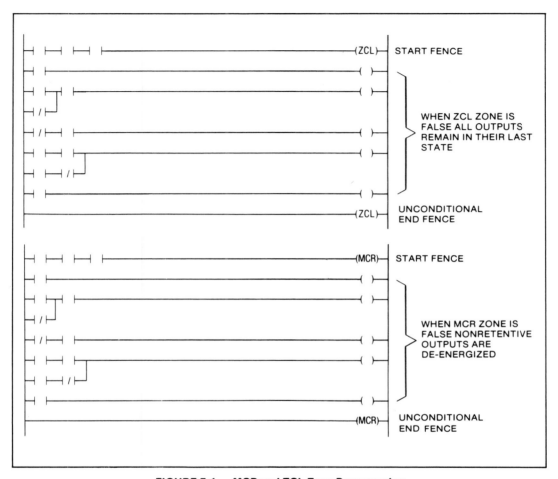

FIGURE 7-1 — MCR and ZCL Zone Programming

Most electromechanical devices have a response time longer than the Processor scan time. Thus, data to and from these devices need not be updated ahead of the normal I/O scan.

7.2.1 Scan Sequence

The Mini-PLC-2 Processor scan sequence can be divided into 2 parts (Figure 7-2):

• I/O Scan

• Program Scan

Upon power up, the Processor begins the scan sequence with the I/O scan. During the I/O scan, data from input modules is transferred to the Input Image Table. Data from the Ouput Image Table is transferred to the output modules.

After completing the I/O scan, the Processor begins the program scan. Here, all User Program instructions are generally scanned and executed in the order they were entered.

The I/O scan and program scan are performed one after the other. The time required to complete both scans is typically 23 msec for 900 instructions.

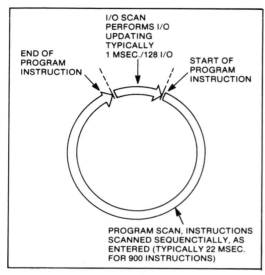

FIGURE 7-2 — Scan Sequence

Typically 23 msec may pass before I/O data is updated in a 1K system. The purpose of IMMEDIATE I/O instructions is to interrupt the program scan to update a word of critical input data or to transfer a word of critical data from the Output Image Table to the module in advance of the normal update sequence.

7.2.2 IMMEDIATE INPUT Instruction

The IMMEDIATE INPUT instruction [I] updates one word of the Input Image Table data in advance of the normal scan sequence (Figure 7-3). The Image Table word represents one Module Group in the I/O Chassis.

The IMMEDIATE INPUT instruction is programmed in the condition area of the ladder-diagram rung. The IMMEDIATE INPUT instruction can be considered as always TRUE; it is always executed whether or not other rung conditions allow logic continuity.

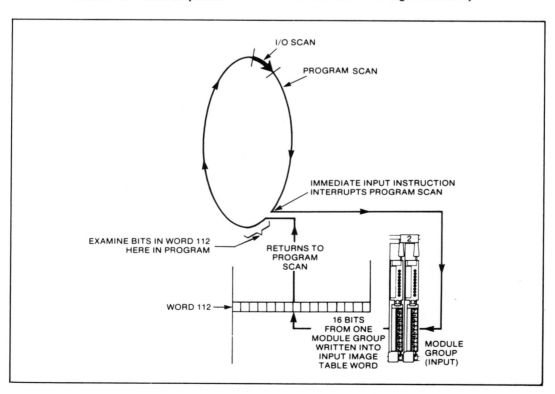

FIGURE 7-3 — IMMEDIATE INPUT Instruction

Program the IMMEDIATE INPUT instruction only when necessary. The need depends on both the response time of the input devices and modules being used, and on the position in the program of the rungs examining these inputs. It is best to program the IMMEDIATE INPUT instruction just before input instructions addressed to the applicable Module Group are examined.

7.2.3 IMMEDIATE OUTPUT Instruction

The IMMEDIATE OUTPUT instruction (IOT) updates one Module Group with data from one Output Image Table word ahead of the normal scan sequence (Figure 7-4).

The IMMEDIATE OUTPUT instruction is programmed as an output instruction in the ladder-diagram rung. This instruction is executed when rung conditions allow logic continuity. Unconditional programming can also be used to cause the Module Group to be updated during each program scan.

Program the IMMEDIATE OUTPUT instruction only when necessary. This depends on the response time of output modules and devices, and on the position of the rungs addressing the applicable Module Group.

The IMMEDIATE OUTPUT instruction should be programmed just after the rungs that con-

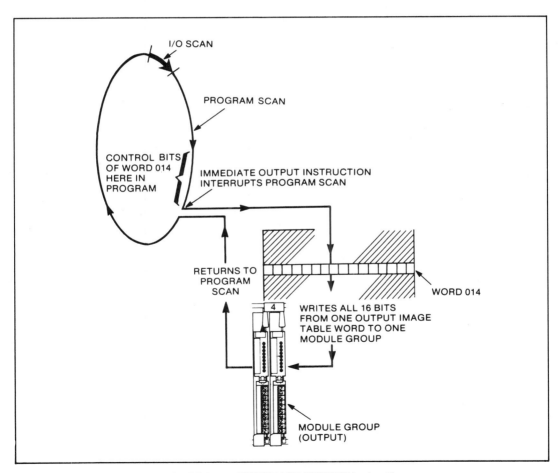

FIGURE 7-4 — IMMEDIATE OUTPUT Instruction

trol the bits in the addressed Output Image Table word.

In PC applications, this instruction only gives a slight advantage when entered near the end of the program scan; since the output data will soon be updated in the I/O scan. This instruction is best applied when entered near the middle of the User Program.

7.3 PROGRAMMING OUTPUT OVERRIDE AND I/O UPDATE INSTRUCTIONS

Instructions are programmed from the Industrial Terminal keyboard with the Processor in PROGRAM mode. When entered, they will be displayed as intensified and blinking with the reverse-video cursor positioned on the first digit of the default word address. The instruction will continue to blink until the word address is entered.

Refer to Table 7-1 for a summarized description of these instructions.

TABLE 7-1 — Output Override and I/O Update Instructions

Note: The MCR and ZCL boundary instructions have no word address. The word addresses XXX of the IMMEDIATE INPUT and OUTPUT instructions are limited to the Input and Output Image Tables respectively.			

KEYTOP SYMBOL	INSTRUCTION NAME	DISPLAY	DESCRIPTION
-(MCR)-	MASTER CONTROL RESET	-(MCR)-	Two MCR instructions are required to control a group of outputs. The first MCR instruction is programmed with input conditions to begin the zone. The second MCR instruction is programmed unconditionally to end the zone. When the first MCR rung is FALSE, all outputs within the zone, except those forced ON, latched ON, or any other retentive output will be de-energized. Do not overlap MCR zones, or nest with ZCL zones. Do not share common outputs between MCR zones.
-(ZCL)-	ZONE CONTROL LAST STATE	-(ZCL)-	Two ZCL instructions are required to control a group of outputs. The first ZCL instruction is programmed with input conditions to begin the zone. The second ZCL instruction is programmed unconditionally to end the zone. When the first ZCL rung is FALSE, outputs in the zone will remain in their last state. Do not overlap ZCL zones, or nest with MCR zones.
-[I]-	IMMEDIATE INPUT	XXX -[I]-	Processor interrupts program scan to update Input Image Table with data from the corresponding module group. It is updated before the normal I/O scan and executed each program scan.
-(IOT)-	IMMEDIATE OUTPUT	XXX -(IOT)-	When rung is TRUE, Processor interrupts program scan to update module group with data from corresponding Output Image Table word address. It is updated before the normal I/O scan and executed each program scan when the rung is TRUE. Can be programmed unconditionally.

Section 9
OPERATING INSTRUCTIONS

9.0 GENERAL

This section contains the operating instructions that are used to move through the program and perform a variety of functions. The instructions are grouped by function and are summarized in Tables 9-1 through 9-6 at the end of this section. They include:

- Data Table Adjustment (Table 9-1)
- Addressing
- Editing Functions (Table 9-2)
- Directories (Table 9-3)
- Search Functions (Table 9-4)
- Troubleshooting Aids (Table 9-5)
- Clear Memory Functions (Table 9-6)

9.1 DATA TABLE ADJUSTMENT

After the size of the Data Table has been determined as described in Section 8.4, the SEARCH 50 function is used to adjust the Data Table. The following display will appear when the [SEARCH] [5][0] keys are pressed in PROGRAM mode.

DATA TABLE ADJUSTMENT

Number of Input/Output Racks	2
Number of Timers/Counters	040
Data Table Size	128

NOTE: The default value of 2 I/O racks will be displayed. It is not a user-entered value for the Mini-PLC-2.

The Data Table is factory configured to 128 words for 1 I/O rack and 40 timers/counters. The Data Table can be reduced in 2-word increments to a minimum of 48 words if no equivalent timers/counters are selected.

The number of equivalent timers/counters to be entered is prompted by a reverse video cursor. When this number has been entered, the Industrial Terminal will compute and enter the Data Table size.

Anytime the Data Table is reduced in size, the Processor searches for instructions in those areas. If an instruction exists in an area to be deleted, the change will not be allowed and the following message will be displayed: "INSTRUCTION EXISTS IN DELETED AREA." To display the rung that is preventing the change, press [SEARCH]. At that time, the decision can be made whether to keep or delete the instruction.

Press [CANCEL COMMAND] to terminate the Data Table Adjustment display.

The instructions for adjusting the Data Table are summarized in Table 9-1.

9.1.1 Memory Layout Display (1770-T3 Industrial Terminal)

The SEARCH 54 function displays a diagram of the areas of memory including the Data Table, User Program, Message Area and the unused memory. The number of words in each area is indicated in decimal numbers.

Press [SEARCH][5][4] to initiate this display, and press [CANCEL COMMAND] to terminate this display.

9.2 ADDRESSING

The ladder diagram instructions are entered with the Processor in the PROGRAM mode. When entered, they are displayed as intensified and blinking to indicate cursor position and that information is needed.

When entering addresses and data, the reverse-video character cursor can be manipulated to the left and right using the [←] and [→] keys to make corrections. The character cursor cannot be moved to the left past the first digit. If the character cursor is moved off the instruction address to the right, the instruction will be entered. It will stop blinking but will remain intensified until the next instruction is pressed or the instruction cursor is moved to the right.

Any time a digit being entered is not within the proper limits, the message "DIGIT OUT OF RANGE" will be displayed. The cursor will

TABLE 9-1 — Data Table Adjustment

FUNCTION	MODE	INDUSTRIAL TERMINAL	KEY SEQUENCE	DESCRIPTION
Data Table Adjustment	PROGRAM	Any ∎	[SEARCH] [5][0] [Numbers]	Enter the number of equivalent timers/counters. The Industrial Terminal displays the size of the reduced Data Table.
Processor Memory Layout	Any	1770-T3	[SEARCH] [5][4]	Displays the number of words in the Data Table Area, User Program Area, Message Area and unused memory.
Either			[CANCEL COMMAND]	To terminate.

∎ 1770-T1, -T2 or -T3

remain in the same position until a valid digit is entered.

9.3 EDITING

Changes to an existing program can be made through a variety of editing functions when the Processor is in PROGRAM mode. Instructions and rungs can be added or deleted; addresses, data and bits can be changed; a rung left incomplete due to an interruption while programming can be located and corrected. The Editing instructions are summarized in Table 9-2.

9.3.1 Inserting an Instruction

Only non-output instructions can be inserted in a rung. There are two ways of doing this.

One way is to press the key sequence [INSERT] [Instruction] [Key sequence of Address]. The new instruction will be inserted after the cursor's present position. If an instruction is to be entered at the beginning of a rung, the cursor must be positioned on the output instruction of the previous rung. However, if the cursor is on the END statement, the instruction will be inserted in the position preceding the cursor.

The other way to insert an instruction is to press the key sequence [INSERT] [←] [Instruction] [Key sequence of Address]. The new instruction will be inserted before the cursor's present position.

If, at any time, the memory is full, the instruction cannot be entered and a "MEMORY FULL" message will be displayed.

9.3.2 Removing an Instruction

Only non-output instructions can be removed from a rung. Output instructions can be removed only by removing the complete rung.

To remove an instruction, place the cursor on the appropriate instruction and press the key sequence [REMOVE] [Instruction]. If the wrong instruction is pressed, an "INSTRUCTIONS DO NOT MATCH" message will be displayed.

Note: Bit values and data of word instructions are not cleared. However, the Input Image Table bits will be rewritten during the next I/O scan except if the instruction had been removed in a forced ON condition. The force function would prevail until removed.

9.3.3 Inserting a Rung

A rung can be inserted anywhere within a program by pressing [INSERT][RUNG] and entering the instructions. The cursor can be positioned anywhere in the previous rung. The new rung will be inserted after the rung which contains the cursor. If it is necessary to remove a newly entered instruction, the rung must be completed first. If the cursor is on the END statement, the [INSERT][RUNG] keys need not be used. The rung can be entered just as in initial program entry.

TABLE 9-2 — Editing Functions

FUNCTION	MODE	INDUSTRIAL TERMINAL	KEY SEQUENCE	DESCRIPTION
Inserting a Condition Instruction	PROGRAM	Any ▣	[INSERT] [Instruction] [Address]	Position the cursor on the instruction that will precede the instruction to be inserted. Then press key sequence.
			or	
			[INSERT] [←] [Instruction] [Address]	Position the cursor on the instruction that will follow the instruction to be inserted. Then press key sequence.
Removing a Condition Instruction	PROGRAM	Any ▣	[REMOVE] [Instruction]	Position the cursor on the instruction to be removed and press the key sequence.
Inserting a rung	PROGRAM	Any ▣	[INSERT] [RUNG]	Position the cursor on any instruction in the preceding rung and press the key sequence. Enter instructions. Editing is prevented until Output is entered.
Removing a rung	PROGRAM	Any ▣	[REMOVE] [RUNG]	Position the cursor anywhere on the rung to be removed and press the key sequence. NOTE: Only addresses corresponding to OUTPUT ENERGIZE, LATCH and UNLATCH instructions are cleared to zero.
Change data of a word instruction	PROGRAM	Any ▣	[INSERT] [Data]	Position the cursor on the word whose data is to be changed. Press the key sequence.
Change the address of a word instruction	PROGRAM	Any ▣	[INSERT] [First Digit] [←] [Address]	Position cursor on a word instruction with data and press [INSERT]. Enter first digit of the first data value of the instruction. Then use the [←] and [→] keys as needed to cursor to the word address or data. Enter the appropriate digits.
Replace an Instruction or change Address of Instructions without Data	PROGRAM	Any ▣	[Instruction] [Address]	Position cursor on the instruction to be replaced or whose address is to be changed. Press the key sequence.

TABLE 9-2 — Editing Functions (cont.)

FUNCTION	MODE	INDUSTRIAL TERMINAL	KEY SEQUENCE	DESCRIPTION
On-Line Data Change	RUN	Any ∎	[SEARCH] [5][1] [Data]	Position cursor on the word whose data is to be changed. Press key sequence. Cursor keys can be used.
			[INSERT]	Press [INSERT] to enter new data into memory.
			[CANCEL COMMAND]	To terminate On-Line Data Change
All Editing Functions	As Applicable	Any ∎	[CANCEL COMMAND]	Aborts the operation at the current cursor position.

∎ 1770-T1, -T2 or -T3

If, by chance, the rung was inserted in the wrong position, it must be completed (press [-( )-][CANCEL COMMAND]) before it can be removed.

If, at any time, the memory is full, a "MEMORY FULL" message will be displayed and more instructions will not be accepted.

9.3.4 Removing a Rung

Removing a rung is the only way an output instruction can be removed. Any rung, except the last one containing the END statement, can be removed.

To remove a rung, position the cursor anywhere on that rung and press [REMOVE] [RUNG].

Note: Only bits corresponding to OUTPUT ENERGIZE, LATCH or UNLATCH instructions addresses are cleared to zero. All other word and bit addresses are not cleared when a rung is removed.

9.3.5 Changing Data of a Word Instruction

The data of any word instruction, except the Arithmetic and PUT instructions, can be changed in the PROGRAM mode without removing and re-entering the instruction. This is done by positioning the cursor on the appropriate word instruction and pressing [INSERT][Data Digits]. When the last digit of the data is entered, the function is terminated and the data is entered into memory. The

function can also be terminated and entered into memory before the last digit is entered by pressing [CANCEL COMMAND]. Also, once the first digit has been entered, the [→] and [←] keys can be used to cursor to any digit in the address or value to make a correction.

9.3.6 Replacing an Instruction or Changing the Address of an Instruction Without Data

To replace one instruction with another, place the cursor on the instruction. Then press [Instruction] [Key Sequence of the Address]. This procedure also can be used when changing the address of an instruction that does not contain data.

9.3.7 On-line Data Change

The lower 12 bits of a word or word instruction excluding Arithmetic and PUT instructions can be changed while the Processor is in the RUN or TEST mode. This is done by positioning the cursor on the appropriate instruction and pressing [SEARCH][5][1]. The message "ON-LINE DATA CHANGE, ENTERING DIGITS" will be displayed near the bottom of the screen. The new digits will be displayed to the right of the message as they are entered. Use the [→] and [←] cursor control keys as needed. After the new data is displayed, press [INSERT] to enter the data into memory.

To terminate this function, press [CANCEL COMMAND].

> **WARNING:** When the address of an instruction whose data is to be changed duplicates the address of other instructions in User Program, the consequences of the change for each instruction should be thoroughly explored beforehand. This is to guard against unexpected machine operation which could result in damage to equipment and/or injury to personnel.

9.4 DIRECTORIES (1770-T3 Industrial Terminal)

Directories have been developed as an aid in using the Industrial Terminal. They list the several functions common to a single multi-purpose key such as the [SEARCH] key. The directories are summarized in Table 9-3.

The Help directory, accessed by pressing the [HELP] key gives a master list of directories and the key sequence to access them. Three other directories that can be accessed from the Help directory are:

- Control functions by pressing [SEARCH] [HELP]

- Record functions by pressing [RECORD] [HELP]

- Clear memory functions by pressing [CLEAR MEMORY][HELP]

The other directories listed in the Help directory cannot be accessed. Certain functions listed in the Control function and Record function directories are not available with the Mini-PLC-2 Processor. If the keys to select any one of these are pressed, the Industrial Terminal will issue a "FUNCTION NOT AVAILABLE WITH THIS PROCESSOR" message.

9.5 SEARCH FUNCTIONS

The Industrial Terminal can be used to search the User Program for a specific instruction or address, the first or last rung, the first or last instruction of a rung or for an incomplete rung using the [SEARCH] key as part of the key sequence. In addition, the Industrial Terminal allows either a single rung or multiple rungs to be displayed. The Search instructions are summarized in Table 9-4.

9.5.1 Search for First Rung

The first rung of the program can be located from any point within the program in any mode of operation by pressing [SEARCH][↑]. This positions the cursor on the first instruction of the program.

TABLE 9-3 — Directories

FUNCTION	MODE	INDUSTRIAL TERMINAL	KEY SEQUENCE	DESCRIPTION
Help Directory	Any	1770-T3	[HELP]	Displays a list of the keys that are used with the [HELP] key to obtain further directories.
Control Function Directory	Any	1770-T3	[SEARCH] [HELP]	Provides a list of all control functions that use the [SEARCH] key.
Record Function Directory	Any	1770-T3	[RECORD] [HELP]	Provides a list of functions that use the [RECORD] key.
Clear Memory Directory	PROGRAM	1770-T3	[CLEAR MEMORY] [HELP]	Provides a list of all functions that use the [CLEAR MEMORY] key.
All Directory Functions	As Applicable	1770-T3	[CANCEL COMMAND]	To terminate

TABLE 9-4 — Search Functions

FUNCTION	MODE	INDUSTRIAL TERMINAL	KEY SEQUENCE	DESCRIPTION
Locate first rung of program	Any	Any ∎	[SEARCH] [↑]	Positions cursor on the first instruction of the program.
Locate last rung of program	Any	Any ∎	[SEARCH] [↓]	Positions cursor on the TEMPORARY END instruction, if present, or the END statement.
Locate first instruction of current rung	PROGRAM	Any ∎	[SEARCH] [←]	Positions cursor on first instruction of the current rung.
Locate output instruction of current rung	Any	Any ∎	[SEARCH] [→]	Positions cursor on the output instruction of the current rung.
Locate rung without an output Instruction	Any	1770-T3	[SHIFT] [SEARCH]	Locates any rung left incomplete due to an interruption in programming.
Locate specific instruction	Any	Any ∎	[SEARCH] [Instruction keys] [Address]	Locates instruction searched for. Press [SEARCH] to locate the next occurrence of instruction.
Locate specific word address	Any	Any ∎	[SEARCH] [8] [Address]	Locates this address in the program (excluding -| |- and -|/|- instructions). Press [SEARCH] to located the next occurence of this address.
Single rung display	Any	Any ∎	[SEARCH] [DISPLAY]	Displays the first rung of a multiple rung display. Press key sequence again to view multiple rungs.

∎ 1770-T1, -T2 or -T3

9.5.2 Search for Last Rung

The last rung of the program (END statement) can be located from any position in the program by pressing [SEARCH][↓]. The cursor will stop at the TEMPORARY END instruction, if present, or the END statement. With the 1770-T3 Industrial Terminal, if the cursor was on the TEMPORARY END instruction, the END statement can be reached by pressing the [SEARCH][↓] keys again. With the 1770-T1 or -T2 Industrial Terminal, the temporary END statement must be removed in order to locate the END statement.

9.5.3 Search for First Instruction of a Rung

With the Processor in the PROGRAM mode, the first instruction of the rung containing the cursor can be located by pressing [SEARCH] [←]. If not in PROGRAM mode, the cursor will move off the screen to the left. To bring it back, press the [←] key. The cursor is displayed by blinking the instruction.

9.5.4 Search for Output Instruction of a Rung

With the Processor in any mode, the output instruction of the rung containing the cursor can be located by pressing [SEARCH][→].

9.5.5 Search For Incomplete Rung (1770-T3 Industrial Terminal)

In the event that an interruption in programming occurred and a rung was inadvertently left without an Output instruction, this rung can be located by pressing the [SHIFT] [SEARCH] keys. The Processor can be in any mode.

9.5.6 Search For Specific Instruction and Specific Address

The procedures for finding a specific instruction or an address are similar. Any instruction in User Program can be located by pressing [SEARCH] [Instruction][Key Sequence of Address]. Any address (excluding those associated with EXAMINE ON and EXAMINE OFF instructions) can be located by pressing the keys [SEARCH][8][Key Sequence of Address]. The address entered is the word address. For the OUTPUT ENERGIZE, LATCH and UNLATCH instructions, the Industrial Terminal will locate all of the bit addresses associated with the word address.

The message "SEARCH FOR" and the entered key sequences will be displayed at the bottom of the screen. The message "EXECUTING SEARCH" will appear temporarily. The 1770-T3 Industrial Terminal will begin to search for the address and/or instruction from the cursor's position. It will look past the TEMPORARY END boundary to the END statement. Then it will continue searching from the beginning of the program to the point where the search began.

A 1770-T1 or -T2 Industrial Terminal will not look past the temporary END statement. It will continue searching from the beginning of the program to the point where it began the search

If found, the rung containing the first occurrence of the address and/or instruction will be displayed as well as the rungs after it. If the [SEARCH] key is pressed again, the next occurrence of the address and/or instruction will be displayed. When it cannot be located or all addresses and/or instructions have been found, a "NOT FOUND" message will be displayed at the bottom of the screen.

This function can be terminated at any time by pressing [CANCEL COMMAND]. All other keys are ignored during the search.

9.5.7 Single Rung Display

Upon power-up, a multiple rung display appears on the screen. A single rung can be viewed by pressing [SEARCH][DISPLAY]. To return to the multiple rung display, press [SEARCH][DISPLAY] again.

9.6 TROUBLESHOOTING AIDS

The following troubleshooting aids are useful during starting-up and when troubleshooting a system:

- Bit Manipulation and Monitor Functions (1770-T3 Industrial Terminal)
- FORCE ON and FORCE OFF Functions
- TEMPORARY END Instruction (1770-T3 Industrial Terminal)
- ERR Message Display

The Troubleshooting aids are summarized in Table 9-5.

9.6.1 Bit Manipulation and Monitor (1770-T3 Industrial Terminal)

Bit Monitor allows the status of all 16 bits of any Data Table word to be displayed. Bit Manipulation allows the status of the displayed bits to be selectively changed or forced, and is useful in setting initial conditions in the data of word instructions.

BIT MONITOR

Bit Monitor can function when the Processor is in any mode. By pressing the key sequence [SEARCH][5][3][Key Sequence of Word Address], the status of all 16 bits of the desired word will be displayed. While the cursor is in the word address field, the [→] and [←] keys can be used to change address digits.

The status of the 16 bits in the next highest or next lowest word address also can be displayed by pressing the [↓] or [↑] keys, respectively. Bit Monitor also can display the status of Force conditions, if any. See 9.6.2 below.

TABLE 9-5 — Troubleshooting Aids

FUNCTION	MODE	INDUSTRIAL TERMINAL	KEY SEQUENCE	DESCRIPTION
Bit Monitor	Any	1770-T3	[SEARCH] [5][3] [Address]	Displays the ON/OFF status of all 16 bits at specified word address and corresponding force conditions if they exist.
			[↑] or [↓]	Displays the status of 16 new bits at the next lowest or highest word address, respectively.
Bit Manipulation	PROGRAM or TEST	1770-T3	[SEARCH] [5][3] [Address]	Displays the ON/OFF status of all 16 bits at specified word address and corresponding force conditions if they exist.
			[→] or [←]	Moves cursor to the bit to be changed.
			[1] or [0]	Enter a "1" to set bit ON or a "0" to set bit OFF.
			See FORCING below	Forcing or removing forces from input bits or output devices.
Either of above			[CANCEL COMMAND]	To terminate.
FORCE ON or FORCE OFF instruction	TEST or RUN	Any ■	[FORCE ON] [INSERT] or [FORCE OFF] [INSERT]	Position the cursor on the Image Table bit or bit instruction to be forced ON or OFF and press the key sequence. The input bit or output device will be forced ON or OFF. ■
Removing a FORCE ON or FORCE OFF instruction	TEST or RUN	Any ■	[FORCE ON] [REMOVE] or [FORCE OFF] [REMOVE]	Position the cursor on the Image Table bit or bit instruction whose force ON is to be removed and press the key sequence.
Removing all FORCE ON instructions	TEST or RUN	Any ■	[FORCE ON] [CLEAR MEMORY]	Position cursor anywhere in program and press key sequence.
Removing all FORCE OFF instructions	TEST or RUN	Any ■	[FORCE OFF] [CLEAR MEMORY]	Position the cursor anywhere in program and press key sequence.
Forced Address Display	Any	1770-T3	[SEARCH] [FORCE ON] or [SEARCH] [FORCE OFF]	Displays a list of the bit addresses that are forced ON or OFF.
				The [SHIFT] [↓] and [SHIFT] [↑] keys can be used to display additional forces.

TABLE 9-5 — Troubleshooting Aids (cont.)

FUNCTION	MODE	INDUSTRIAL TERMINAL	KEY SEQUENCE	DESCRIPTION
Any of the above			[CANCEL COMMAND]	To terminate
Inserting a Temporary END Statement	PROGRAM	1770-T1, -T2	[INSERT] [-|/|-] [8]	Position the cursor on the instruction that will precede the temporary END statement. Press the key sequence and END will be displayed on the screen. The remaining rungs will not be displayed or scanned.
Inserting a TEMPORARY END Instruction	PROGRAM	1770-T3	[INSERT] [T.END]	Position the cursor on the instruction that will precede the TEMPORARY END instruction and press the key sequence. TEMPORARY END will be displayed on the screen. The remaining rungs, although displayed and accessible, are not scanned.
Removing a Temporary END Statement	PROGRAM	1770-T1, -T2	[REMOVE] [-|/|-]	Position cursor on END statement and press key sequence. If temporary, it can be removed. If it cannot be removed, it is an END of program statement.
Removing a TEMPORARY END Instruction	PROGRAM	1770-T3	[REMOVE] [T.END]	Position cursor on the TEMPORARY END instruction and press key sequence.

1 1770-T1, -T2, or -T3

2 When in TEST mode, the Processor will hold outputs OFF regardless of attempts to force them ON.

BIT MANIPULATION

Bit Manipulation can function when the Processor is in Program mode. When in TEST mode, the User Program may override the bit status in the next scan.

The [←] and [→] keys can be used to cursor over to any bit. With the cursor on the desired bit, its status can be changed by pressing the [1] or [0] key. Bit Manipulation also allows the forcing of Image Table bits as described in 9.6.2 below.

To terminate this function, press [CANCEL COMMAND].

WARNING: If it is necessary to change the status of any Data Table bit, be sure that the consequences of the change are thoroughly understood beforehand. If not, unpredictable machine operation could occur directly or indirectly as a result of changing the bit status. Damage to equipment and/or injury to personnel could occur.

9.6.2 FORCE ON and FORCE OFF Functions

The Force functions are used to selectively force an input bit or output device ON or OFF. The Processor must be in the TEST or RUN mode.

The Force functions determine the ON/OFF status of input bits and output devices by overriding the I/O scan. An input bit can be forced ON or OFF regardless of the actual state of the corresponding input device. However, forcing an output terminal will cause the corresponding output device to be ON or OFF regardless of the rung logic or the status of the Output Image Table bit.

From 1 to 16 bits of an Input Image Table word and from 1 to 16 terminals of an Output Module Group can be forced ON or OFF separately or in combination.

NOTE: When in TEST mode, the Procesor will hold outputs OFF regardless of attempts to force then ON, even though the output bit instruction will be intensified.

USING THE 1770-T1, -T2 or 1770-T3 (Series A Rev A) INDUSTRIAL TERMINAL

The same force, ON or OFF, can be applied to any of the bits within a word. However, if a bit in a different Input Image Table word, or a terminal in a different Output Module Group is forced the same way, all previous forces in that word or module group are instantly removed. For example, if any terminal in Output Module Group 013 is forced OFF, and another terminal such as 012/01 is then forced OFF, all force conditions in the Output Module Group 013 will be removed.

Forcing functions can be applied in Ladder Diagam display by placing the cursor on the desired Examine or Energize instruction. After positioning the cursor, any one of the following key sequences can be used for placing or removing a forced condition:

- [FORCE ON][INSERT]
- [FORCE OFF][INSERT]
- [FORCE ON][REMOVE]
- [FORCE OFF][REMOVE]

USING THE 1770-T3 (Series A Rev. B or later) INDUSTRIAL TERMINAL

Simultaneous forcing of bits in different Input Image Table words or terminals in different Output Module Groups is prevented by the

1770-T3 Series A Rev B or later model Industrial Terminal. If attempted, one of the following messages would appear.

SIMULTANEOUSLY FORCING BITS ON (OR OFF) IN TWO INPUT IMAGE TABLE WORDS IS NOT ALLOWED

SIMULTANEOUSLY FORCING BITS ON (OR OFF) IN TWO OUTPUT MODULE GROUPS IS NOT ALLOWED

A bit that is already forced, cannot be forced in the opposite mode. For example, if bit 012/03 is forced ON and an attempt is made to force it OFF, the following message will be displayed.

BIT ALREADY FORCED. EXISTING FORCE MUST BE REMOVED

Forcing functions can be applied using the 1770-T3 Industrial Terminal in either of two ways using: 1) Bit Manipulation/Monitor display of an I/O word or b) Ladder Diagram display of User Program. By pressing the key sequence [SEARCH][5][3][Key sequence of Address], the bit status and force status of the 16 corresponding input bits or output terminals of the desired word can be displayed. The [→] and [←] keys can be used to cursor over to the desired bit. Or, in the Ladder Diagram display, forcing can be applied by placing the cursor on an Examine or Energize instruction.

ALL MODELS

When in TEST mode, the Processor will hold outputs OFF regardless of attempts to force them ON even though the output bit instructions will be intensified.

In every mode except the PROGRAM mode, the ON or OFF status of a forced bit will appear beneath the bit instruction in the rung. In all Processor modes, a "FORCED I/O" message will be displayed near the bottom of the screen when bits are forced ON or OFF.

Note: The ON or OFF status of OUTPUT LATCH/UNLATCH instructions is also displayed below the instruction. However, this is displayed only in PROGRAM mode.

All Force ON or all Force OFF functions can be removed at once in Ladder Diagram Display by pressing either of the following key sequences:

- [FORCE ON][CLEAR MEMORY]

- [FORCE OFF][CLEAR MEMORY]

All force functions will be removed immediately if any of the following conditions should occur: the Industrial Terminal or Processor is disconnected or loses AC power; the [MODE SELECT] key is pressed; or a terminal of a different Output Module Group or a bit in a different Input Image Table word is forced (excluding 1770-T3 Series A Rev B or later).

WARNING: When an energized output is being forced OFF, keep personnel away from the machine area. Accidental removal of Force functions will instantly turn ON the output device. Injury to personnel could result.

9.6.3 Forced Address Display (1770-T3 Industrial Terminal)

A complete list of bit addresses that are forced ON and OFF can be displayed by the Industrial Terminal. Either of the following key sequences can be used.

- [SEARCH][FORCE ON]

- [SEARCH][FORCE OFF]

If all the bits forced ON or OFF cannot be displayed at one time, the [SHIFT] [↓] and [SHIFT][↑] keys can be used to display additional forced bits.

To terminate this display, press [CANCEL COMMAND].

9.6.4 TEMPORARY END Instruction

The TEMPORARY END instruction (or temporary END statement) can be used to test or debug a program up to the point where it is inserted. It acts as a program boundary because instructions below it in User Program are not scanned or operated upon. Instead, the Processor immediately scans the I/O Image Table followed by User Program from the first instruction to the TEMPORARY END instruction, or temporary END statement.

USING THE 1770-T3 INDUSTRIAL TERMINAL

When the TEMPORARY END instruction is inserted, the rungs below it, although visable and accessible, are not scanned. Their con-

tent can be edited, if desired. The displayed section of User Program made inactive by the TEMPORARY END instruction will contain the message "INACTIVE AREA" in the lower right-hand corner of the screen.

The TEMPORARY END instruction can be inserted in either of two ways:

a) Cursor to the last rung of the User Program to be kept active. Position the cursor on the output instruction. Press [INSERT][→][T.END]

b) Cursor to the first rung of the User Program to be made inactive. Position the cursor in the first instruction in the rung. Press [INSERT][←][T.END].

To remove this instruction, position the cursor on it and press [REMOVE][T.END].

To enter a rung after the T.END instruction, place the cursor on the T.END instruction and press [INSERT][RUNG]. Then enter the new rung.

Although more than one TEMPORARY END instruction can be inserted, no rungs will be executed beyond the TEMPORARY END instruction closest to the beginning of the program. The TEMPORARY END instruction uses one word of User Program.

USING THE 1770-T1 OR -T2 INDUSTRIAL TERMINAL

The temporary END statement is similar in function to the TEMPORARY END instruction described above with the following exceptions. When the temporary END statement is inserted, the rungs below it are not shown (nor are they scanned) and no rungs can be entered below it. The temporary END statement will look just like the normal END statement.

The temporary END statement can be inserted in two ways:

a) Cursor to the last rung of the User Program to be kept active. Position the cursor on the output instruction. Press [INSERT][-|/|-][8].

b) Cursor to the first rung of User Program to be made inactive. Position the cursor on the first instruction in the rung. Press [INSERT] [←][-|/|-][8].

To remove this instruction, position the cursor on the END statement and press [REMOVE] [-|/|-]. If it is temporary, it will be removed and the subsequent program instructions will be displayed.

9.6.5 ERR Message for an ILLEGAL OPCODE

An illegal opcode is an instruction code that the Processor does not recognize. It will cause the Processor to fault and will be displayed as an ERR message in the ladder diagram rung in which it occurs. The 4-digit hex value of the illegal opcode is displayed above the ERR message by the 1770-T3 Industrial Terminal. The 1770-T1 or -T2 Industrial Terminal will display the ERR message without the hex value.

If an illegal opcode should occur, the rung containing it can be compared with the equivalent rung in a hard copy printout of the program. A decision must be made either to replace the error with its correct instruction, see paragraph 9.3.6 Replacing an Instruction, or to remove it. The ERR message due to an illegal opcode cannot be removed directly. Instead, remove and replace the entire rung as described in paragraphs 9.3.3 and 9.3.4 Inserting and Removing a Rung. The cause of the problem should be identified and corrected in to correcting the ERR message.

9.7 CLEARING MEMORY

The option of clearing the Data Table, User Program and Messages is available with various CLEAR MEMORY functions. The Clearing Memory instructions are summarized in Table 9-6.

9.7.1 Data Table Clear (1770-T3 Industrial Terminal)

Part or all of the Data Table can be cleared by pressing [CLEAR MEMORY][7][7], entering a start and end word address, and then pressing [CLEAR][MEMORY] again. The Data Table will be cleared between and including these two word addresses.

9.7.2 User Program Clear (1770-T3 Industrial Terminal)

Part or all of the User Program can be cleared by pressing [CLEAR MEMORY][8][8]. The User Program will be cleared from the cursor position to the TEMPORARY END instruction, or the END statement. Neither the Data Table nor Messages are cleared.

9.7.3 Partial Memory Clear

Part of the User Program and the Messages can be cleared by pressing [CLEAR MEMORY] [9][9]. The User Program and Messages are cleared from the cursor position to the end of memory. None of the bits in the Data Table are cleared.

9.7.4 Total Memory Clear

The complete memory can be cleared by pressing [SEARCH][↑] to position the cursor on the first instruction of the program and then pressing [CLEAR MEMORY][9][9]. This resets all the Data Table bits to zero. A total memory clear should be done before entering the User Program.

TABLE 9-6 — Clear Memory Functions

FUNCTION	MODE	INDUSTRIAL TERMINAL	KEY SEQUENCE	DESCRIPTION
Data Table Clear	PROGRAM	1770-T3	[CLEAR MEMORY] [7][7]	Displays a start address and an end address field.
			[Start Address] [End Address]	Start and end word addresses determine boundaries for Data Table clearing.
			[CLEAR MEMORY]	Clears the Data Table within and including addressed boundaries.
User Program Clear	PROGRAM	1770-T3	[CLEAR MEMORY] [8][8]	Clears User Program from the position of the cursor to the END statement or TEMPORARY END instruction. Does not clear Data Table or Messages.
Partial Memory Clear	PROGRAM	Any ∎	[CLEAR MEMORY] [9][9]	Clears User Program and messages from position of the cursor to end of memory. Does not clear Data Table.
Total Memory Clear	PROGRAM	Any ∎	[SEARCH] [↑] [CLEAR MEMORY] [9][9]	Position the cursor on the first instruction of the program. Clears total memory (Data Table, User Program and Messages).

∎ 1770-T1, -T2 or -T3

SELECTED ANSWERS

Chapter 1

Questions
1. noninverting, infinite (ideal)
3. differential, rejects
5. log, multiplication
7. TTL
13. high-pass, low-pass, bandpass, band-reject
17. in phase (0 degrees)

Problems
1. a. -5
3. 250 kHz
7. $V_{ref} = \pm 3$ V
9. a. 0.75, 250 mV
11. 7
    a. $V_o = 0.385$ V
    b. 100 $\mu\Omega$ (with $+10$ V supply)
13. 30 mA, 0.6 mA
15. upper trip voltage = 4.8 V
    upper trip current = 4 μA
    lower trip voltage = 3.6 V
    lower trip current = 3 μA
17. a. $g_m = 3000$ μS
    b. 30 mV
    c. 278 μA

Chapter 2

Problems
1. a. 1429 Hz
    b. 10.17 V
    c. 2034 Hz
    d. 2144 Hz
    e. 0.181 V

3. $R_1 = 10$ kΩ
   $C_1 = 600$ pF
   $V = \pm 10$ V
5. $C_1 = 0.073$ μF
   $R_1 = 10$ kΩ
   $C_2 = 0.76$ μF
7. $C_1 = 0.375$ μF
   $R_1 = 10$ μF
9.   a. 8.125 V
     b. 9.375 V
     c. 5.4 kΩ

Chapter 3

Questions
1. D, S
3. I, D, D
5. I
9. lower

Problems
1. 2095 r/min
3. 972 Ω
5. 5 A, 0.5 Nm, 500 W
9. 0 V, 31.5 A

Chapter 5

Questions
1. series, DC
3. electromagnetic induction, electrical
5. synchronous, starting, rotor
9. synchronous speed

Problems

1. 600 r/min
3. 1800 r/min

Chapter 7

Questions

1. phase, gain
3. zero-voltage, motor
5. PLL, 0.001

Problems

1. 103.6 V
3. **a.** 3.83 V/Hz
    b. 197.4 V
    c. 10,476 W or 140 hp at 1000 r/min

Chapter 8

Questions

1. measurand, position
3. temperature, length
5. digital, photovoltaic
7. magnetoresistor
9. orifice plate, Bernoulli
15. resistance increases, negative, no

Problems

1. **a.** 130°
    b. 107°
3. 0.86 cm
5. 0.256 s
7. 4.014 cm
9. **a.** 165.45 Ω
    b. 180.9 Ω
11. **a.** 3.28 V
    b. 3.48 V

Chapter 9

Questions

1. disturbance
3. open loop, lags, loads
5. set point
7. rate, rate
15. yes
21. no

Problems

1. 1, 2
3. 50%
5. **a.** 3.33 V
    b. 3.33 V

Chapter 10

Questions

1. telemetry
3. analog, digital
5. double polarity
7. position
9. quanta
11. aliasing
13. low-pass filter
15. no

Problems

1. 40 kHz
3. 625 Hz
5. 50 kHz

Chapter 11

Questions

1. transducer, transmission, display, interpretation
3. time division multiplexing
5. frequency, amplitude
7. calibration
21. 22 kHz
23. 2775 Hz

Chapter 12

Questions

1. event-based sequential process
3. control of solids, liquids, environment, and universal systems
5. operator, machine
7. rung, wiring
9. right, left
13. NO, NC

Chapter 13

Questions

1. 1969
3. physically, chemically
5. programming device
7. scan
9. max: 20 MHz, min: 0.1 MHz
11. No
13. I/O
15. random access memory, read only memory
17. PC ladder logic, Boolean statement, mnemonic

Acknowledgments (Continued from copyright page)

mission of General Electric Company; Figure 7.19, National Semiconductor Corporation; Figure 7.24 and 7.26, Airpax Corporation, a North American Philips Company; Figure 7.25, 7.32, 7.33, copyrighted by and reprinted with permission of General Electric Company; Figures 7.34 and 7.35, International Rectifier; Figure 7.36, copyrighted by and reprinted with permission of General Electric Company; Tables 7.2 and 7.3, Westinghouse Electric Corporation.

Chapter 8: Figure 8.7, Hewlett-Packard Company; Figure 8.10, Omega Engineering, Inc.; Figure 8.11, Nanmac Corporation; Figures 8.12, 8.14, Fenwal Electronics, Division of Kidde, Inc.; Figure 8.16, Omega Engineering, Inc.; Figure 8.21, Hewlett-Packard Company; Figure 8.43, Micro Switch, a Honeywell Division; Figures 8.48 and 8.51, Schaevitz Engineering; Figures 8.52, 8.53, and 8.54, Chilton Company, a division of ABC, Inc.; Figure 8.62, Micro Switch, a Honeywell Division; Figure 8.68, National Semiconductor Corporation; Figure 8.75, Hersey Products, Inc., Industrial Measurement Group; Figure 8.82, Metritape, Inc.; Table 8.2, *Temperature Measurement Handbook,* 1983, p. T-15, Omega Engineering, Inc., Stamford, CT; Table 8.3, © 1981 *Instrument and Control Systems,* Chilton Company.

Chapter 9: Figures 9.11 and 9.15, copyrighted by and reprinted with permission of General Electric Company; Figures 9.23 and 9.24, C.A. Norgren Company.

Chapter 10: Figures 10.29, 10.30, 10.31, 10.32, photos by Leslie P. Sheets.

Chapter 11: Figures 11.1, 11.3, 11.5, 11.7, 11.8, 11.9, 11.10, 11.11, adapted from information contained in IRIG 106–80, *Telemetry Standards,* May, 1973, written by the Telemetry Group, Range Commander's Council and published by Range Commander's Council, Secretariat, White Sands Missile Range, NM; Figure 11.19, photo by Leslie T. Sheets; Table 11.1, adapted from information contained in IRIG 106–80, *Telemetry Standards,* May, 1973, written by the Telemetry Group, Range Commander's Council and published by Range Commander's Council, Secretariat, White Sands Missle Range, NM.

Chapter 12: Figure 12.12, photo by Leslie T. Sheets; Figures 12.16 and 12.20, Allen-Bradley Company.

Chapter 13: Figure 13.2, Instruments and Control Systems, Chilton Company, a division of ABC, Inc.; Figures 13.3, 13.4, 13.5, 13.6, 13.21, 13.22, 13.23, 13.25, 13.26, Allen-Bradley Company.

The data sheets have been reproduced courtesy of National Semiconductor Corporation (© 1982) and Texas Instruments, Incorporated.

Appendix C has been reproduced courtesy of Control Engineering-Technical Publishing, a Company of The Dun and Bradstreet Corporation.

Appendix D has been reproduced courtesy of Allen-Bradley Company.

INDEX